Advances in Network Electrophysiology

Makoto Taketani
Michel Baudry

Editors

Advances in Network Electrophysiology

Using Multi-Electrode Arrays

With 222 Illustrations, 13 in Full Color

 Springer

Makoto Taketani
Panasonic Research and Development
Company of America
Cypress, CA 92630
USA
taketanim@us.panasonic.com

Michel Baudry
Department of Biological Sciences
University of Southern California,
    Los Angeles
Los Angeles, CA
USA
baudry@usc.edu

Library of Congress Control Number: 2005925103

ISBN 10: 0-387-25857-4
ISBN 13: 978-0387-25857-7

Printed on acid-free paper.

Printed in Singapore.   (TB/KYO)

9 8 7 6 5 4 3 2 1

springeronline.com

Contents

Contributors

Ashish Ahuja
Department of Electrical Engineering, University of Southern California, Los Angeles, CA, USA

Maura Arsiero
Institute of Physiology, University of Bern, Bern, Switzerland

Kathrin Banach
Department of Physiology, Loyola University, Chicago, IL, USA

Michel Baudry
Neuroscience Program and Department of Biomedical Engineering, University of Southern California, Los Angeles, CA, USA

Theodore Berger
Neuroscience Program and Department of Biomedical Engineering, University of Southern California, Los Angeles, CA, USA

Karl-Heinz Boven
Multi Channel Systems MCS GmbH, D-72770 Reutlingen, Germany

Bruno Buisson
Department of Pharmacology and Toxicology, Trophos SA, Marseille, France

John C. Chang
University of Illinois at Urbana-Champaign, Beckman Institute, Urbana, IL, USA

Laura Lee Colgin
Department of Psychiatry and Human Behavior, University of California, Irvine, USA

Spiros Courellis
Department of Biomedical Engineering, University of Southern California, Los Angeles, CA, USA

Pascal Darbon
Plasticität et Physio-Pathologie de la Montricität, CNRS et Universität de la Mäditerranäe, Marseille, France

Thomas B. DeMarse
Department of Biomedical Engineering, University of Florida, Gainesville, FL, USA

Ulrich Egert
Neurobiology and Biophysics, Institute for Biology III, University of Freiburg, Freiburg, Germany

Michael Fejtl
Multi Channel Systems MCS GmbH, D-72770 Reutlingen, Germany

Ghassan Gholmieh
Department of Biomedical Engineering, University of Southern California, Los Angeles, CA, USA

Michele Giugliano
Brain Mind Institute, Ecole Polytechnique Federale de Lausanne, Switzerland

Larry P. Gonzalez
Department of Psychiatry and Behavioral Sciences, University of Oklahoma Health Sciences Center, Oklahoma City, OK, USA

Kamakshi V. Gopal
Department of Speech and Hearing Sciences and Center for Network Neuroscience, University of North Texas, Denton, TX, USA

Guenter W. Gross
Department of Biological Sciences and Center for Network Neuroscience, University of North Texas, Denton, TX, USA

Elke Guenther
Natural and Medical Sciences Institute, University of Tübingen, Reutlingen, Germany

David Hakkoum
Departement de Neurosciences Fondamentales, Centre Medical Universitaire, Geneva, Switzerland

Martin Han
Department of Biomedical Engineering and Department of Electrical Engineering, University of Southern California, Los Angeles, CA, USA

Marc Olivier Heuschkel
Ayanda Biosystems SA, Lausanne, Switzerland

Thoralf Herrmann
Natural and Medical Sciences Institute, University of Tübingen, Reutlingen, Germany

Prashantha D. Holla
Department of Psychiatry and Behavioral Sciences, University of Oklahoma Health Sciences Center, Oklahoma City, OK, USA

Ken-ichi Honma
Department of Physiology, Hokkaido University Graduate School of Medicine, Sapporo, Japan

Sato Honma
Department of Physiology, Hokkaido University Graduate School of Medicine, Sapporo, Japan

Min-Chi Hsiao
Department of Biomedical Engineering, University of Southern California, Los Angeles, CA, USA

Yousheng Jia
Tensor Biosciences, Irvine, CA, USA; Laboratory of Electrophysiology, Department of Obstetrics and Gynecology, Harbor-UCLA Medical Center, Torrance, CA, USA

Don Kubota
Tensor Biosciences, Irvine, CA, USA

Michael Krause
Tensor Biosciences, Irvine, CA, USA

Hans-Rudolf Lüscher
Institute of Physiology, University of Bern, Bern, Switzerland

Edward O. Mann
University Laboratory of Physiology, Oxford University, Oxford, UK

Vasilis Marmarelis
Department of Biomedical Engineering, University of Southern California, Los Angeles, CA, USA

Ken D. Marshal
Department of Psychiatry and Behavioral Sciences, University of Oklahoma Health Sciences Center, Oklahoma City, OK, USA

Thomas Meyer
Multi Channel Systems GmbH, Reutlingen, Germany

Anand Mohan
Department of Psychiatry and Behavioral Sciences, University of Oklahoma Health Sciences Center, Oklahoma City, OK, USA

Andreas Möller
Multi Channel Systems MCS GmbH, D-72770 Reutlingen, Germany

Dominique Muller
Departement de Neurosciences Fondamentales, Centre Medical Universitaire, Geneva, Switzerland

Wataru Nakamura
Clinic of Pediatric Dentistry, Hokkaido University Hospital, Hokkaido University, Sappow, Japan

Wilfried Nisch
NMI, University of Tübingen, Reutlingen, Germany

Ole Paulsen
University Laboratory of Physiology, Oxford University, Oxford, UK

Jerome Pine
Division of Physics, Mathematics, and Astronomy, California Institute of Technology, Pasadena, CA, USA

Steve M. Potter
Department of Biomedical Engineering, Georgia Institute of Technology, and Emory Univrsity School of Medicine, Atlanta, GA, USA

Ken Shimono
Alpha MED Sciences Co., Ltd., Osaka, Japan

Tetsuo Shirakawa
Center for Advanced Oral Medicine, Hokkaido University Hospital, Hokkaido University, Sappow, Japan

Dong Song
Department of Biomedical Engineering, University of Southern California, Los Angeles, CA, USA

Walid Soussou
Neuroscience Program, University of Southern California, Los Angeles, CA, USA

Esther-Marie Steidl
Department of Pharmacology and Toxicology, Trophos SA, Marseille, France

Alfred Stett
Natural and Medical Sciences Institute, University of Tübingen, Reutlingen, Germany

Luc Stoppini
Biocell Interface SA, Plan-les-Ouates, Switzerland

Jürg Streit
Institute of Physiology, University of Bern, Bern, Switzerland

Makoto Taketani
Panasonic Research and Development Company of America, Cypress, CA, USA

Armand R. Tanguay Jr.
Neuroscience Program, Department of Biomedical Engineering, and Department of Electrical Engineering, University of Southern California, Los Angeles, CA, USA

Anne Tscherter
Clinica Neurologica, Diparimento di Neuroscienze, Università degli Studi di Roma "Tor Vergata," Rome, Italy

Daniel A. Wagenaar
Department of Physics, California Institute of Technology, Pasadena, CA

Zhuo Wang
Neuroscience Program, University of Southern California, Los Angeles, CA, USA

Bruce C. Wheeler
University of Illinois at Urbana-Champaign, Beckman Institute, Urbana, IL, USA

James Whitson
Tensor Biosciences, Irvine, CA, USA

Corina Wirth
Department of Physiology, University of Bern, Bern, Switzerland

Preface

While considerable progress has been made over the last decades in our understanding of electrophysiological processes at the single channel, single synapse and single neuron levels, our understanding of electrophysiological processes at the neuronal network level is still in its infancy. This is in large part due to the technical difficulties of recording electrical activity from large numbers of neurons simultaneously and for prolonged periods of time. Although the first multi-electrode device was built in the mid-1970s, the field of network electrophysiology has only recently started to make significant contribution to our understanding of complex brain operations and functions. These recent advances have been the results of progress in electronic technology, providing for new devices capable of stimulating and recording from large numbers of neurons, as well as advances in the computational methods required to store and analyze the enormous amount of data generated by the new devices. This book is an attempt to review the recent progress in both electronics and computational tools developed to analyze the functional operations of large ensembles of neurons and to provide the readers a sense of the applications made possible by these technological tools. As this field is rapidly growing and evolving, it was difficult to select the contributors and the topics to include in this volume. Instead of being exhaustive, we decided to remain more focus, and to limit the reviews to three general topics.

The first section places the emphasis on the technological development of multi-electrode arrays (MEAs) and related electronics and software. In the first Chapter, Jerome Pine reviews the relatively brief history of MEAs. Chapters 2 and 3 are written by two groups of scientists who have been and continue to be involved in developing commercially available MEA instruments. While Fejtl, Stett, Nisch, Boven, and Möller describe mostly the hardware and software of their instruments in Chapter 2, Whitson, Kubota, Shimono, Jia, and Taketani focus more on the MEA applications and discuss why researchers use MEAs (Chapter 3). This section also includes more recent developments of new MEA devices. Heuschkel (Chapter 4) describes an array of spiky 3D microelectrodes which should improve recording in acute slices by reducing the distance between the cells and the recording electrodes, and should allow better measurement and stimulation conditions than with planar electrode arrays. Hakkoum, Muller and Stoppini (Chapter 5) describe a MEA built on a porous membrane or onto a permeable support that can be used to conduct long-term electrophysiological studies applied specifically to 3-D interface-type organotypic cultures. Soussou et al. (Chapter 6) illustrate the utility and advantages of MEAs in electrophysiological investigations with acute hippocampal slices, while introducing a new generation of conformally designed higher-density MEAs as an adjuvant approach to facilitate and enhance MEA-based research. Finally, Chang and Wheeler (Chapter 7) discuss their attempt to build neuronal networks on MEAs.

The second section of the book reviews a number of applications of the MEA technology to dissociated cell cultures. Dissociated cultures have been favorite specimens for MEAs since the early days of the technology and recent studies using these preparations have provided a much deeper understanding of the properties of neuronal networks. The ability to record and stimulate neuronal activity for long periods of time in cultured neurons and myocytes has provided a unique tool for testing chronic effects of drugs on network physiology. First, Gross and Gopal (Chapter 8) provide evidence that networks prepared from different tissues from the murine CNS have different native activity states and may also differ quantitatively in their pharmacological responses, although these responses remain similar to what is observed in vivo. Potter and colleagues (Chapter 9) describe their interesting technologies that allow recording and stimulation on every electrode of an MEA, and a new closed-loop paradigm that brings in vitro research into the behavioral realm, as they embody their networks in Neurally-Controlled Animats. The whole system of MEA culture plus embodiment thus becomes a "hybrot," because it is a hybrid robot with both living and artificial components. Finally, Guigliano and his collaborators (Chapter 10) review their work related to the analysis and the modeling of the development of neuronal activity in cultured cortical neurons. They make the remarkable observation that it is relatively easy to mathematically capture the essential features of the synaptic interactions in the network and to model the behavior of these networks. Chapter11 is the only non-CNS chapter of this book, where Egert, Banach and Meyer discuss the use of MEAs with cardiac myocytes to understand the dynamics of these special types of networks.

Brain tissues, on the other hand, used to be difficult targets for MEA observation despite the fact that the tissues preserve intact structural relationships between groups of cells and should thus provide fruitful information of brain networks. The third section proves that this is no longer true. The first four chapters of the section review the use of MEAs to identify or classify drugs based on the pattern of modifications of spontaneous or evoked electrical responses elicited in various networks. Gholmieh and colleagues (Chapter 12) summarize their successful efforts to build a hippocampal-based biosensor for neurotoxin detection and classification. They combine the use of MEAs with a mathematical analysis of the input/output functions performed by hippocampal networks while classification is performed by an artificial neural network. Guenther et al. describe a preparation of the vertebrate retina on microelectrode arrays they use to record local electroretinograms in vitro (Chapter 13). They then show that this so-called retinasensor is a suitable in vitro tool to easily and effectively assess effects of pharmacological compounds and putative therapeutics on retinal function. In chapter 14, Gonzalez et al. examine the effects of ethanol on an identified population of neurons within the hippocampus using MEAs for the evaluation of neuronal function. In particular, they show that altered hippocampal cholinergic function follows withdrawal from chronic ethanol treatment. Our own work is reviewed in chapter 15 and stresses the advantages of the Brain-on-a-Chip^{TM} technology for drug evaluation and discovery, as it provides a bridge between biochemical and single-cell testing and behavior by determining the effects of compounds on living slices of brain containing intact networks of neurons.

The last four chapters are directed at understanding the origins and functions of various rhythmic activities in neuronal ensembles and at developing new approaches to better understand the rules governing transformation of signals between inputs and outputs. It is widely believed that these rhythmic activities are the basis for most of higher brain functions. In addition to the advantage of long-term recording described in the previous chapters, the use of MEAs has added spatial information to conventional physiology; in particular detailed two dimensional current source density analysis has provided for the first time information related to the spatio-temporal movement of currents in brain slices, thus opening new views regarding brain network functions. Streit, Tscherter and Darbon discuss the rhythm generation in spinal cultures in chapter 16. Honma, Nakamura, Shirakawa and Honma review their work on circadian rhythm, much longer rhythmic activity in chapter 17. The fast rhythms observed in hippocampus under certain conditions are discussed by Colgin (Chapter 18) and Mann and Paulsen (Chapter 19).

We wish to thank Andrea Macaluso and Krista Zimmer for their continuous support and encouragement to bring this book to completion.

Makoto Taketani
Michel Baudry

I
Development of MEA for Cells, Acute Slices, and Cultured Tissues

1
A History of MEA Development

JEROME PINE

Introduction

For this volume about the current state of the art in MEA electrophysiology, it is valuable to set the stage with an overview of what has come before, extending up to the present, including some interesting antecedents of the work described here and also some descriptions of work that supplements these chapters. The time span is over thirty years, and most readers will not be familiar with all this past work. We believe there is value in knowing the tradition on which we are building.

1.1 Beginnings and Basics

More than 30 years ago, in 1972, Thomas et al. published the first paper describing a planar multielectrode array for use in recording from cultured cells (Thomas et al., 1972). Their introductory paragraph is prescient:

Currently available and rapidly improving techniques permit the in vitro culture of an increasing variety of bio-electrically active tissues and single cells. Perhaps the most interesting questions to be asked of such cultures are those dealing with the development and plasticity of electrical interactions among the cultured elements (tissues or single cells). Exploration of these questions would be greatly facilitated by a convenient non-destructive method for maintaining electrical contact with an individual culture, at a large number of points, over periods of days or weeks. This report describes one approach to the development of such a method.

The multielectrode array that was developed for tests had two rows of 15 electrodes each, spaced 100 µm apart, and was intended for experiments with cultured dorsal root ganglion neurons. The main features of multielectrode arrays to this day are shown in Figure 1.1 from their paper.

The array was on glass, with gold electrodes and leads over an adhesion layer, insulated with photoresist. The electrodes were plated with platinum black to

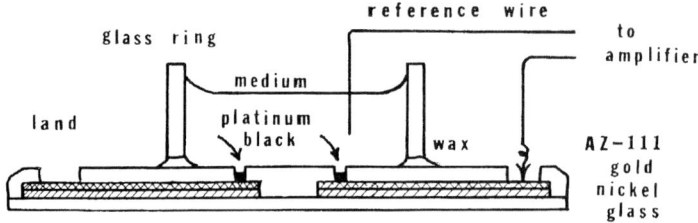

FIGURE 1.1. Schematic diagram of the MEA structure. (From Thomas et al, 1972.)

reduce the impedance of their connection to the culture medium (Gesteland et al., 1959; Robinson, 1968) and were 7 μm square. Standard photolithography methods were used. Initial experiments to try to record from dissociated chick dorsal root ganglion neurons were unsuccessful, and this was ascribed to a confluent glial layer on which the neurons were grown, which insulated them from the electrodes. Turning to dissociated chick myocytes, they found it possible to record robust signals 20 to 1000 microvolts high, after the myocytes had formed a confluent contracting layer over the electrodes. Isolated cells or clumps did not yield measurable signals. It seems likely that the current inflow to initiate myocyte contraction, which can produce extracellular voltages, is over a large area of the cell; and thus for a single cell or small group of cells it does not produce a large enough localized change in potential near an electrode.

Five years later, in 1977, with a very similar introduction to that of Thomas et al., Guenter Gross and his collaborators proposed the idea of a multielectrode array, without knowledge of the previous work (Gross et al. 1977). Their electrodes were gold, insulated with a thermosetting polymer, and were about 10 μm in diameter, deinsulated with UV laser pulses. There were 36 electrodes spaced 100 or 200 μm apart. The initial test reported in 1977 showed recordings from an isolated snail ganglion laid over the electrodes, with single-action potentials having amplitudes up to 3 mv, depending upon the cell size. More specifics for the same preparation were provided in a later paper (Gross, 1979).

The first successful recordings from single dissociated neurons were reported by me in 1980 from a multielectrode array with two parallel lines of 16 gold electrodes, platinized, and insulated with silicon dioxide (Pine 1980). The electrodes were about 10 μm square and 250 μm apart. Though replicating many of the features of the earlier work, I did not know of it while making the array and doing the experiments. Figure 1.2 below shows a scanning micrograph of a typical "fluffy" platinum black deposit on one of the electrodes, which has up to 100 times the surface contact area with the medium as would smooth gold.

The rat superior cervical ganglion neurons used in these experiments had been growing for one to three weeks in culture, and had formed rich interconnected networks. They were grown on a fibrous collagen substrate 3 to 5 μm thick. The cells were about 20 μm in diameter, and recordings were made from 19 cells in

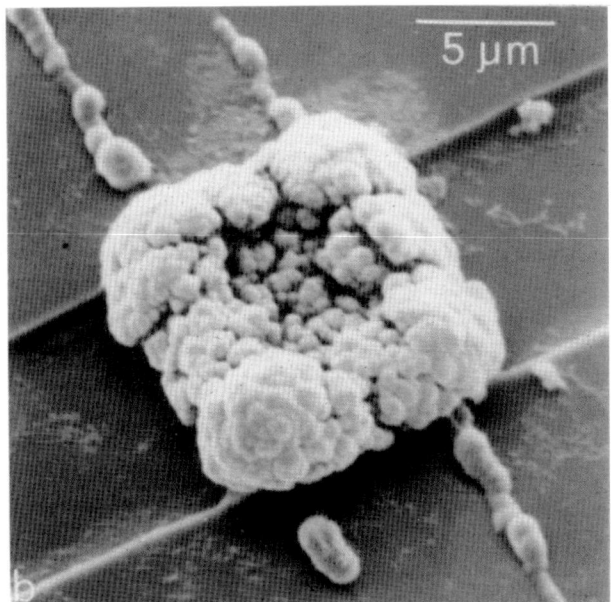

FIGURE 1.2. A platinum black coated electrode. (From Pine, 1980.)

9 cultures, not in contact with electrodes but typically about 25 μm away. Signals were on average 50 μv, with signal-to-noise ratios of 5 to 15:1. It was found that the electrodes could be used for stimulation with a voltage pulse of 0.5 volts and duration of 1 millisecond.

It was important to ask if the results were compatible with what might be expected from known neuron properties. It is easy to show that for a point current sink, or outside a spherical sink, a voltage is generated equal to $I\rho/4\pi r$, where I is the current, ρ is the resistivity of the medium, and r is the distance from the center of the current sink. It is believed that typically a neuron action potential is activated by current flow into the axon hillock, which would represent a point sink (Angelides et al., 1988; Stuart and Sakmann, 1994; Claverol-Tinture and Pine, 2002). At a distance of 25 μm, a current of 16 nanoamperes at the peak of the action potential, which is reasonable, would produce a signal of 50 μv like those observed. These signals are also similar to those observed in vivo, yet somewhat smaller because the culture medium has lower resistivity than brain.

Figure 1.3 shows oscilloscope traces recorded simultaneously from an intracellular stimulating pipette and an extracellular MEA electrode about 25 μm away. The upper intracellular traces show five successively larger short stimuli which elicit action potentials for the two largest. The lower traces show the simultaneous extracellular recordings, from three nonexcitations and two action potentials, so similar they overlap. The difference of a factor of about 1000 in the

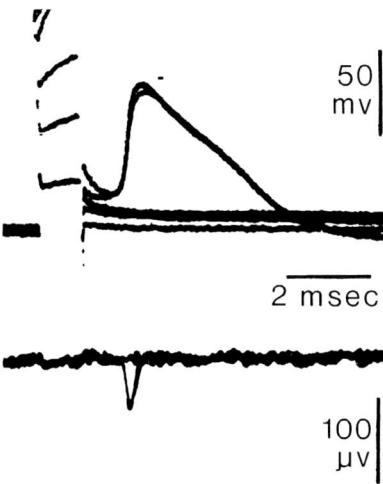

FIGURE 1.3. Five superimposed intracellular oscilloscope traces are at the top. Action potential signals are seen for the largest two. Below are extracellular recordings made simultaneously with the intracellular ones. The two below baseline are from the action potentials, and are too similar to resolve. (From Pine, 1980.)

size of the signals points up the challenge of making good recordings with MEAs, and the shape of the recorded signal shows very well that it comes at the peak of the inward current at the start of the action potential. A great deal of detail about the theory and design issues for MEAs is provided in a book chapter by Kovacs (1994).

1.2   The 1980s

Following up on his earlier work, Gross used his arrays to record from dissociated spinal cord cultures in 1982. Good signals were obtained from spontaneous activity, which was shown to be very temperature dependent below about 30°C, decreasing rapidly to a small value at room temperature. Periodic and aperiodic bursts of activity were seen, which were studied in more detail in followup experiments (Droge et al., 1986.) Figure 1.4 shows some of these culturewide bursts, recorded from three electrodes separated by up to 1 mm. A great variety of patterns was seen, including some very precisely periodic ones.

In 1981, Jobling et al. reported pioneering work with a nine-electrode MEA in which the electrodes were the gates of FET transistors on a silicon chip (Jobling et al., 1981). They demonstrated its effectiveness in recordings from hippocampal slices with good signal-to-noise ratio, while stimulating the slice with a conventional stimulating electrode in a fiber tract. Both slice recording and FET-based MEAs were developed much further by others, but this group did not report any further work.

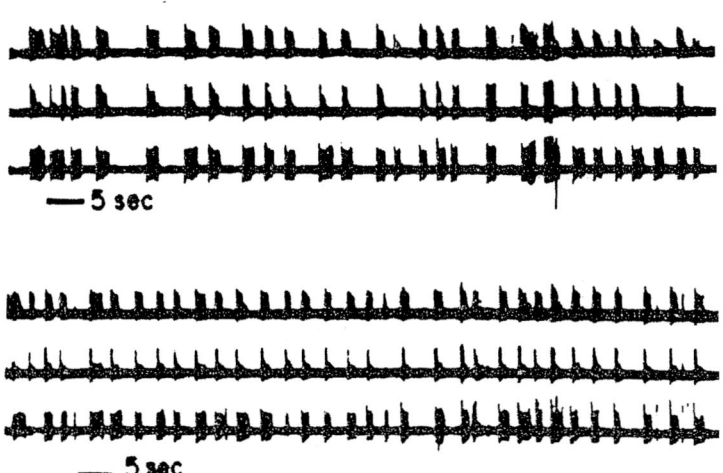

FIGURE 1.4. Simultaneous recordings on three widely spaced electrodes of a monolayer cortical culture, about four weeks in vitro. (From Gross et al., 1982.)

Wheeler and Novak built passive multielectrode arrays tailored to the need of analyzing hippocampal slice activity by performing current source density analyses of field potentials. (Wheeler and Novak, 1986; Novak and Wheeler, 1988). They built an 8 × 4 array of 32 electrodes, 20 μm in diameter and 200 μm on centers, using conventional fabrication techniques, polyimide insulation, and platinization. Figure 1.5 shows in the center a view of a hippocampal slice and the electrode

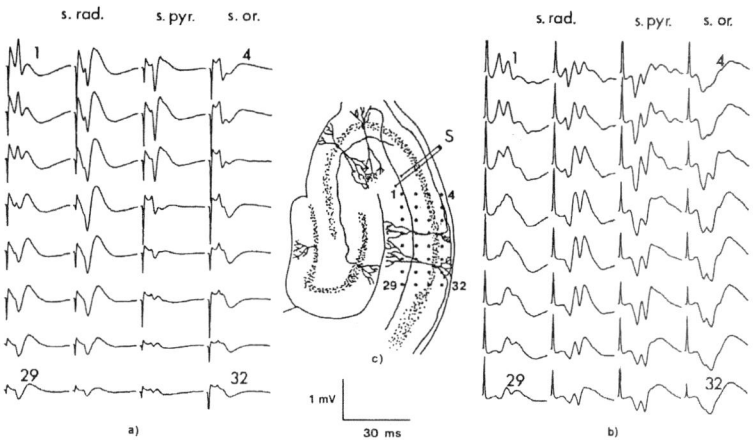

FIGURE 1.5. Responses of a hippocampal slice to a stimulus of the Shaffer collaterals. The center of the figure shows the slice, the stimulating electrode, and the 32 recording electrodes. At left are the normal responses and at right the result when picrotoxin was added to the bath to reduce inhibitory transmission. (From Novak and Wheeler, 1988.)

array. At left are recordings generated by a stimulus to the Shaffer collaterals which synapse on the CA3 neurons. At right, the slice has been treated with picrotoxin, which reduces inhibition. The array can be seen to sample the regions of the dendrites and cell bodies of these neurons. The size and timings of the field potentials can be used to infer the inward or outward currents in the underlying neurons.

MEA data was used for calculating a current source distribution in detail (Wheeler and Novak, 1986). By adding picrotoxin to the slice bath they could create epileptic seizures and use their analysis to infer the current sources and sinks. Figure 1.6 shows the epileptic MEA recordings at left and the corresponding current sources (positive) and sinks (negative) at right. The traces extend for a time of 30 milliseconds. The slant lines in the current source density analysis show a propagation of the activity along the slice with a speed of about 250 μm per millisecond.

Many large invertebrate neurons are "identifiable" by their size and location in ganglia, can be dissected out, and can be used with other identified neurons to form simple networks in culture that replicate some or all of their connections in vivo. The MEA can provide a means for long-term noninvasive communication with such networks for stimulation and recording, much superior to conventional

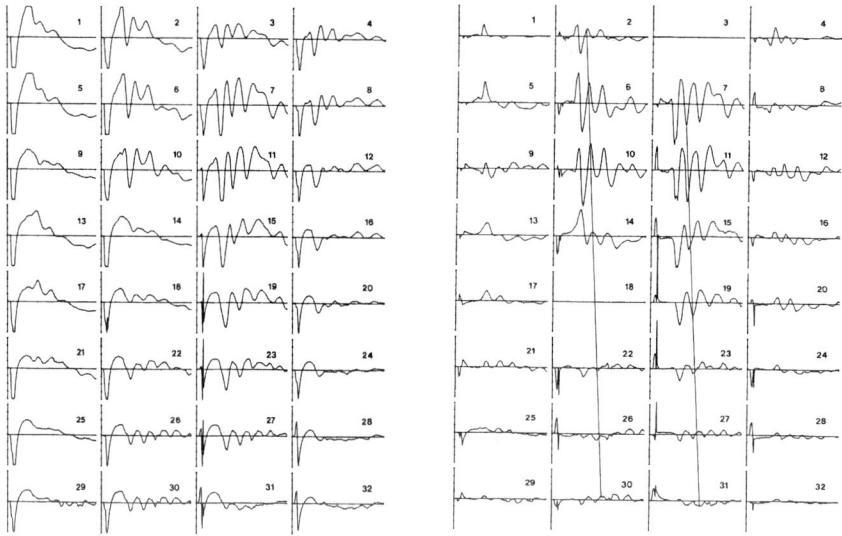

FIGURE 1.6. Epileptiform activity in the hippocampal slice, positioned as in Figure 1.5. At left are the recordings. The time span is 30 msec and the vertical scale spans +2 to −2 mv. At right are the results of a current source density analysis for the locations of the electrodes. The time span is 30 msec and the vertical scale is arbitrary units oriented so that current sources are positive.

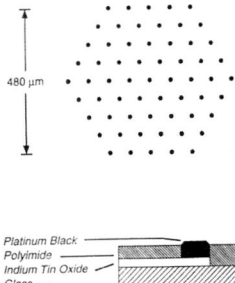
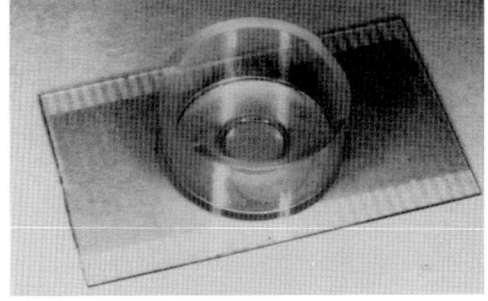

FIGURE 1.7. The 61-electrode array built in the Pine lab. At left, the electrode structure, and at right a completed MEA. (From Meister et al., 1994 and Regehr et al., 1989.)

electrodes. In 1989, Regehr et al. studied invertebrate neurons using a 61-electrode array fabricated originally by Gilbert and Pine (Regehr et al., 1989). This "Pine array" was used in a variety of experiments in the Pine lab and elsewhere. Figure 1.7 shows the layout of the electrodes and their construction at left. They are in a close-packed hexagonal pattern, 70 μm apart. The array was designed to be able to record from any neuron of a low-density culture which is on its area, based on the results of Pine's 1980 experiments.

The array was fabricated on a thin glass coverslip to facilitate observation of cells at high magnification with a short working distance lens and an inverted microscope. The conductors were transparent indium tin oxide, originally introduced by Gross, so as not to interfere with microscopy. The 61 leads end on pads at the edge of the cover slip from where they are connected to external printed circuit-board traces with silicone rubber "zebra connectors". A complete array is shown at the right in Figure 1.7.

The experimenters wanted to explore these possibilities by recording from a variety of invertebrate neurons, from snails, aplysia, and leeches. Not surprisingly, the activity of these large neurons was easy to record with large signals. Sometimes the neurons were right over an electrode and could form a seal over it, and at other times the electrode was in a more extracellular location near a cell body or a process. These varied geometries led to a variety of MEA signals and a few are illustrated in Figure 1.8.

The two traces at the left in Figure 1.8 show recordings from "B19"-identified snail neurons, the top one from an electrode covered over by a cell and the bottom one from an electrode near a process from the cell body. At right are recordings from an aplysia Retzius cell stimulated by an intracellular electrode. At the top is the intracellular recording, and below are, first an MEA electrode near a process, and second an electrode covered over by the cell body. It seems clear that these big cells can sometimes seal over an electrode well enough that capacitative coupling can produce a replica of the intracellular action potential. However, if an MEA electrode is just adjacent, or makes a poor seal, then the recording replicates

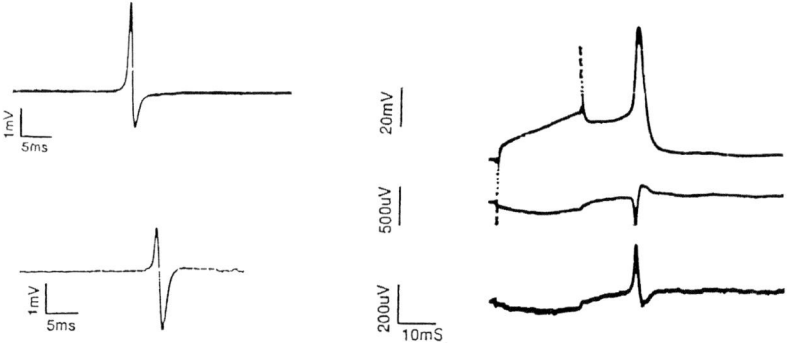

FIGURE 1.8. At left, recordings from B19 identified snail neurons, at top from an electrode covered by a cell, and at bottom from one near a cell. At right, simultaneous recordings from two electrodes, from a stimulated aplysia cell. At top, the intracellular signal; immediately below, a signal from an electrode near the cell; and at bottom the signal from an electrode covered by the cell. (From Regehr et al., 1989.)

the inward current during the action potential. In the paper a detailed model is constructed that can predict the variety of recorded signals.

In 1989, Meister et al. used a Pine lab MEA to study activity from an explanted salamander retina positioned over the array (Meister et al., 1989; Meister et al., 1993). The ganglion cell layer of the retina was in contact with the electrode array and the photoreceptors were illuminated from above by light patterns generated on a CRT screen. Very clean extracellular signals were seen, made larger by the relatively low conductivity of the overlying retinal tissue. The retina remained healthy and responsive for many hours. Figure 1.9 shows at left a schematic view of the array, from which simultaneous recordings from 50 neurons near the electrodes labeled in black were obtained in one experiment. Spike sorting separated multiple

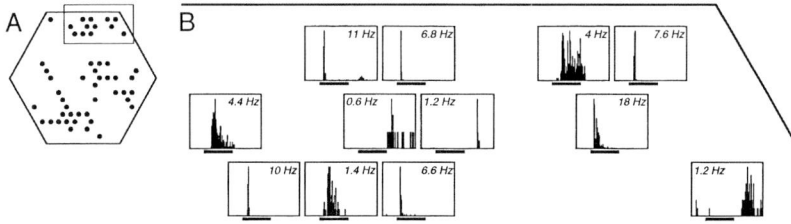

FIGURE 1.9. Multi-neuron signals from a salamander retina. At left is a schematic view of the locations on the array where signals from 50 different neurons were recorded. Twelve typically varied poststimulus time histograms are shown, from a two-second flash indicated by the dark line. (From Meister et al., 1994.)

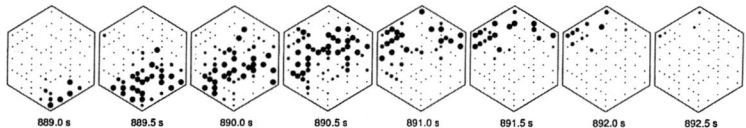

FIGURE 1.10. A spontaneous wave of activity propagating across a five-day postnatal ferret retina. The size of the black circles is proportional to the 0.5 sec firing rate. (From Meister et al., 1991.)

neurons on some electrodes, and in some cases axons produced a trail of recordings extending from a cell body.

The illumination was a two-second flash, uniform over the retina, and the post-stimulus time histograms for a few cells in the upper right region are shown at the right in Figure 1.9. They were derived from 100 successive stimuli at 10-second intervals. The light flash is indicated by the heavy line below the histograms. The variety of responses to such a simple stimulus is striking. Some neurons fire very precisely timed signals, whereas others produce responses that decay with varying times or even remain steady. Cells are seen that respond to the turning off of the flash as well as when it turns on. In later experiments, the retina was illuminated by a multicolored checkerboard pattern changing randomly every 15 milliseconds. By looking backward in time, the "spike-triggered average" stimulus for each neuron was obtained, indicating its receptive field and color sensitivity, simultaneously for all the recorded cells.

The experimental setup used for the salamander experiments was later used with retinas from newborn ferrets and cats, where development of the retinal connections is still taking place (Meister et al., 1991). The animals are blind at birth and for some time afterwards. Very good recordings were again obtained, with no light stimulus, showing spontaneous bursting. When the time and space dependence was analyzed, it was seen that waves of activity passed across the retina, with varying originating points and directions. Figure 1.10 shows the propagation of such a wave across the retina of a five-day postnatal ferret. The size of the black circles is scaled to the firing rate over 0.5-second intervals. The propagation of a wave from lower right to upper left during a 3-second period is clearly evident.

1.3   The 1990s

The MEA extracellular electrode is not adapted to the detection and measurement of subthreshold synaptic potentials. For a synaptic potential of 10 millivolts, the expected recording near a cell body will be much less than one tenth that for an action potential, because it will be generated by a diffuse outward capacitative current from the cell body and neighboring dendrites in contrast to the large current into

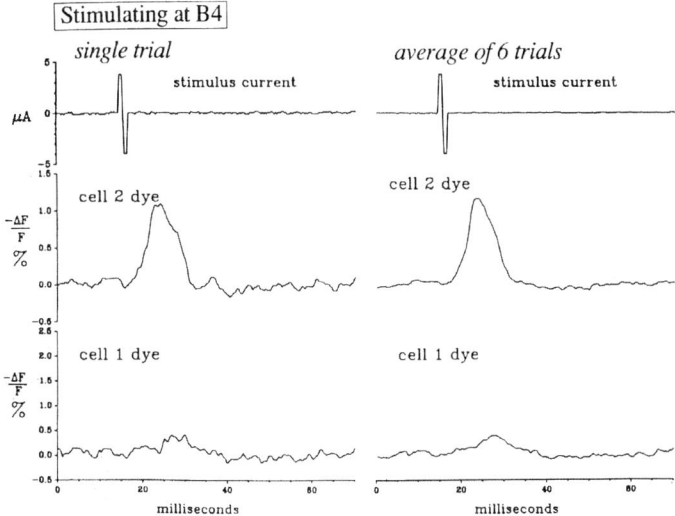

FIGURE 1.11. At left, dye recordings from a stimulated neuron and one that receives synaptic input from it. At right, the same signals, averaged over six trials. (From Chien and Pine, 1991.)

the axon hillock associated with an action potential. Thus, as in vivo, subthreshold synaptically generated potential changes are not expected to be observable. For studies of network development and plasticity this is a serious shortcoming, and motivated a combination of an MEA for stimulation and voltage sensitive dyes for recording. In 1991, Chien and Pine investigated the use of Pine lab MEAs combined with voltage-sensitive dye recording (Chien and Pine, 1991). Because the dyes are directly sensitive to membrane potential, subthreshold signals are only reduced linearly in comparison with the action potential. Nonetheless, their measurement provides a challenge, inasmuch as the dye-recorded signal-to-noise ratio is often small.

Figure 1.11 illustrates the result of one experiment. A neuron in a culture of rat sympathetic neurons was stimulated with an MEA electrode and the postsynaptic potential in a neuron driven by that cell was recorded. At left in the figure is an optical recording of the action potential of the stimulated neuron and of a postsynaptic potential about one tenth as large, while at right the average for 6 trials is shown, and clearly shows the postsynaptic potential. For optical recording from single cultured neurons the dominant noise is shot noise from the photon flux, so that signal averaging improves the signal-to-noise ratio by a factor of the square root of the number of trials. The data show that even when the dye signal is only 1% of the total fluorescence for an action potential, the stimulus-locked synaptic potentials can be cleanly measured. In many instances the dye signal is several times larger and the technique will be much less challenging.

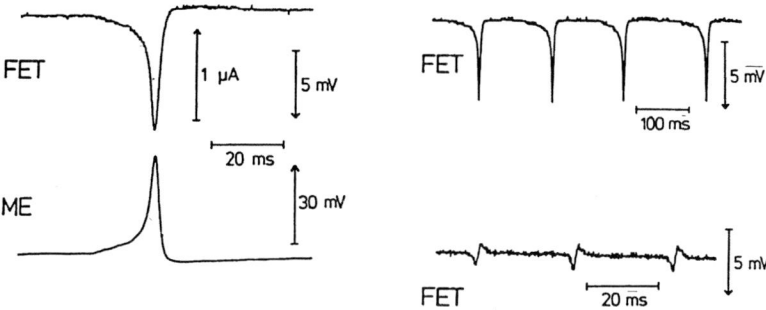

FIGURE 1.12. FET recordings. At left, a simultaneous intracellular and extracellular recording of a spontaneous action potential. At top right, spontaneous activity from a cell making a good seal and below it from a cell making a poor seal. (From Fromherz et al., 1991.)

In 1991, Fromherz and his collaborators investigated the use of a field effect transistor to record action potentials from large aplysia Retzius cells, about 50 μm in diameter (Fromherz et al., 1991). The insulated gate of the FET, about 6 by 10 μm, was completely covered by the cell. Figure 1.12 shows at left the very clean FET recording and the intracellular voltage on a stimulating electrode. At right are two examples of spontaneous activity, one with very good contact between the cell and the FET and one that is like the extracellular voltage which is the derivative of the intracellular potential change. A wide range of signals was seen, and it was hypothesized that they resulted from variations of the contact between the cell and the gate. This began a series of investigations in the Fromherz lab aimed at understanding the FET-neuron interface.

In 1995, Welsh et al. published an account of MEA experiments using the Pine lab MEA which capitalized on the possibility of using it to record from a neural network for long periods of time (Welsh et al., 1995). The suprachiasmatic nucleus in the mammalian brain generates the diurnal circadian rhythm. Suprachiasmatic neurons were dissociated and cultured, and their spontaneous activity monitored for days with the MEA. A surprising result, shown in Figure 1.13, was that the cultured network did not synchronize, but that each neuron independently exhibited oscillation of its activity with an approximately 24-hour period. The figure shows the firing rates of four neurons as a function of time over a three-day period. They are not synchronous, nor do they have exactly identical periods. Thus, each neuron must have its own self-contained circadian rhythm generator. The intracellular machinery for doing this was uncovered eventually by molecular biologists.

Thiébaud et al. reported in 1997 on work designed to improve recording from slices with MEAs (Thiébaud et al., 1997). They constructed perforated silicon substrates that supported an array of electrodes which were either platinum bumps

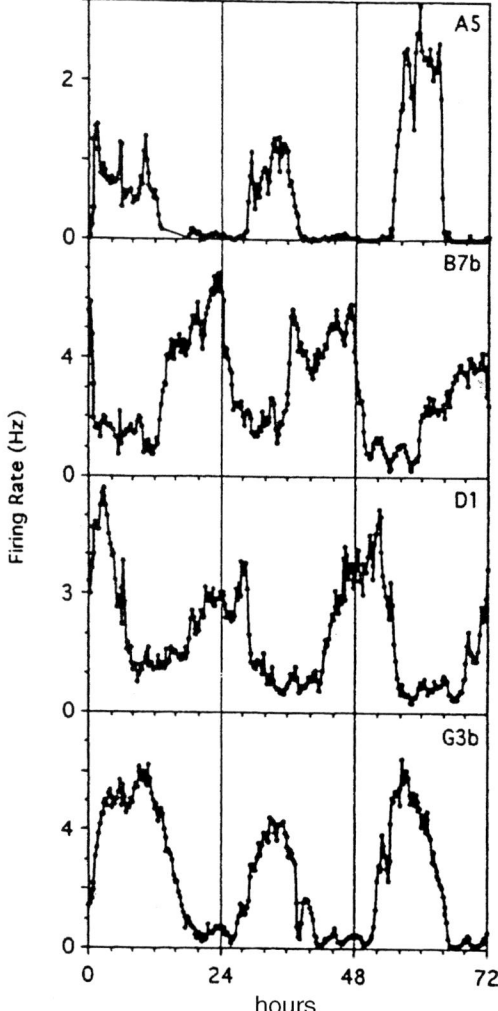

FIGURE 1.13. Spontaneous activity of suprachiasmatic neurons in culture. The diurnal variation of firing rate is large and unsynchronized. (From Welsh et al., 1995.)

or silicon pyramids 45-μm high. The open area of the substrate was 27%. The perforated support was intended to preserve the viability of the slice, and the three-dimensional electrodes to make better contact with the cells of the slice. A reservoir of cell culture medium was connected through the perforated substrate to the slice. They used cultured slices grown for up to 15 days in culture, and they could record for many days. They could also use acute slices for eight hours. A diagram of their system is shown in Figure 1.14.

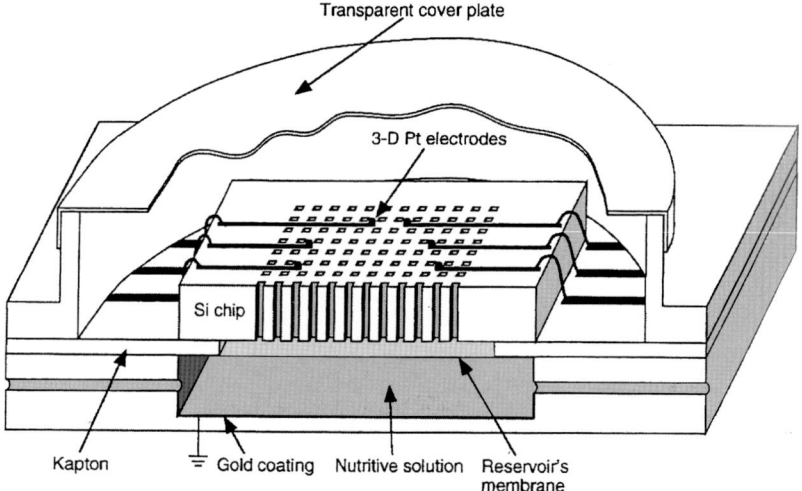

FIGURE 1.14. Schematic view of a slice measurement setup with a perforated silicon substrate. (From Thiébaud et al., 1997.)

The development and plasticity of small neural networks is a long term interest of the Pine lab. It was hoped that network connectivity could be determined at a given time by stimulating each neuron of a network and observing the responses of the others. However, this was made problematic for a conventional MEA by the lack of single neuron specificity for both stimulation and recording. The network of processes over the electrodes made it very difficult to limit the electrode interactions to specific neurons. To address this issue, a silicon-based "neurochip" was fabricated by Maher et al. (1999). The initial version was a 4 × 4 array of wells spaced 100 μm apart on a silicon chip. A thinned area allowed for the creation of wells 15 μm deep designed to contain a single neuron, with an electrode in the bottom of each. A scanning micrograph of one well seen in section after breaking a chip is shown in Figure 1.15. The structure is based on a truncated pyramidal cavity created with anisotropic etching. A gold electrode at the bottom of the well is about 6 μm square.

Initially, the idea was to place neurons into the well through a top hole and have them grow out through small holes at the corners of an overlying silicon nitride cover. For hippocampal neurons, which move along an axon to reach their targets during development, this did not work. The cell body escaped through the axon's corner hole. Therefore long thin tunnels were created for outgrowth of axons and dendrites, too small for the cell body to escape through. In the figure, a thin silicon nitride "umbrella" extends outward from the well, and five raised areas can be seen radiating outward. Under them are tunnels 0.5 μm high and 10 μm wide, through which axons and dendrites grew. Recording and stimulation with these "neurochip" electrodes was easier than for flat arrays. However,

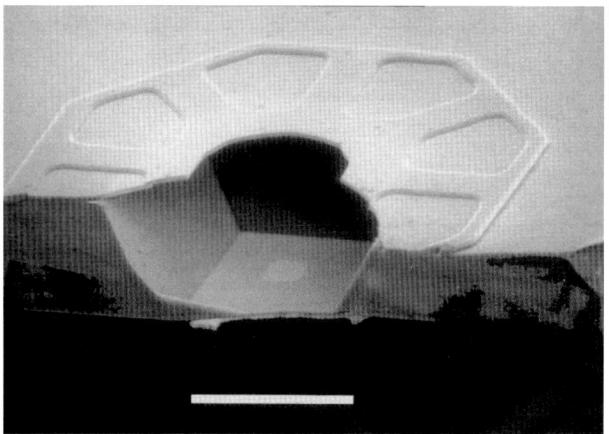

FIGURE 1.15. A neurochip well, broken open to show the structure. The scale bar is 20 microns long. There are radial tunnels under a top silicon nitride layer for outgrowth of axons and dendrites, and a gold electrode in the bottom. (From Maher et al., 1999.)

a serious problem arose from the growth of glial cells (necessary for survival of the cultured neurons) because they grew over the wells and could be reached by neurons inside and used as a path to escape. The neurons seemed in fact to be attracted to the glial cells. Further work on the project was stopped, pending a redesign.

During the late 1990s, groups in Japan at the Matsushita and NTT laboratories, led by Taketani and Kawana, fabricated 64-electrode MEAs for use in slice experiments and with cultures of dissociated cortical neurons (Maeda et al., 1995; Oka et al. 1999). For slices, a rocking device was developed for keeping cultured "organotypic" slice cultures alive over many weeks, so that their development could be observed over time (Kamioka et al., 1997). For cortical cultures, experiments probed plasticity of connections as a result of tetanic stimulation (Jimbo et al., 1999). The striking result was obtained that stimulation of one electrode affected all the neurons driven by it to either enhance or reduce their response. But, for different stimulus electrodes, the response of individual neurons could be either enhanced or reduced. Figure 1.16, originally in color but here in monochrome, illustrates this effect in the graphs across the top or at the right side of the figure.

The top graphs of the figure show how the typical responses to test pulses applied to each of two stimulating electrodes changed after tetanic stimulation by the electrode. All the neurons on a "pathway" from that electrode showed either enhancement or reduction of their firing probability. In contrast, the graph at right, for a typical neuron, shows that its response to stimuli from the 64 electrodes can be either enhanced or reduced by the tetanic stimuli. These unexpected results were

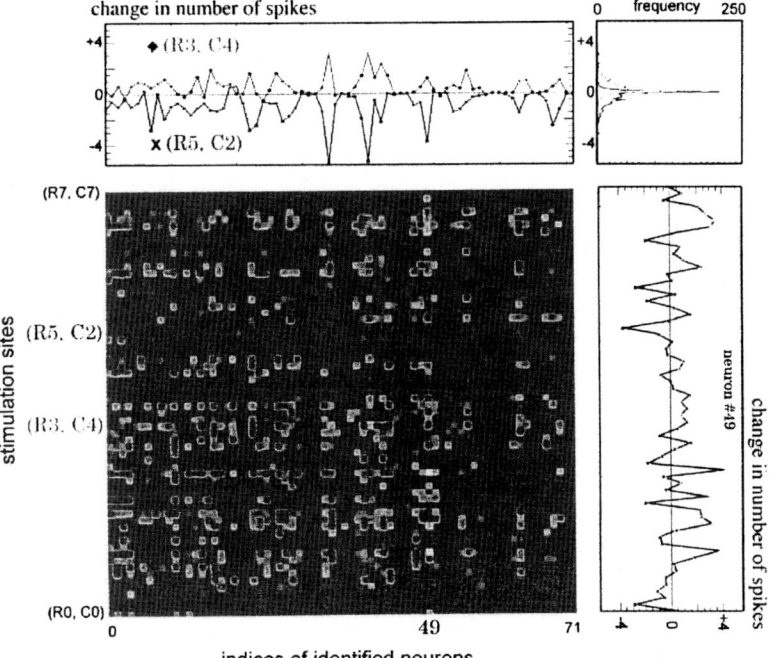

FIGURE 1.16. Response of a dense cortical culture to stimuli at 64 different electrodes, along rows. The firing rates after stimulation for 71 neurons are displayed in columns. Graphs at top and bottom show firing rate changes of responses to stimuli for two rows and of one column after tetanizations of all rows. (From Jimbo et al., 1999.)

part of an ongoing study in the Kawana group of the dynamics of dense cortical cultures grown on MEAs (Maeda et al., 1995).

1.4   The 21st Century

Several of those whose early work has been described above have continued in the period from 2000 to the present. Some of the latest developments are summarized here, without duplicating the work described in the successive chapters of this volume. One significant issue for those trying to study plasticity of cortical networks has been the culturewide bursts that occur at intervals of seconds, in which typically every electrode of the MEA records. Jimbo et al. have studied the dynamics of high-density cortical cultures in detail, and have found that a single stimulus can produce such a burst, even when delivered to a single neuron

Single neuron stimulation

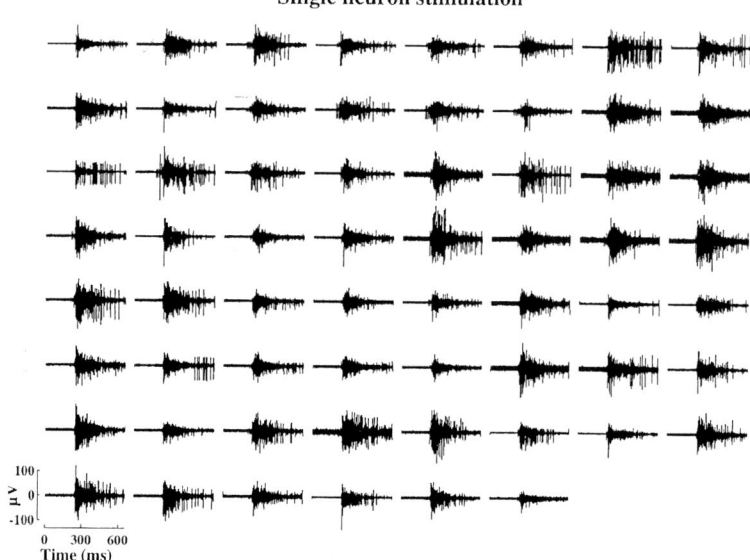

FIGURE 1.17. Dishwide bursting in a dense cortical culture. (From Jimbo et al., 2000.)

(Jimbo et al., 2000). Figure 1.17 shows such culturewide activity, which typically lasts over 100 milliseconds.

The Potter group has studied these dishwide bursts, and is engaged in work aimed at understanding and controlling them (Wagenaar et al., 2001; Madhaven et al., 2003). It is hypothesized that their goal of achieving controlled induction of plasticity in the network connections will be thwarted by these bursts.

Not unlike the Potter group, Shahaf and Maron have been trying to detect learning in dense cortical cultures (Shahaf and Maron, 2001). Their paradigm in one experiment was to stimulate one electrode and identify a neuron that responded with a low probability during a time $50+/-10$ msec afterward. Then, with continuing stimulation, they stopped when two out of ten responses occurred and paused. Then the stimulation was resumed and again stopped when a desired response of two or more out of ten successive stimuli occurred. Figure 1.18 shows an example of "learning" with this paradigm, with an increased response after training during the targeted time window from 40 to 60 msec.

The Fromherz group, continuing work with FET electrode arrays, has developed a structure that embodies a picket fence around each electrode to hold a large invertebrate neuron in place, and has used these structures to study small connected networks (Zeck and Fromherz, 2001). In addition, they have built a closely spaced line of 96-FET electrodes 20 μm apart, and used it for a current source density analysis on a hippocampal slice (Besl and Fromherz, 2002).

Meister and his collaborators have continued studies of retinal signal processing. Studies of motion-sensitive neurons have revealed that the stimulus motion in their

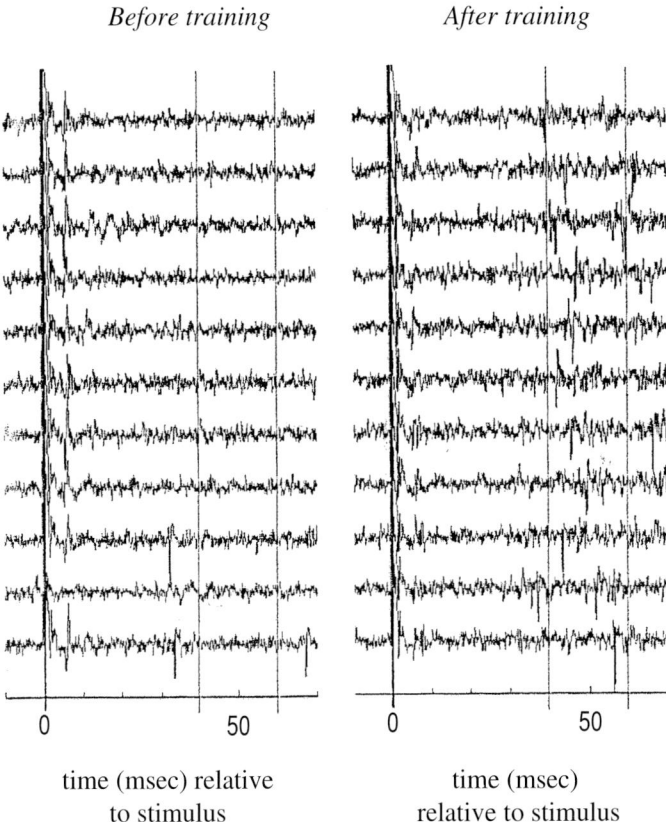

FIGURE 1.18. Learning in a cortical culture. Ten responses at one electrode after stimulation of another electrode are shown at the left. This pair is chosen for poor response in the target period from 40 to 60 msec after the stimulus. At right is shown a "learned" response during this time period for ten trials after training. (From Marom and Shahaf, 2002.)

receptive field must be different from the background motion for them to respond. Figure 1.19 shows raster plots for jittering motion, which is normal for the eye, that dramatically show this effect. The object jittering motion is the same in cases (a) and (b), and the background jitter is in phase in (b) and not in (a) (Olvevczky et al., 2003).

Another study from Meister's lab found that groups of up to seven ganglion cells fire synchronously and that such groups may account for more than 50% of all the spikes recorded from the retina (Schnitzer and Meister, 2003). These patterns convey messages about the visual stimulus far different from what has been inferred from studies of single ganglion cells.

Finally, the Pine lab has returned to the development of the neurochip, utilizing a new fabrication strategy in which "cages" are created on a flat substrate, initially silicon, but possibly glass as well. The technique involves creating cage structures

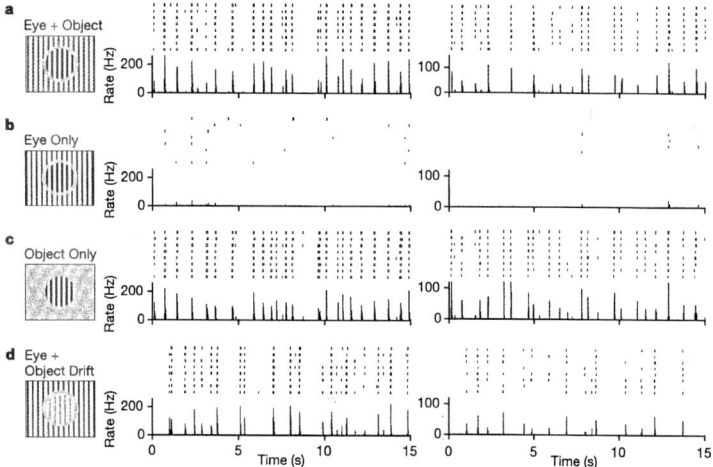

FIGURE 1.19. Raster plots of the response of a motion-sensitive neuron is modified by the background motion of its surround. In traces a,c, and d the motions are not synchronous, whereas in b they are, and there is no response. (From Olveczky et al., 2003.)

of parylene plastic, which allows a great deal of latitude in the shapes of structures, beyond what would be practical with the usual silicon micromachining technology. (Tooker et al., 2004). Figure 1.20 is a schematic view of a cage structures now being investigated. The idea of tunnels for outgrowth, to keep the cell in the cage, is continued from the earlier work. The height of the cage is about 8 μm, but it can be made higher if necessary. The top of the cage is overhanging on the inside, to prevent neurons from climbing out. An additional outward-projecting overhang can be added to keep glial cells from growing over or in. Initial tests have shown normal neuron outgrowth and connected networks developing for cells in a 4 × 4 array of cages.

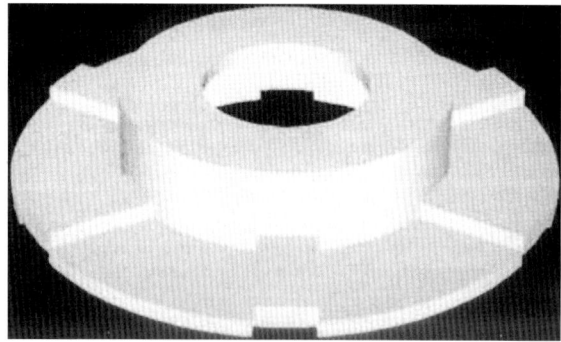

FIGURE 1.20. Schematic view of a parylene neuron cage. (From Tooker et al., 2004.)

Summarizing, during thirty years there have been a great variety of explorations of the possibilities provided by MEAs. Some have led to new understanding, and others have built technologies that promise new future knowledge. It is an exciting time for the MEA community.

References

Angelides, K.J., Elmer, L.W., Loftus, D., and Elson, E. (1988). Distribution and lateral mobility of voltage-dependent sodium channels in neurons. *J. Cell Biol.* 106: 1911–1925.

Besl, B. and Fromherz, P. (2002). Transistor array with an organotypic brain slice: field potential records and synaptic currents. *Eur. J. Neurosci.* 15: 995–1005.

Chien, C.B. and Pine, J. (1991). Voltage-sensitive dye recording of action potentials and synaptic potentials from sympathetic microcultures. *Biophys. J.* 60: 697–711.

Claverol-Tinture, E. and Pine, J. (2002). Extracellular potentials in low-density dissociated neuronal cultures. *J. Neurosci. Meth.* 117: 13–21.

Droge, M.H., Gross, G.W., Hightower, M.H., and Czisny, L.E. (1986). Multielectrode analysis of coordinated, multisite, rhythmic bursting in cultured CNS monolayer networks. *J. Neurosci.* 6: 1583–1592.

Fromherz, P., Offenhausser, A., Vetter, T., and Weis, J. (1991). A neuron-silicon junction: A Retzius cell of the leech on an insulated-gate field effect transistor. *Science* 252: 1290–1293.

Gesteland, R.C., Howland, B., Lettvin, J.Y., and Pitts, W.H. (1959). Comments on micoelectrodes. *Proc. IRE* 47: 1856–1862.

Gross, G.W. (1979). Simultaneous single unit recording in vitro with a photoetched laser deinsulated gold multi-microelectrode surface. *IEEE Trans. Biomed. Eng.* 26: 273–279.

Gross, G.W., Reiske, E., Kreutzberg, G.W., and Mayer, A. (1977). A new fixed-array multimicroelectrode system designed for long-term recording of extracellular single unit activity in vitro. *Neurosci. Lett.* 6: 101–105.

Jimbo, Y., Kawana, A., Parodi, P., and Torre, V. (2000). The dynamics of a neuronal culture of dissociated cortical neurons of neonatal rats. *Biol. Cybern.* 83: 1–20.

Jimbo, Y., Robinson, H.P.C., and Kawana, A.(1993). Simultaneous measurement of intracellular calcium and electrical activity from patterned neural networks in culture. *IEEE Trans. Biomed. Eng.* 40: 804–810.

Jimbo, Y., Tateno, T., and Robinson, H.P.C. (1999). Simultaneous induction of pathway-specific potentiation and depression in networks of cortical neurons. *Biophys. J.* 76: 670–678.

Jobling, D.T., Smith, J. G., and Wheal, H.V. (1981). Active microelectrode array to record from the mammalian central nervous system in vitro. *Med. Biol. Eng. Comp.* 19: 553–560.

Kamioka, H., Jimbo, Y, Charlety, P.J., and Kawana, A. (1997). Planar electrode arrays for long-term measurement of neuronal firing in cultured cortical slices. *Cellular Eng.* 2: 148–153.

Kovacs, G.T.A, (1994). Introduction to the theory, design, and modeling of thin-film microelectrodes for neural interfaces. In: Stenger, D.A. and McKenna, T.M., eds., *Enabling Techniques for Cultured Neural Networks.* Academic Press, San Diego, pp. 121–165.

Maeda, E., Robinson, H.P.C., and Kawana, A. (1995). The mechanisms of generation and propagation of synchronized bursting in developing networks of cortical neurons. *J. Neurosci.* 15: 6834–6845.

Mahavan, R., Wagenaar, D.A., and Potter, S.M. (2003). Multisite stimulation quiets bursts and enhances plasticity in cultured networks. *Society for Neuroscience Annual Meeting*, Abstract 808.14.

Maher, M.P., Pine, J., Wright, J., and Tai, Y.-C. (1999). The neurochip: A new multielectrode device for stimulating and recording from cultured neurons. *J. Neurosci. Meth.* 87: 45–56.

Meister, M., Pine, J., and Baylor, D.A. (1989). Multielectrode recording from the vertebrate retina. *Invest. Ophthalmol. Vis.* 30(Suppl.): 68.

Meister, M., Pine, J., and Baylor, D.A. (1994). Multi-neuronal signals from the retina – acquisition and analysis.*Neurosci. Meth.* 51: 95–106.

Meister, M., Wong, R.O.L., Baylor, D.A., and Schatz, C.J. (1991). Synchronous bursts of action potentials in ganglion cells of the developing mammalian retina. *Science* 252: 939–943.

Novak, J.L. and Wheeler, B.C. (1988). Multisite hippocampal slice recording and stimulation using a 32 element microelectrode array. *J. Neurosci. Meth.* 23: 149–159.

Oka, H., Shimono, K., Ogawa, R., Sugihara, H., and Taketani, M. (1999). A new planar multielectrode array for extracellular recording: Application to hippocampal acute slice. *J. Neurosci. Meth.* 93: 61–67.

Olveczky, B.P. , Baccus, S.A., and Meister, M. (2003). Segregation of object and background motion in the retina. *Nature* 423: 401–408.

Pine, J. (1980). Recording action potentials from cultured neurons with extracellular microcircuit electrodes. *J. Neurosci. Meth.* 2: 19–31.

Regehr, W.G., Pine, J., Cohan, C.S., Mischke, M.D., and Tank, D.W. (1989). Sealing cultured neurons to embedded dish electrodes facilitates long-term stimulation and recording. *J. Neurosci. Meth.* 30: 91–106.

Robinson, D. A. (1968). The electrical properties of metal electrodes. *Proc. IEEE* 56: 1065–1071.

Schnitzer, M.J. and Meister, M. (2003). Multineural firing patterns in the signal from eye to brain. *Neuron.* 37: 499–511.

Shahaf, G. and Maron, S. (2001). Learning in networks of cortical neurons. *J. Neurosci.* 21: 8782–8788.

Stuart, G.J. and Sakmann, B. (1994). Active propagation of somatic action potentials into neocortical pyramidal cell dendrites. *Nature* 367: 69–72.

Thiébaud, P., de Rooij, N.F., Koudelka-Hep, M., and Stoppini, L. (1997). Microelectrode arrays for electrophysiological monitoring of hippocampal organotypic slice cultures. *IEEE Trans. Biomed. Eng.* 44: 1159–1163.

Thomas, C.A., Springer, P.A., Loeb, G.E., Berwald-Netter, Y., and Okun, L. M. (1972). A miniature microelectrode array to monitor the bioelectric activity of cultured cells. *Exp. Cell Res.* 74: 61–66.

Tooker, A., Meng, E., Erickson, J, Tai, Y.-C., and Pine, J. (2004). Development of biocompatible parylene neurocages. *IEEE EMBS Meeting*, 2004.

Wagenaar, D.A., DeMarse, T.B., Potter, S.M., and Pine, J. (2001). Development of complex activity patterns in cortical networks cultured on microelectrode arrays. *Ann. Mtg. Soc. Neurosci.* Abstr. 922.3.

Welsh, D.K., Logothetis, D.E., Meister, M., and Reppert, S.M. (1995). Individual neurons dissociated from rat suprachiasmatic nucleus express independently phased circadian firing rhythms. *Neuron.* 14: 697–706.

Wheeler, B.C., and Novak, J.L. (1986). Current source density estimation using microelectrode array data from the hippocampal slice preparation. *IEEE Trans. Biomed. Eng.* 33: 1204–1212.

Zeck, G. and Fromherz, P. (2001) Noninvasive neuroelectric interfacing with synaptically connected snail neurons immobilized on a semiconductor chip. *Proc. Nat. Acad. Sci.* 98: 10457–10462.

2
On Micro-Electrode Array Revival: Its Development, Sophistication of Recording, and Stimulation

MICHAEL FEJTL, ALFRED STETT, WILFRIED NISCH,
KARL-HEINZ BOVEN, AND ANDREAS MÖLLER

Introduction

Network activity of electrically active cells such as neurons and heart cells underlies fundamental physiological and pathophysiological functions. Despite the well-known properties of single neurons, synapses, and ion channels to exhibit long-term changes upon electrical or chemical stimulation, it is believed that only a concerted effort of many cells make up what it is commonly experienced in humans as self-awareness. In particular, higher brain functions such as associative learning, memory acquisition and retrieval, and pattern and speech recognition depend on many neurons acting synchronically in space and time. Moreover, pathophysiological conditions such as epilepsy, Alzheimer's disease, or other psychological mental impairments have been shown to rely on many neurons to form one of the latter states.

Thus many researchers have been seeking a multi-channel approach to bridge the gap in understanding single-cell properties and population coding in cellular networks. Despite the pioneering work by Thomas et al. (1972), Wise and Angell (1975), and Gross (1979) a remarkable step forward in Micro-Electrode Array (MEA) applications has been achieved only over the last ten years or so, particularly due to the lack of affordable computing power and commercial MEA systems. In this chapter we describe the technology of the most common MEA chips manufactured by the NMI (Natural and Medical Sciences Institute, Reutlingen, Germany), its various design approaches, and current developments. Furthermore, the multi-channel recording and analysis system around the MEA chip developed by and available from MULTI CHANNEL SYSTEMS (MCS, Reutlingen, Germany) together with new developments for multi-site stimulation are described. Additionally, emphasis is given towards an in-depth understanding of the physical prerequisites for adequate extracellular stimulation and recording using MEA. This in turn has led to new approaches in MEA electrode and insulation material as well as new developments in artifact suppression by using digital electronic feedback circuits implemented in a 60-channel MEA amplifier.

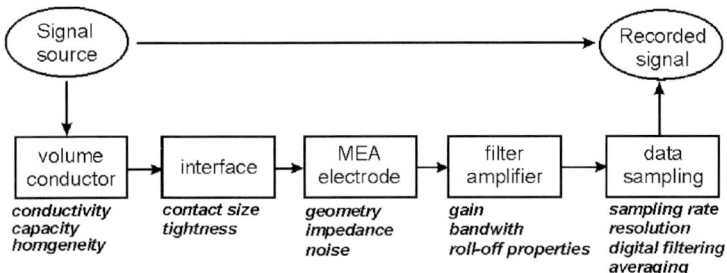

FIGURE 2.1. Pathway showing what parameters are involved in shaping the recorded signal coming from an original cellular signal source.

2.1 Theoretical Considerations of MEA Extracellular Stimulation and Recording

In principle, a MEA is a two-dimensional arrangement of voltage probes designed for extracellular stimulation and monitoring of electrical activity of electrogenic cells, either isolated or in neuronal, muscle, and cardiac tissue. If an analysis of the performance of these probes and its transfer properties is to be performed the electrical characteristics of the main components of the entire system have to be considered (Figure 2.1): (i) the cellular signal sources and the tissue allowing spread of ionic current; (ii) the contact between the cell and the tissue, respectively, and the electrodes; (iii) the substrate and the embedded microelectrodes; and (iv) the external hardware with stimulators and filter amplifiers connected to the electrodes.

2.1.1 Single-Cell Recording with Planar Electrodes

When recording from cells cultured on the MEA surface, individual cells may contact the planar electrodes as shown in Figure 2.2. The cell body is partially covering the electrode surface, and the free electrode area is in contact with the external saline and connected to ground. The amplifier connected to the conducting lane records the sum of the potentials at the surface of the free electrode and the surface of the electrode covered by the membrane. Neglecting the low resistance R_b of the bath solution above the free electrode, the relation between the voltage at the contact pad V_{pad} and in the cleft between cell membrane and the electrode V_J is given by the frequency-independent relation:

$$\frac{V_{pad}}{V_J} = \frac{C_{JE}}{C_E + C_{sh}} \approx \frac{a_{JE}}{a_E}$$

where C_{JE} is the capacity of the covered electrode area of size a_{JE}, C_E the capacity of the entire electrode area of size a_E, and C_{sh} is the shunt capacity of the connecting lane. Given that $C_{sh} \ll C_E$, the amplitude of the recorded signal depends linearly on the ratio of the covered electrode area and the entire electrode area.

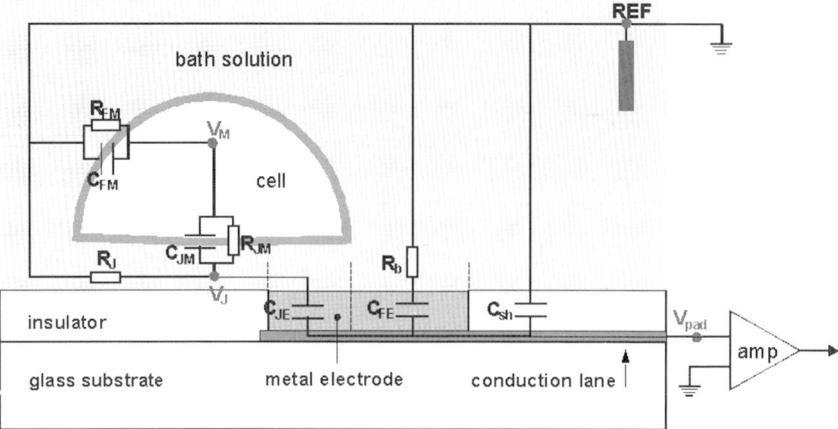

FIGURE 2.2. Extracellular recording of single-cell activity with planar electrodes. Given the electrical circuit the voltage picked up by the amplifier between the contact pads and the reference electrode can be calculated.

The lesson that can be learned from this simplified consideration is the following. In combination with an ideal insulation of the connecting lanes exhibiting neglecting shunt capacity and the use of ideal bandpass filters (infinite input impedance, low cut-off frequency of the highpass, high cut-off frequency of the lowpass), MEA electrodes can be operated as frequency-independent voltage-followers for the currentless monitoring of cellular signals. More advanced considerations on single-cell contacts are given by Buitenweg et al., who used a geometry-based finite-element model for studying the electrical properties of the contact between a passive membrane (Buitenweg et al., 2003), a membrane containing voltage-gated ion channels (Buitenweg et al., 2002), and a planar electrode, respectively.

2.1.2   Tissue Recording with MEAs

The spatial distribution of the voltage in a thin layer of a conductive tissue sheet above the surface of the electrodes of the MEA is recorded with respect to the reference electrode located in the bath solution (Figure 2.3). The signal sources that generate the field potential are compartments of single cells, for example, dendrites or axon hillocks. The electrical activity, be it either spontaneous or evoked by chemical or physical stimulation, spreads within the cellular compartments and from cells to cells via synaptic connections. This spread of excitation within cells and the tissue is always accompanied by the flow of ionic current through the extracellular fluid. Related to the current is an extracellular voltage gradient that varies in time and space according to the time course of the temporal activity as well as the spatial distribution and orientation of the cells.

The recordings may exhibit slow field potentials as well as fast spikes arising from action potentials. The passive spread of cellular signals in tissue slices has

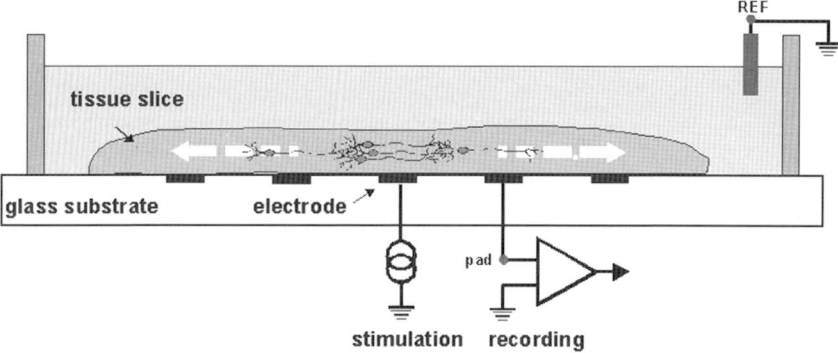

FIGURE 2.3. Stimulation and recording of electrical activity in tissue slices with a MEA. The substrate-integrated planar electrodes can be used both for stimulation and recording.

been investigated by Egert et al. (2002). They could detect spike activity with MEA electrodes at distances of up to 100 μm from a neuron in an acute brain slice. Typically, signal sources are within a radius of 30 μm around the MEA electrode center.

2.1.3   Extracellular Electrical Stimulation with MEA

The MEA electrodes are also used for extracellular electrical stimulation by applying either current or voltage impulses to the electrodes. In principle, the equivalent circuit is the same as for recording when the amplifier is replaced by the stimulation source.

Application of voltage to the electrodes charges the capacity of the electrical double layer of the metal–electrolyte interface. This leads to fast, strong, but transient, capacitive currents with opposite sign at the rising and falling edges of voltage pulses resulting in transient hyperpolarization and depolarization of cellular membranes (Fromherz and Stett, 1995; Stett et al., 2000) This is similar to the effect of brief biphasic current pulses commonly used for safe tissue stimulation (Tehovnik, 1996). In both cases, however, membrane polarization of the target neurons is primarily affected by the voltage gradient generated by the local current density and tissue resistance in the vicinity of the cells. Thus the stimulation efficacy depends on the effective spread of the injected current within the electrode–tissue interface and within the tissue. It is customary to express the stimulation strength in charge injected per pulse, often standardized to the geometric area of the stimulation electrode. The charge injected with current pulses depends only on the amplitude and duration of the pulse, whereas in the case of controlled voltage pulses the charge additionally depends on the tissue resistance and the capacity of the electrode–tissue interface.

For electrochemical reasons, the voltage at the electrodes should always be controlled and as low as possible. Hence microelectrodes should offer a high charge-injection capacity, a parameter that describes the limit of a MEA electrode to be charged without leading to an irreversible electrochemical reaction at the electrode/electrolyte interface.

2.2   MEA Design: Current State and Future Layouts

Micro-electrode arrays were developed in the early 1970s by several research groups (for a historical perspective, see Potter, 2001). Initially, MEAs were mainly made out of Au as the electrode material. Inasmuch as planar Au electrodes have a high impedance, a common practice is to platinize Au electrodes in order to reduce the impedance and thus achieve a better S/N ratio. However, Pt-treated Au electrodes are not stable over a long time period due to the degradation of the Pt-layer and thus they have to be replatinized in order to be used again. To overcome this disadvantage the NMI set out to develop a new MEA with a low and long-term stable electrode impedance.

The standard MEA electrode now is made of TiN by plasma-enhanced chemical vapor deposition (PECVD), and the insulator is made of silicon nitride (Si_3N_4). The PECVD process of a Ti target under a nitrogen atmosphere leads to a fractal deposition of a TiN electrode on the MEA (Figure 2.4). Due to the nanocolumnar 3-D structure the overall surface area of the TiN electrode is much increased compared to a standard 2-D Au or Pt-electrode with the same electrode diameter (Haemmerle et al., 1994; Nisch et al., 1994). The high surface area yields an increase in the overall capacitance and thus leads to a reduced noise level of the electrode, and allows a continuous and reliable electrical stimulation, even over a time period of several weeks (van Bergen et al., 2003). We routinely observe a noise level of less than $+/-10$ µV, measured with a 30-µm MEA electrode at a frequency cut-off of 1 Hz to 3 kHz and a sampling rate of 25 kHz. Increasing demand for specific MEA layouts based on specific biological questions has prompted the NMI for an ongoing development

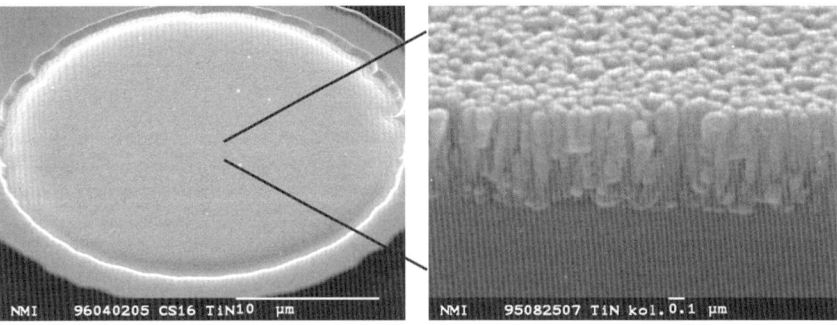

FIGURE 2.4. A single TiN MEA electrode is shown at µm and nm resolution. Note the nanocolumnar structure of the electrode, revealed by the REM image on the right.

of MEAs with custom-designed layouts geared towards specific applications. So far, specific MEA layouts have been designed for the following applications.

2.2.1   The Standard Line of MEA

MEAs come in a pattern of 8×8 or 6×10 electrodes. They are used for acute brain slices, single-cell cultures, and organotypic preparations, and are made out of Ti (titanium) or ITO (indium tin oxide) leads and titanium nitride (TiN) electrodes with a diameter of either 10 or 30 μm. Noteworthy to mention is the fact that MEAs can be optionally delivered with an internal reference electrode and different culture rings/chambers are available to accommodate everyone's need for either acute recordings or long-term cultures, or even combining patch/intracellular approaches with MEA recordings. The insulation is made out of Si_3N_4 in all cases.

2.2.2   Thin MEA

A special approach towards combining MEA recording and imaging has been achieved with the introduction of the "Thin"-MEA. In particular, high-power objectives with a high numerical aperture usually have a very low working distance on the order of only several hundred micrometers. Thus inverted microscopes using such high-power lenses are not able to image through standard MEA due to the thickness of 1 mm. To circumvent this problem MEA have been constructed using cover slip glass. This so-called "Thin"-MEA has a thickness of only 180 μm, and the conductive leads and contact outer pads are made of ITO. The Thin-MEA is mounted on a ceramic support to prevent breakage and can be readily used in combination with high-power objectives (Eytan et al., 2004).

2.2.3 2×30 MEA

This design is intended for studying local responses at a high spatial resolution while at the same time looking at functional connectivity between two organotypic slices placed next to each other. The 60 electrodes are split into two groups. Each group is composed of a 6×5 electrode pattern, and the groups are separated by 500 μm. Within a group the electrode spacing is just 30 μm and the electrode diameter is 10 μm. This allows an unparalleled insight into local connectivity and interconnectivity over a wide range of 500 μm. Moreover, these MEAs are being used to record multi-unit activity, for instance, in slices of the retina. Here, the small interelectrode distance plays a special role as a two-dimensional multi-trode sensor, giving rise to improved spike separation. In principle, the activity of a single neuron is picked up by more than one MEA electrode due to the small inter-electrode distance. Because the distance of the MEA electrodes to a particular cell varies slightly, neighboring MEA electrodes record a slightly different waveform from the very same cell at the same time point. Thus this multi-dimensional fin-gerprint-like pattern identifies a single cell more precisely than conventional spike sorting methods based simply on a one-dimensional waveform analysis (Segev and Berry, 2003).

2.2.4   Flex MEA

Multi-channel recordings in vivo and in semi-intact preparations require a different approach to utilize MEA. Here a flexible 6 × 6 MEA layout based on polyimide has been constructed. It is currently being used to record surface electrocorticograms of the somatosensory cortex in rats and to electrically stimulate the same region via the flex-MEA (Molina-Luna et al., 2004). Conductive leads and outer pad contacts are made out of Au, and the 32-channel flex-MEA electrodes are made out of TiN and have a diameter of 30 μm and spacing of 300 μm, respectively. In addition to the 32 recording electrodes, 2 indifferent and 2 large ground electrodes are incorporated in this design. The polyimide is perforated to allow better attachment of the tissue to the array as well as axonal growth.

2.2.5   High-Density MEA

For various reasons, spatial resolution is important when conduction velocity or synaptic delays are to be measured precisely over a long distance. Given 60 electrodes there is either a very good spatial resolution (small interelectrode distance) and a small recording area, or vice versa. Hence a MEA with a high density of electrodes covering a large area should alleviate these problems. A first step towards such a MEA is currently under development. The layout is comprised of 256 electrodes in a square grid pattern utilizing a 100-μm interelectrode distance in the center and 200 μm in the periphery, yielding a total recording area of about 2.8 × 2.8 mm (Figure 2.5). The chip is wire-bonded to a standard PC chip socket. Thus the chip can be easily mounted in an industry standard socket holder.

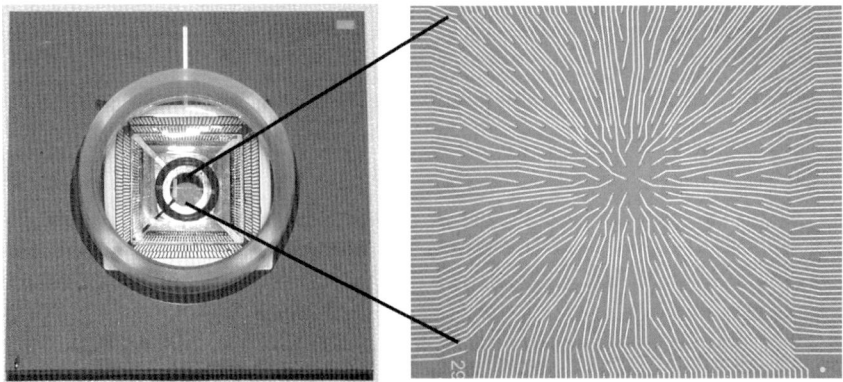

FIGURE 2.5. A high-density MEA is wire-bonded to a standard IC socket commonly used in personal computers. The layout of the high-density HD-MEA is shown on the right.

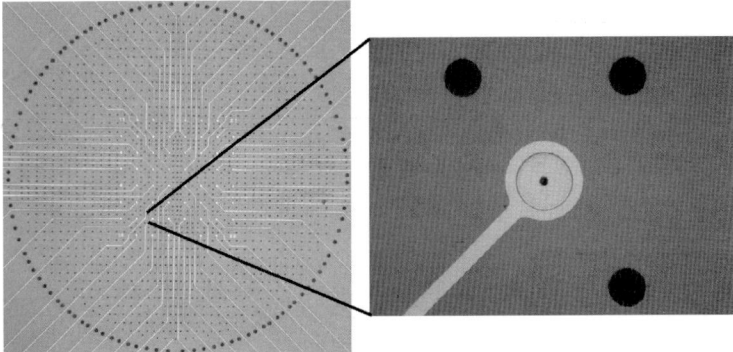

FIGURE 2.6. A MEA with holes in the substrate. The ring enclosing the actual MEA field consists of perfusion holes with 50-μm diameter. The openings in between the MEA electrodes are 20 μm in diameter and negative pressure enhances the contact of the tissue onto the substrate. The hole in the center of the MEA electrode is 5 μm.

2.2.6   Perforated MEA for Tissue Recording

A prerequisite for extracellular recordings with good signal-to-noise ratio is a tight contact between tissue and the electrodes. Working with acute brain slices, positioning and fixation of the slices on the electrode array are often hardly achieved and the distance between the electrode surface and intact cell layers is too large. To avoid these difficulties, we developed MEAs with numerous openings in the substrate (Figure 2.6). By applying negative pressure to these openings from the bottom side it is possible to position and fixate the slices on the electrode side of the MEA substrate. Negative pressure also enhances the contact between tissue and electrodes and therefore the magnitude of extracellular recorded signals is enhanced. The evoked potentials from acute brain slices recorded with this new MEA reach up to 3 mV$_{P-P}$ (preliminary data). Additionally, the openings can be used to perfuse the slice from the bottom side.

2.3   The MEA60 System

The MEA60 System was originally introduced in 1996 by MULTI CHANNEL SYSTEMS (MCS) and many researchers throughout the world are using the system with a variety of biological preparations and for diverse applications. For a more detailed description and applications with the MEA60 system we refer to various chapters in this book. The main components are described briefly, inasmuch as in-depth information about the system has been published and can be found elsewhere (www.multichannelsystems.com).

The 60-channel amplifier has a compact design (165 × 165 × 19 mm), and due to the surface-mounted technology (SMD) of pre- and filter amplifiers the complete electronic circuit and amplifier hardware was built into a single housing. This ensures optimal signal-to-noise ratio of the recording, because no further cables are necessary other than a single SCSI-type cable connecting the amplifier to the data acquisition card. This results in an overall low noise level of the complete amplifier chain ($\times 1200$, 12-bit resolution, 10 to 3 kHz) of $+/-3$ μV, which is well within the $+/-5$ to10 μV noise level of a MEA TiN electrode. Given the low noise of the recording system single units in the lower range of 20 to 30 μV can be readily detected (Granados-Fuentes et al., 2004).

Standard PC technology is used as the backbone of high-speed multi-channel data acquisition. The data acquisition card is based on PCI-bus technology and allows the simultaneous sampling of up to 128 channels at a sampling rate of 50 kHz per channel. Three analog channels and a digital I/O port are accessible, allowing the simultaneous acquisition of analog data such as current traces from a patch clamp amplifier or temperature together with the MEA electrode data. The digital I/O port features trigger IN/trigger OUT functionality. This is an important feature when, for instance, a stimulator is set up to elicit a stimulation pulse to one or more MEA electrodes once a physiological parameter such as the amplitude of a field potential or the spike rate has reached a certain and user-defined threshold. This circuit allows for recurrent feedback stimulation in the system, most notably used in studies revealing developmental properties in neuronal networks (DeMarse et al., 2001; Eytan et al., 2003; Bakkum et al., 2004).

Historically, electrophysiologists are used to standard 19-in. hardware racks where they can plug in amplifiers, oscilloscopes, filters, and so on. Usually this results in lots of cables and wiring prone to catch up noise. Hence the goal was to create a virtual and purely software-based rack with the most common instruments implemented in a digital way. In essence, if one starts to think about software for multi-channel data handling and analysis, what comes immediately to one's mind is the amount of data. Given a 50 kHz sampling rate and 128 channels, several gigabytes of raw data could be acquired in only one hour of recording. Thus the software was actually designed to not get what you see. On the contrary, the data to be recorded is strictly defined by the user, and so is the content of the various oscilloscope-like displays.

The concept of data streams is the core concept of the MC_Rack software. Here, tools can be chosen for displaying data, for digital filtering, for extracting spikes out of raw data, for analyzing the slope/amplitude of an evoked response, and/or calculating the spike rate. They can be used independently of each other, and multiples of the same instrument can be implemented and always stay independent, creating their own data streams. Each time you plug in an instrument in the virtual rack a new data stream is created. These data streams and the MEA electrode channels can now be independently selected and shown on the monitor and/or stored on the disk. Thus the user has full flexibility in terms of what he or she wants to see on the screen and what should be stored on the hard disk. In this way data reduction is achieved and only the important information is stored.

2.4   Multi-Site Stimulation and Artifact Suppression

A common problem in MEA stimulation is to address several MEA electrodes in parallel to deliver a current or voltage pulse to the tissue while retaining the recording properties of the MEA electrodes. Although a few custom-made solutions have been presented (Jimbo et al., 2003; Wagenaar and Potter, 2004), a commercial system readily available to the public has only recently been introduced by MCS. The MEA1060-BC amplifier in combination with the MEA_Select software enables the user to address any of the available MEA electrodes as stimulation sites simply by mouse-click. Thus, one, ten, or even all sixty MEA electrodes can be selected for stimulation. Two distinct stimulation patterns can be fed into the amplifier and readily distributed among the MEA electrodes.

Moreover, in order to retain the recording properties of the MEA electrodes selected for stimulation an electronic Blanking Circuit (BC) has been incorporated into the amplifier. This blanking circuit utilizes electronic switches to actively decouple all MEA electrodes from the main amplifier input stage during the time course of stimulation. The ON and OFF states of the switches are driven by the rising and falling phase of a TTL pulse which is supplied by the stimulator and simultaneously fed into the amplifier. The total time the switches need to be active depends on the actual stimulus strength and waveform. The user can define the time for activating the switches in the MEA_Select software. Although artifact suppression at nonstimulated MEA electrodes retains the recording property less than 1 msec past stimulation, the stimulated MEA electrodes need more time to discharge, and even in the case of using biphasic pulses in current mode to actively discharge a MEA electrode, it was shown that it takes more than 1 msec before the stimulated MEA can be used again for recording. Wagenaar and Potter (2004) reported the recording of spike activity 40 msec to 160 msec past stimulation on stimulated MEA electrodes. In our hands we did observe similar values but we strongly emphasize that more complicated stimulation patterns other than simple square wave pulses may lead to a reduction in discharge time, rendering the MEA electrodes for recording on a much faster time scale. Wagenaar et al. (2004) have also studied current and voltage square wave stimulation patterns in detail. They argue that although the negative phase of a current pulse is more effective than the positive flank in eliciting a biological response, a positive then negative going controlled voltage pulse is even more effective (Wagenaar et al., 2004). However, stimulation patterns other than biphasic square wave patterns have not been investigated.

In general, the blanking circuit is built into a DC-coupled headstage with a gain of $\times 50$, and a filter amplifier with a gain $\times 20$ is added after the headstage. Thus a total amplification similar to the standard MEA1060 amplifier is achieved. A schematic drawing of the blanking circuit in the headstage is depicted in Figure 2.7.

In order to provide a real-time feedback approach to MEA recording a new four- and eight-channel stimulator has been developed. Here a 12-Mbps fast download via USB provides an instantaneous stimulus sequence to selected MEA electrodes.

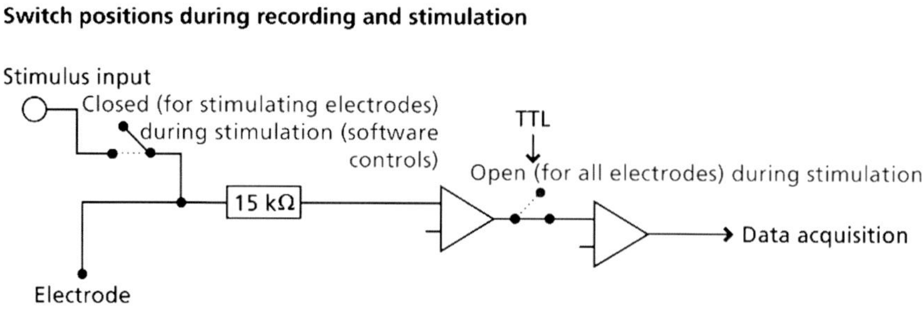

Switch positions during recording and stimulation

FIGURE 2.7. The diagram shows the principal operation of the blanking circuit. A TTL pulse operates the opening and closing of electronic switches to actively decouple the preamplifiers in the headstage off the main filter amplifier circuit.

Even more, once a particular stimulus sequence has been preloaded to the stimulator and waits for a trigger to be sent to the MEA, a real-time feedback can be achieved, because the time necessary to send a preloaded stimulus sequence to the MEA electrodes is 40 µs at most. The online capability of the new stimulator series allows a continuous alteration of the stimulus patterns on all channels independently, making it ideal for arbitrary waveforms to be used for stimulating a neuronal network on a MEA through selected stimulation sites.

2.5   Up-Scaling MEA Systems

The demanding necessity in basic research and pharmaceutical applications to record from more than one MEA simultaneously has posted us to develop a scaleable MEA60 system. In essence, up to four amplifiers can now be hooked up to a single computer and 120 out of a potential 240 MEA electrodes can be recorded. This is particularly useful if statistical questions are being addressed in terms of network activity studying circadian rhythms (Van Gelder et al., 2003; Granados-Fuentes et al., 2004), or applying compounds to see if the substance causes a similar effect in all four MEA preparations. Because automation is a clear and defined goal of MCS, developments are under way to enhance automation of recording from multiple MEAs. In particular, compound application and data analysis as well as the addition of flag conditions, for instance, the question if a response is stable over time, will be automated to enhance throughput in the pharmaceutical industry.

2.6   Outlook and Conclusions

Remarkable progress in MEA recording has been made over the last five years or so, both in hardware design and software features. However, customer feedback always results in new ideas and approaches to a more sophisticated and integrated system. There are at least two important routes to be taken in the future. First, 60 electrodes seem a lot, but on the other hand this results in a reduced spatial resolution. As an example, Current Source Density (CSD) analysis is a popular analytical tool to define active current flow into a cell caused by synaptic activation (current sinks) and passive recurrent flow due to the basic law of closed current loops (current sources). A one-dimensional CSD analysis requires a constant conductivity throughout the medium in order to yield meaningful results. The greater the distance between two recording sites the greater is the potential error in estimating the conductivity values. Thus the closer the recording sites the better the result of a CSD analysis will be. Hence the high-density MEA will lead to a better estimation of current sinks and sources while at the same time recording over a wider range will be achieved compared to a MEA with 60 electrodes with the same interelectrode distance.

Secondly, with the number of recording channels increasing a new concept of data acquisition and transfer is needed and by the same token portability is a debated issue. The general concept is to amplify and digitize the data streams directly in the amplifier housing. Then the USB port will be used for a fast data transfer directly to the personal computer. In general, a lab-in-a-box MEA system will comprise a new hardware and software concept which, in our hands, has at least these advantages: (1) transferring digital data, even over a longer distance, is much less error prone than transferring amplified analog data; (2) utilizing the USB port allows the use of high-end laptop computers, making this system truly portable; and (3) incorporating the stimulator and blanking circuit into the MEA system will provide additional sophistication and user-friendliness. A single box then will incorporate all of the functionalities which up to now are separated and only available by distinct standalone products. Thus this new system will provide a ready-to-go portable MEA recording and stimulation workstation, in particular when MEA applications are increasingly developed as multi-recording and stimulation assays used for drug screening in the pharmaceutical industry (Stett et al., 2003).

To conclude, the basic foundations of extracellular stimulation and recording have led to new concepts in multi-channel MEA recording in basic research and the pharmaceutical industry. Several new MEA layouts are now readily available and the first important steps have been made to provide standardized commercial systems, so that users can communicate and form a platform to discuss their results based on an established technology, similar to patch clamp approaches. Nevertheless, more sophistication is needed and this will result in portable systems using MEA with more electrodes, giving rise to a higher spatial resolution and improved data collection and analysis.

References

Bakkum, D.J., Shkolnik, A.C., Ben-Ary, G., Gamblen, P., DeMarse, T.B. and Potter, S.M. (2004). Removing some 'A' from AI: Embodied Cultured Networks. In: Lida, F., Steels, L., and Pfeifer, R. eds., *Embodied Artificial Intelligence,* Springer-Verlag, Berlin.

Buitenweg, J.R., Rutten, W.L., and Marani, E. (2002). Modeled channel distributions explain extracellular recordings from cultured neurons sealed to microelectrodes. *IEEE Trans Biomed Eng.* 49:1580–1590.

Buitenweg, J.R., Rutten, W.L., and Marani, E. (2003). Geometry-based finite element modeling of the electrical contact between a cultured neuron and a microelectrode. *IEEE Trans Biomed Eng.* 50: 501–509.

DeMarse, T.B., Wagenaar, D.A., Blau, A.W., and Potter, S.M. (2001). The neurally controlled animat: Biological brains acting with simulated bodies. *Autonomous Robots* 11: 305–310.

Egert, U., Heck, D., and Aertsen, A. (2002). 2-Dimensional monitoring of spiking networks in acute brain slices. *Exp Brain Res.* 142: 268–274.

Eytan, D., Brenner, N., and Marom, S. (2003) Selective adaptation in networks of cortical neurons. *J Neurosci.* 23(28): 9349–9356.

Eytan, D., Minerbi, A., Ziv, N.E., and Marom, S. (2004). Dopamine-induced dispersion of correlations between action potentials in networks of cortical neurons. *J Neurophysiol.* 92(3): 1817–1824.

Fromherz, P. and Stett, A. (1995). Silicon-neuron junction: Capacitive stimulation of an individual neuron on a silicon chip. *Phys Rev Lett.* 75: 1670–1673.

Granados-Fuentes, D., Saxena, M.T., Prolo, L.M., Aton, S.J., and Herzog, E.D. (2004). Olfactory bulb neurons express functional, entrainable circadian rhythms. *Eur J Neurosci.* 19(4): 898–906.

Gross, G.W. (1979). Simultaneous single unit recording in vitro with a photoetched laser deinsulated gold multimicroelectrode surface. *IEEE Trans Biomed Eng* 26(5): 273–279.

Haemmerle, H., Egert, U., Mohr, A., and Nisch, W. (1994) Extracellular recording in neuronal networks with substrate integrated microelectrode arrays. *Biosens Bioelectron.* 9(9–10): 691–696.

Jimbo, Y., Kasai, N., Torimitsu, K., Tateno, T., and Robinson, H.P. (2003). A system for MEA-based multisite stimulation. *IEEE Trans Biomed Eng.* 50(2): 241–248.

Molina-Luna, K., Buitrago, M.M., Schulz, J.B., and Luft, A.R. (2004). Thin-film microelectrode array for motor cortex mapping. Progr. Nr. 190.24.2004 Abstract/Itinerary Planner, Washington, DC, *Soc Neurosci* 2004.

Nisch, W., Bock, J., Egert, U., Haemmerle, H., and Mohr, A. (1994) A thin film microelectrode array for monitoring extracellular neuronal activity in vitro. *Biosens Bioelectron,* 9(9–10): 737–741.

Potter, S.M. (2001). Distributed processing in cultured neuronal networks. In: Nicolelis, M.A.L. (ed.), *Progress in Brain Research, Vol. 130: Advances in Neural Population Coding*, Elsevier Science B.V., pp. 49–62.

Segev, R. and Berry II, M.J. (2003). Recording from all of the ganglion cells in the retina. *Soc Neurosci Abstr.* 264: 11.

Stett, A., Barth, W., Weiss, S., Haemmerle, H., and Zrenner, E. (2000). Electrical multisite stimulation of the isolated chicken retina, *Vision Res.* 40: 1785–1795.

Stett, A., Egert, U., Guenther, E., Hofmann, F., Meyer, T., Nisch, W., and Haemmerle, H. (2003). Biological application of microelectrode arrays in drug discovery and basic research. *Anal Bioanal Chem.* 377(3): 486–495.

Tehovnik, E.J. (1996). Electrical stimulation of neural tissue to evoke behavioral responses. *J Neurosci Methods.* 65: 1–17.

Thomas, C.A. Jr, Springer, P.A., Loeb, G.E., Berwald-Netter, Y., and Okun, L.M. (1972). A miniature microelectrode array to monitor the bioelectric activity of cultured cells. *Exp Cell Res.*, Volume 74(1), pp. 61–66.

Van Bergen, A., Papanikolaou, T., Schuker, A, Moeller, A., and Schlosshauer, B. (2003). Long-term stimulation of mouse hippocampal slice culture on microelectrode array. *Brain Res Protoc.*, 11(2): 123–33.

Van Gelder, R.N., Herzog, E.D., Schwartz, W.J., and P.H. Taghert (2003). Circadian rhythms: In the loop at last. *Science* 300: 1534–1535.

Wagenaar, D.A., and Potter, S.M. (2004). A versatile all-channel stimulator for electrode arrays, with real-time control. *J Neural Eng.* 1: 39–45.

Wagenaar, D.A., Pine, J., and Potter, S.M. (2004). Effective parameters for stimulation of dissociated cultures using multi-electrode arrays. *J Neurosci Methods* 38(1–2): 27–37.

Wise, K.D. and Angell, J.B. (1975) A low-capacitance multielectrode probe for use in extracellular neurophysiology. *IEEE Trans Biomed Eng.* 22(3): 212–219.

3
Multi-Electrode Arrays: Enhancing Traditional Methods and Enabling Network Physiology

JAMES WHITSON, DON KUBOTA, KEN SHIMONO, YOUSHENG JIA, AND MAKOTO TAKETANI

Introduction

Early research in the field of multi-electrode arrays (MEAs) was largely concerned with the development of MEA hardware (see Chapter 1 for a review). This research lay the ground work that enabled commercial entities, such as Panasonic (Oka et al., 1999) and Multi Channel Systems (Egert et al., 1998), to develop and manufacture the first MEA-based instruments to be sold in large numbers. Current estimates place the number of instruments sold at over 250 worldwide. The proliferation of these instruments within the electrophysiology community has led to a variety of new applications. This chapter presents a sampling of these applications meant to outline some of the major application categories currently associated with MEAs.

In what follows, MEA applications are divided into two broad categories: experiments that can be performed with traditional non-MEA instrumentation but are enhanced by the use of MEAs; and experiments that can only be performed using MEAs because they depend upon one or more of their unique characteristics. Within each of these broad categories a variety of protocols, tissue types, tissue preparations, and measurements is presented. The examples are largely drawn from users of the Panasonic MED64 System due to the authors' unique access to these researchers; however, the broad outline offered here should apply to users of other MEA instrumentation as well.

3.1 Enhanced Traditional Methods

Traditional slice electrophysiology employs individually implanted glass or metal electrodes to stimulate and record from brain slice tissue. These experiments typically measure spikes, evoked field potentials, or spontaneous field potentials under either static or interface conditions within a slice recording chamber. The tissue samples used in these experiments may derive from a wide variety of brain regions, and they may derive from adult animals for acute testing (lasting hours) or from very young animals for culture testing (lasting days or even weeks).

Over the years, improvements made to MEA hardware have enabled them to perform the same types of experiments as traditional electrophysiology instrumentation. The first MEAs developed in the 1970s were used to record spiking behavior in both acute and culture tissue preparations (Gross et al., 1977; Gross, 1979). Later improvements would add stimulation capabilities (Jobling et al., 1981). However, the relatively high impedance of the early electrodes did not permit the recording of field potentials. The first MEA field potential recordings were achieved by Novak and Wheeler (1988), who succeeded in recording field potentials for two to four hours before the electrodes failed. This early success lay the groundwork for the later development of improved electrodes with the sort of fidelity and survival time required to implement all the major categories of traditional slice electrophysiology (Oka, 1999). However, although modern MEAs can in general perform all the major types of traditional experiments, they are not always the best choice for a particular application.

Determining MEA applicability requires careful consideration of their specific strengths and weaknesses. Ignoring for the moment their potential to enable entirely new types of experiments, their major strengths from a traditional perspective include: (1) the ability to gather data from multiple sites in parallel as if running multiple experiments in a single slice; (2) the ability to change stimulation and recording sites very quickly among those available in the array; (3) the ability to easily set up "within-slice" controls by taking advantage of the many available electrodes; and (4) avoiding the need to place multiple electrodes individually by hand. Limitations include: (1) in most cases, smaller amplitude recordings as compared to traditional instrumentation because the electrodes are not inserted inside the tissue; (2) the electrodes cannot be moved independently because they are arranged in a fixed pattern; and (3) the sensitivity to fluid level fluctuations under "interface" conditions is often greater than that seen with traditional instrumentation.

The sample applications that follow illustrate the use of MEAs for each of three major categories of traditional electrophysiology: spike recording, evoked field potentials, and spontaneous field potentials. They also illustrate the use of MEAs for the study of a variety of different tissue types and preparations, and many demonstrate drug testing. For each application shown, the reasoning behind the choice of MEA is discussed.

3.1.1   Spike Recording

Acute Hypothalamic Slices

The hypothalamus is a key brain structure known to regulate a variety of autonomic and hormonal functions. As such, it is the focus of research aimed at understanding and treating a variety of diseases. Some researchers are pursuing the study of hypothalamus using MEAs (Welsh et al. 1995; Honma et al. 1998). Given that the hypothalamus contains a variety of nuclei, a single slice placed on top of an MEA affords researchers easy access to multiple nuclei (Figure 3.1), enabling them to stimulate and record from multiple regions within a single slice. For spike recording experiments, this means the researcher can record more spiking cells

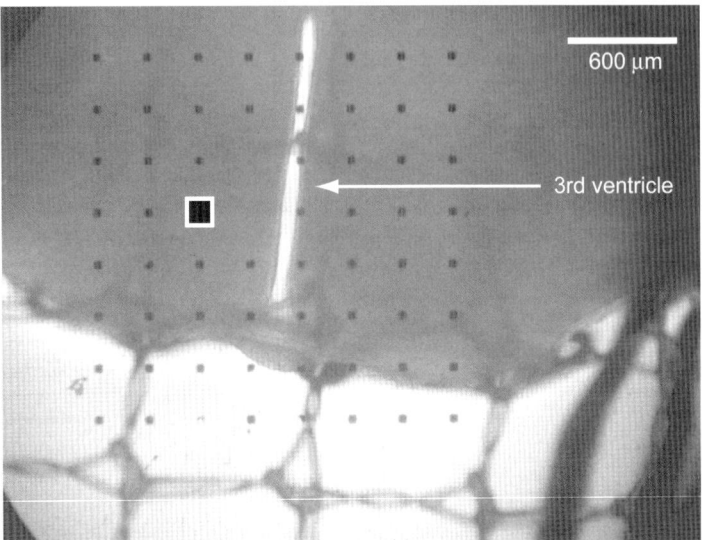

FIGURE 3.1. Micrograph of a coronal rat brain slice, bregma −3.14mm. A 64-electrode array covers the periventricular nucleus, the arcuate nucleus, the median eminence, and the part of the ventro-medial and the dorso-medial hypothalamic nucleus. The white arrow indicates the third ventricle, and the black-filled square the site of the electrode used for stimulation.

per slice, and he or she can easily discover and record from different cell types producing different behaviors in parallel.

An example of just such a case is provided by a recent hypothalamic obesity drug study (unpublished results, Jia, 2004*). Figure 3.2 shows a rat hypothalamus slice placed on top of an 8 × 8 multi-electrode array and several single unit recordings that were taken at the same time from different indicated locations in the slice. After recording a stable spike rate baseline for each of the units, 0.1 μM Ghrelin was added to the bath. Some units reacted with an increase in firing rate, and others instead decreased their firing rate. Previous work in the hypothalamus predicts this effect and attributes it to different cell types. These expected but opposite responses serve as a within-slice control condition helping to validate the results gathered from each slice.

Hypothalamic slices on MEAs have proven effective for testing obesity drugs. Figure 3.3 compares the results obtained with 5-HT and d-FEN (the "Fen" half of the weight loss drug known commercially as "Fen-Phen"). The results show that the two compounds have similar effects as expected: both are known to suppress appetite in behavioral studies. For this application the MEA offered key advantages. On average, four to five units were recorded per slice without the need to implant

---

* Data courtesy Dr. Yousheng Jia, Tensor Biosciences, Irvine, CA 92612, USA.

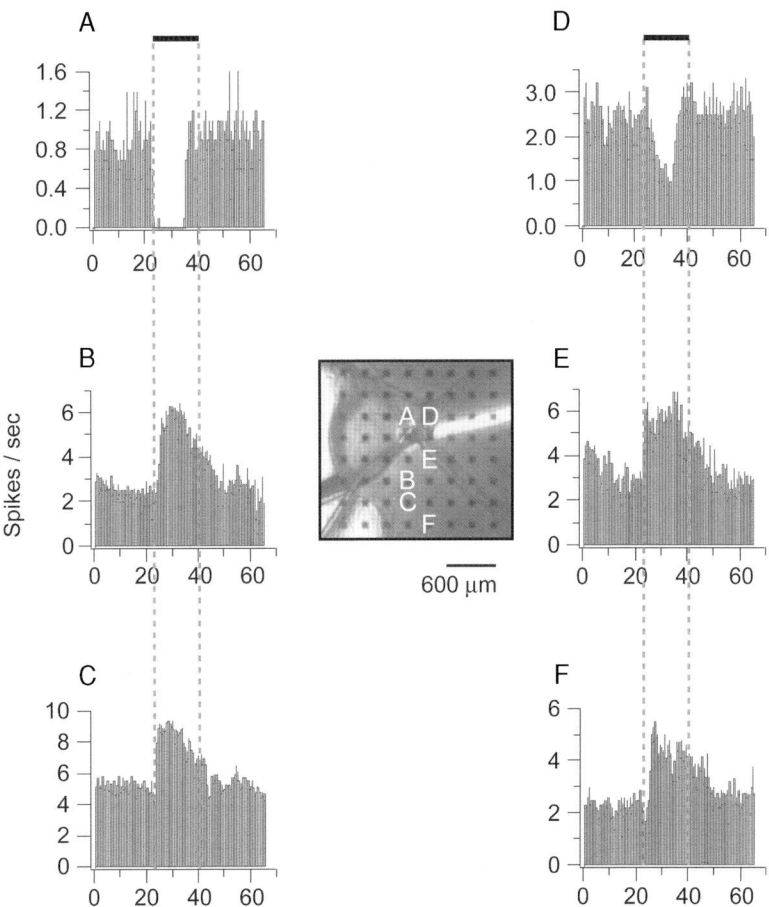

FIGURE 3.2. A hypothalamic slice with multiple spiking unit recordings from a MEA. The central image is a micrograph taken of a hypothalamic slice on top of an array of 64 electrodes arranged in an 8 × 8 grid. Each of the small black squares is an electrode. Six different electrodes produced spiking single units. These are indicated on the micrograph in white letters A–F. Each of the six electrodes has a correspondingly lettered spike frequency plot (spikes/second) indicating the change in spiking behavior observed during baseline, Ghrelin application, and washout. (Courtesy of Tensor Biosciences.)

and move multiple electrodes by hand, as would be necessary to achieve this many unit recordings with traditional instrumentation. And the recording of cells with opposite drug reactions helped to validate the results obtained from each slice.

Dissociated Dorsal Root Ganglia Cell Cultures

In humans, sensory axons innervating the torso and limbs originate from the dorsal root ganglia (DRG). Its cells serve to relay sensory information from muscles, skin, and joints to the spinal cord. Much of the research in this area concerns the treatment

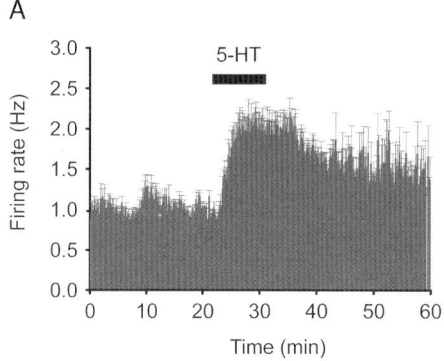

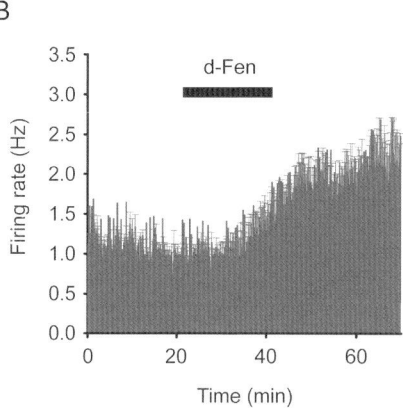

FIGURE 3.3. The excitatory effect of 5-HT and d-FEN on single units in rat arcuate nucleus. (A) The spiking rate of multiple single units is plotted before and after the application 5HT; (B) and also before and after the application of d-FEN. In both case, the horizontal black bar below the drug name indicates the duration of drug application. (Courtesy of Tensor Biosciences.)

of pain. In particular, the inhibition of capsaicin-induced current may provide a basis for reducing capsaicin receptor-mediated nociception (Chen et. al., 2004). Recent work using cultured DRG neurons on a MEA demonstrates the feasibility of studying spiking DRG neurons at multiple sites in parallel.

A MEA with cultured DRG neurons spread broadly across its 64 electrodes is shown in Figure 3.4. Spiking behavior is seen at three of the electrodes after applying 1 μM Capsaicin. Here, the primary advantage of the MEA is through-put and labor reduction: one simply spreads the neurons broadly over the array, waits for the culture to mature, and then examines all 64 sites in parallel to find good spiking neurons. Probably not all sites will have neurons worth recording, but even if just a few are found, this approach is still much faster than using traditional instrumentation to search by moving one or more implanted electrodes by hand.

A

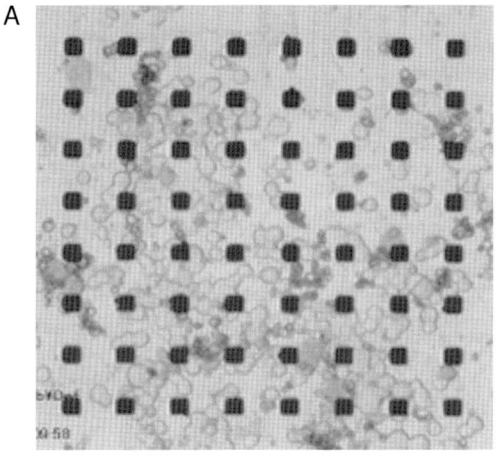

B

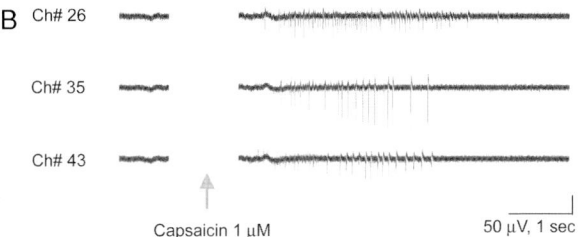

Capsaicin 1 µM   50 µV, 1 sec

FIGURE 3.4. Cultured DRG neurons showing Capsaicin induced spiking. (A) Micrograph of cultured rat DRG neurons after three days on MED-P515A probe (150 µm spacing). (B) Spontaneous responses were recorded from each electrode. Capsaicin (1 µM) elicited repetitive spikes for several seconds. (Courtesy of Alpha MED Sciences, Co., Ltd.)

Co-Cultured Organotypic Septo-Hippocampal Slices

In contrast to the dissociated cell culture of the previous section, the micrograph shown in Figure 3.5 depicts an organotypic co-culture. The tissue slice at the top of the micrograph is a slice taken from a neonatal rat (+9 days), and the slice at the bottom is from the septum (+5 days). In the intact brain, the septum sends cholinergic axonal projections into the hippocampus that are thought to help govern rhythmic activity in the hippocampus. After being co-cultured for 19 days, spike recordings are taken using the two electrodes circled in Figure 3.5.

Figure 3.6 shows how spiking behavior is affected by the application of physostigmine and atropine: the former increases the spike rate over baseline, whereas the latter decreases it. Because physostigmine is a cholinesterase inhibitor and atropine is a muscarinic receptor inhibitor, these results strongly suggest that working cholinergic synapses have formed in the hippocampus due to an influx of septal projections. In other words, in vivo septo-hippocampal anatomy is being

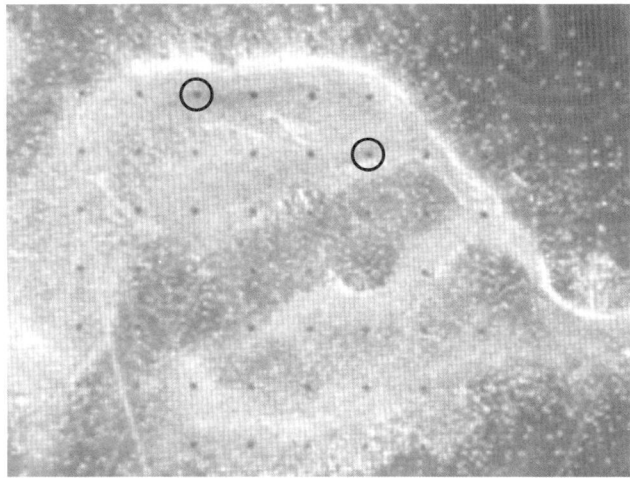

FIGURE 3.5. Micrograph of septo-hippocampal co-culture. Septum slices (bottom half, from 5-day-old rats) were co-cultured with hippocampal slices (upper half, from 9 day-old rats) on the Panasonic MED-P545 probes for 19 days (450 μm interelectrode distance). Spontaneous activity was measured from the two electrodes shown circled.

mimicked by the co-culture. The use of a MEA with transparent wiring here helps make tracking the health of the culture visually very easy.

3.1.2   Evoked Field Potentials

Acute Hippocampal Slices

The hippocampus is a key player in the formation of memories. As such, it is the focus of intense research aimed at treating Alzheimer's and other memory-related disorders. It is also very well suited for slice studies owing to the planar organization of projections among its subfields. Figure 3.7 summarizes its major projections systems and the regions of interest to electrophysiologists. For each subfield, there is a different preferred method of stimulation and a different expected response. The study of these responses can reveal much about the mechanisms underlying the function of a compound.

A number of researchers are pursuing the study of the hippocampus with MEAs (Novak and Wheeler, 1988; Egert et al. 1998; Shimono et al. 2000). Figure 3.8 shows a case where an AMPA reuptake inhibitor is applied to a hippocampal slice, and altered responses are seen under four different conditions at three locations in the slice (two different stimulation patterns are applied to CA1). Although these experiments could also be performed with traditional instrumentation, there are a few key advantages gained through the use of MEAs here. There is increased throughput owing to the ability to stimulate and record from multiple sites, and the different responses expected in the various regions act as within-slice controls that help validate the viability of the slice and the experimental conditions.

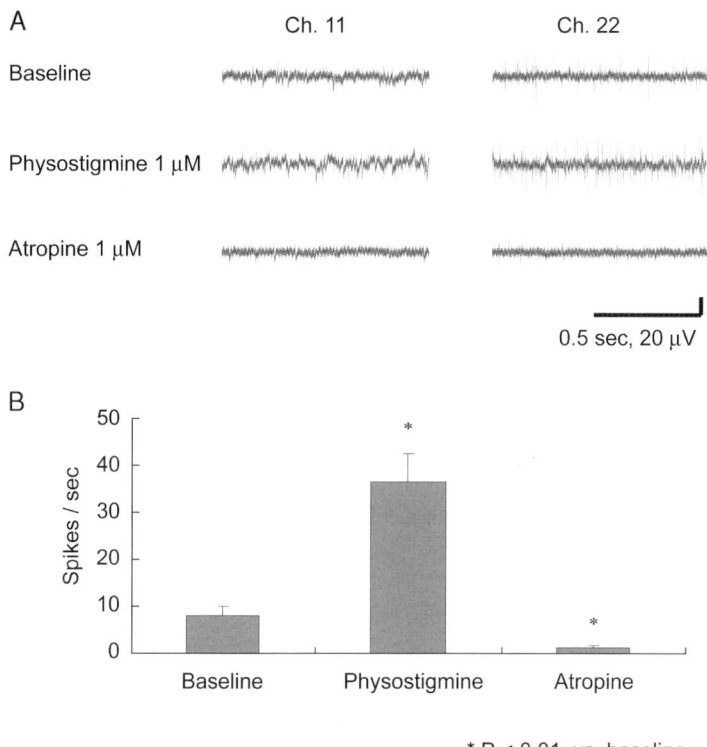

FIGURE 3.6. Evaluation of drug effect on co-cultured septo-hippocampal slice. (A) Physostigmine 1 μM, a cholinesterase inhibitor, increased activities in both CA1 and CA3. Atropine 1 μM, a muscarinic receptor antagonist, blocked activity. (B) Quantification of the effects of physostigmine and atropine. Results are means ± SEM of data obtained in 2 slices, 4 recording electrodes in hippocampus area in each slice.

Acute Spinal Cord Slices

Research focused on the spinal cord concerns both functional questions as well as treatments for various health problems, among them paralysis and pain. Several studies of spinal cord slices employ MEAs (Tscherter et al., 2001; Streit et al., 2001; Darbon et al., 2004; Legrand et al., 2004). Figure 3.9 demonstrates the viability of using an MEA to study spinal cord tissue. A MEA is placed on the dorsal horn of a rat spinal cord slice. Upon stimulation, the expected responses are seen among the electrodes adjacent to the stimulation sites.

Acute Heart Slices

Among electrophysiologists, heart research is largely concerned with the oscillatory behavior of the neurons that control heart muscle contractions. This work can lead to a better understanding of heart function and treatments for diseases such

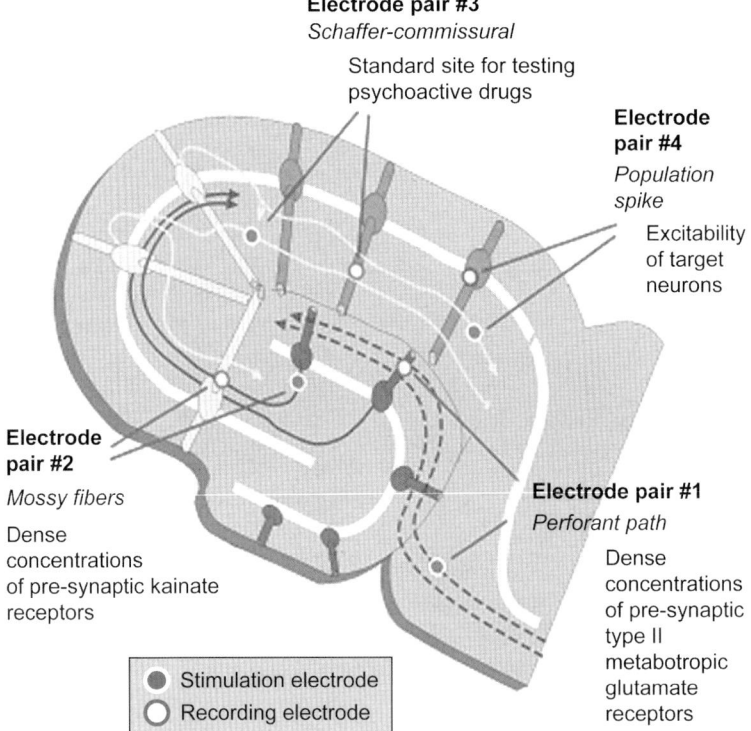

Electrode pair #3
Schaffer-commissural
Standard site for testing
psychoactive drugs

Electrode pair #4
Population spike
Excitability of target neurons

Electrode pair #2
Mossy fibers
Dense concentrations of pre-synaptic kainate receptors

Electrode pair #1
Perforant path
Dense concentrations of pre-synaptic type II metabotropic glutamate receptors

● Stimulation electrode
○ Recording electrode

FIGURE 3.7. Stimulating and recording sites on hippocampal tri-synaptic circuit. Axonal projections travel among the subregions. The order of the projections is as follows: enthorinal cortex to DG (electrode pair #1) to CA3 (electrode pair #2) to CA1 (electrode pair #3) to subiculum (electrode pair #4). For each of the regions shown with a stimulation and recording electrode pair, there is a characteristic response expected for a particular stimulation pattern. (Courtesy of Tensor Biosciences.)

as arrhythmia. A number of researchers employ MEAs in this field (Feld et al., 2002; Lu et al., 2004). Figure 3.10 demonstrates the use of a MEA to study a drug effect on heart tissue. Here ventricular tissue is placed on top of the array and evoked responses are recorded using 100 mM Quinidine during baseline, washin, and washout. The drug increases the latency and duration of the evoked action potentials.

3.1.3 Spontaneous Field Potentials

Acute Hippocampal Slice Oscillations

Rhythmic oscillations are a fundamental feature of brain physiology. Indeed, disruptions to brain wave activity (electroencephalograms) have been used clinically

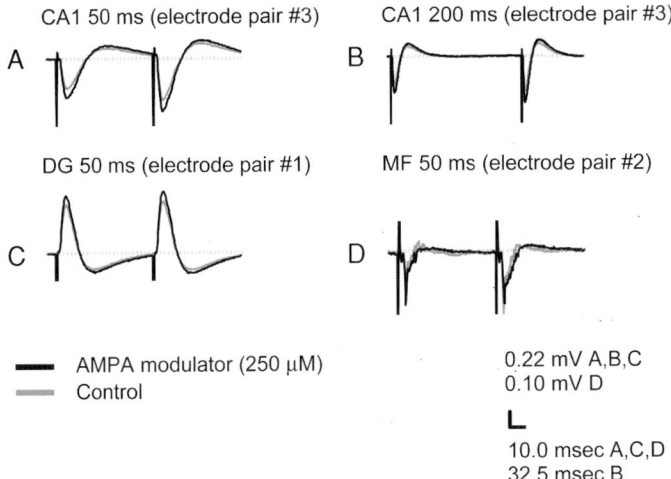

CA1 50 ms (electrode pair #3)   CA1 200 ms (electrode pair #3)

A   B

DG 50 ms (electrode pair #1)   MF 50 ms (electrode pair #2)

C   D

━━ AMPA modulator (250 µM)   0.22 mV A,B,C
━━ Control   0.10 mV D

                                  L
                                  10.0 msec A,C,D
                                  32.5 msec B

FIGURE 3.8. Drug effects on paired-pulse field EPSPs in various areas of rat hippocampus. The gray trace is before and black trace is after the application of a 250 µM AMPA modulator. (A) Shows response from CA1 for a 50 msec paired-pulse stimulation; (B) Shows response from CA1 for a 200 ms paired-pulse stimulation; (C) Shows response from DG for a 50 msec paired-pulse stimulation; (D) Shows response from CA3 for a 50 msec paired-pulse stimulation. (Courtesy of Tensor Biosciences.)

for decades to diagnose brain damage and disease in vivo. Now researchers are using hippocampal slices and MEAs to study this oscillatory behavior in vitro (Shimono et al., 2000; Paulsen, 2003). In an intact brain, cholinergic inputs from the septum to the hippocampus govern hippocampal rhythmic activity. In a slice, carbachol can be used to artificially induce cholinergic rhythms in the hippocampus. Figure 3.11 shows the varied effects of carbachol within each subfield of the hippocampus. A steady beta rhythm (10 to 30 Hz) is seen most prominently among the apical dendrites of CA1 and CA3. In Figure 3.12, using a larger MEA to record from a broader area reveals the varied effects of carbachol for two different anatomical structures: the rhythmic oscillations of entorhinal cortex are found to be higher in frequency than those of the hippocampus. Notice that the use of MEAs here enables the easy observation of rhythms at multiple slice locations in parallel.

Synchronized Cardiac Muscle and Stem Cell Culture Activity

Stem cell research holds great potential for improving both our understanding of brain function and the treatment options for diseases such as Parkinson's. Some stem cell researchers are turning to MEAs for their electrophysiology work (unpublished results, Kodama et al., 2004). Figure 3.13 provides an example: a MEA is shown with a partition dividing its left and right sides. Cultured neonatal cardiac

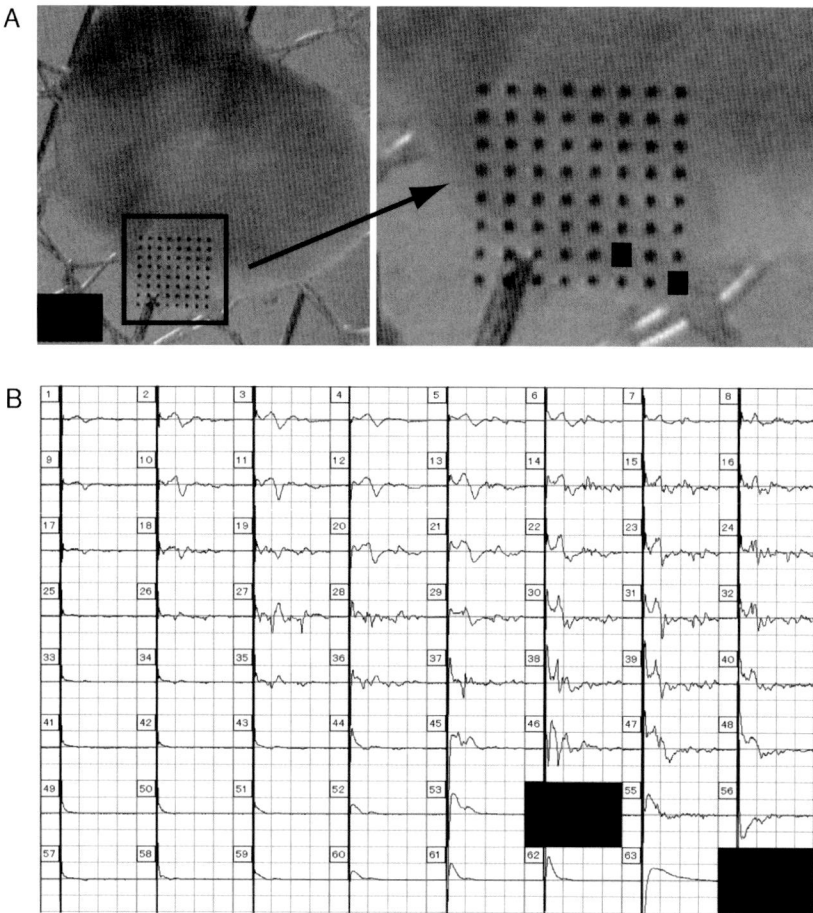

FIGURE 3.9. Evoked response from rat spinal cord slice. (A) Micrographs of a slice from rat spinal cord placed on a Panasonic MED-P210A probe (100 μm inter-polar distance) and centered on dorsal horn (left) and close-up of outlined region (right). (B) Bipolar stimulation is delivered using electrodes located at the edge of dorsal root (marked with large black squares). (Courtesy of Alpha MED Sciences Co., Ltd.)

muscle cells have been grown on the left side, and embryonic stem cell derived cardiac muscle cells have been grown on the right side. Part A of Figure 3.14 shows the activity obtained from the separated cells. Both sides are clearly alive and active. Part B of Figure 3.14 shows that after removing the partition and waiting for three days, synchronous activities appear among the stem cells on the right side. This study suggests that cardiac muscle cells derived from embryonic stem cells can create "electrical syncytium" with intact cardiac muscle cells. Similar work has also been done by Egashira et al. (2004). The use of a MEA for this application makes it quite easy to locate sites of synchronized activity on each side of the partition line.

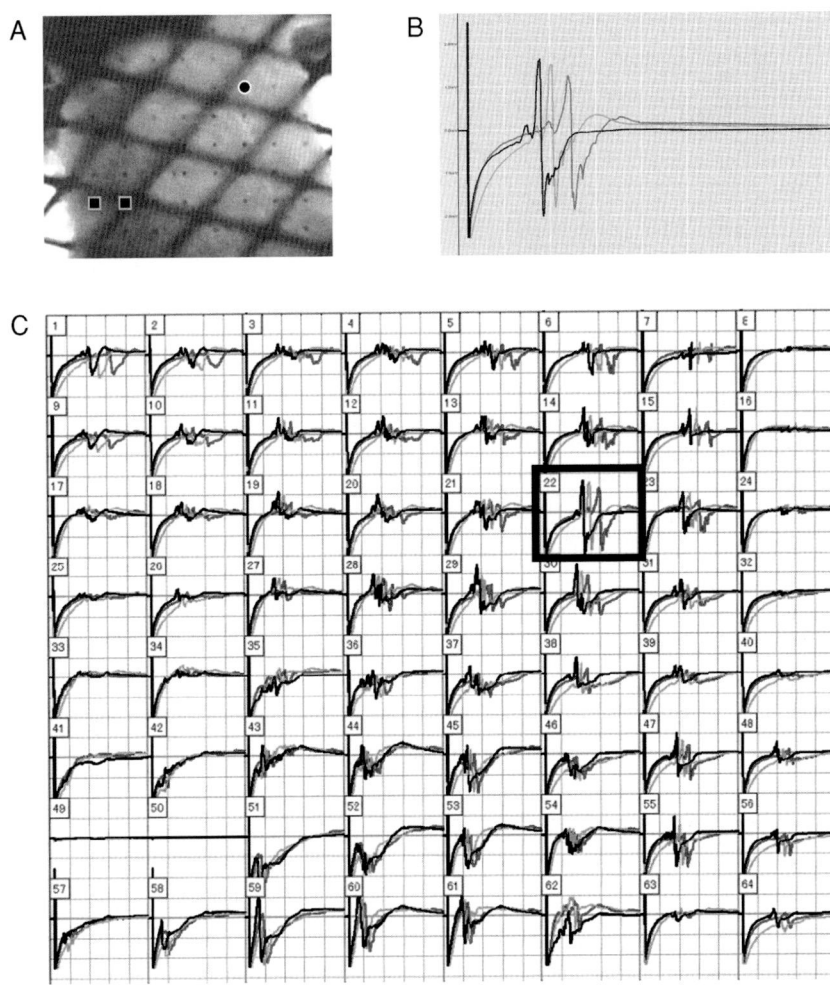

(1.0 mV, 20 ms/div)

FIGURE 3.10. Drug effect on acute ventricular muscle slice. (A) Micrograph of adult rat left ventricular muscle slice with 250 μm thickness. The two filled black squares near the lower left denote the stimulation sites while the black circle near the upper right the recording site of interest—electrode 22. (B) Pacing response at electrode 22 evoked by 100 μA electric stimulation to the two electrodes: the leftmost black trace is without drug, the rightmost medium gray trace is with drug, and the middle light gray trace is washout. In the presence of 100 μM Quinidine, the latency and the duration of the action potential were prolonged. (C) The responses from all 64 electrodes with electrode 22 outlined in black. (Courtesy of Alpha-MED Sciences, Co., Ltd.)

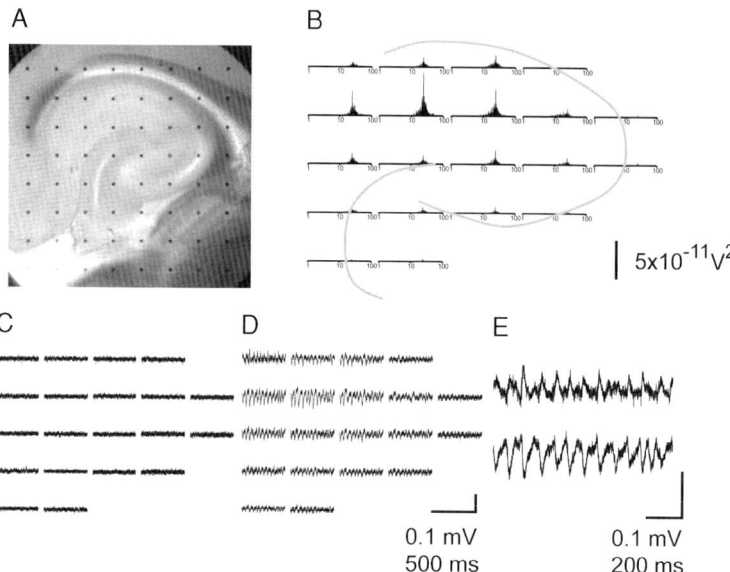

FIGURE 3.11. Distribution of carbachol-induced beta waves within the hippocampus. (A) Micrograph of hippocampal slice on an MEA. (B) Spectra of carbachol-induced spontaneous activity at 20 electrodes on a logarithmic scale from 1 to 100 Hz. Activity is seen in the 10 to 30 Hz frequency range, primarily in apical dendrites of fields CA1 and CA3 (calibration bar: 5×10^{-11} V^2). (C) Activity in baseline conditions. (D) Activity measured after infusion of 50 μM carbachol (calibration bars for (C) and (D): 0.1 mV; 500 msec). (E) Reversal of polarity across the cell body layer of field CA1 (calibration bars: 0.1 mV; 250 msec). (Figure copyrighted 2000 by the Society for Neuroscience.)

3.1.4   Summary of MEA Advantages for Traditional Physiology

MEAs enable the performance of many new types of experiments that cannot be performed at all with traditional instrumentation (see Section 3.2). However, as a potential replacement for traditional instrumentation in the performance of traditional experiments, MEAs, like any new tool, have their strengths and weaknesses.

The potential advantages of MEAs over traditional instrumentation for various combinations of tissue preparations and slice chamber conditions are summarized in Table 3.1. For each combination, a list of the potential advantages is given. The lists use the terms: "site sel" which refers to speeding up the selection of stimulation and recording sites that yield stable baseline responses; "parallelism" refers to the potential to essentially run multiple traditional experiments in one slice; and "sterile" refers to the potential to increase the sterility of slice conditions, which is only important for cultures, by growing them directly on MEAs, storing them under closed conditions, and avoiding repeated electrode insertions and extractions. Next to the list items are plus, minus, and equal signs indicating

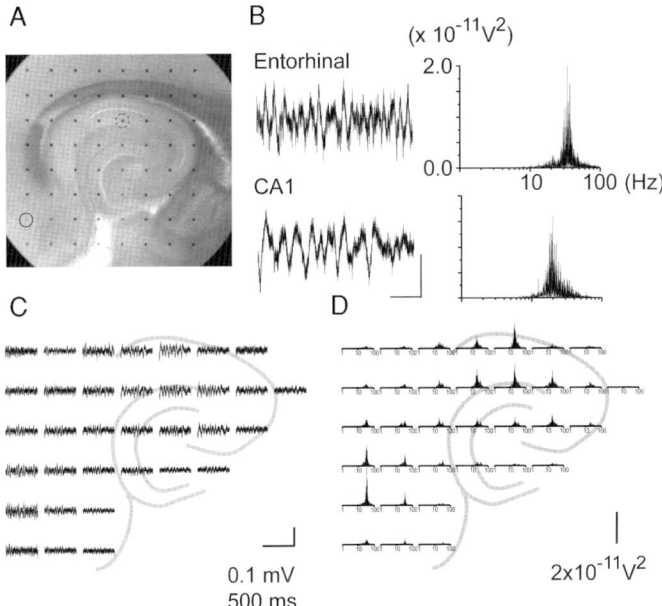

FIGURE 3.12. Two distinct carbachol-induced rhythms in hippocampus and cortex. (A) Micrograph of a cortico-hippocampal slice on a broad array. The solid circle in the lower left denotes an entorhinal cortex electrode, the dotted circle in the upper middle a CA1 electrode. (B) Sample activity in response to infusion of carbachol (50 µM) from the selected electrodes in entorhinal cortex and field CA1 (left) and power spectra for recordings over 3 seconds at these two sites (right). (Left calibration bars: 50µV, 100 msec; right charts share the same calibration.) Although field CA1 exhibits betalike rhythm centered around 20 Hz, carbachol elicits higher frequency (35 to 40 Hz) activity in entorhinal cortex (top right spectrum). (C) Distribution of representative activity in the slice (calibration bars: 0.1 mV, 500 msec). (D) Distribution of low-pass (0 to100 Hz) filtered power spectra in the slice (calibration bar: 2×10^{-11} V^2). (Figure copyrighted 2000 by the Society for Neuroscience.

when there is an advantage, disadvantage, or no gain for MEAs over traditional setups. A plus/minus combination means the results obtained will depend on the specific application. An equal/minus combination means they are either equal or there are disadvantages to using MEAs over traditional instrumentation.

A review of the table entries reveals that some combinations of experimental conditions and tissue preparations more clearly favor the use of MEAs than others. The most potentially advantageous combination is "dissociated cell cultures" and "static and submerged" conditions, where one would expect faster selection of stable recording sites, many parallel experiments (in the best case one per electrode), and increased culture sterility if desired. The least potentially advantageous combination is "acute slices" and "perfused and interface" conditions. Here, the careful tuning of the fluid level that may be necessary to achieve a stable baseline

Partition

NCM   ESCM

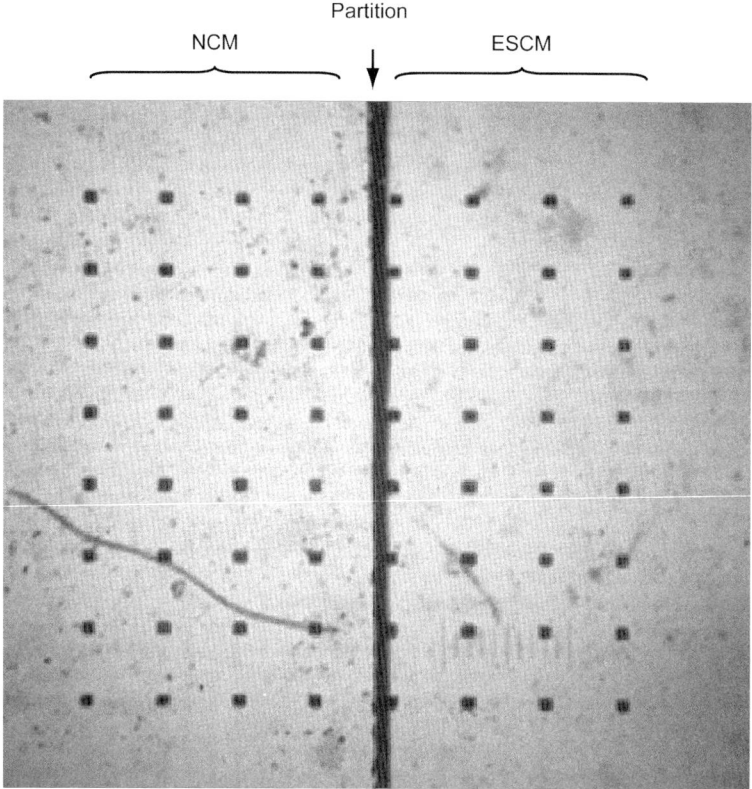

FIGURE 3.13. Detailed view of the microelectrode array area containing cardiac muscle cells and embryonic stem cells. Left side: neonatal cardiac muscle cells (NCM). Right side: embryonic stem cell cardiac muscle cells (ESCM). A physical partition divides the two sides at the beginning of the experiment.

under interface conditions can slow the site selection process; however, the impact of this issue depends mostly on the skill of the practitioner, so with training it can be mitigated.

At the risk of stating the obvious, the following conditions should be met by an experiment before undertaking it with MEAs.

• Because it is impossible to move any of the electrodes independently, one needs an array size and shape that conforms to the stimulation and recording electrode placements needed for the experiment;
• The speed advantages offered by MEAs increase with decreasing placement accuracy requirements, because there is an inherent tradeoff between the time taken to place a slice on an array and placement accuracy;
• Given that the recordings made with a MEA often have smaller response amplitude than traditional instrumentation, it is important to make sure that the signal-to-noise ratio is sufficient to gather viable data.

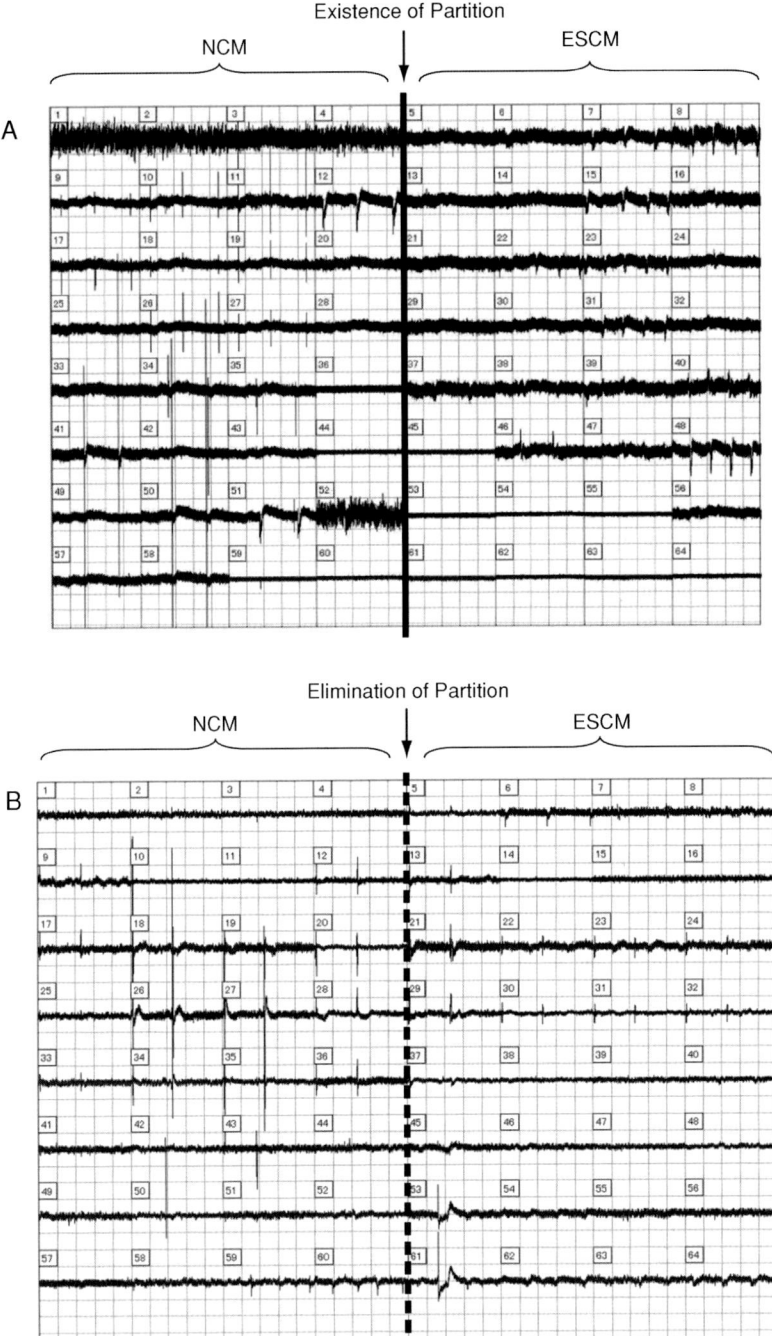

FIGURE 3.14. Activity changes of cardiac muscle cells derived from embryonic stem cells. (A) Spontaneous field potentials from 64 sites in the presence of partition. Independent activities observed in both left and right side cultured cells. (B) Electrical responses from 64 sites in the absence of partition. (Courtesy of Drs. Lee and Kodama, Nagoya University, Japan.)

TABLE 3.1. Potential MEA advantages over traditional setups.

	Acute slices	Dissociated cell cultures	Organotypic cultures
Perfused and submerged	+ site sel +/− parallelism	No data	No data
Perfused and interface	= /− site sel +/− parallelism	No data	No data
Static and submerged	+ site sel +/− parallelism	+ site sel + parallelism + sterile	+ site sel +/−parallelism + sterile
Static and interface	+ site sel +/− parallelism	No data	No data

3.2   New Methods: Network Physiology

This section presents a sampling of MEA applications that are difficult or impossible to perform using traditional instrumentation. These applications rely upon various unique characteristics of MEAs and are typically concerned with the study of network-level phenomena. Many make use of the two-dimensional (2-D) nature of the MEA grid to measure network phenomena in new ways, such as 2-D current source density and phase maps. Others use MEAs to grow networks in culture form directly on the multi-electrode array, which enables stimulation and recording from specific locations in the slice day after day; with traditional culture preparations on membranes the electrodes cannot easily be maintained in specific locations for long periods.

3.2.1   Two-Dimensional Current Source Density Analysis

Discovering Functional Anatomy in the Hippocampus

For decades there has been a great deal of interest in modeling the functional behavior of the hippocampal trisynaptic circuit (Marr 1971; McNaughton and Morris, 1987; Treves and Rolls, 1992; Jarrard, 1993; O'Keefe and Recce, 1993; Buzsaki and Chrobak, 1995; Muller, 1996; Eichenbaum, 1997; Nadel and Moscovitch, 1997; Squire and Zola, 1998; Tulving and Markowitsch, 1998; Lisman, 1999). These efforts have been aided by traditional anatomical and physiological studies that have revealed a great deal about its various components. However, without 2-D recordings of activity patterns using many electrodes, it is difficult to directly determine the extent and pattern of activity propagation across large sections of tissue. In many cases, modelers have had to deduce these properties from indirect evidence.

Work by Shimono et al. (2002) shown in Figure 3.15 demonstrates how an MEA can be employed to reveal the functional anatomy of a region: here axonal inputs to the apical dendrites of the CA1 region are stimulated at various points along the proximal-distal axis. 2-D current source density (2-D–CSD) analysis shows where current sinks develop in response to stimulation at each location. A cumulative analysis of these activity patterns (Figure 3.16) reveals that inputs to CA1 at different levels along the proximal-distal axis have varying levels of

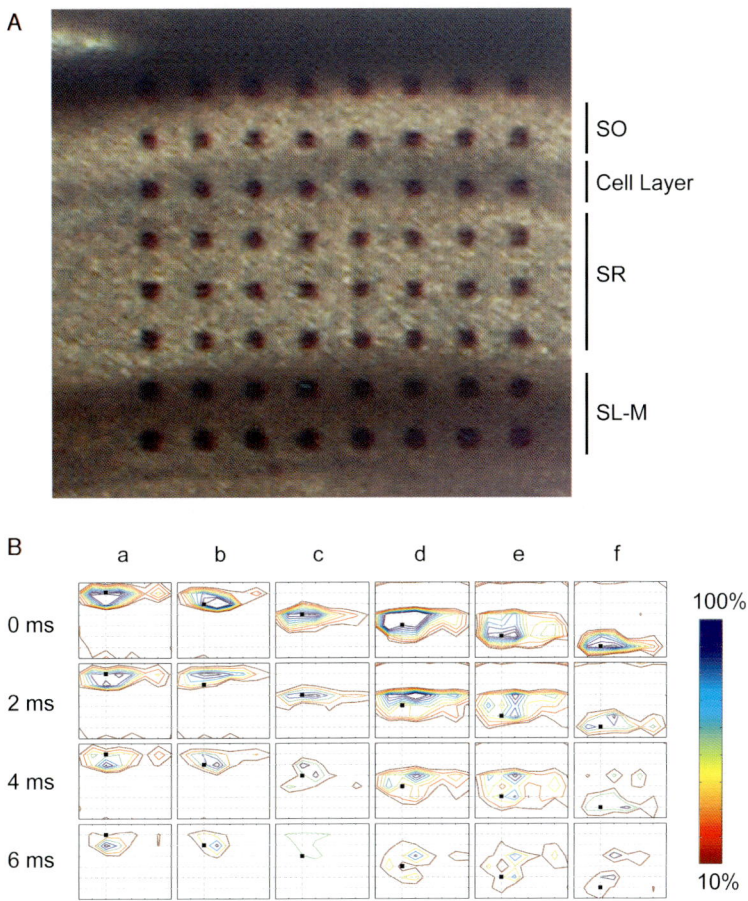

FIGURE 3.15. Computed current sinks shifted along the proximo-distal axis of the CA1 region. (A) Micrograph of a hippocampal slice placed on the microelectrode array with inter-electrode spacing of 100 μm and centered on apical region of field CA1. (B) The current sinks evoked by stimulation in stratum oriens (second row of the array) are shown in column a. The current sinks evoked in the cell layer—third row of the array—are shown in column b. The current sinks evoked in the stratum radiatum (fourth, fifth, and sixth rows of the array) are shown in columns c, d, and e. The current sinks evoked by the stimulation in the stratum lacunosum/moleculare (seventh row of the array) are shown in column f. In each contour map the filled black square denotes the site of the stimulation electrode. (Reprinted from Shimono et al., 2002, with permission from Elsevier.)

influence on the apical dendritic field: the more distal stimulations were found to be broader than the proximal ones, suggesting the existence of proximally directed collaterals. The lateral influence of the projections, moving parallel to the cell body layer, was essentially bandlike, suggesting that CA1 pyramidal cells are activated in a largely uniform way by these inputs. Note that these results could not be obtained without the use of a 2-D MEA to enable calculation of 2-D current source densities.

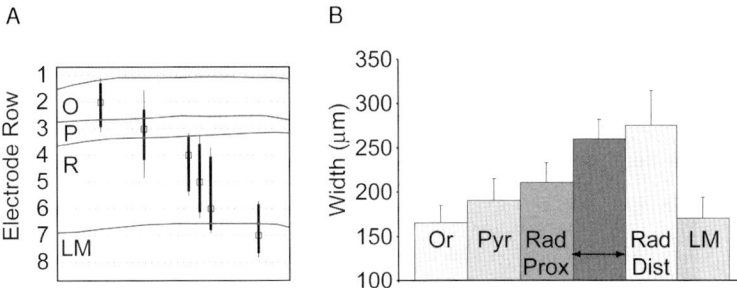

FIGURE 3.16. Statistical analysis of normalized current sinks along the proximo-distal axis. A) micrograph depicting typical electrode placement relative to slice. Single rows of electrodes provide coverage of laminar regions s. oriens, s. pyramidale, and s. lacunosum-moleculare, while three rows span the proximo-distal expanse of s. radiatum. Stimulation by electrodes in each of these rows, shown by black boxes (not drawn to scale), results in sinks whose average proximal and distal extents are shown by thick black bars. Thin black bars represent standard deviations. Results for regions s. oriens through s. lacunosum-moleculare proceed left to right to allow side by side comparison and do not correspond to electrode columns. B) Average proximo-distal widths of current sinks by lamina. Sinks within s. oriens are relatively small, remaining constrained by the cell boundary layer. As stimulation proceeds apically, sinks increase in width. (Reprinted Shimono et al., 2002, with permission from Elsevier.)

The work discussed above employed the technique known as current source density analysis. Through a mathematical transformation of field potential data (Nicholson and Freeman, 1975; Nicholson and Llinas, 1975), an estimate of current source densities can be derived. When the transform is extended to operate on MEA-acquired 2-D field potential data (Shimono et al., 2000), a clear picture of where high and low concentrations of currents into and out of neuronal regions is revealed. If this technique is applied to every timeslice of recorded data, dynamic network-level phenomena become apparent. In short, 2-D–CSD analysis is an example of a new class of tools created to help interpret two-dimensional data and will increase our understanding of the underlying computations of neuronal circuits. The next two subsections introduce other efforts that employ this tool.

Rhythmic Oscillations in the Hippocampus

Previously, MEAs were shown to add value to the study of rhythmic oscillations in the hippocampus by enabling the easy observation of carbachol induced rhythms at multiple sites in parallel (see Section 3.1.3). However, the spectral analysis of the data gathered across these sites was performed independently per site. The results shown in Figure 3.17 contain a series of 2-D–CSD analyses that can only be performed by combining the readings obtained across a 2-D matrix of equally spaced electrodes. The series reveals the pattern current sources and sinks that emerge as a result of carbachol-induced rhythmic oscillations.

From an arbitrarily chosen starting point, a sink appears in the apical dendrites of the border between fields CA3 and CA1, with an associated source across the

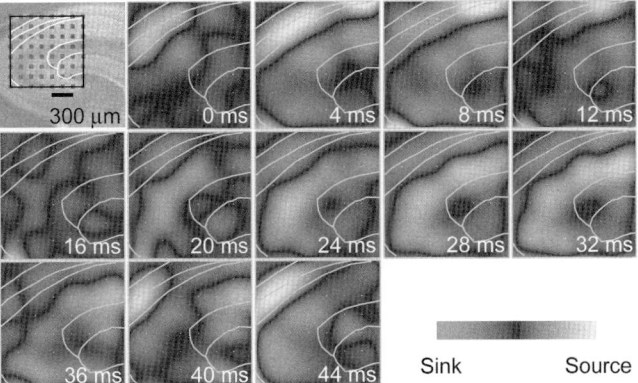

FIGURE 3.17. Current source density analyses of carbachol-induced activity in hippocampus. Each frame shows the instantaneous computed current source density in the region of the electrode array. Taken together they reveal a repeating pattern of sink/source formation across the region. (Figure copyrighted 2000 by the Society for Neuroscience.)

cell boundary layer in the basal dendrites. Within a few milliseconds, an additional focal sink has appeared in apical CA1 with a corresponding basal source. The fields merge and intensify, then dissipate after about 12 msec. After a brief interim during which activity is not distinguished from background, a source appears in the apical dendrites at about 20 msec, with a corresponding sink in the basal dendrites. These expand and intensify before dissipating by roughly 20 msec later (40 msec), after which an apical sink reappears to reinitiate the cycle. The cycle repeats indefinitely (as indicated in the next figure), with an approximate frequency of 25 Hz.

Atrial Pacing

Previously a MEA was shown to provide a viable platform for studying evoked responses in heart tissue (Section 3.1.3) however, the 2-D nature of the array was used only to speed up the experiment. In contrast, Figure 3.18 shows a 2-D–CSD analysis of sink/source propagation in response to a stimulation delivered at two electrodes in the center of the slice. A series of snapshots of the activity is shown at 5 msec intervals represent a single cycle of pacing oscillatory activity, which begins with sink formation primarily in the upper left quadrant, replaced by a source in roughly the same location 15 msec later. This kind of analysis requires the MEAs a grid of equally spaced electrodes.

3.2.2   Phase Maps of Planar Signal Propagation

Ventricular Pacing

Another form of 2-D analysis that can be performed using a MEA is a phase map. Figure 3.19 shows a phase map obtained by stimulating ventricular tissue at two

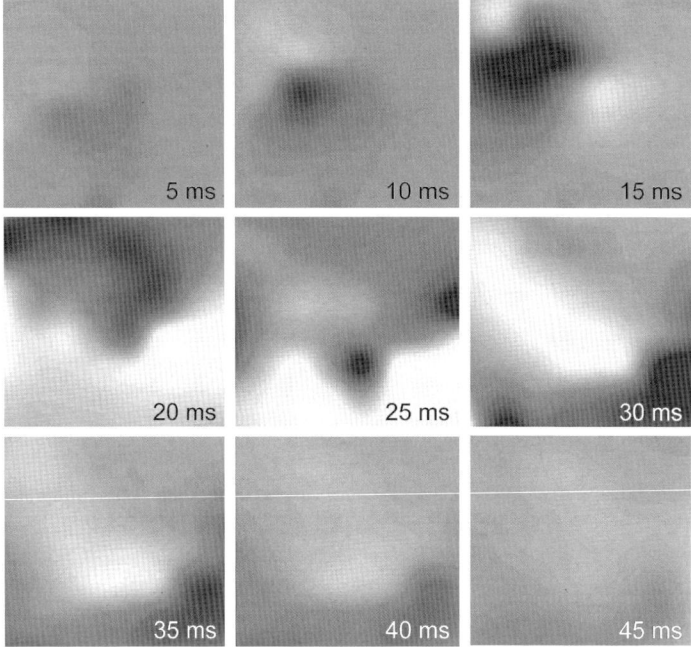

FIGURE 3.18. Propagation of atrial pacing activity. Each time frame shows a computed two-dimensional current source density at a particular time. The times after the stimulation, delivered to the center of the slice, are shown in each panel. Positive potentials are white and negatives are black. (Courtesy of Dr. H.Yeh, Mackay Memorial Hospital, Taiwan.)

stimulation sites (part A) and observing the signal propagation delays (part B). A phase map is a contour plot where the contour lines indicate specific signal propagation delays. In this particular example, the contour lines are plotted at 5 msec intervals. The earliest contour line (5 msec) is closest to the stimulating electrodes in the upper right quadrant of part B. From here, the signal does not propagate across the tissue at a uniform speed; rather, it propagates more quickly through the lower half of the figure than the upper half. This sort of analysis, though possible on a smaller scale (fewer electrodes) using traditional instrumentation, is made very simple over large regions by the use of a MEA.

Atrial Pacing

Similar to the result just seen for ventricular tissue, signal propagation in atrial tissue can be studied using MEAs. Figure 3.20 shows a case where two sites on the left side of the slice are stimulated (the black squares in part A). The resulting phase map is shown in part B. Unlike the phase map from the ventricular tissue, signals propagate from the stimulation site to the right in a very uniform wavelike manner.

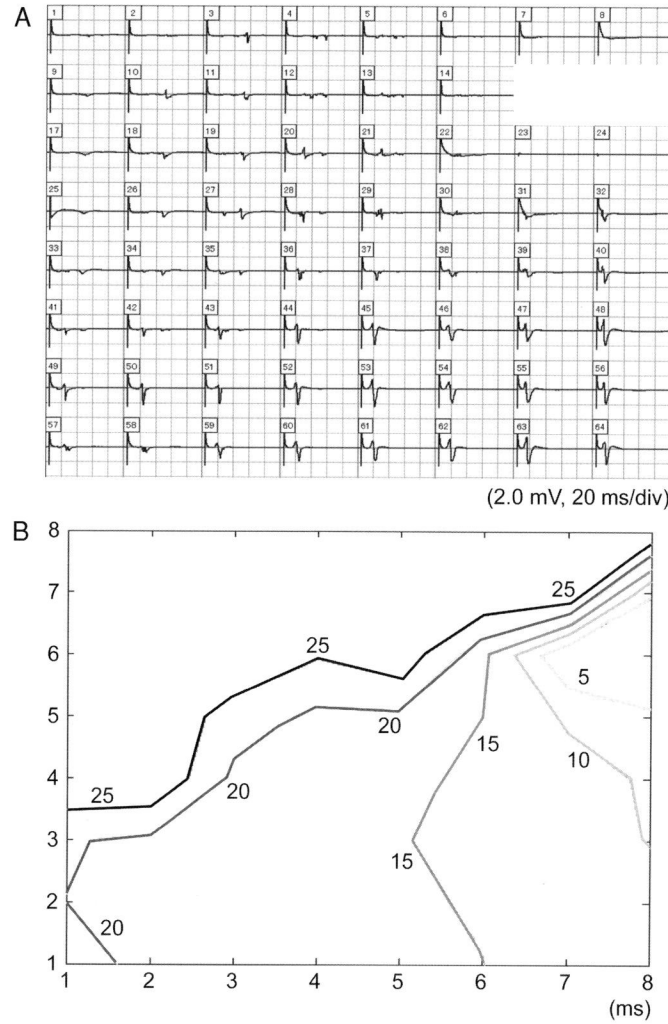

FIGURE 3.19. Evoked response from guinea pig ventricular muscle. (A) Pacing responses (2.0 mV, 20 msec/div) evoked by electrical stimulation (100 μA) to the two electrodes on Panasonic MED probe (marked with black rectangles). (B) Phase map of the pacing activity. Each contour shows the latency of the evoked responses from stimulation. (Courtesy of Dr. Tsubone, Graduate School of Agriculture and Life Sciences, University of Tokyo, Japan.)

3.2.3   Long-Term Site-Specific Culture Studies

LTP Lasting Days in Organotypic Hippocampal Cultures

Organotypic cultures grown directly on the MEA surface were introduced in Section 3.1.1 in the form of septo-hippocampal co-cultures. There, the use of MEAs simplified recording and helped lower the risk of infection; however,

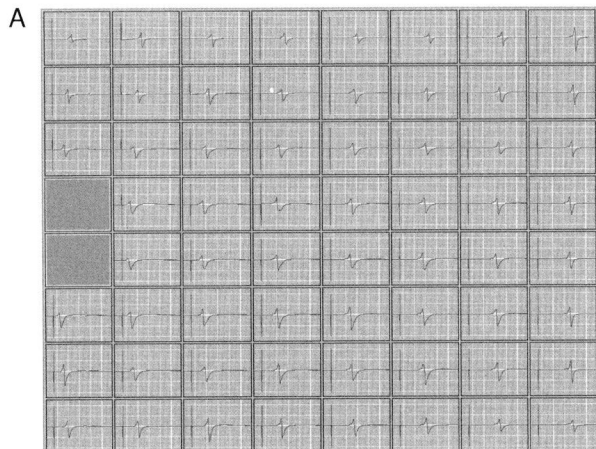

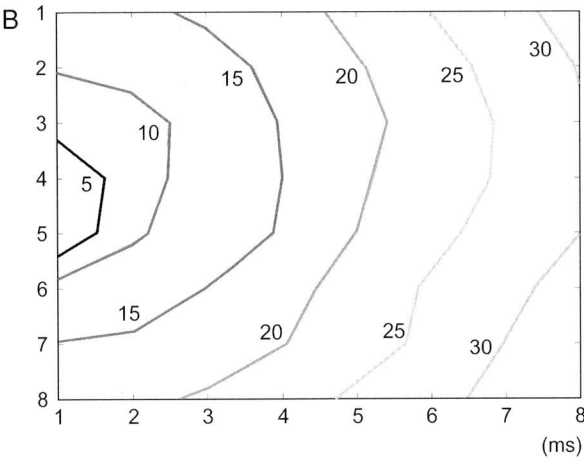

FIGURE 3.20. Primary myocyte culture. (A) Pacing responses were evoked by electric stimulation to the two electrodes on MED probe: the black squares indicate the stimulation sites. (B) Phase map of the pacing activity. Each contour shows the latency in milliseconds of the evoked responses from stimulation. (Courtesy of Drs. Lee and Kodama, Nagoya University, Japan.)

the same results could have been gathered using traditional instrumentation and cultures grown on membranes. Figure 3.21 presents an experiment that can only be performed using a culture grown on a MEA.

The experiment is an LTP study conducted over a two-day period using organotypic hippocampal cultures. Field EPSPs evoked by Schaffer collateral fiber stimulation were recorded in field CA1 before and after tetanus stimulation. High-frequency stimulation intended to potentiate the pathway was delivered during the first hour of the experiment. The maximum amplitude of field

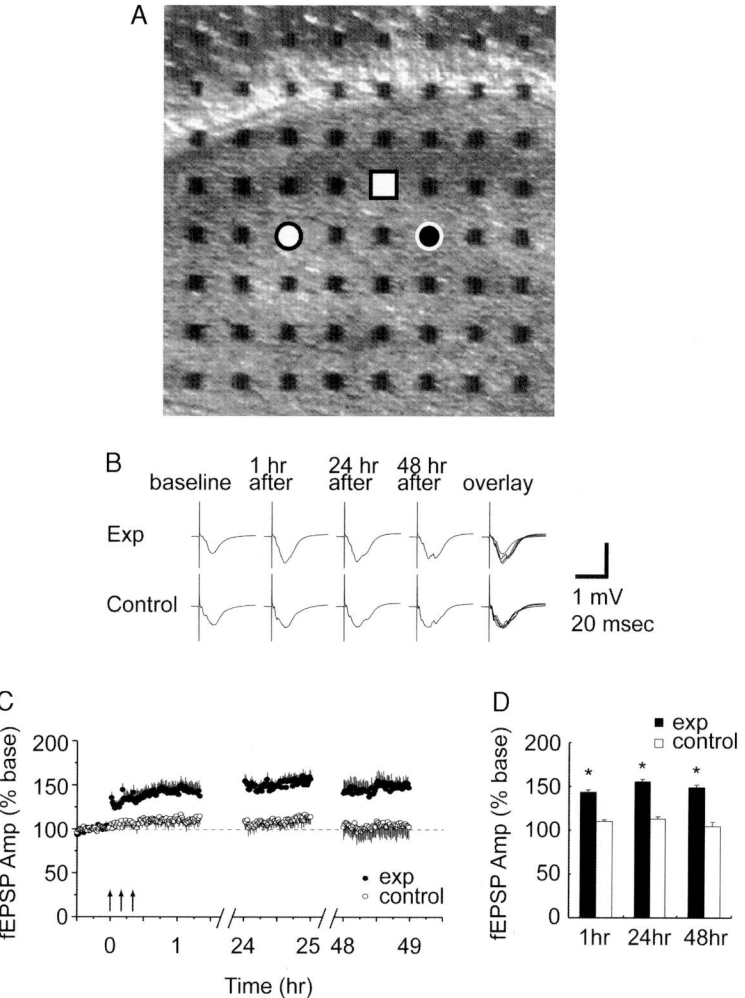

FIGURE 3.21. Long-lasting recording of long-term potentiation in cultured hippocampal slices. (A) Micrograph of a hippocampal slice cultured on an MED probe. The recording electrode is indicated by a white square, and stimulation electrodes are indicated by a white-filled circle (tetanized pathway (exp)) and a black-filled circle (control pathway). (B) Field EPSPs evoked by Schaffer collateral fiber stimulation were recorded in field CA1 before and after tetanus stimulation. (C) Summary graph of long lasting LTP recording. (D) Averaged LTP amplitudes at different intervals.

EPSPs was determined and calculated as a percentage of averaged baseline values (means ± S.E.M., $n = 8$). A statistically significant amount of potentiation was observed throughout the two-day experiment. Figure 3.22 helps to confirm that this effect is indeed LTP by applying a well-known LTP blocker, APV, and observing that LTP is indeed blocked; however, after APV washout it can still be induced.

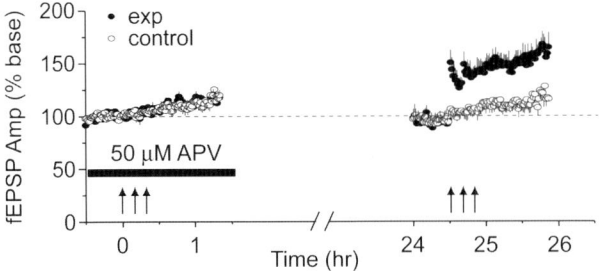

FIGURE 3.22. Reversible blockade of LTP by APV in cultured hippocampal slices. High-frequency stimulation was applied to one pathway (exp) and another unstimulated pathway received low-frequency stimulation (control) in the presence of APV (50 μM). Slices were returned to the incubator in culture medium without APV, and were tested the following day. Another high-frequency stimulation was applied to the same pathway and responses measured for another 2 hours. Each point represents the mean ± S.E.M. ($n = 6$) of the relative amplitude of fEPSPs in both pathways. Arrows indicate time of high frequency stimulation.

This experiment can only be performed under conditions that enable the same exact axons in a given slice to be stimulated over a period of days, because LTP is a synapse-specific phenomenon. Under normal traditional conditions, if a culture is grown on a membrane, the stimulating and recording electrodes will have to be removed periodically to move the slice into an incubator. It is virtually impossible to reinsert the stimulating electrodes such that they will activate exactly the same axons, day after day. However, if the culture is grown on top of a MEA, the position of the electrodes relative to the tissue is fixed, even when it is placed in an incubator.

3.3   Combining MEAs and Fluorescent Microscopy

Another application enabled by MEAs is the combining of multi-electrode electrophysiology and fluorescent microscopy. Microscopy systems offer excellent spatial resolution of activity and can track molecular-level events such as calcium influx into a neuron. MEAs have excellent temporal resolution, and can stimulate slices, triggering events that can be imaged through digital microscopy with a fluorescent dye. MEAs make this possible by allowing a microscope unobstructed access to the slice while at the same time stimulating and recording from the slice. Traditional instrumentation makes this difficult because manipulators and other equipment tend to block or obscure the microscope's access to the slice. A number of researchers are pursuing this methodology (unpublished results, Kilborn et al., 2002[†]). Figure 3.23 shows in part A a micrograph of a hippocampal culture, and

---

[†] Data courtesy Drs. Ken Shimono, Alpha MED Sciences Co., Ltd., Osaka, Japan; and Karl Kilborn, Intelligent Imaging Innovations, Inc., Santa Monica, CA, USA.

A   B

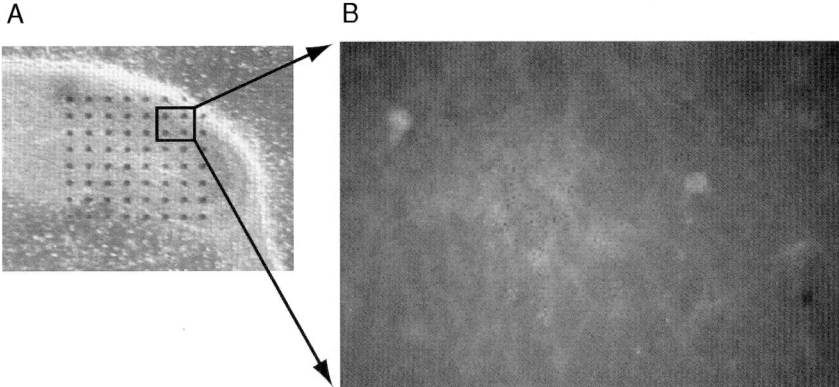

FIGURE 3.23. Micrographs of hippocampal slice and multi-cellular calcium imaging. (A) Hippocampus slice cultured on the MED probe (14 days in culture). (B) Pyramidal layer in the same slice labeled with Fura-2. (Courtesy of Intelligent Imaging Innovations.)

in part B an image of the slice taken with the microscope using a dye for calcium imaging. Figure 3.24 shows in part A the field EPSPs obtained with burst stimulation of the slice. Part B shows the rapid elevation of somatic calcium in dye-loaded cell bodies with the onset of each burst. By one second after each burst, the calcium level returned to near baseline. Early work in this field was done by Jimbo et al. (1993).

3.4   The Future: Commercial Drug Discovery

MEAs are making inroads into the pharmaceutical industry. Already, ten out of the top twenty pharmaceutical companies own and use MEA instruments. Most of these are being used to perform traditional types of experiments, often tied to drug safety testing. The enhanced traditional experiments made possible by the use of MEAs fit well with the needs of the drug discovery industry because these tend to increase throughput and reduce labor by simplifying experiments (see Section 3.1).

This is especially true if the hardware and software that are used to run the experiments are also designed to help optimize throughput. Among Panasonic MED64 users in the pharmaceutical industry, having an instrument that makes the selection of different stimulating electrodes very quick and easy has proven very beneficial. Also, the use of wizard-style software (e.g., the "Performer" software from Panasonic) to guide practitioners step by step through traditional protocols is believed to have increased throughput and reduced training time for operators.

Tensor Biosciences has taken these principles a step further with the creation of a drug screening system called the DS-MED. It employs custom software, hardware, and incubators that have all been designed and built to streamline drug screening experiments and increase throughput. Figure 3.25 shows a working

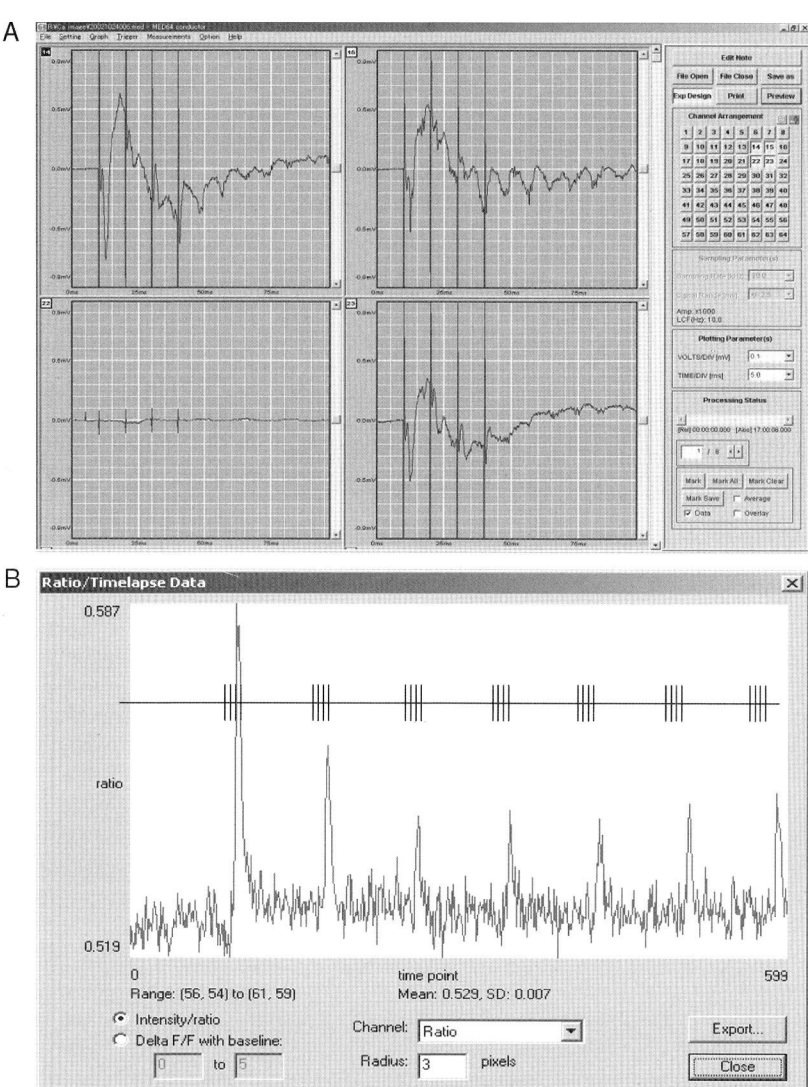

FIGURE 3.24. Simultaneous multi-cellular calcium imaging and multi-site electrophysiological recording. (A) Field potential was recorded with MED64 System at 4 sites in the pyramidal layer of hippocampus slice culture during the burst stimulation. The traces show the responses to a single 4-pulse stimulation. (B) Fluorescence intensity was simultaneously measured in the same pyramidal layer of hippocampus slice culture during the burst stimulation using a 3i digital microscopy workstation. (Courtesy of Intelligent Imaging Innovations.)

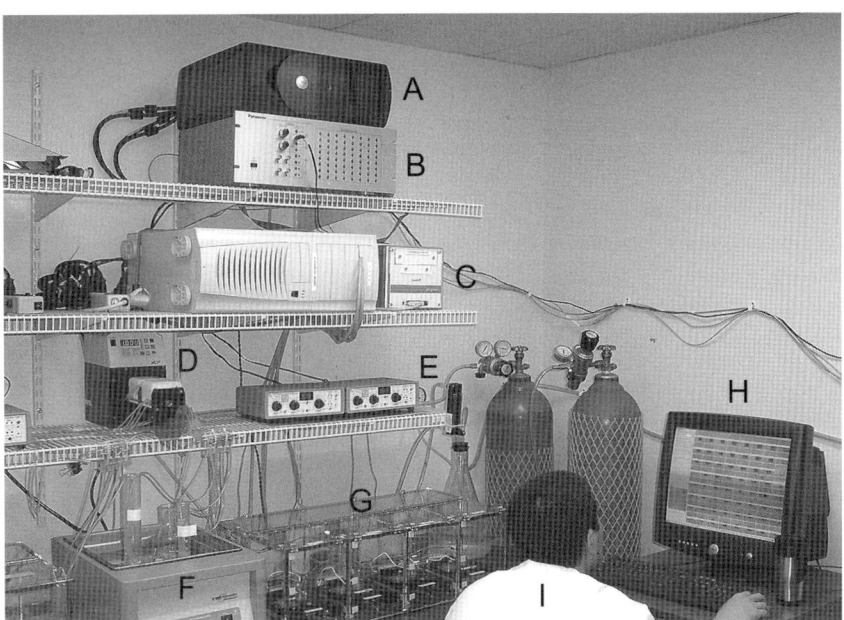

FIGURE 3.25. The DS-MED system for drug screening with MEAs. (A) A PC running the software that controls all the hardware and runs experiments. (B) A Panasonic MED64 Amplifier. (C) Custom multiplexing hardware that enables running multiple interleaved experiments in parallel. (D) An eight-line pump for perfusion. (E) Two inline heaters that handle two lines each. (F) A water bath containing ACSF and test compounds. (G) A custom four-chamber incubator with four MEAs running independent experiments. (H) A computer display showing the control software. (I) Lab technician Frank Tsuji. (Courtesy Tensor Biosciences.)

prototype capable of running four MEA experiments in parallel. The custom hardware, labeled C in the figure, is a multiplexing system that enables a single MED64 amplifier to run multiple interleaved experiments, each with its own MEA and automatic stimulation electrode switching (no patch cables). The custom software, shown at H, controls all the experiments from a single interface, and allows each one to start and stop independently, which is key to optimizing throughput because anytime a slice goes bad it can be replaced, which would not be possible with batch processing. The custom incubator, labeled G, lets the operator access any of four MEAs quickly and easily without leaving her chair.

Looking beyond enhanced traditional experiments for safety testing, the use of MEA technology for drug discovery purposes can also serve the need for more realistic information-rich tissue models in drug discovery. Currently most neurobiological drug discovery programs jump from single-cell screening to whole-animal behavioral testing. In so doing they skip the network level of analysis, that is, the analysis of drug effects on intact networks of neurons. Behavior is the product of networks, not single neurons. Single-cell studies, though extremely useful for initial screening, are ultimately lacking in realism and relevance. And behavioral studies, though invaluable for safety testing, rely on indirect measures of physiological activity that are subject to a great deal of interpretation in the context of treatment quality evaluation.

A more ideal approach would appear to be the use of in vivo multi-electrodes in combination with behavioral experiments. Here drug effects observed on a network-level activity can be correlated with behavioral effects in real-time. However, this approach has very limited throughput. One of the more exciting potential uses for MEAs is to employ them as higher-throughput versions of these experiments. By focusing on the replication of desirable physiological effects established with the multi-electrode in vivo method, and then parallelizing testing with hardware such as the DS-MED, higher-throughput network-level drug testing might become a reality.

References

Buzsaki, G., and Chrobak, J.J. (1995). Temporal structure in spatially organized neuronal ensembles: A role for interneuronal networks. *Curr. Opin. Neurobiol.* 5: 504–510.

Chen, S.C. and Wu, F.S. (2004). Mechanism underlying inhibition of the capsaicin receptor-mediated current by pregnenolone sulfate in rat dorsal root ganglion neurons. *Brain Res.* 1027: 196–200.

Darbon, P., Yvon, C., Legrand, J.C., and Streit, J. (2004). I_NaP underlies intrinsic spiking and rhythm generation in networks of cultured rat spinal cord neurons. *Eur. J. Neurosci.* 20: 976–988.

Egashira, K., Nishii, K., Nakamura, K., Kumai, M., Morimoto, S., and Shibata, Y. (2004). Conduction abnormality in gap junction protein connexin45-deficient embryonic stem cell-derived cardiac myocytes. *Anat. Rec.* 280A: 973–979.

Egert, U., Schlosshauer, B., Fennrich, S., Nisch, W., Fejtl, M., Knott, T., Mueller, T., and Haemmerle, H. (1998). A novel organotypic long-term culture of the rat hippocampus on substrate-integrated multielectrode arrays. *Brain Res. Pro.t* 2: 229–242.

Eichenbaum, H. (1997). How does the brain organize memories? *Science* 277: 330–332.

Feld, Y., Melamed-Frank, M., Kehat, I., Tal, D., Marom, S., and Gepstein, L. (2002). Electrophysiological modulation of cardiomyocytic tissue by transfected fibroblasts expressing potassium channels: A novel strategy to manipulate excitability. *Circulation* 105: 522–529.

Gross, G.W. (1979). Simultaneous single unit recording *in vitro* with a photoetched laser deinsulated gold multimicroelectrode surface. *IEEE Trans. Biomed. Eng.* 26: 273–278

Gross, G.W., Rieske, E., Kreutzberg, G.W., and Meyer, A. (1977). A new fixed-array multi-microelectrode system designed for long-term monitoring of extracellular single unit neuronal activity *in vitro*. *Neurosci. Lett.* 6: 101–106.

Honma, S., Shirakawa, T., Katsuno, Y., Namihira, M., and Honma, K. (1998). Circadian periods of single suprachiasmatic neurons in rats. *Neurosci. Let.t* 250: 157–160.

Jarrard, L.E. (1993). On the role of the hippocampus in learning and memory in the rat. *Behav. Neural. Biol.* 60: 9–26.

Jimbo, Y., Robinson, H.P.C., and Kawana, A. (1993). Simultaneous measurement of intracellular calcium and electrical activity from patterned neural networks in culture. *IEEE Trans. Biomed. Eng.* 40: 804–810.

Jimbo, Y., Robinson, H.P.C., and Kawana, A. (1998). Strengthening of synchronized activity by tetanic stimulation in cortical cultures: Application of planar electrode arrays. *IEEE Trans. Biomed. Eng.* 45: 1297–1304.

Jobling, D.T., Smith, J.B., and Wheal, H.V. (1981). Active microelectrode array to record from the mammalian central nervous system *in vitro. Med. Biol. Eng. Comput.* 19: 553–560.

Legrand, J.C., Darbon, P., and Streit, J. (2004). Contributions of NMDA receptors to network recruitment and rhythm generation in spinal cord cultures. *Eur. J. Neurosci.* 19: 521–532.

Lisman, J.E. (1999). Relating hippocampal circuitry to function: Recall of memory sequences by reciprocal dentate-CA3 interactions. *Neuron.* 22: 233–242.

Lu, Z.J., Pereverzev, A., Liu, H.L., Weiergräber, M., Henry, M., Krieger, A., Smyth, N., Hescheler, J., and Schneider, T. (2004). Arrhythmia in isolated prenatal hearts after ablation of the Cav2.3 (α1E) subunit of voltage-gated Ca2+ channels. *Cell Physiol. Biochem.* 14: 11–22.

Marr, D. (1971). Simple memory: A theory for archicortex. *Proc. R. Soc. Lond. B Biol. Sci.* 272: 23–81.

McNaughton, N. and Morris, R. G. (1987). Hippocampal synaptic enhancement and information. *Trends Neurosci.* 10: 408–415.

Muller, R. (1996). A quarter of a century of place cells. *Neuron* 17: 813–822.

Nadel, L. and Moscovitch, M. (1997). Memory consolidation, retrograde amnesia and the hippocampal complex. *Curr. Opin. Neurobiol.* 7: 217–227.

Nicholson, C. and Freeman, J.A. (1975). Theory of current source-density analysis and determination of conductivity tensor for anuran cerebellum. *J Neurophysiol* 38: 356–368.

Nicholson, C. and Llinas, R. (1975). Real time current source-density analysis using multi-electrode array in cat cerebellum. *Brain Res.* 100: 418–424.

Novak, J.L. and Wheeler, B.C. (1988). Multisite hippocampal slice recording and stimulation using a 32 element microelectrode array. *J. Neurosci. Meth.* 23: 149–159.

Oka, H., Shimono, K., Ogawa, R., Sugihara, H., and Taketani, M. (1999). A new planar multielectrode array for extracellular recording: Application to hippocampal acute slice. *J. Neurosci. Meth.* 93: 61–67.

O'Keefe, J. and Recce, M.L. (1993). Phase relationship between hippocampal place units and the EEG theta rhythm. *Hippocampus* 3: 317–330.

Paulsen, O. (2003). Network oscillations studied using planar multielectrode arrays. *The Second Network Physiology Symposium, satellite symposium of The Society for Neurosciences 33rd Annual Meeting, New Orleans, Louisiana, USA.*

Ren, D. and Miller, J.D. (2003). Primary cell culture of suprachiasmatic nucleus. *Brain Res. Bull.* 61: 547–553.

Shimono, K., Brucher, F., Granger, R., Lynch, G., and Taketani, M. (2000). Origins and distribution of cholinergically induced beta rhythms in hippocampal slices. *J. Neurosci.* 20: 8462–8473.

Shimono, K., Kubota, D., Brucher, F., Taketani, M., and Lynch, G. (2002). Asymmetrical distribution of the Schaffer projections within the apical dendrites of hippocampal field CA1. *Brain Res.* 950: 279–287.

Squire, L.R. and Zola, S.M. (1998). Episodic memory, semantic memory, and amnesia. *Hippocampus* 8: 205–211.

Streit, J., Tscherter, A., Heuschkel, M.O., and Renaud, P. (2001). The generation of rhythmic activity in dissociated cultures of rat spinal cord. *Eur. J. Neurosci.* 14: 191–202.

Thomas, Jr, C.A., Springer, P.A., Loeb, G.E., Berwald-Netter, Y., and Okun, L.M. (1972). A miniature microelectrode array to monitor the bioelectric activity of cultured cells. *Exp. Cell. Res.* 74: 61–66.

Treves, A. and Rolls, E.T. (1992). Computational constraints suggest the need for two distinct input systems to the hippocampal CA3 network. *Hippocampus* 2: 189–199.

Tscherter, A., Heuschkel, M.O., Renaud, P., and Streit, J. (2001). Spatiotemporal characterization of rhythmic activity in rat spinal cord slice cultures. *Eur. J. Neurosci.* 14: 179–190.

Tulving, E., and Markowitsch, H.J. (1998). Episodic and declarative memory: Role of the hippocampus. *Hippocampus* 8: 198–204.

Welsh, D.K., Logothetis, D.E., Meister, M., and Reppert, S.M. (1995). Individual neurons dissociated from rat suprachiasmatic nucleus express independently phasedcircadian firing rhythms. *Neuron* 14: 697–706.

4
Development of 3-D Multi-Electrode Arrays for Use with Acute Tissue Slices

MARC OLIVIER HEUSCHKEL, CORINA WIRTH, ESTHER-MARIE STEIDL, AND BRUNO BUISSON

4.1 Acute Tissue Slices On Multi-Electrode Arrays

Multi-site recording of neuronal activity in freshly prepared slices provides an ideal approach towards understanding the function of neuronal processing in complex networks. Such an approach has given elegant results with native retina where ganglionic neurons could be recorded by planar Multi-Electrode Arrays (MEA; Meister et al., 1991, 1994). Other brain regions display some 2-D-organized networks, but these structures need to be sectioned prior to use with multi-electrode arrays. Some of these preparations (hippocampus, cerebellum, cortex, and spinal cord) have been used for many years with conventional glass electrodes that could be placed directly on active living cells. The dead cell layer of mechanically cut slices constitutes, however, an insulation layer between living neurons and the electrodes of planar arrays, preventing the achievement of a good signal-to-noise ratio (SNR). The recent development of 3-D tip-shaped electrode arrays now offers the technical opportunity to record from acute slices with unprecedented spatiotemporal resolution.

4.1.1 Acute Tissue Slices

Acute tissue slice experimentation is of great interest in basic research and industry for purposes of unraveling network properties, providing an insight into excitatory and inhibitory relationships between different cells or cell groups, as well as for testing drugs in discovery and safety pharmacological processes.

The major advantage of using acute tissue slices compared to organotypic tissue slice cultures and dissociated cell cultures is that the native network connectivity is preserved for at least several hours after sample preparation. However, tissue slices still represent an incomplete network, as many connections to cells that were located outside the slice preparation are cut, reducing the inputs and outputs to a limited cell network. As most experiments with acute tissue slices are generally made within eight hours after sample preparation, it can be expected that there will be no major modification of cell connectivity; that is, there is preservation of the network characteristics that were originally present in the animal. This is not the case for organotypic slice cultures, where the network connectivity changes because of synaptic rearrangement during the culturing period of 10 to 14 days. Moreover, as experiment duration is typically short, problems associated with long-term tissue slice nutrition on a solid substrate are avoided. This is a major limitation of organotypic slice cultures where solution renewal is poor at the bottom of the slice, the area that is in direct contact with the electrodes (Thiebaud et al., 1997, 1999).

For safety pharmacology applications, acute slices represent, prior to in vivo tests, the most physiological approach to reveal potential drug-induced side effects. In addition, adverse effects can be evaluated in reference to known compounds. Because many slices can be produced with a single brain structure (and a single animal), new chemical entities can be evaluated in ascending dose–response protocols at a high number of electrodes with a powerful resolution. This will economize on the use of animals compared to conventional tests performed in vivo. When compared to reference compounds, these tests can offer a high degree of precision for diagnostics.

4.1.2   Dead-Cell Layer Problem: Definition

Since their first introduction (Thomas et al., 1972; Gross, 1979), MEA biochips have been widely used for monitoring neuronal activity in in vitro preparations. Local Field Potentials (LFP) as well as single-unit action potentials have been recorded successfully in dissociated spinal cord preparations (Gross et al., 1993; Streit et al., 2001), cortical cell cultures (Jimbo et al., 1993; Maeda et al., 1995; Gross et al., 1997), acute retina preparations (Litke and Meister, 1991; Hammerle et al., 1994; Meister et al., 1994), and cardiac myocytes cultures (Israel et al., 1984; Connolly et al., 1990). In all these preparations, active neurons (i.e., the effective sources of signals) were in direct contact with the electrode surface. Similarly, acute retina preparations present a clean tissue surface that adheres to the electrodes establishing a tight contact with the electrode surface. Under such conditions, even small biological signals can be detected with a reasonable signal-to-noise ratio and can be extracted out of the noise (Litke and Meister, 1991; Meister et al., 1991, 1994).

The use of MEA biochips with organotypic slice cultures (Stoppini et al., 1997; Egert et al., 1998; Thiebaud et al., 1999; Tscherter et al., 2001) also allowed recording of LFP as well as of single spike events. For acute tissue slice experiments, the main technique used is the patch clamp technique (Brockhaus et al., 1993; D'Angelo et al., 1998; Ziak et al., 1998). However, patch is limited to studying only a small number of cells in parallel, whereas other alternative techniques, such as optical imaging techniques (Yuste et al., 1999), are limited by dye phototoxicity, which dramatically reduces experiment duration.

Only a few groups have reported monitoring of acute tissue slices on MEA biochips. Evoked epileptiform field potentials in the mV range have been recorded (Novak and Wheeler, 1986, 1988, 1989) from rat hippocampus slices in an environment similar to that in a standard interface chamber, which is known to improve signal levels. Only small LFP amplitudes recorded from acute rat hippocampus slices on conventional MEA biochips have been reported (Oka et al., 1999). Recently, Egert et al. (2002) published results on the monitoring of spontaneous activity in acute cerebellar slices on planar MEA biochips with amplitudes in the 100 μV range. There, it was also indicated that the first active cell layer inside the acute cerebellar slice was located about 15 to 30 μm from the slice border, allowing establishment of a reasonably good contact with the electrodes.

The consequence of the small responses recorded in the past has limited use of acute tissue slices on MEA biochips in the scientific community and in industry. As it would be highly interesting to use acute tissue slices on MEA biochips, it was apparent that further investigation was needed to reveal the cause of the low signal response. A close look at sample preparation procedures for acute tissue slice experimentation may explain this problem as discussed below.

• During slice preparation, the tissue is cut into slices (thickness of 250 to 500 μm) using a tissue chopper. Depending on the type of tissue used (brain, spinal cord, etc.) and its specific network structure, it is very important that the cutting orientation and/or plane are the same as the one where neurons have their major dendritic trees and axons. Otherwise, most cell connections of the network will be cut, thereby drastically reducing biological activity.
• Cutting the tissue injures or kills the cells located at the border of the slice and leads to an edema. Dead cells release intracellular potassium and neurotransmitter that can mediate neurotoxicity. For these reasons, there would only be a few active cells within the first 50 μm at the border of the slice. This so-called "dead-cell layer" forms an electrically passive layer generating a parasitic shunt between planar recording electrodes and active cells inside the tissue slice. Due to the decay of electrical potential with the square of the distance from the edge of the slice, only a small number of neurons are sampled and consequently small signal amplitudes are typically obtained. In comparison to intracellular signals recorded using glass microelectrodes, the signal amplitudes obtained are 100 to 1000 times smaller. Thus, it is important to reduce the distance between active cells in the slice and recording electrodes.
• When the slice is placed onto the MEA biochip, it has to be stuck so that it cannot move anymore. A tight seal resistance between electrodes and active cells in the tissue is difficult to achieve due to the dead-cell layer. Furthermore, a small gap (up to 100 nm) is always present between tissue and electrodes, which is filled with saline solution. This results in current leakage causing loss of part of the signal. Hence, the use of a surface coating for enhancing tissue attachment to the MEA biochip is very important.

In summary, small responses obtained by conventional MEA biochips are primarily due to poor contact between active cells in the slice and electrodes.

4.1.3   Dead-Cell Layer Problem: Solutions

In order to overcome the dead-cell layer problem, the most widely used solution is the organotypic slice culture technique, where the dead cells at the border of the tissue slice are eliminated during a culture period of about two weeks. After this period, active cells have moved closer to the recording electrodes, thus enhancing electrode coupling. In this case, biological activity can be recorded (Stoppini et al., 1997; Thiebaud et al., 1997; Tscherter et al., 2001) allowing single-action potential and LFP to be measured. However, as mentioned above, the native tissue connectivity is lost in organotypic cultures because of synaptic rearrangement.

Another possibility to overcome the dead-cell layer problem is to penetrate the slice with recording electrodes in order to reduce as much as possible the distance between recording electrodes and active cells. This can be achieved by using three-dimensional electrode arrays. Sharp tip-shaped electrodes are required to penetrate the first layers of dead and injured cells at the slice border without significantly dilacerating the tissue. Good slice penetration is very important, as it may also enhance cell coupling to the electrodes and reduce leakage currents.

The remaining problem is to make tissue adhesion on the electrode similar for all types of cell/tissue cultures. Many different approaches for enhancing cell and/or tissue adhesion have been devised, which include the use of adhesion proteins and chemical matrices (Adler et al., 1985; Makohliso et al., 1998; Bledi et al., 2000; Branch et al., 2000). The choice of an optimal adhesion promoter depends on the type of cells or tissue studied and the techniques employed. It was found that polyethyleneimine (PEI) dissolved in distilled water provides a good surface coating and enhances the adhesion of acute tissue slices (Oka et al., 1999; Heuschkel et al., 2002; Wirth and Luscher, 2004). Another procedure used to achieve good contact between the tissue and the MEA biochip is the employment of a U-shaped platinum wire with a grid of nylon threads that is placed on top of the tissue slice.

4.1.4   Advantages of 3-D Multi-Electrode Arrays

It was first postulated that 3-D electrodes with a height of between 50 and 100 μm should improve recording in acute slices by reducing the distance between cells and recording electrodes and should allow better measurement and stimulation conditions as compared to planar electrode arrays.

If we compare the two electrode configurations presented in Figure 4.1, some advantages of the 3-D electrode configuration can be listed.

• The main advantage of 3-D electrodes is tissue slice penetration, which reduces the distance between electrodes and active living neurons, as demonstrated with histological studies with organotypic tissue slices (Kristensen et al., 2001). Simulations and measurement experiments without tissue also showed that larger signal amplitudes could be recorded with a 3-D electrode as compared to a planar electrode when the signal source is located at a given distance from the substrate ground level (Heuschkel et al., 2002).

• 3-D electrodes present a geometrical advantage with an increased surface. It has been calculated that the area of a 3-D electrode is about twice as large as planar electrodes with same base dimensions (Heuschkel et al., 2002). This increase

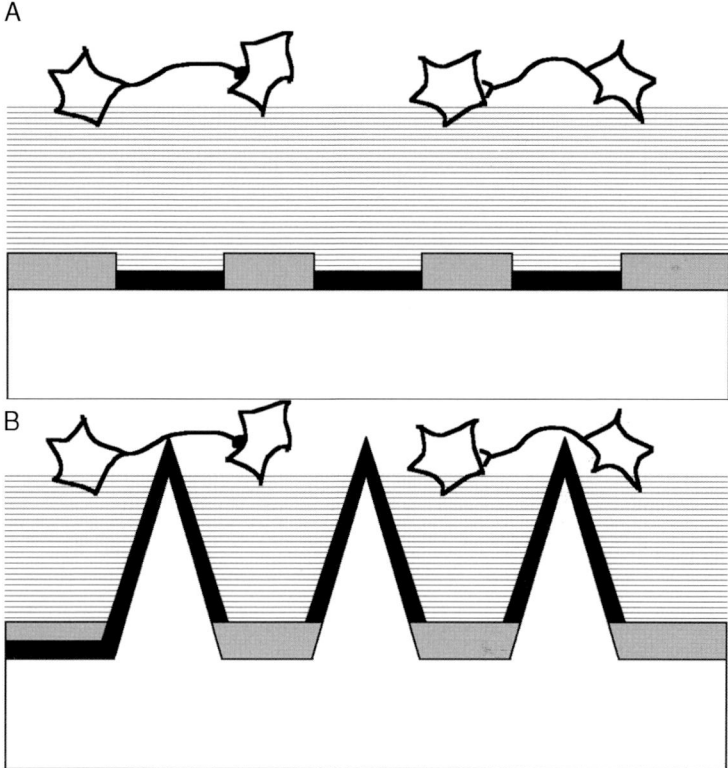

FIGURE 4.1. Schematic cross-section of tissue slice/electrode configuration. (A) Note the layers of dead cell (horizontal lines) lie between the active cells (on top) and the electrodes (black). (B) In the case of a three-dimensional electrode configuration, the distance between active cells and electrodes is reduced due to tissue penetration allowing better measurement conditions.

in surface area reduces electrode impedance, which increases recorded signal amplitudes.
• 3-D electrodes also present a geometrical advantage when used for electrical stimulation. As the electrodes are located at the top of sharp tips, the electrode shape allows generation of a higher electrical field around the electrode during stimulation. Furthermore, due to a larger electrode surface, the electrode can store a larger electrical charge and thus can deliver a higher safe injected charge.

4.2   Multi-Electrode Array Fabrication Technologies

In this section, MEA biochip fabrication is described. A more detailed description of 3-D microstructure realization on glass substrates is also presented.

4.2.1 Multi-Electrode Array Fabrication Concept

Most commercially available MEA biochips are based on large glass substrates with dimensions of about 5 cm × 5 cm, micro-fabricated electrodes made of various materials, an electrode insulation layer, and a glass ring building a culture chamber of about 1.5 ml. As large glass substrates are used for MEA fabrication, only one single MEA biochip can be realized out of one single glass substrate. Consequently, each MEA biochip is produced individually, thus generating high micro-fabrication costs.

Furthermore, due to high device costs, MEA users try to reuse the biochip as many times as possible in order to reduce costs per experiment. This implies that after each experiment the MEA biochip will undergo a cleaning and sterilization procedure generating some electrode degradation. This is difficult to evaluate as it depends on the previous experiment and history of each MEA biochip. It has been reported that MEA biochips have been reused several times, however, the overall experimental conditions degrade with each MEA reuse until the electrode quality becomes unacceptable and the MEA biochip is finally discarded. An ideal MEA biochip would be low-cost and disposable as the costs per experiment would be low and good electrode quality would be obtained due to single use.

The major advantage of micro-fabrication technologies used in the micro-electronics industry as well as for MEA biochip fabrication is the reduction of chip size due to minimization of feature sizes. This overall chip size reduction allows the generation of hundreds or even thousands of components on one single substrate. By multiplying the number of components on each processed substrate, the cost per component is dramatically reduced.

Thus, the fabrication of low-cost MEA disposables could be envisaged, permitting widespread use of MEA technology. For example, if dimensions of a typical MEA biochip could be reduced to 1 cm × 1 cm and circular glass substrates with a diameter of 10 cm would be used, then it would be possible to fabricate 57 MEA biochips on one single substrate. Compared to one MEA biochip per substrate, the cost advantage would be significant. However, the limitations of chip size reduction would be the large number of connections on a very small surface that would have to be connected to an external data acquisition system, and the reduced size of the effective culture chamber workspace making cell cultures and tissue handling more complicated.

The MEA fabrication concept employed by Ayanda Biosystems is a compromise between the current commercially available MEA biochips and disposables. MEA biochips based on 2.1 cm × 2.1 cm or 1.5 cm × 1.5 cm glass chips are used, allowing the realization of 12 and 24 MEA biochips (see Figure 4.2) on a circular glass substrate (diameter of 10 cm), respectively. These device sizes still allow building culture chambers similar to the currently available MEA biochips without significantly reducing the access to the recording/stimulation electrodes. An approach based on Printed Circuit Boards (PCB) is used to interface MEA chips to data acquisition systems. The main advantages of the PCB technology are its low cost and its layout flexibility that allows easy adaptation of the MEA biochips to any external data acquisition interface.

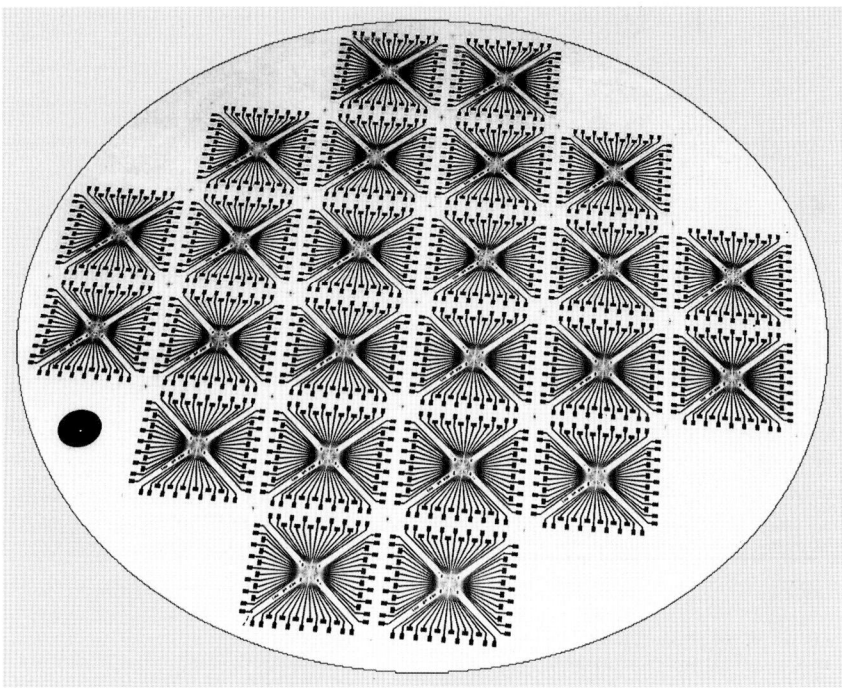

FIGURE 4.2. Micrograph of a glass substrate (diameter 10 cm, thickness 700 μm) including 24 micro-fabricated 3-D MEA chips. The MEA chip dimension is 1.5 cm × 1.5 cm.

Thus, the MEA biochip itself is an assembly of a micro-fabricated glass chip and a PCB covered with a culture chamber made of a glass ring (see Figure 4.3). The MEA chip is glued onto the PCB at electrode connection pads. The culture chamber and the sealing of the MEA biochip are made using a glass ring that is embedded in silicone sealant. This fabrication concept also allows easy changing of the electrode layout, that is, the possibility to have several different electrode layouts that can be used with the same PCB, due to a standardization of the chip/PCB connections. On the other hand, the assembly of the MEA chip onto a PCB presents several minor disadvantages.

An aperture in the PCB, whose dimensions are smaller than the glass chip, defines the workspace for the device. Furthermore, the sealing of the biochip using silicone sealant will contribute to workspace reduction. As an example, for a MEA biochip based on a 1.5 cm × 1.5 cm glass chip, the resulting workspace corresponds to a circle with a diameter of 8 mm, which is sufficient for most MEA applications. If a larger workspace is required, then larger MEA chip dimensions have to be chosen.

The MEA biochips are sensitive to high temperatures due to differing thermal expansion coefficients of the different used materials. Thus, thermal treatments at high temperature (above 80°C) have to be avoided or require very slow temperature changes in order to prevent contact breakage. As a consequence, these

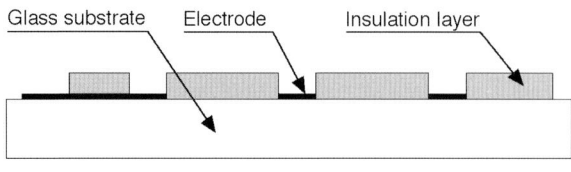

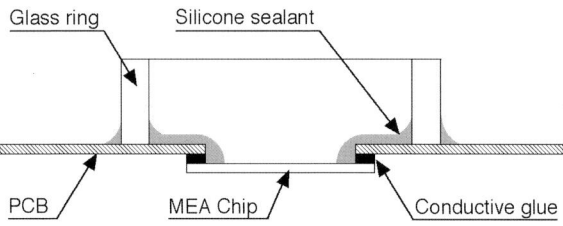

FIGURE 4.3. Schematic cross-section of a MEA chip (top) and of MEA biochip (bottom). The MEA chip is composed of a glass substrate, patterned metal electrodes, and an epoxy insulation layer. The MEA chip is glued under a printed circuit board (PCB) for interfacing to signal amplification and data acquisition system. A glass ring, which is fixed on top of the PCB, and MEA biochip sealing using silicone rubber builds a watertight culture chamber with a volume of approximately 1.5 ml.

MEA biochips are not suited for autoclave sterilization. However, there are many other very effective available sterilization procedures such as UV light, gamma-ray irradiation, or heating the MEA biochip at 56°C for six to eight hours. For acute tissue slices, MEA biochip sterilization is not required.

In summary, micro-fabrication techniques allow production of MEA biochips with significant cost reduction due to smaller MEA chip dimensions. This is at present not sufficient to turn MEA biochips into single-use disposables. Nevertheless, there are still technological improvements that can be done in order to achieve this goal in future.

4.2.2   Materials and Fabrication Technologies

4.2.2.1   Multi-Electrode Array Materials

The choice of materials for MEA biochip fabrication is very important as the electrical and the experimental conditions may be influenced by a poor material choice. The major material characteristics areas follows.

- *Biocompatibility:* It is very important that all materials that will be in contact with the cell culture or tissue do not show any toxic effect, and do not affect the cell "well-being" while in the MEA biochip. Furthermore, the materials should allow good adhesion of the biological preparations.

- *Good electrode characteristics:* The electrical characteristics of the electrodes should allow measurement of small signal amplitudes with a good signal-to-noise ratio.
- *Transparency:* The MEA biochip should be as transparent as possible in order to enhance observation using an inverted microscope. Furthermore, if electrodes can be made from transparent material, optical (potentiometric dyes) and electrical measurement can be combined in one single experiment.
- *Cost:* Low-cost materials should be used in order to reduce MEA biochip fabrication costs.

The material characteristics necessary to produce a high-quality MEA biochip are discussed below. In addition, they are compared with materials commonly used for MEA biochips.

MEA Substrate Materials

Today, most MEA biochips using metal electrodes are based on glass substrates (Hammerle et al., 1994; Meister et al., 1994; Gross et al., 1997; Oka et al., 1999; Heuschkel et al., 2002). Other materials, such as polymers (Boppart et al., 1992; Stoppini et al., 1997) and silicon (Fromherz et al., 1991; Offenhausser et al., 1997) were also used previously as MEA substrate material.

On silicon substrates, electrodes have to be isolated from the substrate as silicon is a semiconductor. Silicon material is also about ten times more expensive than glass. It is best suited for field-effect transistor monitoring devices, integrated signal amplification and filtering of signal, and the addition of specific features, such as holes through the substrate (Thiebaud et al., 1997), for example.

The main advantages of glass substrates are transparency, good chemical and electrical isolating characteristics, and low cost. As most researchers use inverted microscopes, the transparency of the MEA biochip is of importance for cell culture/tissue observation during experimentation. The cells attached to metallic electrodes cannot be easily observed with an inverted microscope due to metal opacity.

MEA Electrode Materials

Metals such as gold, platinum, platinum black, titanium nitride, and indium-tin oxide (ITO) are commonly used as electrode materials. As each of these materials has different electrical characteristics, the choice of electrode material depends on the application and specific requirements.

For example, if very low noise levels are required for monitoring biological activity due to very small signal amplitudes, porous materials such as platinum black (Oka et al., 1999) and titanium nitride (Egert et al., 1998) are best due to their low impedance. For specific cell patterning, gold electrodes present some advantages with specific cell-promoting chemistries (Nam et al., 2004). For a combination of optical and electrical recordings, ITO electrodes are best due to optical transparency of the material allowing fabrication of fully transparent MEA biochips (Kucera et al., 2000).

The most commonly used electrode material is platinum black due to its good electrical characteristics (low noise level). However, the deposition of platinum black requires a supplementary time-consuming plating process. Furthermore, platinum black displays a weak adhesion to the underlying electrode that often leads to material removal during MEA biochip cleaning procedures.

A good compromise between good electrical characteristics and an inexpensive fabrication process is the use of metallic platinum as an electrode material. The noise level obtained with planar 40 μm × 40 μm electrodes is in the range of 20 μV (Heuschkel et al., 2002).

MEA Insulation Materials

The choice of material for the electrode insulation layer is important because it will have an impact on parasitic capacitances between the electrode leads and the culture medium (conductive saline solution). In order to reduce parasitic capacitances, the material should have a low dielectric constant or have a large thickness.

Silicon nitride and silicon dioxide (also often combined) are widely used as insulation materials. These materials are deposited as thin-film layers (<1μm) on top of the MEA biochip and are patterned afterwards by photolithography/etching procedures. The small thickness of these materials is sometimes a problem because it is generally insufficient to reduce the parasitic capacitances.

Polymers such as polyimide and SU-8 epoxy are also used as insulation layers. The main advantage of using polymers is that they can be deposited with larger thickness (several μm) than obtainable with silicon-based insulation layers. Furthermore, the polymer materials are available in a photosensitive format making their patterning simple and low cost. Polyimide can be used as a MEA substrate and insulation layer at the same time for the realization of flexible MEA biochips.

SU-8 epoxy has also been used as insulation layer because it can be patterned easily by photolithography. It is also fully transparent, does not have any toxic effect on cells, and has a high chemical stability due to its highly cross-linked structure.

MEA Chip Interfacing Materials

In order to adapt the MEA chip to an external data acquisition system, the MEA chip is mounted onto a printed circuit board. The PCB material is composed of an epoxy and glass fiber matrix, and is often covered by an epoxy photosensitive resist. Several experiments of cell cultures on PCBs and epoxy-covered PCBs showed that cells could survive for a few days. Thus, the main disadvantage of standard PCB is the apparent material toxicity. Therefore, it needs to be isolated from the cell culture or tissue by using, for example, silicone rubber. The main advantages of PCB material are its low costs and its layout flexibility.

MEA Culture Chamber Materials

A glass, plastic, or silicone rubber ring is a simple way to generate a culture chamber on top of the MEA biochip. As the MEA biochip has to be sealed in order to remain watertight, silicone rubber such as Sylgard silicone DC 184 from Dow Corning can also be used for culture chamber and PCB insulation.

4.2.2.2 Multi-Electrode Array Micro-Fabrication Technologies

The technologies used for MEA fabrication are based on standard silicon (CMOS) microelectronic components manufacturing methods. Therefore, the MEA biochips are mainly fabricated in a cleanroom environment. The main techniques used are photolithography, metal deposition, and wet chemical etching. The developed 3-D MEA biochips were fabricated using fabrication techniques that are presented briefly below (see Figure 4.4 for fabrication process flow). Detailed descriptions of processes can be found elsewhere (Sze, 1985; Madou, 1997; Rai-Choudhury, 1997). The complete MEA biochip fabrication process-flow has also been described in detail elsewhere (Heuschkel, 2001).

Photolithography

This technique allows patterning of a light-sensitive resist on a substrate, and is commonly used in micro-fabrication. It allows fabrication of etching and/or deposition mask on a substrate for thin-film layer patterning. There are two different kinds of photosensitive resist: the positive-tone resists and the negative-tone resists.

A photolithography process is based on the following steps. The resist is first deposited on the substrate (spin coating) and heated in order to evaporate solvents. A solid layer remains with a thickness of one to several micrometers. The resist is then exposed to UV light through a metal mask (the pattern was defined on that mask). The resist is then developed. For a negative-tone resist, the exposed areas will remain on the substrate but for a positive-tone resist the exposed areas will be dissolved. In turn, the resist can then be used as mask for etching an underlying layer.

Another application of photosensitive resist is use as a mask for metal deposition. A negative-tone resist is first deposited and patterned on a substrate. Then a metal layer is deposited onto the substrate and the resist-covered areas. The resist is then dissolved in developer, the overlying metal layer being removed at the same time. The substrate areas that were free of resist are the only areas covered by the remaining metal. This method of metal patterning is known as "lift-off process".

Metal Deposition Techniques

Metal deposition can be done using mainly two different techniques: metal evaporation or metal sputtering. A thin film layer of a metal (thickness between 20 and 100 nm) such as chromium, titanium, or tantalum is first deposited onto the substrate to enhance adhesion. Finally, metals such as, for example, gold or platinum are deposited on top of the adhesion metal during the same process. Metal layer thickness in the range of 100 nm to 1 μm can be deposited on the substrate.

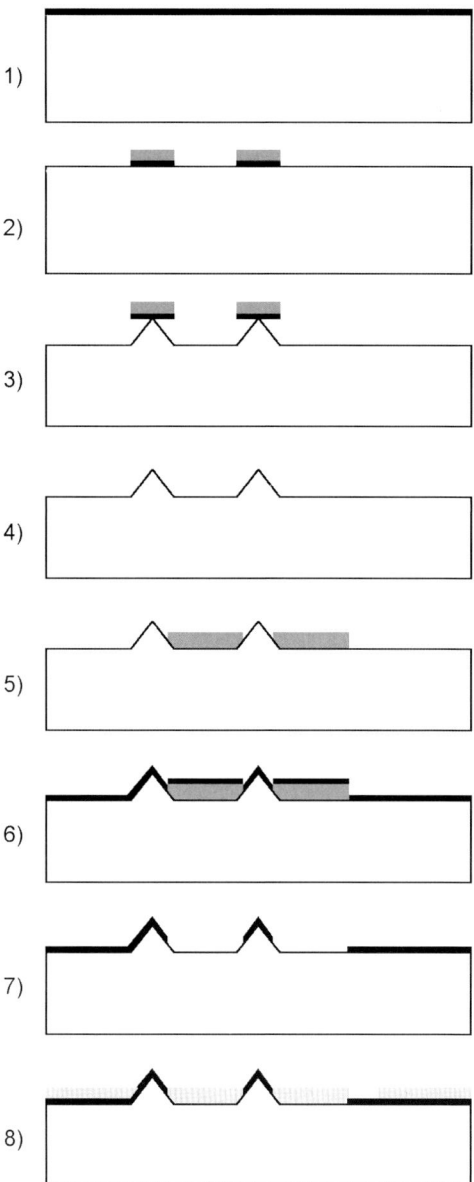

FIGURE 4.4. Schematic diagram of 3-D MEA fabrication process. (1) The first process step is the realization of a mask for substrate etching. A soda–lime glass substrate is covered with a 100-nm chromium layer deposited by sputtering. (2) It is patterned using a negative tone resist and an adequate etching solution. (3) The glass substrate is then etched in HF solution in order to obtain glass tips. (4) The etching process is stopped right after mask detachment allowing fabrication of sharp tips. (5) The next process consists of the electrode deposition using a lift-off process. A negative tone resist is coated onto the substrate and patterned. There is no resist remaining at electrode and electrode wire locations. (6) Electrode material is sputtered everywhere on substrate. (7) The resist is then dissolved in acetone solvent releasing the electrode pattern. (8) Finally, the substrate is covered with SU-8 epoxy insulation layer that is patterned by photolithography.

Wet Chemical Etching

Wet chemical etching is a standard technique used for silicon, metal, and glass etching. Patterned substrates with the material to etch are placed into an adequate etching solution until etch pattern is achieved. The case of glass etching is discussed in more detail in the next sections.

Back-End Technologies

Back-end technologies such as substrate dicing allowing chip release and chip-packaging techniques are done outside of the micro-fabrication environment. During MEA biochip fabrication, the chip has to be assembled onto a PCB for interfacing purposes. This step is done using screen-printing (Lambert, 1974; Swerdlow, 1985). Conductive glue is screen-printed through a stencil mask on the metal pads of the PCB. Placing the two parts together and glue hardening complete the gluing process.

4.2.3   Fabrication of Tip-Shaped Glass Microstructures

4.2.3.1   State of the Art of 3-D Multi-Electrode Arrays

The first 3-D MEA biochips described in literature (Thiebaud et al., 1997, 1999, 2000) were fabricated on silicon substrates for use with organotypic tissue cultures. 3-D electrodes were first made by electroplating platinum hillocks with a height of 20 µm (Thiebaud et al., 1997) and afterwards by wet chemical etching of bulk silicon substrate in KOH solution with a tip-height of about 47 µm (Thiebaud et al., 1999, 2000). The main weakness of the electrodeposited platinum hillocks was the mechanical stability of the hillock base, leading to easy breakage. On the other hand, the silicon tips showed good mechanical properties and allowed improved monitoring of organotypic slice culture activity.

The authors (Heuschkel et al., 2002) have previously described 3-D MEA biochips based on glass substrates for use with acute tissue slices and have compared it with similar planar MEA biochips. The fabrication process of these 3-D tip-shaped microstructures is described in detail below.

4.2.3.2   The Nature of Glass

Glass is the product of melted crystalline materials at elevated temperatures, which have subsequently been cooled to rigid condition without crystallization. The chemical composition is largely inorganic, with silica (SiO_2) being the most important material. Silicate glasses are composed of three-dimensional molecule networks, the basic structural unit being a silicon–oxygen tetrahedron in which a silicon atom is bonded to four oxygen atoms. Some other atoms such as sodium are ionically bonded to oxygen and disrupt the continuity of the network because some oxygen atoms are no longer bonded to two silicon atoms. Inorganic oxides are also incorporated into silicate glasses. Elements such as boron are introduced into the network and replace silicon atoms, and are referred to as network formers.

Mono- or divalent cations such as calcium, magnesium, and sodium that do not enter the network but form ionic bonds with nonbridging oxygen atoms are referred to as network modifiers.

There are two different glasses that are commonly used in micro-fabrication: a borosilicate glass known commercially as Pyrex (Corning, glass code 7740) and soda–lime glass commonly known as float glass or window glass. In Pyrex glass, which has a low alkali oxide composition, vitreous boric oxide with a boric–oxygen triangle network replaces parts of the three-dimensional network. Its main advantage is its thermal expansion coefficient that is similar to various metals and silicon. This characteristic allows glass/silicon bonding and is the reason for its broad use in sensor and micro-fluidics applications. The soda–lime glasses are the oldest and the most widely used glasses. They contain silica and alkali oxides (principally sodium oxide) and lime (mostly calcium and magnesium oxide). Usually, a small amount of alumina is included in the glass formulation to improve working characteristics and chemical durability. Glass compositions and main characteristics have been described in more detail elsewhere (McLellan and Shand, 1984; Scholze, 1991).

4.2.3.3   Wet Chemical Etching of Glass

The high chemical resistance of glass to almost all chemicals makes it difficult to structure. However, HF immediately dissolves the main component, silica, according to the following chemical reaction.

$$SiO_2 + 6\,HF \rightarrow H_2SiF_6 + 2\,H_2O$$

In reality, HF glass etching processes are more complicated because they are affected by side-reaction products due to the other glass components settling at the glass surface in the form of insoluble hexafluorosilicates or fluoride salts. These precipitates tend to impede the etching process. Network modifiers such as Na_2O, K_2O, and CaO are incorporated into the glass forming nonbridging oxygen atoms, and act by partially breaking its three-dimensional structure. Cations from the network-modifying oxides are bonded ionically to the silicate network and are therefore relatively mobile. In aqueous solutions, cations such as Na^+ and Ca^{2+} are leached out of the glass near the surface and replaced by H_3O^+ by ion exchanges, which forms SiOH (silanol) groups in the silica structure. The formation of this superficial hydrated silica film precedes the dissolution of the silica structure.

Dissolution of silica is characterized by a two-part etching mechanism (Suire, 1971; Monk et al., 1993): first, H^+ will open the SiO_2 network at the surface, and then the fluorine species F^- and HF_2^- react with the silicon to form SiF_4 which is soluble in water and forms H_2SiF_6. The adsorption processes of HF molecules, HF_2^- and H^+ ions, determine the reaction rate. Thus, adding more H^+ (e.g., by adding hydrochloric acid) to the etching solution enhances the etch rate. On the other hand, the creation of silanol bonds at the surface of the SiO_2 seems to be limiting the etching reaction at high HF concentrations. It has been reported previously that bath agitation induces only little variation on silica etching rate (Liang and Readey, 1987; Monk et al., 1993).

The low solubility of alkaline earth and lead hexafluorosilicates generated during the glass etching process causes compound precipitation, particularly when soda–lime glasses are etched in HF solutions. The hexafluorosilicates (Na_2SiF_6, K_2SiF_6, $MgSiF_6$, Ca_2SiF_6, etc.) are more or less soluble in water but not soluble in HF solution. In order to enhance process homogeneity, glass substrates can be rinsed regularly in water. The etching process is less effective because the hexafluorosilicates protect the underlying glass from the etching solution, reducing the global etch-rate. Figure 4.5 illustrates the process of hexafluorosilicate generation at the glass surface. A more detailed description of this phenomenon can be found in the literature (Suire, 1971; Spierings, 1993).

4.2.3.4   Masks for Tip-Shaped Glass Structures

In order to generate a tip protruding from glass substrate, a mask for protecting the glass locally is required. The mask geometry will have an impact on final tip-shape, as the mask is under-etched until mask removal. In Figure 4.6, the glass-tip fabrication using wet chemical etching in HF solution is depicted. First, the glass is etched vertically. Then, the mask is under-etched more and more with continuing etching process. Finally, the mask is fully under-etched and is released into the HF solution. This is the moment at which the etching procedure has to be stopped in order to preserve a sharp tip-shape.

Because most photosensitive resists present poor chemical stability and poor adhesion to substrate in HF solutions, the best glass-masking materials are metallic layers. Unfortunately, most metals deposited as an adhesion layer (e.g., titanium, tantalum, and chromium) are etched in HF solutions. On the other hand, copper, platinum, silver, and gold are not etched in HF solution. However, several tests have shown that chromium deposited at high temperature (300°C) has an improved adhesion to glass substrates and presents a good resistance to HF solution when covered by another metal such as copper or photosensitive resist.

Wet chemical etching of glass in HF solutions is assumed to be an isotropic process; that is, the vertical etching and horizontal mask under-etching rate should be identical. The theoretical shape of the under-etched material ought to be semi-circular (if the mask adheres well to the substrate). In practice, however, this is rarely the case. For Pyrex glass, there is an increased mask under-etching, which is more pronounced with increasing HF concentration. This can be explained by poor adhesion between the deposited chromium layer and Pyrex glass enhancing mask removal at the substrate/mask interface. It results in microstructures with a low aspect ratio. Microtips tend to be blunt. For soda–lime glass, a decreased mask under-etching is usually observed. This effect is due to the formation of hexafluorosilicates at the mask location, which protect the vertical walls, reducing the horizontal glass etch-rate. This results in very sharp tip-shaped structures protruding from the glass surface.

A

B

FIGURE 4.5. (A) Scanning electron microscopy picture of an array of tip-shaped micro-structures halfway through the glass etching process. Metal masks (diameter of 150 μm, spacing 200 μm) covered with photosensitive resist (gray circles) remain attached to the glass substrate until total under-etching is achieved. Hexafluorosilicates fill the space between the masks. (B) Higher magnification picture of the same hexafluorosilicate. It mainly attaches to the masks but seems not to be present at substrate surface. However, it generates a solid barrier at mask area and reduces etching solution renewal at tip locations, resulting in the glass etch-rate decrease.

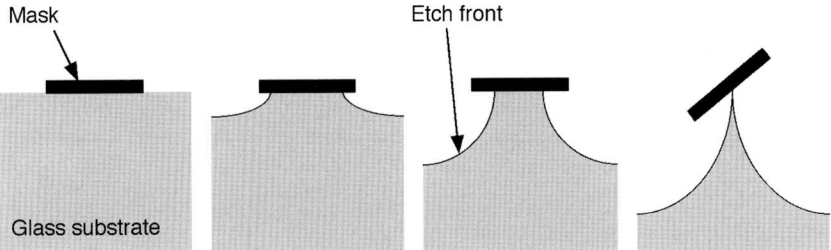

FIGURE 4.6. Schematic of glass etching process in hydro-fluoric acid (HF) solution. Due to isotropic glass etching properties, the mask is under-etched symmetrically until detachment. In order to get very sharp tips the etching process should be stopped right after mask detachment.

Finally, the mask pattern for realization of tip-shaped structures is composed of circles with a diameter of twice the required tip height. As the mask will be under-etched until mask release, the hypothesis of isotropic glass etching was assumed for mask dimension calculation.

4.2.3.5   Tip-Shaped Glass Structures: Morphology Versus Etching Conditions

The morphology of the 3-D glass structures is greatly dependent on etching conditions. Structures made on Pyrex glass have very different shapes depending on HF solution concentration (see Figure 4.7). For low HF concentrations (5%), tips with a circular shape can be achieved, however the morphology of the glass surface is usually rather rough. For higher HF concentrations (49%), mask under-etching is about 3.5 times higher than vertical etching depth, and resulting tips are very flat. However, the glass surface is very smooth.

Structures made on soda–lime glass show different dependencies (see Figure 4.8). Using low HF concentrations (5% and 10%), very sharp and high tips can be fabricated with a smooth surface morphology. For higher HF concentrations (>25%), the glass etch-rate is too fast to prevent hexafluorosilicate precipitation. As a consequence, the etching process cannot be controlled precisely and the coating of the reactive surface by precipitates reduces the homogeneity of the substrate surface. This commonly results in unusable structures.

When comparing the obtained glass etching results, best tip-shaped glass microstructures were obtained using float glass in a 10% HF solution at 20°C, resulting in long and sharp tips with a smooth surface. This fabrication process has therefore been chosen for 3-D MEA biochip fabrication.

4.3   3-D Multi-Electrode Arrays: Results

4.3.1   Obtained MEA Biochips

Various MEA biochips with planar and 3-D electrodes have been developed. Most of these biochips are compatible with the data acquisition system from Multi

A

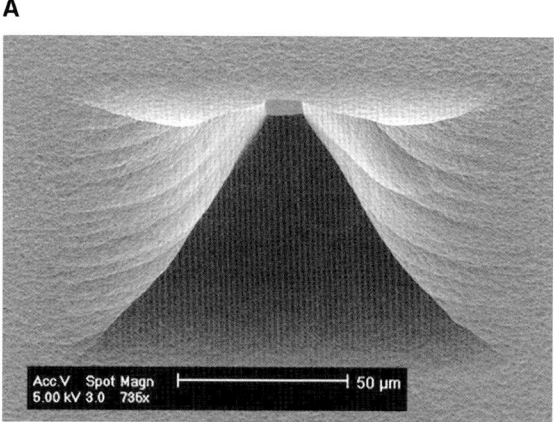

B

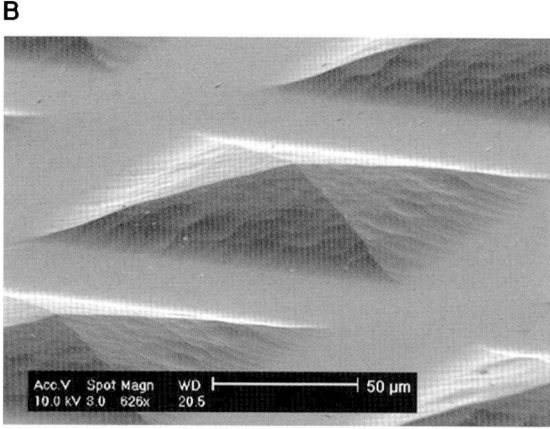

FIGURE 4.7. Scanning electron microscopy pictures of Pyrex tip-shaped micro-structures. A set of square chromium masks was used to generate a pyramidal tip shape. (A) Pyrex tip resulting from etching in a 5% HF solution. The horizontal under-etching is similar to the vertical etching. However, a very rough surface morphology was obtained. Etch-rate was approximately 0.2 μm/min. Note that the top of the tip is flat, which is due to stopping the etching process prior to mask detachment. (B) Pyrex tip resulting from etching in a 49% HF solution. The horizontal etch-rate was found to be four times faster than the vertical etch-rate allowing only fabrication of flat micro-structures. A smooth surface morphology was obtained. Etch-rate was approximately 4 μm/min. (From Heuschkel (2001), with permission.)

Channel Systems. The MEA biochip dimensions are in this case 4.9 cm × 4.9 cm, which is much larger than chip dimensions that are 1.5 cm × 1.5 cm or 2.1 cm × 2.1 cm. A global view of two different MEA biochips is shown in Figure 4.9. Currently, the most used and thus standard MEA electrode arrangement is based on a 60-electrode matrix (8 × 8 without corner electrodes) with an

A

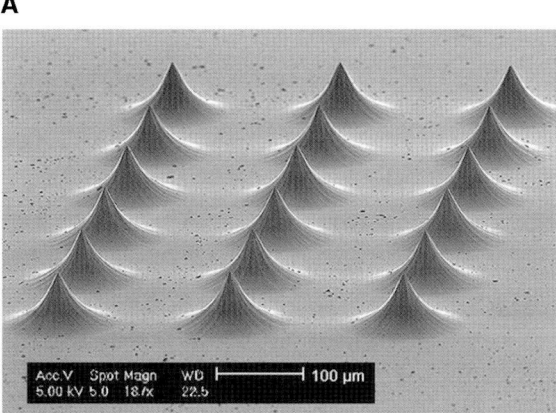

B

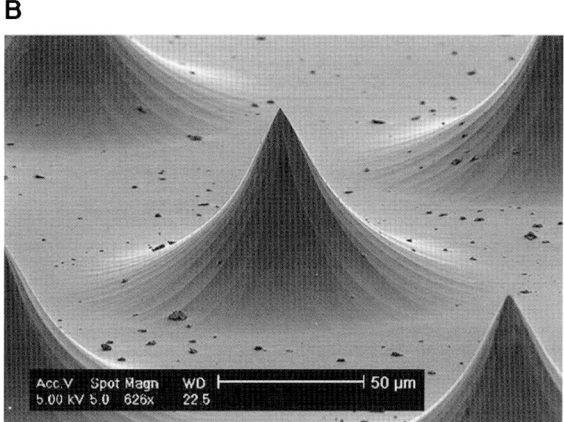

FIGURE 4.8. (A) Scanning electron microscopy pictures of soda–lime glass tip-shaped microstructures etched in a 10% HF solution at 20°C for 40 min. A set of circular chromium masks was used for tip fabrication (diameter 140 μm, spacing 200 μm). The tip height obtained is 70 μm (etch-rate of 2 μm/min). (B) Higher magnification micro-graph of same micro-structure as in (A). The micro-structures have a sharp tip (diameter at top <1 μm) and the glass surface appears to be smooth. (From Heuschkel (2001), with permission.)

electrode spacing of 200 μm and square electrodes with dimensions of 40 μm × 40 μm. However, several other electrode arrangements have also been realized, for example, an 8 × 8 matrix with smaller electrodes and 100 μm spacing, and a 5 × 13 matrix of circular electrodes with a diameter of 40 μm and 140 μm spacing (Wirth and Luscher, 2004), and so on.

Planar electrode MEA biochips have also been produced using a similar fabrication process (without the glass etching procedure). Both types have identical electrode configurations allowing direct comparison between the planar and 3-D MEA biochip variations.

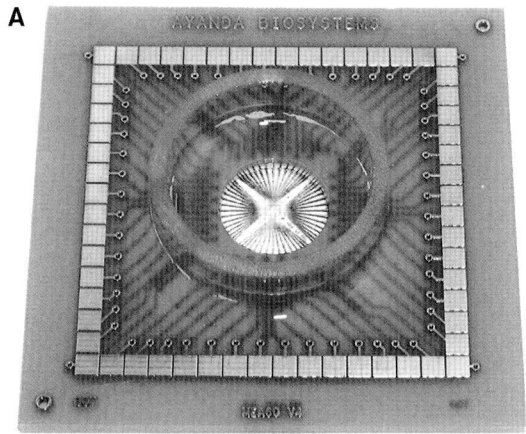

FIGURE 4.9. (A) Aerial view of a 60-electrode 3-D MEA biochip (dimensions of 5 cm ×
5 cm) based on a 1.5 cm × 1.5 cm glass chip. The recording electrodes are placed in an
8 × 8 matrix without corner electrodes, generating a 1.4 mm × 1.4 mm recording area
in the center of the chip. External connection pads are adapted to signal amplification and
data acquisition system from Multi Channel Systems, Germany. (B) Aerial view of another
60-electrode 3-D MEA biochip based on a 2.1 cm × 2.1 cm glass chip. In this MEA biochip,
electrodes are arranged on a 4 × 15 matrix generating a 1 mm × 13 mm recording area in
the center of the chip. Furthermore, additional stimulation (8) and reference (4) electrodes
have been integrated and are connected to external stimulation circuitry and ground via
supplementary external connection pads.

Pictures of 3-D MEA biochip results are presented in Figure 4.10. In
Figure 4.10A an electrode workspace from a standard layout described above is
shown. Figure 4.10B shows another electrode workspace with 68 electrodes on
a 6 × 12 matrix without corner electrodes. In both cases, the electrode height is
about 60 μm. The highest tips already fabricated on MEA biochips have a height
of about 80 μm.

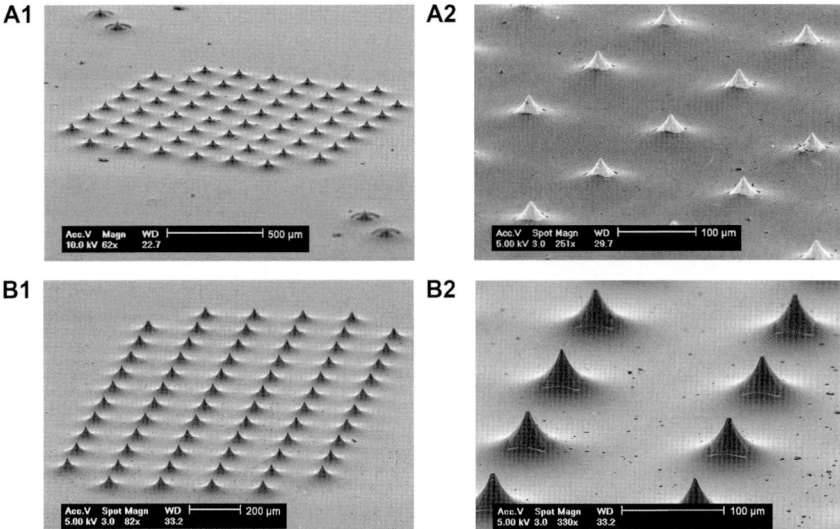

FIGURE 4.10. Scanning electron microscopy pictures of 3-D MEA recording areas. (A) The tip-shaped electrodes are embedded in epoxy insulation layer. MEA electrodes are arranged in an 8×8 matrix without corner electrodes. Tip height is 60 µm, electrode dimensions are 40 µm $\times$ 40 µm, electrode spacing is 200 µm. This layout corresponds to the 3-D MEA biochip presented in Figure 4.9A. (B) MEA electrodes are arranged in a 6×12 matrix without corner electrodes generating a recording area of 1 mm $\times$ 2.2 mm. Tip height is 60 µm, electrode dimensions are 40 µm $\times$ 40 µm, and electrode spacing is 200 µm.

The tip-shape corresponds closely to a semi-circular shape, which was expected for the soda–lime glass etching process. As illustrated in Figure 4.10B, the top of glass tips may present a plateau, which is the consequence of a premature glass etching process termination. As the glass etching is not homogeneous on the whole substrate, it can happen that some masks are not fully under-etched at process end. Ideally, the etching process should be prolonged for a few minutes in order to avoid such flat tips. On the other hand, over-etching the glass tips will result in smaller rounded tips.

Different tip-heights have also been obtained with different glass etching masks. The tip-height is limited by isotropic glass etching characteristics. It was also found that the height of the glass tips is limited by the MEA electrode spacing, and on the other hand the minimal distance (20 µm) required between two neighboring electrode masks that would allow best etching conditions.

4.3.2   Electrical Characteristics

Experimental measurements of electrode properties were obtained from different MEA biochips integrating electrodes made of different materials, sizes, and geometries. For these measurements, impedance (Z) and phase shift (θ) measurements were carried out. The MEA biochips were first filled with a 0.9% NaCl solution. A large platinum counterelectrode was immersed in the saline solution and contacting the PCB connection pads of the MEA biochip completed the electrical circuit. In this configuration, the MEA electrode is small compared to the counterelectrode and its impedance is dominant. A reference signal of 100 mV, 100 kHz was applied to the electrodes. Sweeping the frequency from 100 Hz to 20 kHz provided measurements. Mean values and the relative error of the obtained data were calculated and plotted. The electrode resistance and capacitance were calculated using data from the impedance and phase shift measurement.

The electrode noise level can be measured using a data acquisition system, for example, the MEA60 system from Multi Channel Systems. The MEA biochip is placed into the MEA1060 amplifier and the noise level is displayed from the data-acquisition software. The measurement of the noise level is thus very easy to achieve and is the simplest way to control MEA biochip integrity and electrode quality.

In order to compare the impact of electrode material on the measurement conditions, a first set of experiments using planar MEA biochips made with different electrode material was done (see Figure 4.11). Four different materials commonly used for recording electrodes were chosen for this experiment: platinum, platinum black, indium-tin oxide (ITO), and titanium nitride (TiN).

Platinum electrodes (thickness of 150 nm) were fabricated using a sputtering process. Platinum black electrodes were obtained by platinum electroplating onto platinum electrodes using a solution containing 1% of platinum chloride $H_2PtCl_6/6$ H_2O, 0.01% of lead acetate $Pb(COOCH_3)_2/3$ H_2O, and 0.0025% of HCl (Oka et al., 1999). For good electroplating results, the deposition was done at room temperature with a current density of 30 mA/cm^2 for 2 min. ITO electrodes (resistance <20 Ω/square) were manufactured from commercially available glass substrates covered with a 100 nm thick ITO layer from Merck Balzers. All the above electrodes have identical geometry (square electrode with dimension of 40 µm × 40 µm). TiN electrodes (diameter of 30 µm) were obtained from Multi Channel Systems.

After impedance and phase shift measurement (see Figure 4.11A), the electrode resistance and the electrode capacitance were calculated (see Figure 4.11B) and normalized to 1 mm^2 area, considering that the resistance is inversely proportional to the area and the interface capacitance is directly proportional to it.

The measured curves present a similar shape for all tested electrode materials. At low frequencies, the high electrode impedance and the negative phase shift (close to –90°) indicate that the electrodes act as capacitors. At high frequencies, the electrode coupling becomes resistive (phase shift tends to zero, data not shown) and the impedance (a few kΩ, data not shown) is mainly defined from the electrode spreading resistance and the saline solution resistance.

The results shown in Figure 4.11 illustrate the electrode material characteristics. ITO electrodes present the highest impedance and the smallest capacitance. As a result, ITO proves to be the least desirable material. The resulting noise level is the highest of all measured materials, however, its main advantage remains

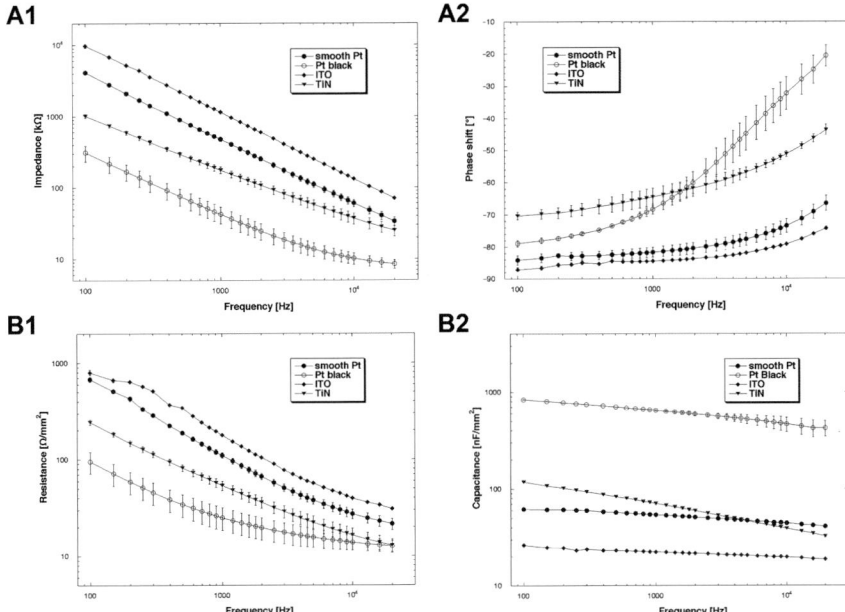

FIGURE 4.11. (A) Impedance Z and phase shift θ versus frequency for planar electrodes made of sputtered platinum (smooth Pt), black platinum (Pt black), indium-tin oxide (ITO), and titanium nitrate (TiN) in a 0.9% NaCl solution. A 100 mV, 100 kHz AC stimulation signal was applied to the electrodes during these measurements. (B) Using measured data, the global electrode resistance and capacitance were calculated. The resistance is defined as $R = Z \cos(\theta)$ and the capacitance is defined as $1/C = (2\pi F) Z \sin(\theta)$, where F is the frequency. Porous materials such as Pt black and TiN, which have a larger electrode surface than smooth metallic electrodes, show the lowest resistance and the largest capacitance. It results in smaller noise level and a larger safe-charge-injection limit. (From Heuschkel (2001), with permission.)

its transparency allowing combination of extracellular electrical monitoring with optical measurement methods.

Smooth platinum electrodes present better electrode characteristics. The noise level of platinum electrodes is good enough for most electrophysiological experiments. Results show that porous materials (Pt black and TiN) have the best electrode properties due to a larger effective area resulting from the high porosity, conveying a smaller electrode resistance and a larger capacitance. This reduces the noise level and increases the current injection capability of the electrodes. One major drawback of Pt black is the poor mechanical stability of the Pt black layer on the underlying platinum electrode. Furthermore, it is cumbersome to deposit the Pt black on an electrode array, as each electrode has to be plated separately. TiN electrodes were deposited under specific deposition conditions that are difficult to reproduce elsewhere.

TABLE 4.1. Noise levels measured using planar electrodes made of different electrode materials and sizes.

Material	Electrode surface	Measured noise level
Indium–tin oxide (ITO)	1600 μm^2	30–40 μV
Platinum	1600 μm^2	20–25 μV
Platinum black	1600 μm^2	5–10 μV
Titanium nitride (TiN)	710 μm^2	10–25 μV

The measured noise levels for the different planar MEA electrodes are presented in Table 4.1. These values can also be obtained by mathematical calculation of the lowpass cut-off frequency and the thermal noise using the data from Figure 4.11. These noise-level values are sufficient for monitoring biological activity as long as the signal-to-noise ratio remains large; that is, amplitudes are in the range of 100 μV to 1 mV.

In order to compare planar electrodes with 3-D electrodes, the electrode impedance and phase shift of both planar and 3-D electrodes (with a tip height of 60 μm) were measured (see Figure 4.12A). The calculated values of the electrode resistance and capacitance are plotted in Figure 4.12B.

In the 3-D arrays, there is a difference in electrode impedance due to an increase of the geometrical electrode surface. The effective 3-D electrode area can be approximated by the lateral surface of a cone (underestimation) or of a pyramid (overestimation) with a basis of 40 μm and a height of 40 μm. This yields an electrode surface between 2809 μm^2 and 3600 μm^2, respectively. Comparing these values to a planar electrode surface of 1600 μm^2, there is an electrode surface increase of a factor between 1.75 and 2.25.

The measured impedance value of the planar electrode is 2.08 times higher than the 3-D electrode impedance, which corresponds closely to the electrode area increase. However, the phase shift yielded similar values in both measurements. The resulting value of the electrode resistance for 3-D MEA electrodes is 2.14 times smaller and the electrode capacitance is 2.08 times larger than for planar electrodes configuration.

The electrode noise is mainly due to the electrode's thermal noise defined by

$$U_{th} = \sqrt{4\,kTRB}$$

where k is the Boltzmann constant ($k = 1.3807 \cdot 10^{-23}$ $J{\cdot}K^{-1}$), T is the temperature, R is the global resistance of the electrode, and B is the electrode bandwidth. According to the thermal noise definition, there should be a noise reduction of a factor 1.44 for the 3-D electrode due to its proportionality to the square root of the electrode resistance. The measured noise levels show a reduction from 20 to 25 μV for planar electrodes to 14 to 17 μV for 3-D electrodes, which corresponds to a decrease of a factor between 1.17 and 1.78. It results that the signal-to-noise ratio is increased for a given signal amplitude due to the electrode geometry variation (larger electrode area). However, the signal-to-noise ratio is dependent on the signal amplitude, which also varies with the electrode configuration.

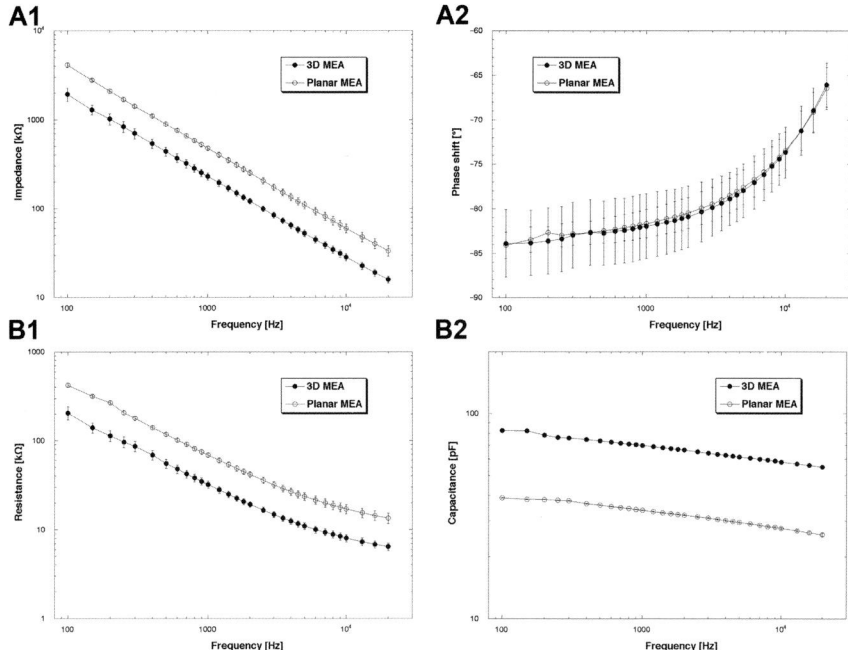

FIGURE 4.12. (A) Impedance Z and phase shift θ versus frequency for planar and 3-D MEA in a 0.9% NaCl solution. A 100 mV, 100 kHz AC stimulation signal was applied to the electrodes during these measurements. (B) Using measured data, the global electrode resistance and capacitance were calculated for both the planar and 3-D electrodes versus frequency. Due to increased electrode area, the 3-D electrodes show a lower resistance and a larger capacitance than planar electrodes. (Reprinted from Heuschkel et al. (2002), with permission from Elsevier.)

Furthermore, the larger capacitance of the 3-D electrodes results in a higher safe-charge-injection limit. This is important for cell or tissue stimulation through the 3-D electrodes because a higher charge can be applied to the electrode before irreversible electrolysis, that is, ohmic current flow occurs.

Electrode configuration simulations and experimental simulation verification results published previously (Heuschkel et al., 2002) have shown that the 3-D electrode configuration should provide better signal recordings than the planar electrode configuration when used with acute tissue slices.

In summary, porous electrode materials provide the best electrode characteristics. It has also been shown that the electrical properties of 3-D tip-shaped electrodes provide better electrical characteristics than planar electrodes. Thus, the MEA biochip with the best electrical characteristics would be a 3-D tip-shaped platinum black electrode. However, we chose to use only smooth platinum because the platinum black deposition process is too cumbersome and more costly.

4.3.3   Typical 3-D Multi-Electrode Array Recordings Using Acute Tissue Slices

In order to illustrate the use of 3-D MEA biochips with acute tissue slices and to show typical signals that can be obtained with such MEA biochips, results obtained using acute rat hippocampal slices are presented below.

Sample Preparation

Experiments were carried out with young Sprague–Dawley male rats (ca. 4 weeks old, 120 g) that were sacrificed by decapitation. Anesthesia was avoided, because most anesthetics do obviously affect the brain and their efficacy depends strongly on the weight of the animal. Animal handling and preparation were carried out according to the Swiss ethical rules for animal experimentation.

The brain was quickly removed and put into ice-cold ($4°C$) saturated with 95% O_2/5% CO_2 artificial cerebrospinal fluid without calcium (calcium-free ACSF composition in mM: NaCl 135, KCl 5, $NaHCO_3$ 15, $MgCl_2$ 1, and glucose 10; pH 7.4). Absence of calcium ions prevents synaptic propagation of action potentials during preparation, reducing cell damage. Transverse hippocampal slices (thickness of 350 μm) were cut with a tissue chopper. Slices were then incubated in normal gazed ACSF solution with 2 mM $CaCl_2$ at room temperature for at least one hour for recovery.

3-D MEA biochips were coated with 0.1% polyethylenimine dissolved in distilled water for two hours, extensively rinsed with distilled water and left to air-dry. Slices were then placed onto the 3-D MEA biochips and positioned to cover the electrode matrix. After positioning of the slice, the surrounding ACSF solution was removed with a pipette. Small filter papers were then used to absorb the solution remaining between the slice and the MEA electrodes promoting tissue adhesion on MEA biochip. Fresh Ringer solution was then quickly introduced in the culture chamber on top of the slice. Figure 4.13 shows a typical acute hippocampal tissue slice on a 3-D MEA biochip. The 3-D MEA biochip was then transferred to the MEA1060 amplifier from Multi Channel Systems and a continuous perfusion rate of 2 to 3 ml/min of ACSF was maintained during the recording session.

Calcium-Free Experiment

The application of a calcium-free ACSF solution is a simple experiment allowing identification of the parts of the signal that belong to the stimulation artifact and induced local field potentials, respectively. As stated above, the absence of calcium blocks synaptic transmission and should leave only the stimulation artifact and eventually presynaptic action potentials. Figure 4.14 shows the results of a calcium-free experiment. The first series of signals (left column) shows evoked local field potentials in response to a stimulation signal (±100 μA, 100 μs duration, 1 pulse every 30 sec) on four different electrodes under normal ACSF solution.

The recordings shown in the middle column were recorded after 4 min of low-calcium ACSF solution perfusion. Only the electrode stimulation artifact remains.

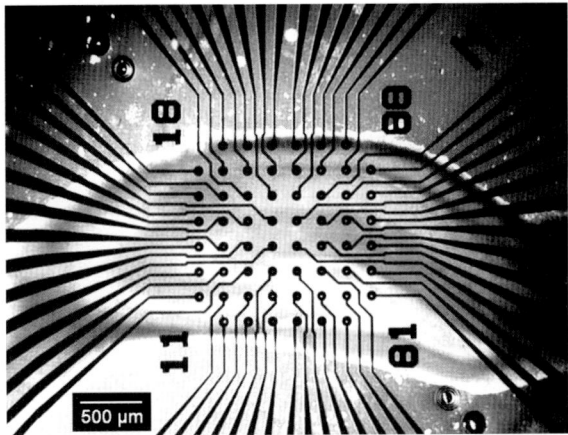

FIGURE 4.13. Light transmission bottom view of typical experimental conditions when using 3-D MEA biochips with acute tissue slices. An acute transverse rat hippocampal slice was plated on a 3-D MEA biochip with an 8 × 8 electrode arrangement. Small white dots at center of electrodes indicate that the top of the 3-D tips present a small plateau. The MEA biochip was coated with polyethylenimine 1% dissolved in distilled water for two hours prior to slice plating procedure.

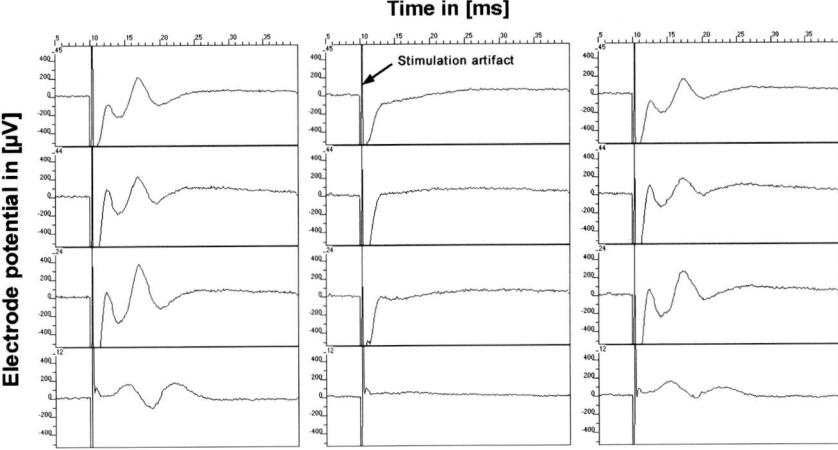

FIGURE 4.14. Application of low calcium solution allows biological signal identification. Left: illustration of electrical stimulation (±100 μA, 100 μs, 1 pulse every 30 sec) and responses from transverse rat hippocampal acute tissue slice from four different electrodes. The measured signals correspond to local field potentials generated by neurons located close to the tip-shaped electrodes. The application of a low calcium solution (0 mM $CaCl_2$) allows inhibition of neuronal activity due to synaptic transmission inhibition. The traces shown in the middle column were recorded after 4 min of low-calcium solution perfusion. The peaks observed at the beginning of each trace represent electrode stimulation artifacts. Right: traces recorded 10 min after low calcium solution washout show biological signal recovery. (From Heuschkel (2001), with permission.)

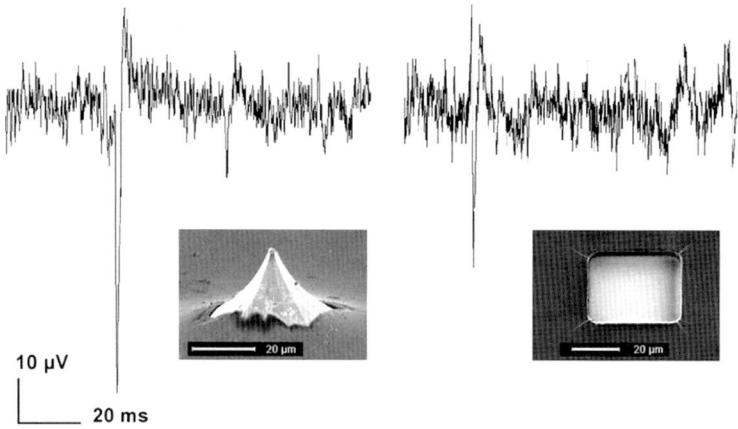

FIGURE 4.15. Typical single-unit action potentials recorded with planar and 3-D electrode configuration in acute tissue slices. The signal amplitude recorded with 3-D electrode configuration is about two times larger than with planar electrode configuration. It is a result of a smaller distance to active electrodes in the slice as well as from better electrode characteristics due to an increase in effective electrode surface.

Note the bottom electrode trace showing a small signal right after the stimulation artifact, presumably due to presynaptic action potentials. This experiment is reversible as shown in the right traces recorded 10 min after washout using standard ACSF solution.

A low concentration solution (1 µM) of tetrodotoxin can also be used to block all neuronal responses. Experiments using TTX showed similar responses than shown in Figure 4.14 (data not shown), but the duration of response recovery was much longer (30 minutes to one hour).

Spontaneous Single-Unit Activity

Single-unit action potentials can also be recorded using MEA biochips. In order to compare signal amplitudes between planar and 3-D electrode configuration, acute hippocampal slices were placed in the MEA biochip. A U-shaped platinum wire with a grid of nylon threads was placed each time on top of the slice. The pressure generated by the nylon grid on the tissue stimulated several neurons to fire action potentials. In several experiments with both planar and 3-D electrode configuration ($n = 5$), typical single-unit action potentials have been recorded. Larger signal amplitudes were always recorded using the 3-D electrode configuration (see Figure 4.15).

4.4   Application Examples

Two application examples illustrate the use of 3-D MEA in basic neuroscience research as well as in safety steps of drug development processes.

4.4.1 Spatiotemporal Evolution of Signal Spread in the Rat Barrel Cortex Assessed with 3-D Multi-Electrode Arrays

4.4.1.1 Introduction

Although fingertips provide important information for humans, rats explore their surroundings with their whiskers. But humans and rats have approximately equivalent capacities in texture discrimination tasks. It is therefore important to know how such sensory data is computed in the somatosensory cortex. Woolsey and Van der Loos (1970) discovered that layer 4 (L4) of the cortical region where whisker information is processed consists of discrete clusters of cell bodies, called "barrels". This cortical area was therefore termed the "barrel cortex". There exists a one-to-one relationship between whiskers and barrels: a single particular whisker activates cells of a single barrel. Since this discovery, many studies gave information about different aspects of the barrel cortex: barrels receive and amplify input from the ventral posterior medial nucleus (Jensen and Killackey, 1987; Diamond, 1995; Hirsch, 1995; Feldmeyer et al., 1999). Axonal and dendritic arbors of barrel neurons are confined to barrel borders (Feldmeyer et al., 1999; Lubke et al., 2000; Petersen and Sakmann, 2000), suggesting that barrels work as functionally independent excitatory neuronal networks (Goldreich et al., 1999; Petersen and Sakmann, 2001; Laaris and Keller, 2002). But it is still not clear if neighboring barrels communicate directly with each other or only through L2/3. L4 cells also target pyramidal cells in L2/3 (Armstrong-James et al., 1992; Feldmeyer et al., 2002) whose axons arborize in L2/3 and in L5 (Gottlieb and Keller, 1997). Excitation in L4 is followed by a rapid and strong feedforward inhibition in L4, limiting spread of excitation both spatially and temporally. However, the precise spatiotemporal behavior of the inhibitory network in L4 and L2/3 has never been described in detail.

A controversial subject is the role of L5/6, generally considered as the main output layer. Cortico-thalamic cells in L6 also send collaterals to L4 (Staiger et al., 1996; Zhang and Deschenes, 1997), and axons of L4 neurons were observed to end in L5/6 (Lubke et al., 2000; Petersen and Sakmann, 2000; Laaris and Keller, 2002). These anatomical observations expand the generally postulated simple "thalamus-L4-L2/3-L5/6"- pathway.

A lot of these studies were based on single-cell recordings. However, in order to understand the mechanisms of signal processing, one has to investigate the neuronal network itself. Therefore, we have turned to 3-D MEA biochips that permit simultaneous recording from 60 electrodes of neuronal network local field potentials (LFP) with a high spatial and temporal resolution. This technique answers questions that have never been satisfactorily explained previously; it was still not clear if neighboring barrels communicate directly with each other or only through L2/3. Nothing was known about the interplay of excitation and inhibition in the supragranular layers of the barrel cortex, and also the role of L6 was not clear.

In a first set of experiments, a single barrel was therefore stimulated and in a second set, stimulation occurred in L6. Stimulation in a barrel led to excitation starting in L4 and spreading to L2/3 and subsequently to L5/6, sometimes activating a neighboring barrel through L2/3. Excitation was followed by inhibition shaping excitation spatially and temporally. Stimulation in L6 resulted in an early ascending inhibition and an excitatory L6–L4–L6 loop.

4.4.1.2   Materials and Methods

3-D MEA biochips consisting of 60 tip-shaped electrodes have been used for these experiments. A specific electrode arrangement with the following properties has been designed: an electrode arrangement in a 5×13 matrix, an electrode diameter of 40 µm, an electrode spacing of 140 µm or 200 µm (center to center), a tip height of 42 µm, and an electrode impedance between 400 and 600 kΩ. The 3-D MEA biochip was coated with 0.1% polyethyleneimine dissolved in distilled water for two hours, rinsed with distilled water, and left to air-dry. Cortical slices with a thickness of 350 µm from 12 to 20 day-old Wistar rats with clearly visible barrels (see Figure 4.16A) were placed on the 3-D MEA biochips for recording. The LFP at each electrode was recorded against a reference electrode placed in the bath.

ACSF-filled double-barrel micro-pipettes with tip diameters of 5 to 10 µm were used for extracellular stimulation. Constant amplitude voltage pulses (0.5 to 5 V, 150 µs, 0.3 Hz) were delivered in the center of the barrels, respectively in L6. Data acquisition was done using a MEA1060 amplifier and a data acquisition card from Multi Channel Systems.

The trace of a single channel consisted of a positive signal immediately following the stimulus artifact that was often followed by a broader negative peak (see Figure 4.16B). As the negative peak disappeared in the presence of 10 µM bicuculline, it must have been due to $GABA_A$ receptors (not shown). The positive peak vanished after adding 10 µM CNQX and 50 µM D-AP5 and must therefore have been due to AMPA and NMDA receptors (not shown). In order to visualize the spatial spread of signals, gray-scale coded activity maps representing the instantaneous voltage amplitude of the LFP recorded by each MEA electrode were generated (see Figure 4.16C). White always codes for positive signal amplitudes, which stand for excitation, and black codes for negative signal amplitudes, which stand for inhibition.

4.4.1.3   Spatiotemporal Evolution of Activity After Single Barrel Stimulation

Stimulation in a single barrel led to a stereotypical spatiotemporal evolution of the activity. Excitation was visible in the stimulated barrel 1.8 ± 0.6 ms ($n = 43$) after stimulation. This short-lasting excitation (on average 1.8 ± 1.2 ms, $n = 43$) was replaced in 60% of the experiments by a slowly increasing inhibitory signal. Before excitation in the barrel was replaced by inhibition, it spread within a short time interval from the stimulated barrel to L2/3 above the stimulated barrel in 95% of the experiments (see Figure 4.17A). Although excitation respected the columnar organization delineated by the barrel borders in L4 and lower L2/3 immediately above the barrel, it became subsequently broader in L2/3 (see Figure 4.17A). In the upper L2/3, it spread horizontally and crossed the borders of the cortical column, designing a mushroomlike activation in L2/3. Inhibition started in the stimulated barrel 3.4 ± 1.6 ms ($n = 23$) after stimulation and spread from L4

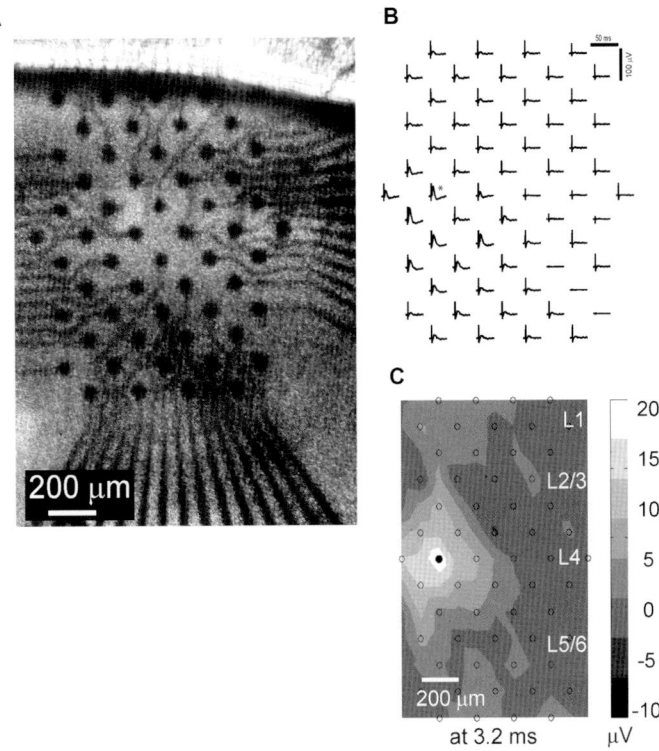

FIGURE 4.16. (A) Small piece of barrel field cortex mounted on a 3-D MEA biochip. The barrels are barely visible as bright clusters in row 5–7 of the MEA biochip. (B) Stimuli-induced local field potentials recorded with the MEA electrodes (averaged over 20 sweeps). The stimulus artifact indicates the timing of the stimulus delivered at the location indicated by the asterisk. Three damaged electrodes in the right lower corner of the MEA biochip were switched off. (C) Spatial activity map of the postsynaptic signals of the recording shown in (B), 3.2 ms after stimulation. Black dot: stimulation location.

to the lower L2/3 following excitation. Inhibition in lower L2/3 was observed in all experiments.

Although excitation of L4 was terminated 3.6 ± 1.8 ms ($n = 43$) after stimulation, the depolarizing signal in upper L2/3 was active for as long as 15 ms (9.6 ± 3.0 ms, $n = 30$) after stimulation. The inhibitory signal in lower L2/3 was visible for up to 14.7 ± 3.4 ms ($n = 34$).

In 12% of the experiments, excitation in a neighboring barrel could be observed (5.1 ± 1.1 ms) after stimulation (see Figure 4.17B). In all these cases, the neighboring barrel was activated through L2/3 and not by direct activation. In two of these five experiments, a symmetrical activation (although with different delays) of both neighboring barrels could be observed.

In 25% of the experiments, excitation propagated first from the stimulated barrel in L4 to the lower L2/3 and less than one millisecond later to L5/6 (see

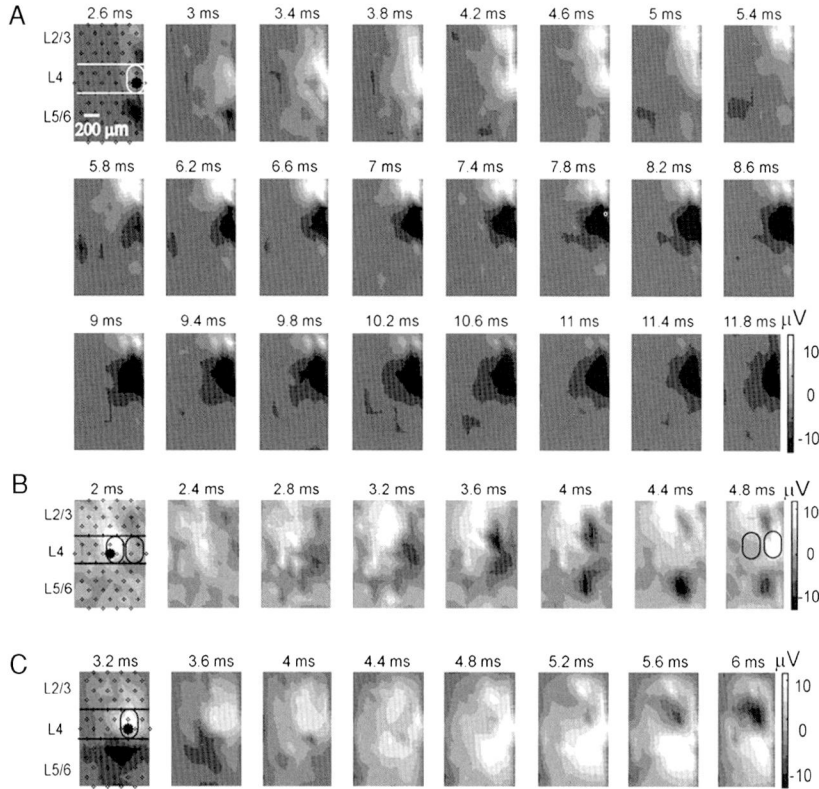

FIGURE 4.17. Spatiotemporal evolution of activity after single barrel stimulation. (A) Time series of activity maps spaced at 0.4 msec intervals. Excitatory signals were initiated in layer 4 (L4) in the stimulated barrel and then spread to L2/3, followed by inhibition flowing from the stimulated barrel to the lower L2/3. (B) Consecutive activity maps of a different experiment after stimulation in a barrel (black dot) illustrating activation of a neighboring barrel through an L4–L2/3–L4 pathway. (C) Same as in (B), but illustrating activation of a neighboring barrel as well as of L5/6.

Figure 4.17C). Excitation remained for 8 ± 1.5 ms in L5/6 and did not respect the columnar organization.

4.4.1.4   Temporal Evolution of the Activity Map After Direct Stimulation of L6

In an additional set of experiments, L6 to L4 connections were investigated by stimulating L6. Figure 4.18 illustrates a typical example of the spatiotemporal evolution of the activity after L6 stimulation. In 13 of 15 experiments, stimulation in L6 resulted in a very early wave of inhibition starting in L6, already 1 msec after stimulation. In about 1.5 ms after stimulation, a first excitation wave was visible in L4, usually confined to a barrel border. This excitation was extremely short because it was suppressed by the rising inhibition from L5/6. This inhibition wave rose from L5/6 to L4 and further to lower L2/3 before it was replaced in L4 by an

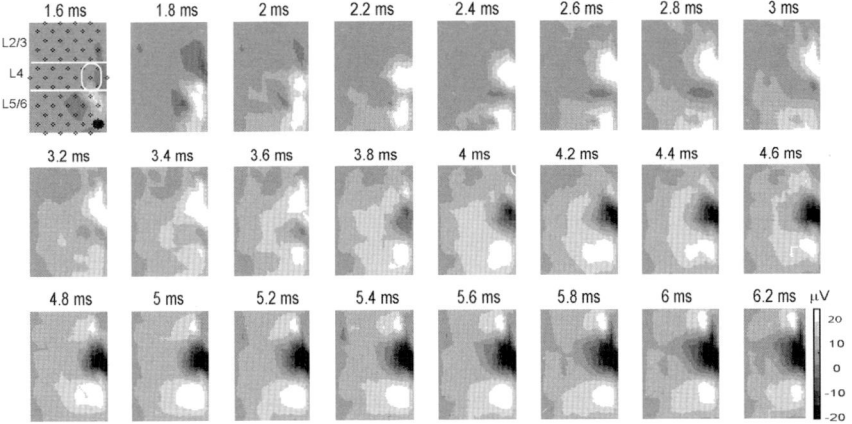

FIGURE 4.18. Response pattern after stimulation in layer 6 (L6). Time series of activity maps showed a very early inhibition ascending from L5/6 to a barrel in L4. It is followed by appearance of excitation flowing from L6 to L4, subsequently to L2/3 and simultaneously backwards to L6. Black dot: stimulation position.

excitation wave (starting 2.2 ± 0.9 ms in L6, $n = 14$), which propagated from L6 to L4 within about 2 ms, activating sometimes a single barrel and sometimes L4 but in a more diffuse manner. Similarly as after direct stimulation in L4, activation in L4 was short-lasting (1.9 ± 1.1 ms, $n = 14$). From L4, excitation spread to lower L2/3 as described in the previous section. Simultaneously, excitation flowed back from L4 to L6, sometimes through a neighboring barrel, fanning out into a broad band of excitation in L6. Excitation in L4 was followed by strong inhibition that remained active for several ms (5.5 ± 2.7 ms, $n = 14$) but barely spread to supragranular layers.

The time course of the positive response revealed a clear L6–L4–L6 loop in all experiments. Similar to direct barrel stimulation, the spatiotemporal evolution of the inhibitory signal was more stereotypical compared to the excitatory signals after stimulation of L6.

4.4.1.5 Discussion and Conclusions

The direct observation of the spatiotemporal spread of excitation and inhibition through all laminae of the cortex provides a dynamic view of the interdependence of excitation and inhibition in cortical columns. The earliest excitatory response remains confined to the barrel respecting the barrel borders, and excitation of neighboring barrels was only observed via L2/3 excitation, demonstrating that barrels are truly functionally independent units. Excitation moves to L2/3 and subsequently to L5/6, illustrating beautifully the L4-L2/3–L5/6 pathway.

The strong and stereotypical inhibition starting in L4 and spreading to L2/3 allows only a brief window of excitability in these layers, underlining how inhibition shapes excitation not only in the temporal, but also in the spatial domain.

The early negative signal spreading from L6 to L4 suggests feedforward inhibition, which may be responsible for end-inhibition in L4 as observed in the visual cortex (Bolz and Gilbert, 1986; Hirsch, 1995). An excitatory L6–L4–L6 loop was found in the visual cortex too (Stratford et al., 1996; Tarczy-Hornoch et al., 1999), and the previously described findings are an indication that such a feedforward loop also exists in the barrel cortex. It may contribute to a strong excitation in L4 and provide L6 with feedback information on the state of excitation in L4.

The technique of 3-D MEA biochips applied to acute cortical slices establishes therefore a new innovative technique, which permits visualization of network behavior of acute brain slices. This work has been published recently in more detail elsewhere (Wirth and Luscher, 2004).

4.4.2   Pharmacological Safety of Lead Compounds Assessed with 3-D Multi-Electrode Arrays

4.4.2.1   Introduction

Drug development strategies aim at reducing time and cost for optimizing and marketing new chemical entities. As an alternative to "classical" target screenings, Trophos has introduced the use of primary neuronal cultures as a complex and integrated system to screen new chemical entities for neurodegenerative disorders. This approach provides the main advantage to integrate the complete complexity of neuronal (patho)physiology and therefore to select more powerful compounds on in vitro "sick" neurons. From this point of view the cell is considered as a "black box" and the target(s), mechanism(s) of action, and possible adverse effects need to be elucidated after the screening processes. Extensive pharmacological profiling of lead compounds could be performed through binding studies on a large scale of receptors, transporters, channels and enzymes that could suggest a potential side effect, but further functional studies will be needed to confirm and to document an hypothesis. To speed up the lead selection process, Trophos has introduced early safety tests to evaluate possible adverse effects of new chemical entities on the main neuronal properties: action potential propagation, propagation, synaptic transmission, and synaptic plasticity. These functional evaluations are performed using acute slices of rat brain recorded on 3-D MEA biochip.

4.4.2.2   Preparation of Acute Hippocampal and Cerebellar Slices

Experiments were carried out with 17 to 21 day-old rats (Sprague–Dawley). Animals were housed and used in accordance to the French and European directives for animal care. The rats were sacrificed by fast decapitation, without previous anaesthesia. The brain was quickly removed and soaked in ice-cold oxygenated buffer with the following composition in mM: KCl 2, NaH_2PO_4 1.2, $MgCl_2$ 7, $CaCl_2$ 0.5, $NaHCO_3$ 26, glucose 11, and saccharose 250. Sodium was omitted in order to prevent cell swelling. Hippocampus or cerebellar slices (thickness of 300 μm) were cut with a McIlwain tissue chopper. Slices were carefully separated with a painting brush and incubated at room temperature for at least one hour in artificial cerebrospinal fluid (ACSF) of the following composition in mM: NaCl 126, KCl

3.5, NaH_2PO_4 1.2, $MgCl_2$ 1.3, $CaCl_2$ 2, $NaHCO_3$ 25, and glucose 11, bubbled with carbogen (95% O_2, 5% CO_2). It is of value to note that in cerebellar sagittal and parasagittal slices the large Purkinje neurons dendritic tree is preserved but that the parallel fibers network is disrupted.

4.4.2.3   Electrophysiological Recordings of Hippocampal or Cerebellar Slices on 3-D Multi-Electrode Arrays

One slice is positioned in the center of the 3-D MEA biochip to cover the 60 electrodes field (1,4 mm^2) without any coating procedure. After positioning the slice, the surrounding solution is fully removed with tissue paper in order to assure a tight adhesion between the slice and the electrodes. A net ballast with a U-shaped platinum wire (where perpendicular hair pieces are stuck like rungs) is carefully placed on the slice to immobilize it. The MEA is then quickly transferred to a MEA1060 amplifier stage (Multi Channel Systems) on a microscope and the slice is continuously perfused with oxygenated ACSF (3 ml/min at 37°C) during the whole recording session. The bath is connected to the ground electrode.

For evoked fEPSPs in hippocampal slices, one of the electrodes visually identi-fied in the CA3 field or in the CA3/CA1 border is chosen as an "active" electrode (to stimulate the Schaeffer collaterals) and it is disconnected from its amplifier. A monopolar biphasic current pulse (-300 μA 50 μs and $+300$ μA 50 μs) is then injected through the "stimulating" electrode every 30 sec. Evoked simultaneous responses (fEPSPs) can usually be recorded from 6 up to 12 electrodes of the CA1 field stratum radiatum, corresponding to the propagation of the stimuli along the Schaeffer collateral pathways. Responses are recorded at a 5 kHz sampling rate every 30 sec with the MC-Rack software (Multi Channel Systems). Data (fEPSPs amplitudes, in μV) are stored on the computer disk until offline analysis.

LTP was induced using a standard protocol (100 Hz LTP). Briefly, a tetanic stimulation (using the same current amplitude as the one employed to evoke fEP-SPs) is applied for 1 sec at 100 Hz and repeated twice at 20 sec intervals. fEPSPs recording after LTP induction is performed for an additional 20 min period with one pulse-stimulation every 30 sec.

Molecules to be tested are directly dissolved from stock-concentrated DMSO solutions into the ACSF (bubbled with 95% O_2, 5% CO_2), so that the final con-centration of DMSO is less than 0.1%.

In the offline analysis, fEPSPs amplitudes recorded at different places of the CA1 stratum radiatum region are normalized, averaged, and plot as a function of time before and after LTP induction. Spikes amplitude and spikes rate were determined with the MC-Rack software (Multi Channel Systems).

4.4.2.4   Spontaneous Action Potentials

Spontaneous action potentials can be recorded in cultured neurons (cortical neu-rons, motoneurons, etc.) after seven to eight days of culture either with standard glass electrodes or with MEA biochips. We have chosen to use acute cerebellar slices of young rats (P16–P21) to record spontaneous spikes of Purkinje neurons (see Figure 4.19A), because such a preparation is immediately available if young

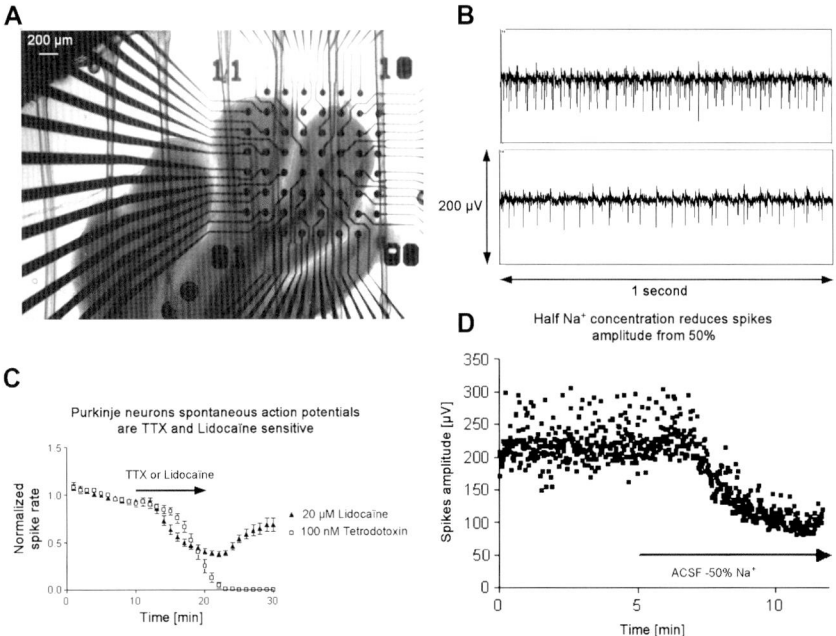

FIGURE 4.19. (A) Bottom view of a typical cerebellar slice on a 3-D MEA biochip. Two lobules cover the 60 electrodes. The layer of Purkinje neurons is close to the interface between the dark inner zone and the clear external zone (granule cell layers). (B) A very regular spiking activity corresponding to the firing of a Purkinje neuron under control conditions (top) and following a 10 min exposure to 20 μM lidocaïne (bottom) recorded at a single electrode. (C) Inhibition of spontaneous spikes by classical sodium channel antagonists. Data points correspond to mean values ± SEM (6 electrodes from 2 slices, lidocaine; 8 electrodes from 1 slice, TTX). (D) As expected, spike amplitude is reduced by 50% when the extracellular sodium concentration is decreased by 50%. Data points correspond to raw spikes amplitudes as a function of time.

rats are accessible. Purkinje neurons spontaneously fired action potential at a frequency of about 30 to 40 Hz (see Figure 4.19B) and spikes were inhibited by classical sodium channel antagonists such as lidocaine and tetrodotoxin (see Figure 4.19C). The "pure" sodium component of the spikes is illustrated by the 50% decrease of their amplitude, which is observed when the sodium concentration is reduced by half in the bath solution (see Figure 4.19D). Thus, potential adverse effects at sodium channels could be revealed with this type of preparation. However, because Purkinje neurons are integrated into a complex neuronal network, modification of spike frequency could also result from an interaction of the tested compound with other channels and/or cells within the cellular network.

4.4.2.5   Synaptic Transmission

Hippocampal slice preparations are widely used for studying excitatory and inhibitory transmission as well as synaptic plasticity. The hippocampal slice is of optimal size for recordings performed on a 3-D MEA biochip (see Figure 4.20A). When stimulating Schaeffer collaterals in the CA3 region, one can record fEP-SPs at electrodes located in the CA1 region. Pharmacological investigations reveal that fEPSPs are strongly inhibited by 50 μM CNQX, which indicates a major glutamatergic transmission through AMPA/kainate receptors (see Figure 4.20B). The fEPSPs amplitude is reduced by 50% when extracellular sodium is decreased by half, which is consistent with a 50% decrease of AMPA/kainate currents and as a consequence a 50% decrease of field-evoked potentials (see Figure 4.20B). When the slice is bathed with a "low calcium–high magnesium" solution, a classical protocol to decrease neurotransmitter release (Goda and Stevens, 1994; Wachowiak et al., 2002), again fEPSPs are strongly inhibited (see Figure 4.20B). These data confirm that fEPSPs correspond mainly to the activation of glutamatergic synapses at CA3–CA1 synapses. Nevertheless, we have been able to observe that a fraction of fEPSPs result from an excitatory $GABA_A$-mediated neurotransmission with HCO_3^- ion as the main permeable ion (data not shown).

To illustrate the usefulness of this recording technique, the effects of two widely used drugs have been investigated in the hippocampal slice preparation. The first one is Cyclosporin A (an immunosuppressant), a molecule known to modulate both calcineurin and the mitochondrial permeability transition pore (Khaspekov et al., 1999; Levy et al., 2003). As illustrated in Figure 4.20C, Cyclosporin A is able to potentiate fEPSP and to strongly increase LTP amplitude (dose-dependent effect). These experiments reveal that pharmacological profiles can be performed with good time and concentration resolution. It is of value to recall that Cyclosporin A normally does not cross the blood–brain barrier (BBB). Therefore, the primary effects of this molecule are very unlikely to occur when the BBB is not disrupted.

The second drug investigated is tamoxifen (an anti-estrogen used in breast cancer), a molecule known to target estrogen receptors, the mitochondrial permeability transition pore (Custodio et al., 1998; Hoyt et al., 2000) and $GABA_A$ receptors (Chesnoy-Marchais, 2003). Brain concentration of tamoxifen can reach 0.5 μM in tamoxifen-treated patients. At 1 μM, tamoxifen slightly increases fEPSPs and LTP amplitude (see Figure 4.20D). Thus, it is possible that tamoxifen's effects on neuronal tissue result from its ability to modulate synaptic transmission.

4.4.2.6   Safety Pharmacology on 3-D MEA: Conclusions

In conclusion, recording of hippocampal slices on 3-D MEA biochips permits revealing potential beneficial or adverse effects of molecules on synaptic transmission. Moreover, this preparation allows the pharmacological evaluation of new compounds in reference to known drugs.

4.5   Conclusions and Outlook

When compared to classical planar electrodes MEA biochips, 3-D tip-shaped MEA biochips offer the main advantage of overcoming the problem of the dead cell

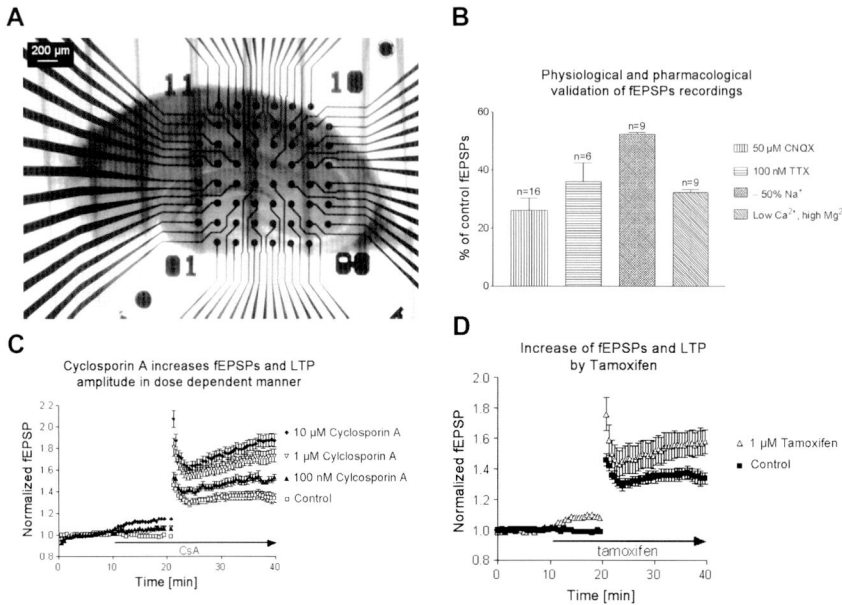

FIGURE 4.20. (A) Light transmission view from bottom of a rat hippocampal slice on a 3-D MEA biochip. The main hippocampal regions (dentate gyrus and CA1, CA2, and CA3) can be clearly identified. (B) Pharmacological and physiological modulation of fEPSPs. CNQX (a selective AMPA/kainate receptor antagonist) and TTX (a selective inhibitor of "fast" sodium channels) both decrease fEPSP amplitude by inhibiting glutamatergic synaptic transmission and action potential propagation, respectively. When the extracellular concentration of Na^+ is reduced by 50%, fEPSP amplitude is decreased also by 50%. This might result from a 50% reduction of the electrochemical gradient through AMPA/kainate receptor channels and, as a consequence, a 50% decrease of postsynaptic membrane depolarization. When synaptic transmission is depressed by lowering extracellular Ca^{2+} and increasing extracellular Mg^{2+}, fEPSPs are strongly inhibited too. Numbers of recording electrodes are indicated between brackets above the bars (mean ± SEM). (C) Time and dose-dependent effect of Cyclosporin A on fEPSPs before and after 100 Hz LTP induction (∼20 min). Data points correspond to mean values ± SEM (40 electrodes from 7 slices, 10 μM CsA; 30 electrodes from 5 slices, 1 μM CsA; 30 electrodes from 6 slices, 0.1 μM CsA; 33 electrodes from 9 slices, control). (D) Effect of 1 μM tamoxifen on fEPSPs and 100 Hz LTP (induced at ∼20 min). Data points correspond to mean values ± SEM (16 electrodes from 4 slices, 1 μM tamoxifen; 33 electrodes from 9 slices, control.)

layer in acute slices and this constitutes a technological breakthrough in this field. MEA recordings currently constitute the only technical approach available for investigating the overall activity of a complex neuronal network. This is feasible due to increased electrode area and the resulting overall reduction of impedance and background noise level. Measurements performed with 3-D MEA biochips show signal shapes comparable to recordings from conventional planar electrode arrays, however, signal amplitudes are improved significantly.

It should be recalled that extracellular electrodes will never provide single-cell resolution such as the one observed with the patch-clamp technique. This high-resolution technique is, however, confronted with another major problem when applied to the recording of native neurons in slices: the access resistance. Small currents, such as the ones evoked by the opening of ligand-gated or voltage-gated channels at very fine and distal neurites (located at tens of microns from the neuronal cell body) are hardly detectable because of the high intracellular access resistance when the patch electrode is attached to the cell body. This is in addition to space-clamp problems. In contrast, if a 3-D tip-shaped electrode is in close contact with high-density dendritic neuronal trees, it has the ability to record the sum of very small field potentials that could hardly be "seen" with conventional patch electrodes. Then, a 3-D tip-shaped electrode could record the sum of small synaptic currents. The 3-D MEA recording technique will certainly have an impact on the electrophysiological data from acute slices that are obtained with conventional glass electrode techniques.

The fabrication concept used at Ayanda Biosystems allows the manufacture of 3-D MEA biochips at reduced costs compared to other commercially available MEA biochips. The authors believe that currently available fabrication technologies would allow MEA biochips to become single-use consumable in a near future. However, there are still several technological problems to be addressed before low-cost MEA biochips can be introduced to the market.

Polymer replication technologies such as plastic injection or hot embossing can also be used for 3-D MEA biochip fabrication. These techniques could be an alternative to wet chemical glass etching process. Its advantage would be replication of homogeneous tip-shaped structures on a plastic substrate at a lower cost. The move to such fabrication technologies implies high development costs. Therefore, it would be applicable only when a large demand for low-cost 3-D MEA biochips exists.

The two application examples illustrate the power of 3-D MEA recording to unravel neuronal network properties and to address pharmacological and toxicological questions in drug discovery and safety pharmacology. These promising results should provide a lever towards the broader use of multi-electrode arrays in electrophysiology research.

Acknowledgments. The authors would like to thank Rebecca Pruss (Trophos), Scott Gilbert (Crystal Vision Microsytems), and Solomzi Makohliso (Ayanda Biosystems) for comments and text editing of the manuscript.

References

Adler, R., Jerdan, J., and Hewitt, A.T. (1985). Responses of cultured neural retinal cells to substratum-bound laminin and other extracellular matrix molecules. *Dev. Biol.* 112(1): 100–114.

Armstrong-James, M., Fox, K., and Das-Gupta, A. (1992). Flow of excitation within rat barrel cortex on striking a single vibrissa. *J. Neurophysiol.* 68(4): 1345–1358.

Bledi, Y., Domb, A.J., and Linial, M. (2000). Culturing neuronal cells on surfaces coated by a novel polyethyleneimine-based polymer. *Brain Res. Protoc.* 5(3): 282–289.

Bolz, J. and Gilbert, C.D. (1986). Generation of end-inhibition in the visual cortex via interlaminar connections. *Nature* 320(6060): 362–365.

Boppart, S.A., Wheeler, B.C., and Wallace, C.S. (1992). A flexible perforated microelectrode array for extended neural recordings. *IEEE Trans. Biomed. Eng.* 39(1): 37–42.

Branch, D.W., Wheeler, B.C., Brewer, G.J., and Leckband, D.E. (2000). Long-term maintenance of patterns of hippocampal pyramidal cells on substrates of polyethylene glycol and microstamped polylysine. *IEEE Trans. Biomed. Eng.* 47(3): 290–300.

Brockhaus, J., Ilschner, S., Banati, R.B., and Kettenmann, H. (1993). Membrane properties of ameboid microglial cells in the corpus callosum slice from early postnatal mice. *J. Neurosci.* 13(10): 4412–4421.

Chesnoy-Marchais, D. (2003). Potentiation of glycine responses by dideoxyforskolin and tamoxifen in rat spinal neurons. *Eur. J. Neurosci.* 17(4): 681–691.

Connolly, P., Clark, P., Curtis, A.S., Dow, J.A., and Wilkinson, C.D. (1990). An extracellular microelectrode array for monitoring electrogenic cells in culture. *Biosens. Bioelectron.* 5(3): 223–234.

Custodio, J.B., Moreno, A.J., and Wallace, K.B. (1998). Tamoxifen inhibits induction of the mitochondrial permeability transition by $Ca2+$ and inorganic phosphate. *Toxicol. Appl. Pharmacol.* 152(1): 10–17.

D'Angelo, E., De Filippi, G., Rossi, P., and Taglietti, V. (1998). Ionic mechanism of electroresponsiveness in cerebellar granule cells implicates the action of a persistent sodium current. *J. Neurophysiol.* 80(2): 493–503.

Diamond, M.E. (1995). Somatosensory thalamus of the rat. In Diamond, I.T., ed., *The Barrel Cortex of Rodents.* Plenum, New York, pp. 189–219.

Egert, U., Heck, D., and Aertsen, A. (2002). Two-dimensional monitoring of spiking networks in acute brain slices. *Exp. Brain Res.* 142(2): 268–274.

Egert, U., Schlosshauer, B., Fennrich, S., Nisch, W., Fejtl, M., Knott, T., Muller, T., and Hammerle, H. (1998). A novel organotypic long-term culture of the rat hippocampus on substrate-integrated multielectrode arrays. *Brain Res. Protoc.* 2(4): 229–242.

Feldmeyer, D., Egger, V., Lubke, J., and Sakmann, B. (1999). Reliable synaptic connections between pairs of excitatory layer 4 neurones within a single 'barrel' of developing rat somatosensory cortex. *J. Physiol.* 521(Pt 1): 169–190.

Feldmeyer, D., Lubke, J., Silver, R.A., and Sakmann, B. (2002). Synaptic connections between layer 4 spiny neurone-layer 2/3 pyramidal cell pairs in juvenile rat barrel cortex: Physiology and anatomy of interlaminar signalling within a cortical column. *J. Physiol.* 538(Pt 3): 803–822.

Fromherz, P., Offenhausser, A., Vetter, T. and Weis, J. (1991). A neuron-silicon junction: A Retzius cell of the leech on an insulated-gate field-effect transistor. *Science* 252(5010): 1290–1293.

Goda, Y. and Stevens, C.F. (1994). Two components of transmitter release at a central synapse. *Proc. Nat. Acad. Sci. U. S. A.* 91(26): 12942–12946.

Goldreich, D., Kyriazi, H.T., and Simons, D.J. (1999). Functional independence of layer IV barrels in rodent somatosensory cortex. *J. Neurophysiol.* 82(3): 1311–1316.

Gottlieb, J.P. and Keller, A. (1997). Intrinsic circuitry and physiological properties of pyramidal neurons in rat barrel cortex. *Exp. Brain Res.* 115(1): 47–60.

Gross, G.W. (1979). Simultaneous single unit recording in vitro with a photoetched laser deinsulated gold multimicroelectrode surface. *IEEE Trans. Biomed. Eng.* 26(5): 273–279.

Gross, G.W., Harsch, A., Rhoades, B.K. and Gopel,W. (1997). Odor, drug and toxin analysis with neuronal networks in vitro: Extracellular array recording of network responses. *Biosens. Bioelectron.* 12(5): 373–393.

Gross, G.W., Rhoades, B.K., Reust, D.L., and Schwalm, F.U. (1993). Stimulation of monolayer networks in culture through thin-film indium-tin oxide recording electrodes.*J. Neurosci. Meth.* 50(2): 131–143.

Hammerle, H., Egert, U., Mohr, A., and Nisch, W. (1994). Extracellular recording in neuronal networks with substrate integrated microelectrode arrays. *Biosens. Bioelectron.* 9(9–10): 691–696.

Heuschkel, M.O. (2001). *Fabrication of Multi-Electrode Array Devices for Electrophysiological Monitoring of In-Vitro Cell/Tissue Cultures*, Series in Microsystems, Volume 13. Hartung-Gorre Verlag, Konstanz, Germany.

Heuschkel, M.O., Fejtl, M., Raggenbass, M., Bertrand, D., and Renaud, P. (2002). A three-dimensional multi-electrode array for multi-site stimulation and recording in acute brain slices. *J. Neurosci. Meth.* 114(2): 135–148.

Hirsch, J.A. (1995). Synaptic integration in layer IV of the ferret striate cortex. *J. Physiol.* 483(Pt 1): 183–199.

Hoyt, K.R., McLaughlin, B.A., Higgins, D.S., Jr., and Reynolds, I.J. (2000). Inhibition of glutamate-induced mitochondrial depolarization by tamoxifen in cultured neurons. *J. Pharmacol. Exp. Ther.* 293(2): 480–486.

Israel, D.A., Barry, W.H., Edell, D.J., and Mark, R.G. (1984). An array of microelectrodes to stimulate and record from cardiac cells in culture. *Am. J. Physiol.* 247(4 Pt 2): H669–674.

Jensen, K.F. and Killackey, H.P. (1987). Terminal arbors of axons projecting to the somatosensory cortex of the adult rat. I. The normal morphology of specific thalamocortical afferents. *J. Neurosci.* 7(11): 3529–3543.

Jimbo, Y., Robinson, H.P., and Kawana, A. (1993). Simultaneous measurement of intracellular calcium and electrical activity from patterned neural networks in culture. *IEEE Trans. Biomed. Eng.* 40(8): 804–810.

Khaspekov, L., Friberg, H., Halestrap, A., Viktorov, I., and Wieloch, T. (1999). Cyclosporin A and its nonimmunosuppressive analogue N-Me-Val-4-cyclosporin A mitigate glucose/oxygen deprivation-induced damage to rat cultured hippocampal neurons. *Eur. J. Neurosci.* 11(9): 3194–3198.

Kristensen, B.W., Noraberg, J., Thiebaud, P., Koudelka-Hep, M., and Zimmer, J. (2001). Biocompatibility of silicon-based arrays of electrodes coupled to organotypic hippocampal brain slice cultures. *Brain Res.* 896(1–2): 1–17.

Kucera, J.P., Heuschkel, M.O., Renaud, P., and Rohr, S. (2000). Power-law behavior of beat-rate variability in monolayer cultures of neonatal rat ventricular myocytes. *Circ. Res.* 86(11): 1140–1145.

Laaris, N. and Keller, A. (2002). Functional independence of layer IV barrels. *J. Neurophysiol.* 87(2): 1028–1034.

Lambert, F. (1974). La sérigraphie. *EMI.* 182: 59–63.

Levy, M., Faas, G.C., Saggau, P., Craigen, W.J., and Sweatt, J.D. (2003). Mitochondrial regulation of synaptic plasticity in the hippocampus. *J. Biol. Chem.* 278(20): 17727–17734.

Liang, D.-T. and Readey, D.W. (1987). Dissolution kinetics of crystalline and amorphous silica in hydrofluoric-hydrochloric acid mixtures. *J. Am. Ceramic Soc.* 70: 570–577.

Litke, A. and Meister, M. (1991). The retinal readout array. *Nuclear Inst. Meth. Phys. Res.* A310: 389–394.

Lubke, J., Egger, V., Sakmann, B., and Feldmeyer, D. (2000). Columnar organization of dendrites and axons of single and synaptically coupled excitatory spiny neurons in layer 4 of the rat barrel cortex. *J. Neurosci.* 20(14): 5300–5311.

Madou, M. (1997). *Fundamentals of Microfabrication*. CRC Press, New York.

Maeda, E., Robinson, H.P., and Kawana, A. (1995). The mechanisms of generation and propagation of synchronized bursting in developing networks of cortical neurons. *J. Neurosci.* 15(10): 6834–6845.

Makohliso, S.A., Giovangrandi, L., Leonard, D., Mathieu, H.J., Ilegems, M., and Aebischer, P. (1998). Application of Teflon-AF thin films for bio-patterning of neural cell adhesion. *Biosens. Bioelectron.* 13(11): 1227–1235.

McLellan, G.W. and Shand, E.B. (1984). *Glass Engineering Handbook*. McGraw-Hill, New York.

Meister, M., Pine, J., and Baylor, D.A. (1994). Multi-neuronal signals from the retina: Acquisition and analysis. *J. Neurosci. Meth.* 51(1): 95–106.

Meister, M., Wong, R.O., Baylor, D.A., and Shatz, C.J. (1991). Synchronous bursts of action potentials in ganglion cells of the developing mammalian retina. *Science* 252(5008): 939–943.

Monk, D.J., Soane, D.S., and Howe, R.T. (1993). Determination of the etching kinetics for the hydrofluoric acid/silicon dioxide system. *J. Electrochem. Soc.* 140: 2339–2346.

Nam, Y., Chang, J.C., Wheeler, B.C., and Brewer, G.J. (2004). Gold-coated microelectrode array with thiol linked self-assembled monolayers for engineering neuronal cultures. *IEEE Trans. Biomed. Eng.* 51(1): 158–165.

Novak, J.L. and Wheeler, B.C. (1986). Recording from the Aplysia abdominal ganglion with a planar microelectrode array. *IEEE Trans. Biomed. Eng.* 33(2): 196–202.

Novak, J.L. and Wheeler, B.C. (1988). Multisite hippocampal slice recording and stimulation using a 32 element microelectrode array. *J. Neurosci .Meth.* 23(2): 149–159.

Novak, J.L. and Wheeler, B.C. (1989). A high-speed multichannel neural data acquisition system for IBM PC compatibles. *J. Neurosci. Meth.* 26(3): 239–247.

Offenhausser, A., Sprossler, C., Matsuzawa, M., and Knoll, W. (1997). Field-effect transistor array for monitoring electrical activity from mammalian neurons in culture. *Biosens. Bioelectron.* 12(8): 819–826.

Oka, H., Shimono, K., Ogawa, R., Sugihara, H., and Taketani, M. (1999). A new planar multielectrode array for extracellular recording: Application to hippocampal acute slice. *J. Neurosci. Meth.* 93(1): 61–67.

Petersen, C.C. and Sakmann, B. (2000). The excitatory neuronal network of rat layer 4 barrel cortex. *J. Neurosci.* 20(20): 7579–7586.

Petersen, C.C. and Sakmann, B. (2001). Functionally independent columns of rat somatosensory barrel cortex revealed with voltage-sensitive dye imaging. *J. Neurosci.* 21(21): 8435–8446.

Rai-Choudhury, P. (1997). *Handbook of Microlithography, Micromachining, and Microfabrication*, Volume 1 & 2. SPIE Optical Engineering Press, Bellingham, WA.

Scholze, H. (1991). *Glass: Nature, Structure, and Properties*. Springer Verlag, New York.

Spierings, G.A.C.M. (1993). Wet chemical etching of silicate glasses in hydrofluoric acid based solutions. *J. Mater. Sci.* 28: 6261–6273.

Staiger, J.F., Zilles, K., and Freund, T.F. (1996). Distribution of GABAergic elements postsynaptic to ventroposteromedial thalamic projections in layer IV of rat barrel cortex. *Eur. J. Neurosci.* 8(11): 2273–2285.

Stoppini, L., Duport, S., and Correges, P. (1997). A new extracellular multirecording system for electrophysiological studies: Application to hippocampal organotypic cultures. *J. Neurosci. Meth.* 72(1): 23–33.

Stratford, K.J., Tarczy-Hornoch, K., Martin, K.A., Bannister, N.J., and Jack, J.J. (1996). Excitatory synaptic inputs to spiny stellate cells in cat visual cortex. *Nature* 382(6588): 258–261.

Streit, J., Tscherter, A., Heuschkel, M.O., and Renaud, P. (2001). The generation of rhythmic activity in dissociated cultures of rat spinal cord. *Eur. J. Neurosci.* 14(2): 191–202.

Suire, J. (1971). Réactions entre le verre et l'acide fluorhydrique. *Silicates Industriels* 36: 73–79 and 101–104.

Swerdlow, R.M. (1985). *The Step-by-Step Guide to Screen-Process Printing.* Prentice-Hall, Englewood Cliffs, NJ.

Sze, S.M. (1985). *Semiconductor Devices: Physics and Technology.* John Wiley & Sons, New York.

Tarczy-Hornoch, K., Martin, K.A., Stratford, K.J., and Jack, J.J. (1999). Intracortical excitation of spiny neurons in layer 4 of cat striate cortex in vitro. *Cereb. Cortex.* 9(8): 833–843.

Thiebaud, P., Beuret, C., de Rooij, N.F., and Koudelka-Hep, M. (2000). Microfabrication of Pt-tip microelectrodes. *Sens. Actuat. B. Chem.* 70: 51–56.

Thiebaud, P., Beuret, C., Koudelka-Hep, M., Bove, M., Martinoia, S., Grattarola, M., Jahnsen, H., Rebaudo, R., Balestrino, M., Zimmer, J., and Dupont, Y. (1999). An array of Pt-tip microelectrodes for extracellular monitoring of activity of brain slices. *Biosens. Bioelectron.* 14(1): 61–65.

Thiebaud, P., de Rooij, N.F., Koudelka-Hep, M. and Stoppini, L. (1997). Microelectrode arrays for electrophysiological monitoring of hippocampal organotypic slice cultures. *IEEE Trans. Biomed. Eng.* 44(11): 1159–1163.

Thomas, C.A., Jr., Springer, P.A., Loeb, G.E., Berwald-Netter, Y., and Okun, L.M. (1972). A miniature microelectrode array to monitor the bioelectric activity of cultured cells. *Exp. Cell Res.* 74(1): 61–66.

Tscherter, A., Heuschkel, M.O., Renaud, P., and Streit, J. (2001). Spatiotemporal characterization of rhythmic activity in rat spinal cord slice cultures. *Eur. J. Neurosci.* 14(2): 179–190.

Wachowiak, M., Cohen, L.B., and Ache, B.W. (2002). Presynaptic inhibition of olfactory receptor neurons in crustaceans. *Microsc. Res. Tech.* 58(4): 365–375.

Wirth, C. and Luscher, H.R. (2004). Spatiotemporal evolution of excitation and inhibition in the rat barrel cortex investigated with multielectrode arrays. *J. Neurophysiol.* 91(4): 1635–1647.

Woolsey, T.A. and Van der Loos, H. (1970). The structural organization of layer IV in the somatosensory region (SI) of mouse cerebral cortex. The description of a cortical field composed of discrete cytoarchitectonic units. *Brain Res.* 17(2): 205–242.

Yuste, R., Lanni, F., and Konnerth, A. (1999). *Imaging Neurons: A Laboratory Manual.* Cold Spring Harbor Laboratory Press, New York.

Zhang, Z.W. and Deschenes, M. (1997). Intracortical axonal projections of lamina VI cells of the primary somatosensory cortex in the rat: A single-cell labeling study. *J. Neurosci.* 17(16): 6365–6379.

Ziak, D., Chvatal, A., and Sykova, E. (1998). Glutamate-, kainate- and NMDA-evoked membrane currents in identified glial cells in rat spinal cord slice. *Physiol. Res.* 47(5): 365–375.

5
Electrophysiological Monitoring of Hippocampal Slice Cultures Using MEA on Porous Membrane

DAVID HAKKOUM, DOMINIQUE MULLER, AND LUC STOPPINI

Introduction

The development of brain slice preparations and particularly slice culture models that allow maintaining brain networks alive under in vitro conditions for several days, weeks, and even months have brought important new possibilities to study physiological and pharmacological mechanisms in the central nervous system (CNS) (Gahwiler, 1987; Heimrich and Frotscher, 1993; Finley et al., 2004). The simplicity of preparation and maintenance of slice cultures and their versatility in terms of physiological models makes them interesting for large-scale testing of pharmacological compounds. These preparations indeed offer the possibility to control the tissue environment, while conserving in vivo-like characteristics including organization of neuronal networks anatomy, three-dimensional aspect, developmental and functional properties of neurons, or reactivity of the tissue to injury or toxicity (Stoppini et al., 1991). In addition many slices can be obtained from the same animal thereby decreasing the number of animals needed for each experiment. Brain slices and slice cultures have thus been increasingly used over the years as models in studies of synaptic plasticity (Toni et al., 2001; Brager et al., 2002), epilepsy (Chen et al., 2004; Aptowicz et al., 2004; Silva et al., 2003; Kovacs et al., 2002), ischemia (Cater et al., 2003; Cronberg et al., 2004; Perez Velazquez et al., 2003; Barth et al., 1996), regeneration (Lee et al., 2002; Teter et al., 1999; McKinney et al., 1999; Stoppini et al., 1993), demyelination (Roth et al., 1995; Notterpek et al., 1993; Ghoumari et al., 2003; Demerens et al., 1996), toxicity (Shimono et al., 2002; Khaspekov et al., 2004; Xu et al., 2003), apoptosis (Lee et al., 2003; Keynes et al., 2004; Vis et al., 2002)) degenerative diseases (Brendza et al., 2003; Bendiske and Bahr, 2003; Duff et al., 2002; Hay et al., 2004; Murphy and Messer, 2004; Sherer et al., 2003), or even psychiatric and mental disorders (Kawasaki and Tsutsui, 2003).

Another interesting aspect of these preparations is that they are amenable to various methodological approaches. The most widely used readout for functional studies has probably been electrophysiology, through extracellular or intracellular patch clamp techniques (Aptowicz et al., 2004; Leutgeb et al., 2003; Yu et al., 2003; Alix et al., 2003; Thomas et al., 1998b), but protocols for all sorts of analyses have been successfully applied to slices or slice cultures, including recently live confocal imaging (Kasparov et al., 2002; Dailey and Waite, 1999; Schwartz and Yu, 1995; Miller et al., 1993), transfection methodologies (McAllister 2004; Ibrahim et al., 2000; Thomas et al., 1998a; Bergold et al., 1993; Ehrengruber et al., 2001; Morrison, III et al., 2000a), or genomic analyses (Morrison, 3rd, et al., 2000b; Fan and Tenner, 2004; Morl et al., 2002).

An important advantage in this respect of slice cultures is the possibility to continuously monitor within the same piece of tissue and for several days the changes associated with development, reaction to injury, or long-term pharmacological treatments. One way to do this is through electrophysiological recordings and the use of multiple electrode arrays (van Bergen et al., 2003; Shimono et al., 2002). As these long-term analyses are extremely important for various pharmacological testing procedures such as toxicity analyses, screening for drugs promoting remyelination, regeneration, or synaptic network remodeling, we undertook to develop a long-term recording system, based on the use of multiple electrode arrays applied to interface-type organotypic slice cultures, in which analyses could be carried out in incubatorlike conditions while fully controlling the perfusion medium applied to the tissue. In previous work, we had developed a multi-recording device (Stoppini et al., 1997) for the analysis of spontaneous and evoked electrical activity in organotypic cultures. We have extended this approach and now designed a new type of array onto the surface of a porous membrane or onto a permeable support. This biological/electronic interface, developed within the context of a small company, Biocell Interface, is described here. In particular, we provide evidence for the applicability of this recording interface for analyses of nerve regeneration.

5.1   The Neurosensor Interface System

5.1.1   Description of the BioCell Neurosensor Interface System

The Neurosensor Interface System (NIS) is based on the use of a multi-electrode recording array designed to conduct long-term electrophysiological studies applied specifically to 3-D interface-type organotypic cultures. The system is illustrated in Figure 5.1; it is composed of three main interconnected parts: a console unit which contains the micro-electrode array. This console is connected to a Biocell Interface Signal Conditioner (BCI-100) for stimulation, amplification, and digitalization of recorded signals, and the whole system is hooked up to and controlled by a computer system.

The console unit is a thermoregulated electronic module that houses the removable BioCell interface system. This interface is composed of a disposable cartridge that includes the multi-electrode array maintained in a sandwich between two chambers: the upper chamber, which is the gas perfusion chamber for maintaining the tissue at the interface, and the lower one, which is the solution perfusion chamber (Figure 5.1b). Cultures are grown on a precut and porous disk of

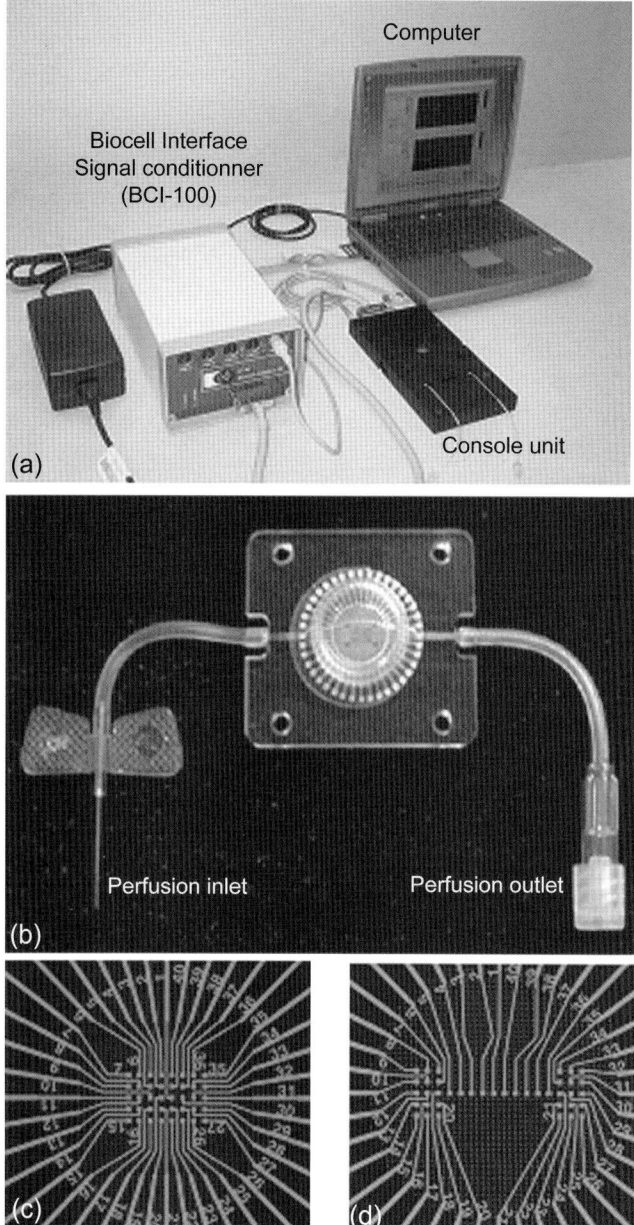

FIGURE 5.1. The Neurosensor Interface System (NIS; arrows) is composed of a console unit connected to the Biocell Interface Signal Conditioner (BCI-100; arrowhead) itself connected to a computer that monitors the experimental parameters (a). The console unit is a thermoregulated electronic module that houses the removable Biocell NIS. The Biocell NIS (b) is a disposable cartridge that contains the multi-electrode array (40 electrodes) built on a porous membrane between two chambers: the gas perfusion chamber and the perfusion chamber. The design of the electrodes can be easily modified depending upon the specific requirements of the experiment or tissue under analysis. Two different designs of multiple electrode arrays are illustrated ((c) and (d)).

membrane (Confetti, 10 mm diameter) that is deposited in the upper chamber on top of the micro-electrode array. The array is built on porous support so that the tissue remains in contact with the solution perfused underneath. The two chambers, the array, and the tissue are then assembled and inserted into the console unit, where the temperature can be precisely controlled.

The multi-electrode array is a network of 40 micro-electrodes 30 ìm thick and made of pure gold by plasma evaporation (Figures 5.1c and d); each electrode is insulated by biocompatible photoresist. This electrode network is embedded on a permeable and transparent membrane so that perfusion solution from the lower chamber can reach the tissue placed on the array. The position of the electrodes can be visualized and adjusted using a dissecting microscope or directly through the camera integrated within the connector. An inlet and outlet tubing system allows gas (to the upper chamber) and medium (to the lower chamber) to be perfused through the appropriate chambers at defined rates. Each electrode can be assigned as stimulating, recording, or earth/ground electrode. They also have variable gains. A golden metallic film layer arranged around the lower chamber makes the reference electrode. The stimulation and triggering can be performed either by the computer or by an external stimulator.

The Biocell Interface Signal Conditioner receives signals from the network of electrodes in contact with the tissue. A software utility allows the user to easily configure and set up the conditioner. Parameters that can be controlled include: the selection of acquisition channels, adjustment of the impedance of electrodes, conversion to AC or DC recording, amplification gain, adjustable offset, and analog/digital conversion. All commands for setting experimental and electrophysiological recording parameters as well as for data analysis are computerized.

The computer and software system includes a board that drives acquisition of digitized signals and stimulation. A data acquisition software package (Multiplexor) allows stimulation of up to 8 pairs of electrodes and recording from up to 8 electrodes among the 40-electrode matrix. Processing software allows online analysis of acquired data. A graphical software tool, compatible with Windows, allows users to configure and set up the conditioner. Every configuration can be saved on disk for further use. This software can be used to control the electrode parameters such as: amplitude of stimulation pulses (0 to 4092 mV), pulse duration (100 μs to 1 sec), pulse rate (1 pulse/day to 100 pulses/sec), temperature of the experiment, and perfusion parameters.

The online analysis software is composed of several panels for the visualization and analysis of the activity collected by the eight selected channels. One panel is for spontaneous activity (Figure 5.2a), a second one for evoked activity (Figure 5.2b), and the third panel is used for online quantification of response parameters such as amplitudes, latency, or slope (Figure 5.2c).

5.1.2 Properties of the BioCell-Interface System

This multi-electrode array system was designed for large-scale applications and long-term analyses such as required for drug testing applied to various models

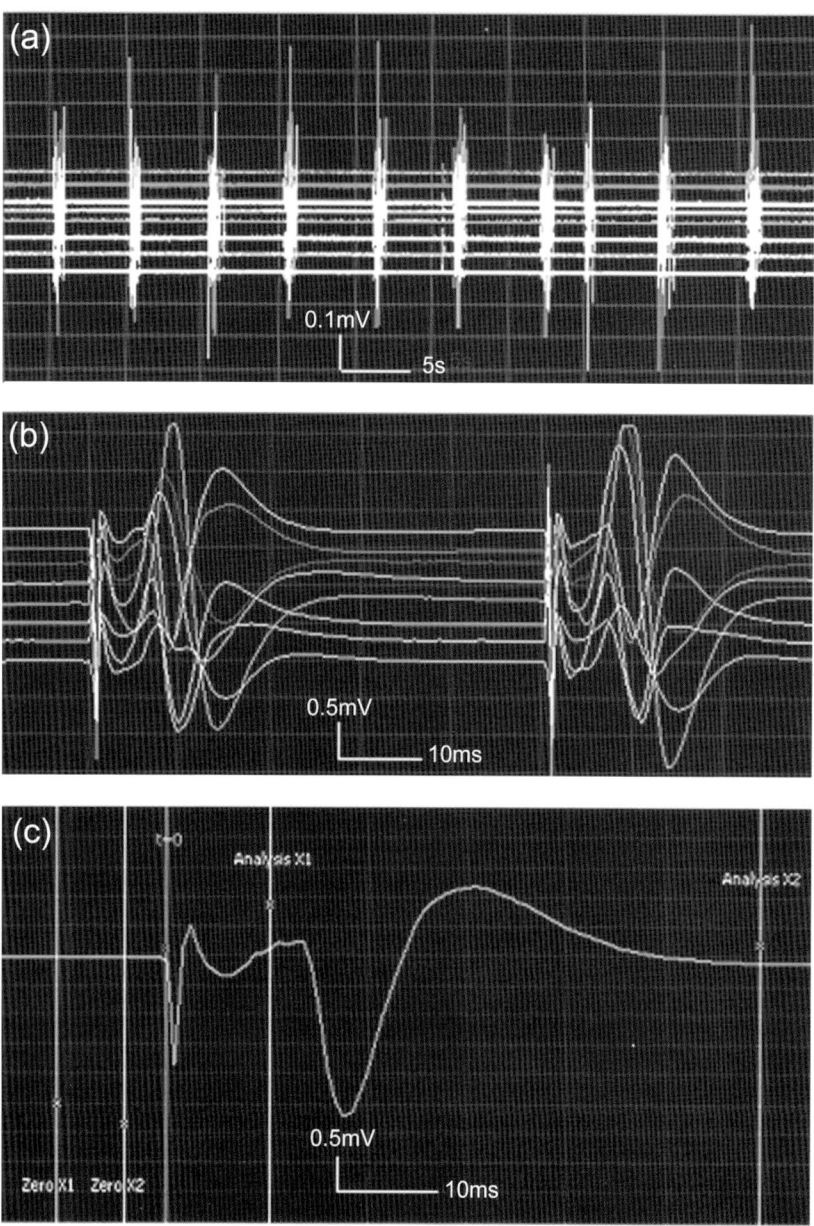

FIGURE 5.2. Software panels used for the analysis of electrical activity. The BioCell software combines different windows that allow the manipulator to visualize and analyze individually all selected channels. In the upper panel (a) are shown the eight recording channels which in this case display spontaneous epileptic activity induced by a strong electrical stimulation. The intermediate panel (b) illustrates the evoked synaptic activity recorded by eight selected channels in response to paired-pulse stimulation of CA3. The lower panel (c) is the trace analysis window for online analysis of the parameters of evoked responses.

developed in organotypic slice cultures. An objective was to have a simple system, easy to use, that could be handled. The main advantages include:

- Use of low cost, disposable cards that exclude possible artifacts coming from residual products from previous experiments and that can be prepared at the same time from the same animal to be run in parallel using multiple simultaneous systems.
- Possibilities to apply different patterns of stimulation in different locations for several days and to record activity simultaneously from several electrodes over prolonged periods of time.
- Possibility to continuously perfuse the tissue with culture medium or artificial CSF solutions, while keeping the cultures in a sterile environment, with control of the temperature and no need for a CO_2 incubator.

5.2   Pharmacological Applications

5.2.1   Stability of Evoked Activity

One important objective for the design of this multi-electrode recording system was to be able to perform analyses of synaptic activity over prolonged periods of time in a simple and reproducible manner. For this the design of the recording chamber was adapted so as to maintain incubatorlike conditions at a much smaller scale. Figure 5.3 illustrates recordings obtained in the CA1 area of hippocampal organotypic slice cultures by stimulation of CA3 neurons. Over a period of six hours in this experiment, synaptic activity could be elicited without major changes in the size and characteristics of responses. In another protocol that we use regularly, synaptic activity is recorded for periods of 20 minutes a day over several consecutive days. Figure 5.3b shows the mean amplitude of the responses recorded in this way on the eight available channels for a period of seven days in one experiment. The stimulation intensity was maintained constant throughout the experiment. Considering the numerous factors susceptible to affect excitability or survival of cultures over this period of time, the stability obtained in this experiment is of particular interest and clearly indicates that the quality of the survival of the slice cultures maintained in the recording chamber has been optimized. The possibility to reach such stability constitutes therefore an important advantage for carrying precise pharmacokinetic or toxicological studies of a given molecule on prolonged periods of time.

5.2.2   Drug Testing

The main type of applications for which this system was designed was to perform pharmacological analyses of the effects of compounds on excitability and synaptic transmission when applied to models that require long-term monitoring of activity. An important condition therefore is to be able to apply drugs to the tissue through an efficient perfusion system that does not alter stability. At the same time, the amount

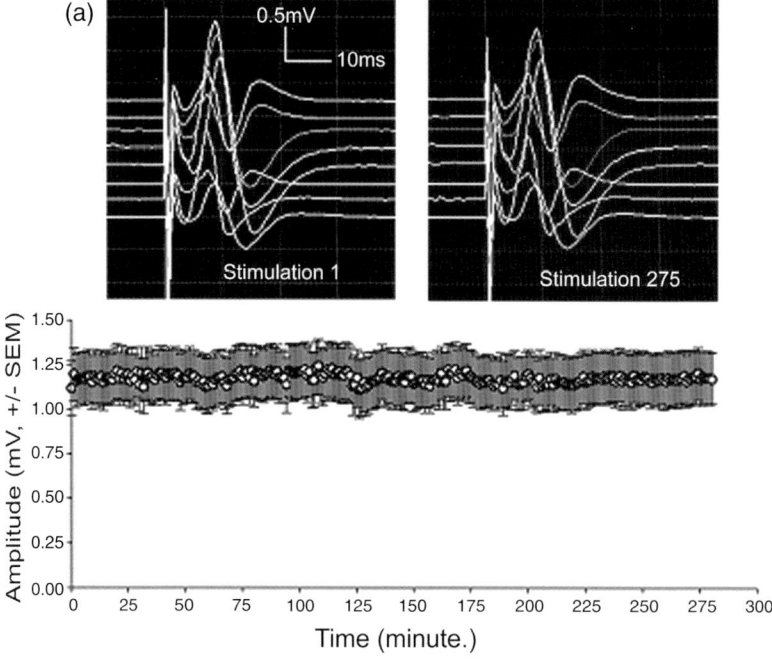

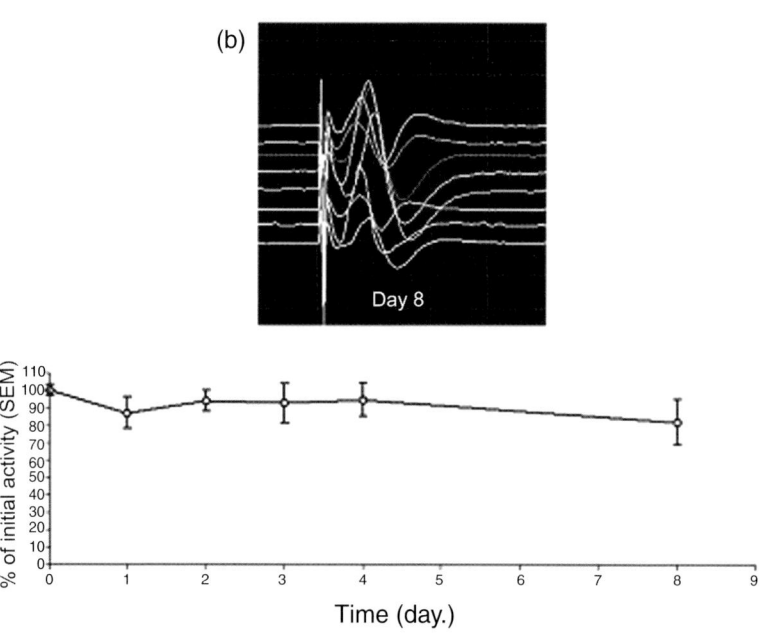

FIGURE 5.3. Stability of evoked activity. Illustration of the synaptic responses amplitude recorded in CA1 hippocampal organotypic slice cultures on eight channels over a period of several hours. The lower panel illustrates another protocol in which synaptic responses were recorded for 20 minutes each day over a period of eight days and represented as the average amplitude (±SEM) obtained during each recording daily session (b).

of medium to be perfused has to remain as small as possible as the quantities of substances to be tested are usually limited. A compromise has therefore to be found for each condition so as to be able to exchange the medium in the chamber, but without excessive perfusion volumes and while keeping the recording stability. Figure 5.4 illustrates the results of classical pharmacological experiments performed using drugs that affect the excitability or mechanism of synaptic transmission. In Figure 5.4a, two antagonists of glutamate receptors, NBQX (10 micromolar), a blocker of AMPA-type receptors, and D-AP5 (50 micromolar), an NMDA receptor blocker, were applied to the tissue at a perfusion rate of 10 ml/min. As expected these antagonists block excitatory synaptic transmission. The effect was obtained within a few minutes of drug application, which is about as fast as that obtained when testing drugs in regular submerged or interface chambers using tissue slices. Also washout of the drugs resulted in a complete recovery with a very comparable time course, indicating an efficient perfusion system that does affect stability of recording despite the fact the array is built onto a porous membrane.

Figure 5.4b is another illustration of the same type of experiments, but carried out with TTX, an antagonist or sodium channel that prevents generation of the action potential. Although the acute effect of TTX could easily be monitored in these experiments, we curiously noticed that applications for longer periods of time clearly affected the survival of the tissue. Washout of the compound after one to two days of application reproducibly resulted in partial recovery of synaptic activity. This example illustrates therefore the possible interest of this system for toxicological analyses of the effect of various compounds on brain tissue survival or functionality.

5.3   Use of the Neurosensor Interface for Studies of Nerve Regeneration

Stoppini and collaborators, using rat hippocampal slice cultures, showed that sectioning of the Schaffer collaterals (i.e., the axons of CA3 pyramidal neurons connecting the dendrites of CA1 cells) resulted in a sprouting and regrowth of lesioned axons that re-established synaptic connections between the two areas and that this capacity to regenerate was age dependent (Stoppini et al., 1997). By recording the amplitude of the activity elicited across the lesion in the days following the section, they were able to show a progressive recovery that occurred within a few days in one- to two-week-old tissue, but required weeks or was only partial in older cultures. The approach used, however, made it difficult to follow recovery within the same slice for the required period of time.

Furthermore, as this property may be of interest for the development of proregenerative drugs, we investigated the possibility of using the BioCell multi-electrode array system to monitor regeneration in this paradigm. Figure 5. illustrates the progressive changes in amplitude of the responses evoked across the lesion as a function of time. This experiment was performed by selecting four electrodes

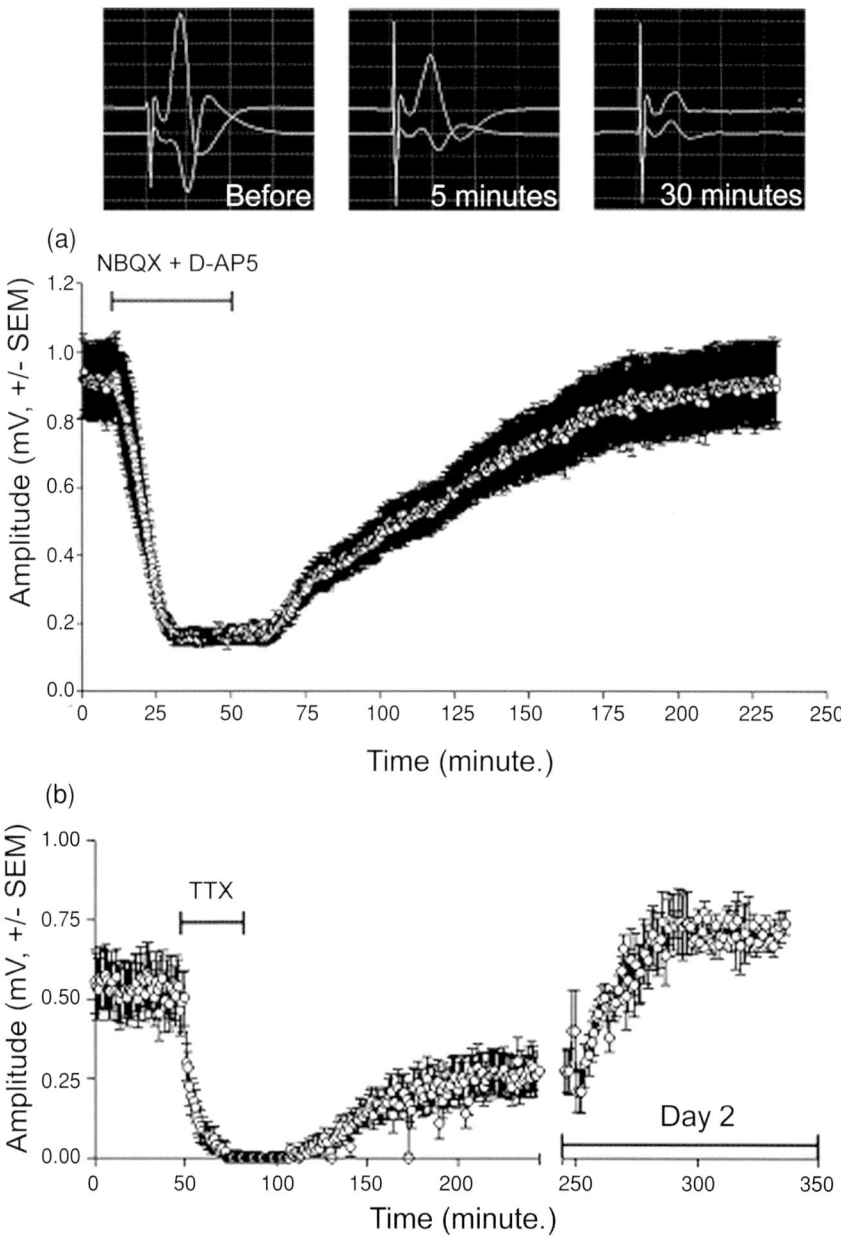

FIGURE 5.4. Pharmacological applications. (a) Illustration of an experiment in which 50 microM of D-AP5 and 10 microM of NBQX were applied to the perfusion medium at a rate of 10 ml/min. Note the time course of activity blockade and recovery illustrating the possibility to rapidly exchange the medium of the chamber, with low volumes and slow perfusion rates without affecting stability. (b) Other example illustrating the acute blockade of excitation by TTX (1 μM).

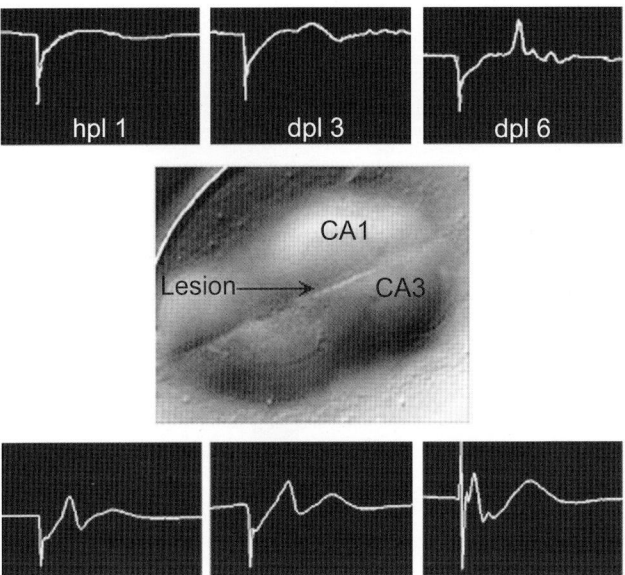

FIGURE 5.5. Paradigm used for studies of nerve regeneration using the Biocell Neurosensor Interface. Slice cultures were cut using a razor blade between CA3 and CA1 areas. Continuous monitoring of activity shows a complete elimination of evoked synaptic responses across the lesion immediately after the lesion and aa progressive recovery of the amplitude of the evoked responses within the days postlesion (center and right upper panels). In contrast, activity recorded within the CA3 region is not modified over time indicating the good viability of the tissue following the section.

located in the CA1 area for the recording of the synaptic responses evoked by the regenerating axons and four others close to the site of stimulation, but within the CA3 area, to verify the viability of the tissue.

These experiments were also carried out using mice rather than rats in order to be able to take advantage of the existence of possible transgenic animals. The slice culture was then cut with a sterile razor blade and synaptic activity across the cut monitored over several days. The results so obtained confirmed the initial observation made in rats, demonstrating the capacity of CA3 cells to sprout through the lesion and to form new synaptic contacts in the CA1 field. They also provided evidence for a similar age dependency of regeneration (Figure 5.6). Although evoked responses across the lesion recovered about 60% of their initial amplitude in young cultures after 10 days (10 to 15 days old), the recovery was only 20% in the older ones. In contrast, the electrodes localized close to the site of stimulation but within the CA3 area remained constant over the entire recovery period independently of the age of the cultures, showing that the tissue remained viable even in the older slices and that the defect in regeneration was not due to a potential damage of older cells. These experiments indicate therefore that the BioCell interface system is of potential

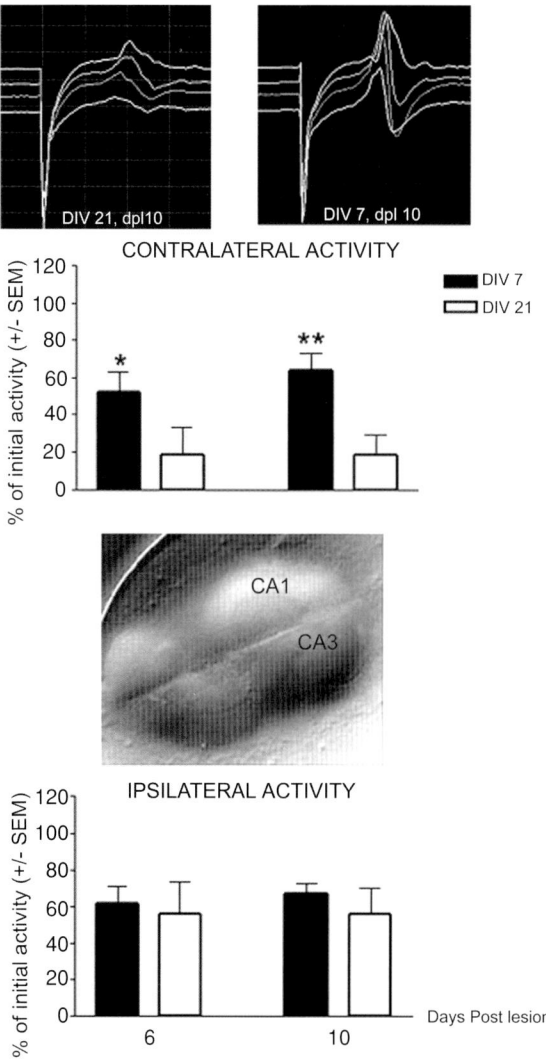

FIGURE 5.6. Age-dependent capacity to regenerate across a lesion in mouse hippocampal cultures. The graph illustrates the percentage of recovery of activity across the lesion obtained 6 days after the lesion in mouse slice cultures in which the lesion was carried out either 7 days or 21 days after explantation. Again the synaptic activity elicited in the CA3 areas does not change over time, indicating that the defects of regeneration observed in older tissue are not due to a greater susceptibility or damage of CA3 cells.

interest for the development of large-scale testing of drugs that could have application in promoting regeneration. It allows continuous and reproducible monitoring of activity even in the case where specific manipulations of the tissue are required.

5.4   Conclusions

We described in this chapter the main characteristics of a new multi-electrode array system that make it of interest for the development of drug-testing approaches. As we showed here, this system is relatively simple and easy to handle; it is character-ized by an excellent stability of recordings, capacity to easily run pharmacological and toxicological experiments by exchanging the perfusion medium and monitor-ing effects on synaptic transmission or tissue viability for several days or weeks, and the possibility to apply this approach to even more complex physiological models such as lesion-induced regeneration or tissue re-myelination.

In comparison to other or previous multi-electrode arrays, the main advantage of the Neurosensor Interface System is probably the design based on a low-cost, disposable cartridge and membrane array that renders its use simple, reliable, and compatible for simultaneous large-scale applications. In contrast, one disadvantage might be the relatively limited number of electrodes available for simultaneous recordings of activity and the difficulty to exploit the system in such a way as to perform single-cell spike recordings or unit recordings. The objective however, when designing this array was to privilege a development compatible with large-scale screening of compounds rather than analysis of signal integration or single-cell properties within a complex network.

The system has therefore been developed to ensure versatility and applicability to numerous different biological models susceptible to be of relevance for drug development.

Acknowledgements. This work was supported by grants from Biocell Interface and Serono Pharmaceutical Research Insitute (Geneva).

References

Alix, P., Winterer, J., and Muller, W. (2003). New illumination technique for IR-video guided patch-clamp recording from neurons in slice cultures on biomembrane. *J. Neurosci. Meth.* 128(1–2): 79–84.

Aptowicz, C.O., Kunkler, P.E., and Kraig, R.P. (2004). Homeostatic plasticity in hippocam-pal slice cultures involves changes in voltage-gated Na$^+$ channel expression. *Brain Res.* 998(2): 155–163.

Barth, A., Barth, L., Morrison, R.S., and Newell, D.W. (1996). bFGF enhances the pro-tective effects of MK-801 against ischemic neuronal injury in vitro. *Neuroreport.* 7(9): 1461–1464.

Bendiske, J. and Bahr, B.A. (2003). Lysosomal activation is a compensatory response against protein accumulation and associated synaptopathogenesis—An approach for slowing Alzheimer disease? *J. Neuropathol. Exp. Neurol.* 62(5): 451–463.

Bergold, P.J., Casaccia-Bonnefil, P., Zeng, X.L., and Federoff, H.J. (1993). Transsynaptic neuronal loss induced in hippocampal slice cultures by a herpes simplex virus vector expressing the GluR6 subunit of the kainate receptor. *Proc. Nat. Acad. Sci. U. S. A.* 90(13): 6165–6169.

Brager, D.H., Capogna, M., and Thompson, S.M. (2002). Short-term synaptic plasticity, simulation of nerve terminal dynamics, and the effects of protein kinase C activation in rat hippocampus. *J. Physiol.* 541(Pt 2): 545–559.

Brendza, R.P., Simmons, K., Bales, K.R., Paul, S.M., Goldberg, M.P., and Holtzman, D.M. (2003). Use of YFP to study amyloid-beta associated neurite alterations in live brain slices. *Neurobiol. Aging* 24(8): 1071–1077.

Cater, H.L., Chandratheva, A., Benham, C.D., Morrison, B., and Sundstrom, L.E. (2003). Lactate and glucose as energy substrates during, and after, oxygen deprivation in rat hippocampal acute and cultured slices.*J. Neurochem.* 87(6): 1381–1390.

Chen, S.F., Huang, C.C., Wu, H.M., Chen, S.H., Liang, Y.C., and Hsu, K.S. (2004). Seizure, neuron loss, and mossy fiber sprouting in herpes simplex virus type 1-infected organotypic hippocampal cultures. *Epilepsia* 45(4): 322–332.

Cronberg, T., Rytter, A., Asztely, F., Soder, A., and Wieloch, T. (2004). Glucose but not lactate in combination with acidosis aggravates ischemic neuronal death in vitro. *Stroke* 35(3): 753–757.

Dailey, M.E. and Waite, M. (1999). Confocal imaging of microglial cell dynamics in hippocampal slice cultures. *Methods* 18 (2): 222–30, 177.

Demerens, C., Stankoff, B., Logak, M., Anglade, P., Allinquant, B., Couraud, F., Zalc, B., and Lubetzki, C. (1996). Induction of myelination in the central nervous system by electrical activity. *Proc. Nat. Acad. Sci. U. S. A.* 93(18): 9887–9892.

Duff, K., Noble, W., Gaynor, K., and Matsuoka, Y. (2002). Organotypic slice cultures from transgenic mice as disease model systems. *J. Mol. Neurosci.* 19(3): 317–320.

Ehrengruber, M.U., Hennou, S., Bueler, H., Naim, H.Y., Deglon, N., and Lundstrom, K. (2001). Gene transfer into neurons from hippocampal slices: comparison of recombinant Semliki Forest Virus, adenovirus, adeno-associated virus, lentivirus, and measles virus. *Mol. Cell Neurosci.* 17(5): 855–871.

Fan, R. and Tenner, A.J. (2004). Complement C1q expression induced by Abeta in rat hippocampal organotypic slice cultures.*Exp. Neurol.* 185(2): 241–253.

Finley, M., Fairman, D., Liu, D., Li, P., Wood, A., and Cho, S. (2004). Functional validation of adult hippocampal organotypic cultures as an in vitro model of brain injury. *Brain Res.* 1001(1–2): 125–132.

Gahwiler, B.H. (1987). Organotypic slice cultures: A model for interdisciplinary studies. *Prog. Clin. Biol. Res.* 253: 13–18.

Ghoumari, A.M., Ibanez, C., El-Etr, M., Leclerc, P., Eychenne, B., O'Malley, B.W., Baulieu, E.E., and Schumacher, M. (2003). Progesterone and its metabolites increase myelin basic protein expression in organotypic slice cultures of rat cerebellum. *J. Neurochem.* 86 (4): 848–859.

Hay, D.G., Sathasivam, K., Tobaben, S., Stahl, B., Marber, M., Mestril, R., Mahal, A., Smith, D.L., Woodman, B., and Bates, G.P. (2004). Progressive decrease in chaperone protein levels in a mouse model of Huntington's disease and induction of stress proteins as a therapeutic approach. *Hum. Mol. Genet.* 13(13): 1389–1405.

Heimrich, B. and Frotscher, M. (1993). Slice cultures as a model to study entorhinal-hippocampal interaction. *Hippocampus* 3 Spec No: 11–17.

Ibrahim, M., Si-Ammour, A., Celio, M.R., Mauch, F., and Menoud, P. (2000). Construction and application of a microprojectile system for the transfection of organotypic brain slices. *J. Neurosci. Meth.* 101(2): 171–179.

Kasparov, S., Teschemacher, A.G., and Paton, J.F. (2002). Dynamic confocal imaging in acute brain slices and organotypic slice cultures using a spectral confocal microscope with single photon excitation. *Exp. Physiol.* 87(6): 715–724.

Kawasaki, H. and Tsutsui, Y. (2003). Brain slice culture for analysis of developmental brain disorders with special reference to congenital cytomegalovirus infection. *Congenit. Anom. (Kyoto)* 43(2): 105–113.

Keynes, R.G., Duport, S., and Garthwaite, J. (2004). Hippocampal neurons in organotypic slice culture are highly resistant to damage by endogenous and exogenous nitric oxide. *Eur. J. Neurosci.* 19(5): 1163–1173.

Khaspekov, L.G., Brenz Verca, M.S., Frumkina, L.E., Hermann, H., Marsicano, G., and Lutz, B. (2004). Involvement of brain-derived neurotrophic factor in cannabinoid receptor-dependent protection against excitotoxicity. *Eur. J. Neurosci.* 19(7): 1691–1698.

Kovacs, R., Schuchmann, S., Gabriel, S., Kann , O., Kardos, J., and Heinemann, U. (2002). Free radical-mediated cell damage after experimental status epilepticus in hippocampal slice cultures. *J. Neurophysiol.* 88(6): 2909–2918.

Lee, P., Son, D., Lee, J., Kim, Y.S., Kim, H., and Kim, S.Y. (2003). Excessive production of nitric oxide induces the neuronal cell death in lipopolysaccharide-treated rat hippocampal slice culture. *Neurosci. Lett.* 349(1): 33–36.

Lee, Y.S., Baratta, J., Yu, J., Lin, V.W., and Robertson, R.T. (2002). AFGF promotes axonal growth in rat spinal cord organotypic slice co-cultures. *J. Neurotrauma* 19(3): 357–367.

Leutgeb, J.K., Frey, J.U., and Behnisch, T., (2003). LTP in cultured hippocampal-entorhinal cortex slices from young adult (P25-30) rats. *J. Neurosci. Meth.* 130(1): 19–32.

McAllister, A.K. (2004). Biolistic transfection of cultured organotypic brain slices. *Meth. Mol. Biol.* 245: 197–206.

McKinney, R.A., Luthi, A., Bandtlow, C.E., Gahwiler, B.H., and Thompson, S.M. (1999). Selective glutamate receptor antagonists can induce or prevent axonal sprouting in rat hippocampal slice cultures. *Proc. Nat. Acad. Sci. U. S. A.* 96(20): 11631–11636.

Miller, L.D., Petrozzino, J.J., Mahanty, N.K., and Connor, J.A. (1993). Optical imaging of cytosolic calcium, electrophysiology, and ultrastructure in pyramidal neurons of organotypic slice cultures from rat hippocampus. *Neuroimage* 1(2): 109–120.

Morl, F., Groschel, M., Leemhuis, J., and Meyer, D.K. (2002). Intrinsic GABA neurons inhibit proenkephalin gene expression in slice cultures of rat neostriatum. *Eur. J. Neurosci.* 15(7): 1115–1124.

Morrison, B., 3rd, Eberwine, J.H., Meaney, D.F., and McIntosh, T.K. (2000a). Traumatic injury induces differential expression of cell death genes in organotypic brain slice cultures determined by complementary DNA array hybridization. *Neuroscience* 96(1): 131–139.

Morrison, B., 3rd, Meaney, D.F., Margulies, S.S., and McIntosh, T.K. (2000b). Dynamic mechanical stretch of organotypic brain slice cultures induces differential genomic expression: Relationship to mechanical parameters. *J. Biomech. Eng.* 122(3): 224–230.

Murphy, R.C. and Messer, A. (2004). A single-chain Fv intrabody provides functional protection against the effects of mutant protein in an organotypic slice culture model of Huntington's disease. *Brain Res. Mol. Brain Res.* 121(1–2): 141–145.

Notterpek, L.M., Bullock, P.N., Malek-Hedayat, S., Fisher, R., and Rome, L.H. (1993). Myelination in cerebellar slice cultures: Development of a system amenable to biochemical analysis. *J. Neurosci. Res.* 36(6): 621–634.

Perez Velazquez, J.L., Kokarovtseva, L., Weisspapir, M., and Frantseva, M.V. (2003). Antiporin antibodies prevent excitotoxic and ischemic damage to brain tissue. *J. Neurotrauma* 20(7): 633–647.

Roth, G.A., Spada, V., Hamill, K., Bornstein, M.B. (1995). Insulin-like growth factor I increases myelination and inhibits demyelination in cultured organotypic nerve tissue. *Brain Res. Dev. Brain Res.* 88(1): 102–108.

Schwartz, R.D. and Yu, X. (1995). Optical imaging of intracellular chloride in living brain slices. *J. Neurosci. Meth.* 62(1–2): 185–192.

Sherer, T.B., Betarbet, R., Testa, C.M., Seo, B.B., Richardson, J.R., Kim, J.H., Miller, G.W., Yagi, T., Matsuno-Yagi, A., and Greenamyre, J.T. (2003). Mechanism of toxicity in rotenone models of Parkinson's disease. *J. Neurosci.* 23(34): 10756–10764.

Shimono, K., Baudry, M., Panchenko, V., and Taketani, M. (2002). Chronic multichannel recordings from organotypic hippocampal slice cultures: Protection from excitotoxic effects of NMDA by non-competitive NMDA antagonists. *J. Neurosci. Meth.* 120(2): 193–202.

Silva, A.P., Pinheiro, P.S., Carvalho, A.P., Carvalho, C.M., Jakobsen, B., Zimmer, J., and Malva, J.O. (2003). Activation of neuropeptide Y receptors is neuroprotective against excitotoxicity in organotypic hippocampal slice cultures. *FASEB J.* 17(9): 1118–1120.

Stoppini, L., Buchs, P.A., and Muller, D. (1991). A simple method for organotypic cultures of nervous tissue. *J. Neurosci. Meth.* 37(2): 173–182.

Stoppini, L., Buchs, P.A., and Muller, D. (1993). Lesion-induced neurite sprouting and synapse formation in hippocampal organotypic cultures. *Neuroscience* 57(4): 985–994.

Stoppini, L., Duport, S., and Correges, P. (1997). A new extracellular multirecording system for electrophysiological studies: Application to hippocampal organotypic cultures. *J. Neurosci. Meth.* 72(1): 23–33.

Stoppini, L., Parisi, L., Oropesa, C., and Muller, D. (1997). Sprouting and functional recovery in co-cultures between old and young hippocampal organotypic slices. *Neuroscience* 80(4): 1127–1136.

Teter, B., Xu, P.T., Gilbert, J.R., Roses, A.D., Galasko, D., and Cole, G.M. (1999). Human apolipoprotein E isoform-specific differences in neuronal sprouting in organotypic hippocampal culture. *J. Neurochem.* 73(6): 2613–2616.

Thomas, A., Kim, D.S., Fields, R.L., Chin, H., and Gainer, H. (1998a). Quantitative analysis of gene expression in organotypic slice-explant cultures by particle-mediated gene transfer. *J. Neurosci. Meth.* 84(1–2): 181–191.

Thomas, M.P., Davis, M.I., Monaghan, D.T., and Morrisett, R.A. (1998b). Organotypic brain slice cultures for functional analysis of alcohol-related disorders: Novel versus conventional preparations. *Alcohol Clin. Exp. Res.* 22(1): 51–59.

Toni, N., Buchs, P.A., Nikonenko, I., Povilaitite, P., Parisi, L., and Muller, D. (2001). Remodeling of synaptic membranes after induction of long-term potentiation. *J. Neurosci.* 21(16): 6245–6251.

van Bergen, A., Papanikolaou, T., Schuker, A., Moller, A., and Schlosshauer, B. (2003). Long-term stimulation of mouse hippocampal slice culture on microelectrode array. *Brain Res. Brain Res. Protoc.* 11(2): 123–133.

Vis, J.C., de Boer-van Huizen, R.T., Verbeek, M.M., de Waal, R.M., ten Donkelaar, H.J., and Kremer, B. (2002). 3-Nitropropionic acid induces cell death and mitochondrial dysfunction in rat corticostriatal slice cultures. *Neurosci. Lett.* 329(1): 86–90.

Xu, G., Perez-Pinzon, M.A., and Sick, T. J. (2003). Mitochondrial complex I inhibition produces selective damage to hippocampal subfield CA1 in organotypic slice cultures. *Neurotox. Res.* 5(7): 529–538.

Yu, T.P., Lester, H.A., and Davidson, N., (2003). Requirement of a critical period of GABAergic receptor blockade for induction of a cAMP-mediated long-term depression at CA3-CA1 synapses. *Synapse* 49(1): 12–19.

6
Mapping Spatio-Temporal Electrophysiological Activity in Hippocampal Slices with Conformal Planar Multi-Electrode Arrays

WALID SOUSSOU, GHASSAN GHOLMIEH, MARTIN HAN, ASHISH AHUJA, DONG SONG, MIN-CHI HSIAO, ZHUO WANG, ARMAND R. TANGUAY JR., AND THEODORE W. BERGER

6.1 Introduction

Over the last two decades, technological advances in the fields of microchip and electronics manufacturing have enabled an increase in the production and use of silicon-based multi-electrode arrays. (Singer, 2000; Morin et al., 2005) These multi-electrode arrays or MEA for short, have come in a variety of shapes and materials, but fall into two broad classes: thin and sharp (implantable) or dish-based (planar). Although many investigations are currently undertaking research in vivo with implantable versions, this chapter focuses on applications of planar MEAs (pMEA), which are very well suited for in vitro experiments with slice or dissociated cells preparations. This chapter illustrates the utility and advantages of pMEAs in electrophysiological investigations with acute hippocampal slices, while introducing a new generation of conformally designed higher-density pMEAs as an adjuvant approach to facilitate and enhance MEA-based research.

Currently, the research being undertaken on pMEAs ranges from studying processes of neuronal plasticity underlying learning and memory, to tracking activity development in networks, and also pharmacological drug screening and testing. These diverse applications can be classified, based on the intricacy of their methodology, into the following nonmutually exclusive categories: (1) MEAs can be used as a multitude of single independent electrodes for rapid high-throughput experiments; (2) the spatial relations between electrode tips can be used synergistically to map electrical activity to tissue location; (3) recording simultaneously from multiple electrodes allows correlation of temporal information, which is not possible with many recordings from single electrodes; (4) the combination of spatial and temporal monitoring reveals the spatiotemporal dynamics of the neuronal network; (5) the ability to maintain cultured preparations on pMEAs allows long-term physiological investigations; and (6) recording and stimulating through the pMEA creates two-way communication with the tissue that is indispensable for investigating and developing neuroprosthetic applications.

High-throughput applications involve sampling several electrodes out of the total number on the MEA and selecting a representative one, or treating subgroups statistically as multiple samples from a homogeneous population. Electrodes within a particular cytoarchitectural region of a slice usually record similar neural responses. This redundancy of observed signals can be used to enhance the statistical significance of results by grouping responses into larger sample sizes. Similar time savings are achieved in cell cultures, where the multitude of electrodes records the activity of numerous cells at the same time, thereby decreasing the number of individual experiments needed to reach a significant population sample. Such high-throughput use of MEAs as biosensors has been applied to drug screening using cell culture (Pine, 1980; Gross et al., 1995) and hippocampal slice rhythmic activity (Shimono et al., 2000). In the first case, drugs are classified according to changes in the firing activity of neuronal cells cultured on MEAs (Gross et al., 1995, 1999). In the second case, changes in the frequency of carbachol-induced theta rhythmic oscillations in hippocampal slices are correlated with specific drug properties (Shimono et al., 2000). In both cases, the MEAs provided multiple sample points in different regions of the network, which enabled either a quick selection of an optimal site or averaging several channels for greater statistical accuracy.

In contrast to using array electrodes as individual and independent streams of data, the spatial arrangement of electrodes can be used to generate spatial maps of the activity in a slice. Any parameter of the recorded potentials can be plotted in a color-coded matrix according to the relative spatial positions of the electrodes in order to generate topographic activity maps. Such spatial activity maps can be matched to a picture of the slice showing the actual electrode positions in order to visualize the activity in relation to the subregions of a slice (Shimono et al., 2000) or map the spatial extent of a response along a network (Jimbo and Robinson, 2000). In addition, if electrodes are close enough to each other, they enable current source density (CSD) analysis, which can elucidate the origins and meaning of the complex field potentials recorded (Wheeler and Novak, 1986).

The ability to simultaneously record from all the MEA electrodes over time enables correlation of activity between different parts of a network in order to study their patterns and plasticity in cell and tissue preparations. The temporal sequence of firing of ensembles of cells can provide information on network states. Beggs and Plenz (2003) analyzed cell bursting avalanches to describe the stability of the network. Jimbo et al. (1999) reported on time-dependent synaptic plasticity in networks of cultured cells in observing that connections between cells that fired within 20 msec before the other were potentiated after tetanus, whereas connections between cells negatively correlated within 20 msec were depressed.

pMEAs combine spatial and temporal information and enable the conversion of static spatial activity maps into dynamic spatiotemporal map sequences. These series of maps can be joined as frames of a movie to visually trace the propagation of spontaneous, evoked, or rhythmic activity across the slice. For example, Novak and Wheeler (1989) studied the temporal propagation of seizure activity, and Shimono et al. (2000) localized and spatiotemporally followed the origin of theta rhythm generated by carbachol in a slice, both using CSD analysis of the signals recorded from the MEA.

The surface of pMEAs is ideal for long-term tissue cultures of both slices and dissociated cells, as it provides a flat, biocompatible, and sterilizable support with embedded electrodes that can continuously monitor culture activity without disrupting the closed system. Longer-term experiments can track changes in activity and plasticity of developing cultures and networks (Gross and Schwalm, 1994; Stoppini et al., 1997; Thiebaud et al., 1997; Jahnsen et al., 1999) under different chronic pharmacological treatments (Shimono et al., 2002).

Although several neuroprosthetics, such as cochlear, cortical (Chapin et al., 1999), and retinal (Humayun et al., 2003) devices rely on implantable in vivo MEA technology, pMEAs still play a major role in understanding network connectivity and dynamics (Meister et al., 1994; Warland et al., 1997). pMEAs are being used as an in vitro testing platform to first characterize the information processing of the target neuronal network, before undertaking in vivo experiments. In our current goal to replace the CA3 hippocampal area with a microchip (FPGA/VLSI) implementation of a nonlinear model of CA3 (Berger et al., 2001), we are using pMEAs to provide a functional proof-of-principle. pMEA experiments allow us first to generate nonlinear models, then to test hardware implementations in order to change parameters and conditions rapidly, cost effectively, and with fewer animals. Similarly, the retinal prosthesis project relies on understanding underlying network dynamics and plasticity of the retina in order to transform incident light into an electrical stimulation pattern that will produce correct visual percepts (Humayun and Weiland, personal communications). Retinal stimulation and recording experiments are thus currently being undertaken on pMEAs to develop a nonlinear mathematical model of the retinal network that will be implemented in the next generation of retinal prostheses (Chichilnisky and Kalmar, 2003; Frechette et al., 2005).

6.2   High-Density Conformal pMEAs

The above-mentioned applications of pMEAs illustrate their versatility and advantages of having multiple electrodes over a single electrode. These advantages could be augmented by increasing the number of electrodes, that is, by increasing the number of sampling points and spatial resolution. Unfortunately, physical as well as technological constraints limit the number of possible electrodes on an MEA. The first obstacle is overcrowding of electrode leads, which can produce unwanted noise and crosstalk between closely spaced lines, and require more complex manufacturing solutions, such as lead stacking or electrode addressing. The second impediment involves connectors, which are difficult to keep small and reliable. The third barrier is the size and cost of signal modulation, digitization, and storage hardware. That is, of course, without mentioning the increased complexity and time consumption of data analysis. This limit on the number of electrodes creates a tradeoff in array design: electrodes can either sparsely

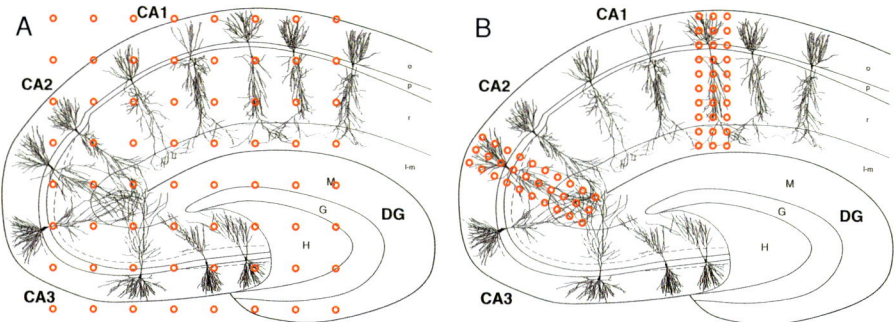

FIGURE 6.1. Electrode positions matched to a drawing of a hippocampal slice with representative cell bodies, axons, and dendrites traced from Nissl stains (Ishizuka et al., 1995). (A) A square 8 × 8 array of electrodes is overlaid on the slice drawing. (B) Two high-density 3 × 10 rectangular arrays whose angles and separation conform to the cytoarchitecture of the slice are overlaid on the CA2 and CA1 pyramidal cells. (Electrode size and spacing not to scale.0

cover a large surface area or be closely spaced (high-density) over a smaller one. Figure 6.1 illustrates the tradeoff due to the constraint on electrode number with two arrangements of electrodes overlaid on a drawing of a hippocampal slice. A low-density square matrix 8 × 8 array (Figure 6.1A) can record field potentials from several different areas of the slice while providing only two and rarely three electrodes along the orientation of a particular cell group. The two more densely packed 3 × 10 subarrays (Figure 6.1B) cover only two pyramidal populations (one in CA1 and another in CA2), but match the correct orientation of the cells' dendritic axis, which allows CSD analysis and multiple points for bipolar stimulation (see below).

This tradeoff has led us to create and test several new high-density conformal pMEAs that are custom designed for our experimental purposes. These pMEAs have several high-density clusters of electrodes whose orientation and location conform to the cytoarchitecture of the slice. The hippocampal slice is an ideal tissue preparation for pMEAs, as its intrinsic two-dimensional trisynaptic excitatory cascade network from dentate gyrus to CA3 and CA1 subregions (DG → CA3 → CA1) is preserved when the hippocampus is sliced along its longitudinal axis. (Andersen et al., 1971) The following experimental examples are designed to illustrate the above-listed applications of pMEAs in electrophysiological experiments on acute hippocampal slices, while emphasizing the advantages of our new conformal high-density designs.

6.2.1 High-Throughput Multiple Independent Sites

We have used pMEAs to build a hippocampal-based biosensor for rapid neurotoxins detection (Gholmieh et al., 2001, 2003). The detection of the chemical compounds is based on a novel quantification method for analyzing short-term plasticity (STP) of the CA1 system in acute hippocampal slices, using random electrical impulse sequences as inputs and population spike (PS) amplitudes as outputs. STP is quantified by the first- and the second-order Volterra kernels using a variant of the Volterra modeling approach. Since they describe the functional state of the biosensor, the Volterra kernels changed differently depending on the chemical compounds that were added to the slice medium, which enabled its classification.

Determining the location of the optimal CA1 population spike response is a critical parameter for this approach. Using 8×8 pMEAs with an interelectrode distance of 150 μm from the University of North Texas (CNNS, Denton, TX, USA), extracellular potentials in the CA1 region were recorded from rat hippocampal slices. All the slices described below were from male Sprague–Dawley rats that were anaesthetized before decapitation. The hippocampi were dissected under cold artificial cerebrospinal fluid (aCSF (in mM): NaCl, 128; KCl, 2.5; NaH_2PO_4, 1.25; $NaHCO_3$, 26; Glucose, 10; $MgSO_4$, 2; ascorbic acid, 2; $CaCl_2$, 2, aerated with a mixture of 95% O_2 and 5% CO_2), and sectioned into 400 μm slices. The tissue was then allowed to rest for at least one hour at room temperature before positioning on the pMEAs and starting the experiments under continuous perfusion of aCSF with reduced $MgSO_4$ (1 mM). Figure 6.2 shows a paired-pulse experiment with a hippocampal slice positioned over the array (A) and potentials recorded at a subset of electrodes of the array (B). In these experiments, slices were stimulated using external bipolar nichrome electrodes, which allow larger current injections than array electrodes. The corresponding 16 recordings spanned *stratum oriens*, *stratum pyramidale*, and the apical dendritic region (*stratum moleculare* and *stratum radiatum*). The optimal recording for our purposes was the one that showed the largest amplitude with a well-defined population spike overriding a positive excitatory postsynaptic potential (EPSP). Once the specific electrode with the corresponding channel was chosen, a random train was delivered to the slice. On several occasions, however, we have been faced with the inability to obtain signals that fit the above criteria due to the scarcity of electrodes and nonoptimal stimulation position.

Faced with the limitations of the currently available pMEAs, we have designed a new pMEA that would optimize CA1 input (stimulation)–output (recording) properties. This pMEA has a 2×8 subarray to stimulate the Schaffer collaterals fibers which constitute the major excitatory inputs to CA1 pyramidal cells, and a 4×12 subarray to record the output of these cells (Figure 6.3A). This novel pMEA, designed to match CA1's cytoarchitecture, eliminates the need to reposition the slice or stimulation electrode for optimizing stimulation and recording locations. The following paragraph describes this optimization protocol.

Hippocampal slices were positioned over the array with the 2×8 subarray under the Schaffer collateral pathway, and the 4×12 subarray under CA1 pyramidal cells (Figure 6.3A). One column of the 2×8 subarray was chosen for stimulation (S column, Figure 6.3B). The eight electrodes in the stimulation column formed seven possible adjacent pair combinations: that is, S1–2, S2–3, ..., S7–8. A testing paired-pulse stimulation (30 msec interval) of 70 to 100 μA was

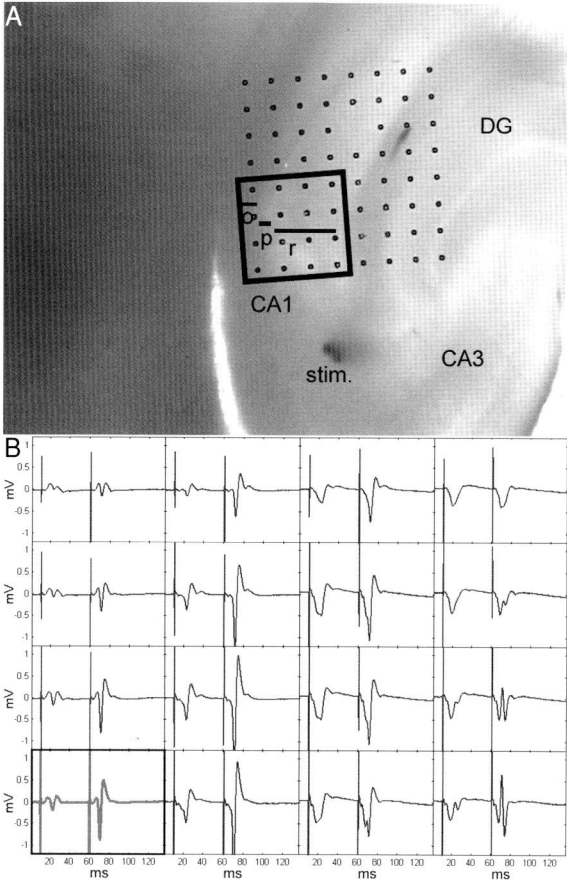

FIGURE 6.2. Recording hippocampal slice evoked responses with a pMEA. (A) Hippocampal slice positioned over the 8×8 array. (B) The corresponding recording from a 4×4 subset. The corresponding 16 recordings span *stratum oriens* (o), the cell body layer and the apical dentritic region (*stratum moleculare* (m), and *stratum radiatum* (r)). The optimal recording for our purposes was the one (highlighted in the lower left quadrant) that showed the largest amplitude with a well-defined population spike overriding a positive excitatory postsynaptic potential (EPSP).

delivered sequentially through each pair to identify the pair that yielded the largest amplitude population spike. That stimulation pair was then used to generate an input/output (I/O) curve, by incrementally increasing the stimulation intensity over a 10 to 250 μA range. The stimulation intensity was then set to yield the half-maximal response, and five sets of paired pulses were delivered again through each of the seven electrode pairs and then the responses were averaged. One column of the 4×12 subarray was used for recording (R column, Figure 6.3B). In this particular experiment, the laminar profile recorded in the R column shows a

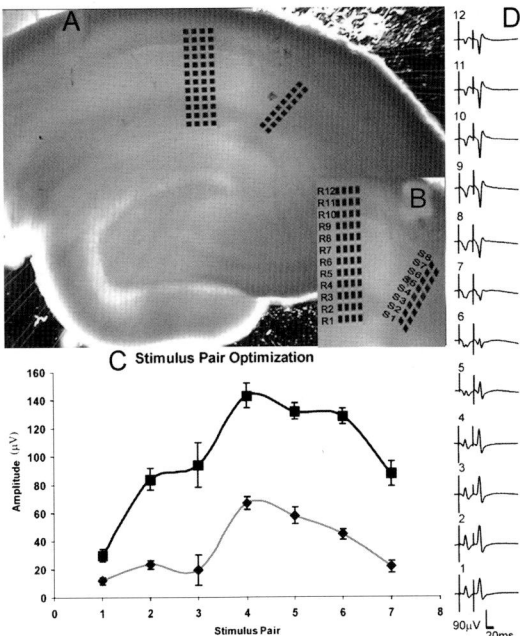

FIGURE 6.3. Optimization of the input–output properties of Schaffer collaterals–CA1: (A) hippocampal slice on a pMEA; (B) blow-up of the recording (R) and stimulation sites (S); (C) population spike amplitude responses for the first (gray) and the second pulses (black) recorded at R11, plotted against the stimulation pair number; (D) laminar profile of FP recordings.

clear pyramidal population spike at R8–12 and the largest dendritic component at R3 and R4 (Figure 6.3D). The recording electrode R11 had the largest discernable population spike. Its stimulation optimization graph is shown in Figure 6.3C where the population spike amplitude is plotted against the different possible stimulation pair combinations. The maximal response for this slice was thus obtained at R11 in response to stimulation from pair S5–S6. Hence, this array allowed us to rapidly localize an afferent pathway, and to optimize the stimulation location to obtain the largest response in the target area.

   We were further interested in selectively stimulating two adjacent afferent pathways to dentate gyrus (DG), in order to compare interactions of their responses in the same slice. For this purpose, we designed the perforant path (PP)-DG-CA3 pMEA, which has three high-density subarrays: two 3 × 7 arrays that span the blades of the DG and one 3 × 6 array to record their mossy fiber/CA3 outputs. The PP is the major excitatory projection into DG and consists of two anatomically and functionally discrete subdivisions, medial and lateral PP. The lateral PP originates in the ventrolateral entorhinal cortex, and terminates in the outer third of the DG molecular layer, whereas the medial PP originates in the dorsomedial entorhinal cortex, and synapses in the middle third of the DG molecular layer (Steward,

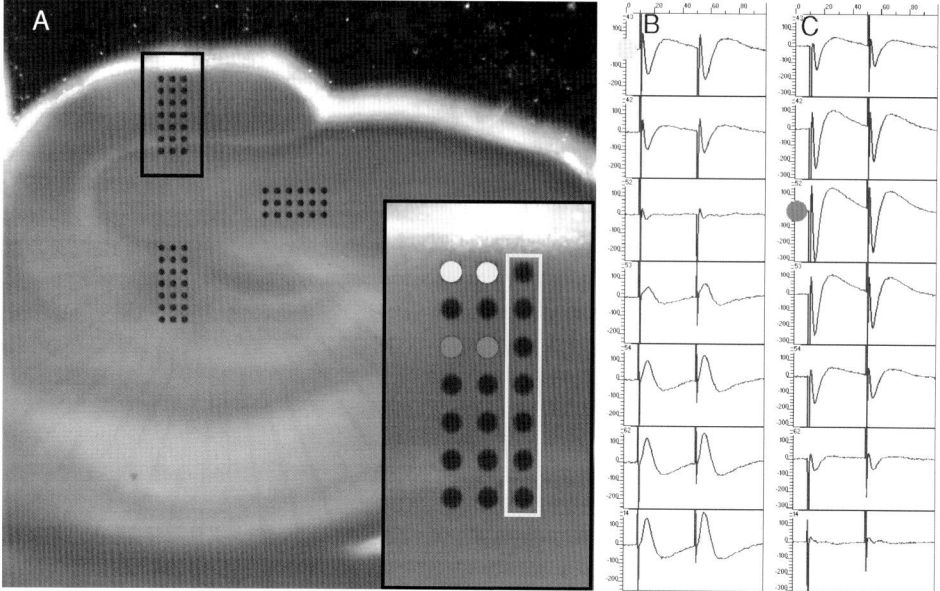

FIGURE 6.4. Paired-pulse facilitation and depression in dentate gyrus: (A) photomicrograph of dentate on a three subarray pMEA. The electrodes highlighted in light and dark gray correspond, respectively, to lateral and medial perforant path stimulation sites, and the white box marks the electrodes whose responses are graphed in B and C; (B) response of a column of electrodes to bipolar paired-pulse stimulation at the light electrodes in lateral PP; (C) paired-pulse depression in response to medial PP stimulation. (B),(C). The interstimulus interval was 50 msec. Y-axis: -250 to 250 μV, X-axis: 100 msec.

1976). Experimentally, the two pathways are distinguished by their anatomical location, and their responses to paired-pulse (pp) stimulation, whereby lateral PP shows facilitation and medial PP responses exhibit depression (McNaughton and Miller, 1984; Dahl et al., 1990). Each of these subpathways encompasses approximately 100 μm, and the electrodes are spaced 50 μm apart, therefore, by sequentially stimulating along a column, each subpathway is activated from at least two rows. We manufactured these pMEAs to fit into the Multi Channel Systems recording setup (MCS, Reutlingen, Germany; Gholmieh G. et al., 2005). The degree of selectivity of a pathway can then be determined by the amount of facilitation or depression observed.

Figure 6.4 shows an example distinguishing between medial and lateral PP in a rat hippocampal slice. After positioning the slice on the array, we sent bipolar stimulations through adjacent electrode pairs in rows, and recorded the responses along the third column. The electrodes highlighted in light gray were used to stimulate the outer third of the PP (B), which corresponds to lateral PP. A 50 msec interval pp produced an increase in the amplitude of the second evoked postsynaptic potential (PSP) compared to the first, hence exhibiting pp facilitation. In contrast, stimulating the same slice through electrodes at medial PP (C, marked in dark gray) revealed pp depression at the same 50 msec interval. The facilitation and depression observed respectively at the anatomical lateral and medial pathways confirm that the two subpathways were indeed selectively excited. Paired-pulse stimulations at the electrodes in between the light and dark gray ones did not produce changes in PSP amplitudes, suggesting that both pathways might be partially stimulated, and canceled each other's effects. This experiment illustrates the ease with which this conformal pMEA, with its closely spaced electrodes was able to selectively activate two adjacent pathways. Such subpathway segregation is extremely difficult to achieve in the same slice with more widely spaced MEAs or inserted twisted-wire stimulating electrodes.

6.2.2 Spatial Mapping of Electrical Activity

In a heterogeneous tissue such as the hippocampal slice, which has several regions populated by different cell types with different interconnectivities, all the electrodes of a pMEA do not record the same signals. The spatial distribution of electrodes determines the subregions that are sampled, which may respond differently to different stimuli. The uniqueness of information carried by each electrode therefore reflects underlying slice cytoarchitecture, and provides information on functional physiology of various subregions. By arranging all electrode responses according to the pMEA topography, we obtain a spatial map of responses. The waveforms in Figures 6.2 through 6.4 are arranged according to their respective positions on the pMEA, and illustrate how responses vary over the slice. In Figure 6.2, the population spike amplitude varies in relation to the distance from the pyramidal cell layer. In Figure 6.4B, a reversal from a negative going potential to a positive one is observed, which indicates good isolation of afferent inputs. These spatial maps show the entire response at every electrode, and although they are comprehensive, visually extracting information from these formats may be hard. Another approach to display spatial maps is to extract the parameter of interest, transform its value into a color code, and then plot it according to its electrode position. By interpolating between adjacent electrode values, a color map is generated, which easily visualizes the topographic distribution of the parameter.

Figure 6.5 illustrates how useful and easy a color spatial map can be for visually presenting a long-term potentiation (LTP) experiment run on a 60 electrode pMEA. In LTP experiments, the amplitude of excitatory postsynaptic potentials (EPSP) is compared before and after high-frequency stimulation (HFS). The EPSP amplitude is therefore extracted and converted into a pseudo-color map that is overlaid on top of a picture of the experimental slice. The spread of the EPSP response is thus quickly visualized as the red area, and the post-HFS (Figure 6.5B) response is notably darker than the pre-HFS, thereby indicating potentiation. The ratio of post-HFS amplitude over pre-HFS amplitude is calculated to determine the percentage of LTP at each electrode. This percentage is color-coded and mapped in Figure 6.5D, which reveals in red the extent of CA1 that exhibited potentiation.

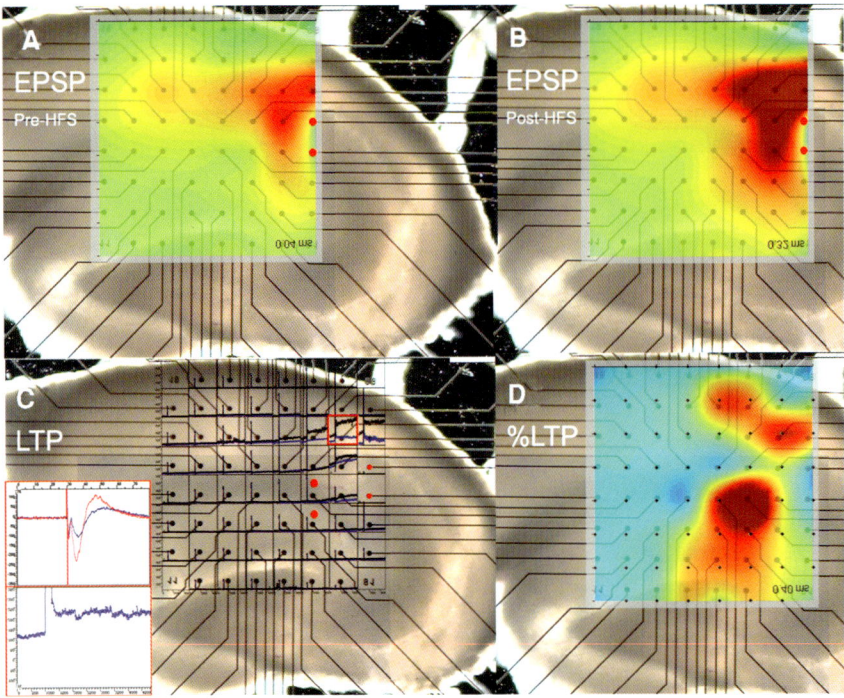

FIGURE 6.5. Extent of LTP is different than extent of response: (A) EPSP map showing CA1 response to Shaffer stimulation (color map reflects interpolation between EPSP amplitudes at recording electrodes, where red indicates 50 mV, dark blue −50 mV, and green 0 mV). (B) Same as in (A) only mapping the response after HFS potentiation. (C) Timecourse of LTP experiment plotting amplitude over time at each electrode location. Top inset shows overlay of EPSP waveforms pre- and post-HFS for the electrode boxed in red. Bottom inset shows that electrode's timecourse (box sizes 500 μV and 80 msec). (D) Spatial map showing extent of spread of potentiation (color map represents interpolation between percent change in amplitude at the recording electrodes between pre- and post-HFS, where red indicates 300%, and dark blue 0%).

This spatial map of LTP is a lot simpler to read than Figure 6.5C, which shows the time course of the LTP experiment at each electrode (we discuss spatiotemporal maps in the next section). It is interesting to observe that not all areas that showed an EPSP in Figures 6.5A and B were actually potentiated in C, and vice-versa.

    The above spatial maps give an excellent picture of what is happening over the entire slice during the experiment; however, they do not provide information about the exact location of activity. There are not enough sampling points to produce accurate information about the differences along the dendritic tree of pyramidal cells. This low spatial resolution is further complicated by the fact that field potentials intermix in the tissue, and their sources may be missed between electrodes. High-density pMEAs sample more points per unit area, and enable

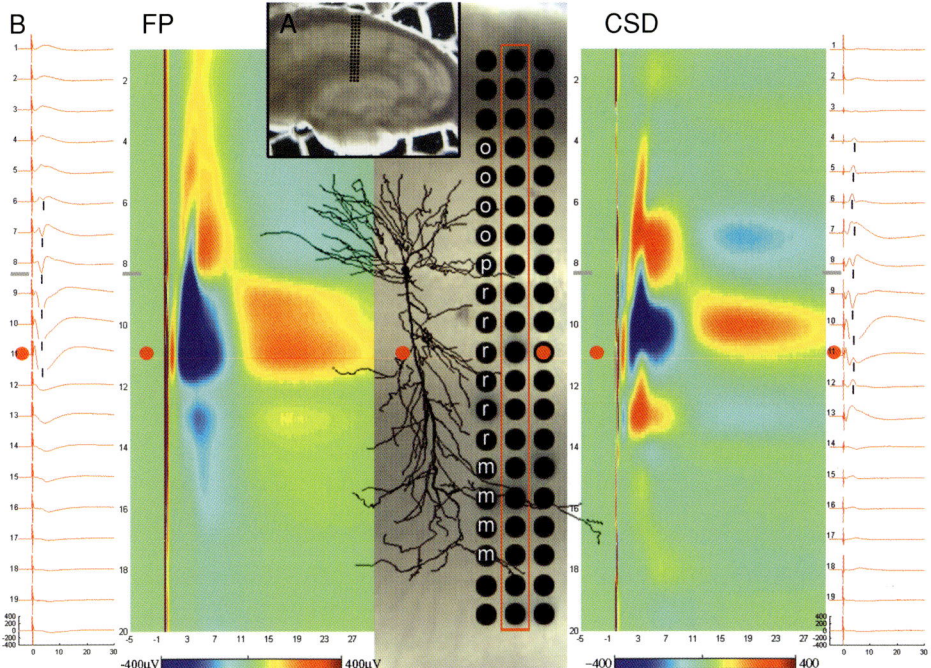

FIGURE 6.6. FP versus CSD in CA1: (A) photomicrograph of a hippocampal slice on the 20×3 pMEA; (B) CA1 laminar profile of the slice in (A) in response to stimulation at the electrode marked in red. The red traces show the recorded FP or the calculated CSD at the middle column (red box), with the numbers corresponding to the electrode position starting from the top. The color map shows the same values converted to a color scale and interpolated between electrodes. The trace drawing of a cell illustrates the relative position of the electrodes with respect to the CA1 pyramidal cells. The horizontal gray bars mark the cell layer. The vertical black lines mark the population spike. The stimulation occurred 10 msec after the beginning of the recording. (x-axis 35 msec, y-axis or color range -400 to $+400$ μV or CSD units. *Strata* are labeled in the electrodes: *oriens* (o), *pyramidale* (p), *radiatum* (r), and *moleculare* (m).)

CSD analysis. (Nicholson and Freeman, 1975; Miyakawa and Kato, 1986) The latter is a spatial filter that reveals sources and sinks of currents, and allows more meaningful interpretation of field potential recordings.

In order to assess the applicability of high-density electrodes for CSD analysis, hippocampal slices were oriented on the 20×3 MEA such that the apical CA1 pyramidal dendrites were parallel to the long axis of the array (Figure 6.6A). Monopolar biphasic stimulations were then delivered through one of the electrodes in a peripheral column (in this case, the right column), and evoked responses were recorded on the 59 other electrodes. The left panel in Figure 6.5B shows field potential (FP) activity recorded at the middle column of electrodes in response to a single stimulation in *stratum radiatum* (marked by a red dot). Positive FPs were observed in *stratum oriens* (above the cell body layer, *stratum pyramidal*), and their reversed negative potentials in *strata radiatum* and *moleculare*. The population spike, seen as a sharp deflection in the waveform, was present on many electrode traces.

The combination of multiple recordings along the pyramidal cell constitutes the FP laminar profile of the CA1 responses. By interpolating the values between adjacent recordings, a topographical map was generated in which voltages were assigned colors. In these color maps, the yellow/red marked the spread of positive EPSPs, and the blue areas delineated negative EPSPs, or the spread of population spikes through positive EPSPs. A CSD of the laminar profile was obtained by applying two-dimensional-CSD analysis to the recorded voltages (Gholmieh G. et al., 2005). The population spike spread was markedly narrower in the 2-D–CSD graphs (right panel, Figure 6.6B). Comparing the FP and CSD (left versus right panel, respectively), color maps showed how CSD also narrowed the longitudinal spread of EPSPs, both the positive and the negative ones. Additionally, CSD analysis unmasked a current source in *stratum moleculare*, which was not visible in the FP. The reversal between the other source and the sink, however, remained at the same location in *stratum pyramidal*.

In summary, spatial activity maps are thus generated by displaying the electrode recording according to their topographical arrangement over the slice. These maps can visualize the spread of a response and the extent of an experimental manipulation's effect. Depending on electrode density, these spatial plots can provide either low resolution information which allows a wide, albeit sparse, coverage of the activity at various slice regions, or at higher density enables CSD analysis for accurate mapping of currents and sources in smaller cellular subregions.

6.2.3   Spatiotemporal Map

The examples above illustrated the power of pMEAs in generating static topographical activity maps by interpolating values between electrodes. The ability to simultaneously record from all electrodes over time, transforms these static maps into dynamic ones that allow visualization of spatiotemporal propagation of activity across a slice.

The propagation of an evoked response from CA3 to CA1 in a hippocampal slice is visualized in Figure 6.7. The mossy fibers of the slice were stimulated by the electrodes marked in red. Each color map illustrates the actual recorded potential at the time point marked in its corner. The sequence of the four frames thus provides a glimpse into the progression of the response over time. A large EPSP is first observed in CA3, as a large negative depolarization color-coded in blue. As the EPSP wanes in CA3, it begins to propagate along the Schaffer collateral fibers towards CA2 and then CA1. From these images, it appears that the deepest depolarization traversed approximately 2 mm in 8 msec, which amounts to 0.25 m/sec. This propagation velocity corresponds to the average value reported for Schaffer collateral fibers by Andersen et al. (2000). The time frames help to visualize easily the propagation of EPSP activity across the slice, which could

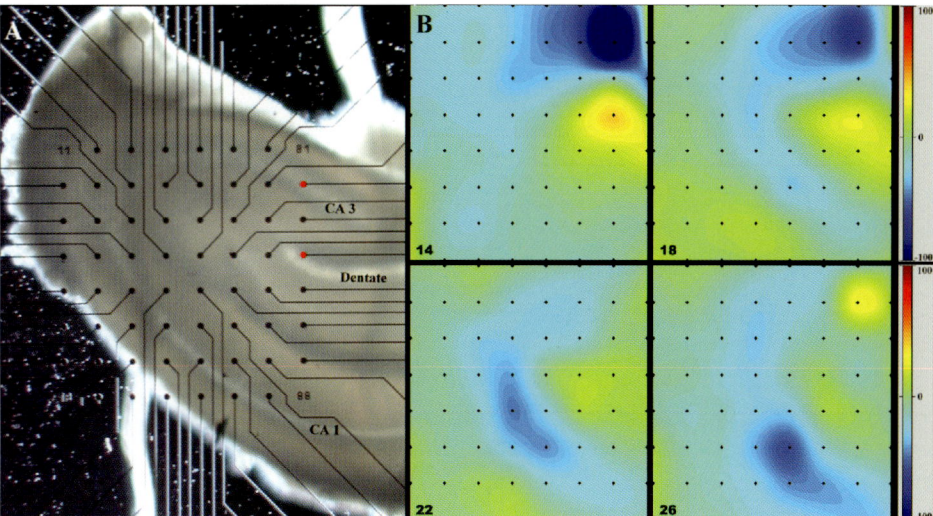

FIGURE 6.7. Propagation of an EPSP along the Schaffer collaterals: (A) photomicrograph of a slice on the pMEA. Red dots mark stimulation site. (B) Four snapshot spatial maps of the recorded potential. The number at the bottom left corner indicates the time elapsed since the stimulation. The pseudo-color range has red at 100 µV and blue at −100 µV.

not be observed at all with single electrode experiments, and would even be hard to discern just by looking at all 60 waveforms.

Although Figure 6.7 traced passive propagation of a postsynaptic potential along the Schaffer collaterals, it is sometimes desirable to trace the activity across synapses in order to see how different subregions are interacting with each other. As part of the biosensor project described in Section 6.2.1, we tested the effects of trimethylolpropane phosphate (TMPP), a byproduct of jet fuel combustion and a $GABA_A$ inhibitor. (Keefer et al., 2001) At 1 mM (Figure 6.8), TMPP disinhibited the slice sufficiently that the monosynaptic population spike in dentate granule cells produced by stimulation in the PP (blue waveforms) was followed by a disynaptic burst of population spikes in CA3. This epileptiform bursting in CA3 slowly spread an EPSP into CA1. Stimulation in CA3 (red waveforms) produced a large monosynaptic population spike in CA3 immediately followed by similar bursting as before and followed by an EPSP in CA1. Finally, stimulation in CA1 (green waveforms) produced a monosynaptic EPSP in CA1 and no activity anywhere else in the slice.

When the responses to the three stimulations were overlaid, propagation across the slice could be clearly visualized by comparing the response delays. The time difference between the first population spike peaks in CA3 triggered by PP and mf stimulations was 12 to 13 msec. These first spikes in CA3 were followed by bursts of spikes 3 to 5 msec apart, and in the case of PP stimulation, the second spike was the largest. The response delay between these two stimulations corresponded

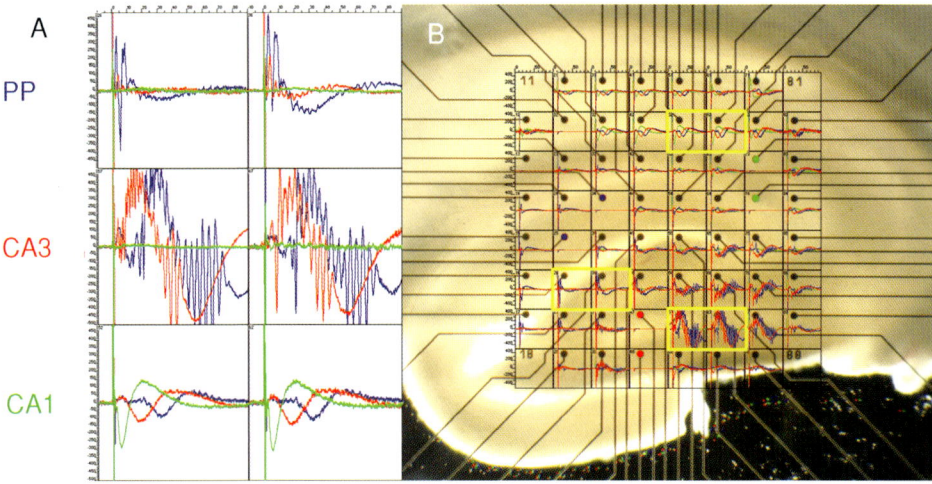

FIGURE 6.8. Effects of 1 mM TMPP application on the hippocampal slice: (A) responses evoked by stimulation at the perforant pathway (PP, blue dots and waveforms), CA3 area (red dots and waveforms), and CA1 (green dots and waveforms); (B) photomicrograph of slice on pMEA overlaid with the responses to the three stimulations. Yellow rectangles show electrodes that are expanded in (A) (*x*-axis 100 msec, *y*-axis 500 μV).

to the EPSP lag observed in CA1 for the same two stimulations (14 msec delay at the troughs). In addition, the EPSP evoked by mf stimulation was delayed 12 msec from the monosynaptic potential elicited by direct stimulation at CA1. These lags reveal that propagation of activity across the slice was facilitated by TMPP. More specifically, TMPP disinhibited the inhibitory neurons in CA3, thus when a sufficient stimulation arrived through the mossy fibers to CA3, the disinhibited pyramidal cells all fired in synchrony (population spike). This population spike spread uninhibited through the recurrent collaterals thereby triggering the observed bursting. All CA3 spikes also traveled along the Schaffer collaterals and produced the observed trisynaptic EPSP in CA1. The delays in CA1 EPSP reflected the number of synapses traversed between the initial stimulation site and the CA1 pyramidal cells. The pMEAs thus enabled not only the localization of the major site of action of TMPP at the CA3 inhibitory interneurons, but also unmasked its facilitatory effect on the propagation of activity throughout the entire trisynaptic pathway of the hippocampal slice.

There appears to be significant variability in the propagation velocity and delays reported in the literature. Using pMEAs, Andersen et al. mapped the angle and distribution of Schaffer collaterals in a sheet-like rat hippocampal preparation, and found a distribution of conduction velocities for different fibers in the hippocampus. They calculated the average conduction velocity of the weakly myelinated Schaffer collateral fibers at 0.25 m/sec, associational fibers averaged 0.39 m/sec, and axons in the fimbria fell into two classes with average velocities of 0.99 and

0.37 m/sec depending on whether they were myelinated (Andersen et al., 2000). Although our results in Figure 6.7 match this value, fibers running parallel to the pyramidal layer were previously reported to have a propagation velocity of 0.3 m/sec in the guinea pig hippocampus. (Andersen et al., 1978) In dissociated hippocampal neuron cultures, conduction velocity was calculated to 0.12 m/sec with multichannel optical recordings with voltage-sensitive dyes, and the synaptic delay was calculated at 1 msec (Kawaguchi et al., 1996).

These differences in velocities are also reflected in the variability of response delays. Our lab has previously reported on trisynaptically evoked population spikes in CA1 from PP stimulations in vivo, where we recorded monosynaptic responses in CA1 at 6 msec, disynaptic ones at 11 msec, and trisynaptic ones at 17 msec, considerably different from the above-reported delays. (Yeckel and Berger, 1995) The monosynaptic and disynaptic responses were postulated to constitute a feedforward circuit, whereby PP feeds directly into CA3 and CA1 (Yeckel and Berger, 1990). Hippocampal trisynaptic response onset delays reported with optical recordings of voltage-sensitive dyes from PP stimulations yielded 5.5 msec delay for granule cells, 6.5 msec for CA3, and 19.2 for CA1. In addition, potentiation by tetanic stimulation decreased the onset latency by 65% at CA1 (Nakagami et al., 1997).

In the isolated guinea pig brain in vitro preparation, Pare and Llinas (1994) measured the population spike latency in CA1 in response to entorhinal cortex (EC) stimulation. They reported spike spatial propagation beyond the transverse axis of the hippocampus, and temporal delays that varied by up to 8 msec depending on the relative recording and stimulation sites. The recorded delays suggested that the topographical relationship between EC and hippocampus, although not spatially confined in lamellae, was preserved in the time domain. They further demonstrated that the spike latency depended on the stimulation intensity, which we also observed in our experiments (data not shown). Furthermore, a 3 msec delay in propagation was observed in CA2 using optical recordings of voltage-sensitive dyes. (Sekino et al., 1997) The transmission from CA3 to CA1 could proceed either directly through the Schaffer collaterals or stop for 3 msec in CA2 before proceeding. The coexistence of these two modes of propagation of activity leads to delays in propagation from CA3 to CA1 that vary from 7 to 6.5 msec. The differences in our reported multisynaptic delays therefore depend on many factors that can be investigated systematically with pMEAs. Of course, higher-density conformal pMEAs could resolve propagation along dendritic trees at higher spatiotemporal resolution, and allow examination of changes of electrical conductance velocity or summation along single cells. In summary, however, the reported propagation velocities and delays point to the power and usefulness of pMEAs for investigations of multisynaptic network activity.

6.2.4   Neuroprosthetic Applications

The pMEAs' ability to stimulate specific regions of a tissue and record its spatiotemporal dynamics makes them useful testbeds for neuroprosthetic devices. We are using our conformal high-density pMEAs to develop a proof-of-concept for a

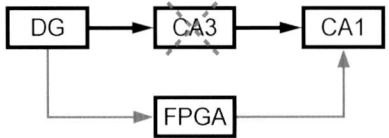

FIGURE 6.9. Diagram of CA3 replacement with a FPGA/VLSI device. The FPGA/VLSI device will replace the functionality of severed CA3, by receiving its inputs and sending out its modeled outputs to CA1.

cortical prosthesis. The goal of this study is to investigate the fundamental science and implement technology that will enable a biomimetic electronic device to be implanted into hippocampus to replace damaged neuronal circuits, and thereby re-establish its original functions. The feasibility of this ambitious endeavor is being first demonstrated in an in vitro hippocampal slice preparation on pMEAs. The proof-of-concept consists of replacing the biological CA3 subregion with an FPGA/VLSI-based model of the nonlinear dynamics of CA3, such that propagation of spatiotemporal patterns of activity in the DG → FPGA/VLSI → CA1 network reproduces that observed experimentally in the biological DG → CA3 → CA1 circuit (Figure 6.9) (Berger et al., 2001; Berger et al., 2005).

Inherent to the success of this project is the development of a stable biological preparation that requires optimization of the anatomical and the electrophysiological conditions. Current pMEAs are not optimized for such an application because the symmetrical distribution of electrodes does not match the cytoarchitecture of hippocampal slices (see above). Therefore, we have designed and built a new generation of pMEAs that allow us to record spatiotemporal activity along the trisynaptic pathway of the hippocampal slice. Based on preliminary results obtained with traditional glass electrodes, we mapped the best recording sites from the DG, CA3, and CA1 as shown in Figure 6.10A and used these data to design a new pMEA (Figure 6.9B).

The pMEA designed to replace CA3 included two different circular pad sizes: 28 µm diameter pads that are grouped in series to form sets of stimulating pads in DG (three at a time) and CA1 (two at a time), and 36 µm diameter pads used for recording in DG, CA3, and CA1. By grouping sets of stimulating pads in series, we were able to achieve a significantly larger pad surface area and correspondingly larger total stimulating current than was possible with single pads, while still maintaining essential conformality to cytoarchitecturally relevant features. The first phase of experiments consisted of recording trisynaptic activity in order to build a mathematical model of the CA3 region. Stimulating the perforant pathway fibers, with 5 µM picrotoxin in the aCSF, elicited population spikes in DG and CA3, and an EPSP in CA1 (Figure 6.3C). The delay between population spikes varied from 8 to 10 msec between DG and CA3, and the EPSP in CA1 was similarly delayed from the spike in CA3. Random interval stimulation trains will be sent through the slice to extract the nonlinear dynamics of the circuit as described in Section 6.1.1., and the generated kernels will be implemented in the FPGA.

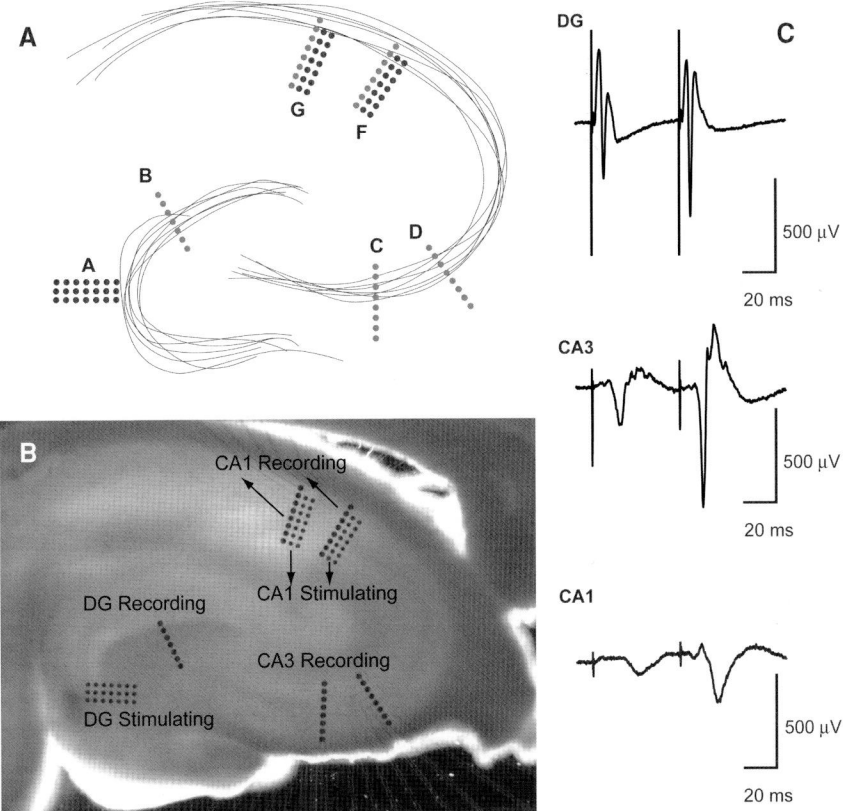

FIGURE 6.10. pMEA for CA3 replacement: (A) the design of conformal multi-electrode array included two different circular pad sizes (A, 3 × 7; F and G, 2 × 7) 28 μm diameter pads with a 50 μm center-to-center spacing grouped in series to form sets of stimulating pads in DG and CA1, and (B, 1 × 7; C, D, F, and G, 1 × 8) 36 μm diameter pads also with a 50 μm center-to-center spacing for recording. The sets are aligned according to rat hippocampal cytoarchitecture covering key input/output regions of DG, CA3, and CA1, thereby allowing complete diagnostic assessment of the nonlinear dynamics of the trisynaptic circuit. (B) Photomicrograph of a slice on the pMEA. (C) Trisynaptic recording from the DG, CA3, and CA1 areas. PP stimulation yielded large populations pikes in DG and CA3 and an EPSP in CA1, all with the appropriate multi-synaptic time delays (scale bars 20 msec and 500 μV).

The second phase of the proof-of-concept involves the severance of CA3–CA1 connections, and replacing the CA3 region dynamics with the FGPA device. The DG responses will then be channeled through the FPGA to the CA1 stimulating electrodes in order to regenerate the biological circuit's output. (Berger et al., 2005) The results from these experiments will allow us to refine our models until we are able to reproduce a CA1 output identical to the one obtained from the intact slice. At that point, we can transfer our acquired knowledge and FPGA to an in vivo testbed. The pMEA is thus serving as a rapid testing tool for the development of our neuroprosthetic application.

6.3 Discussion

Although sequential single-electrode recordings were used diligently to study network connectivity (Andersen et al., 1971; Yeckel and Berger, 1990, 1995), examining spatiotemporal correlations was not possible with this methodology. Multi-electrode array technology offers several advantages over traditional single-electrode recording in the areas of high-throughput testing, spatial mapping of electrical activity, temporal information processing, spatiotemporal activity monitoring, long-term physiological investigations, and neuroprosthetic-driven network dynamics elucidation. Technological advances have increased the production and use of pMEAs, however, the limited number of electrodes has led to the creation of generic arrays consisting of symmetrical matrices of electrodes that suffer from misalignment with tissue cytoarchitecture due to the nonsymmetrical anatomy of the brain. Elliptic (Thiebaud et al., 1999) or circular (Duport et al., 1999) electrode layouts have been designed for hippocampal slices with a single or double layer of electrodes matching the pyramidal cell layer, however these pMEAs are hard to align to the tissue, and do not record dendritic activity. This chapter introduced four high-density pMEAs that are custom designed for specific in vitro investigations with hippocampal slices. These new arrays are designed for input optimization to area CA1, medial and lateral perforant path differentiation, CSD studies, and recording trisynaptic activity.

In all in vitro electrophysiological experiments, considerable care and time is spent locating appropriate stimulation and recording sites. Herein, the first advantage of pMEAs presents itself for high-throughput experiments, by expediting the process of localizing the optimal stimulation and recording sites in a target area. The first example presented above illustrated the ease with which the optimal recording site or several sites could be selected from a geometrically arranged pMEA. By defining the desired recording criteria, users could quickly select the channels they wished to record and analyze. The redundancy of the channels could serve as a backup for weak signals or damaged electrodes, or to increase the statistical sample size by grouping the recordings into area-specific subsets (e.g., dendritic vs. pyramidal). The second example illustrated how the large number of densely packed electrodes in a conformal array can be used to optimize stimulation and recording sites for finer control over the input–output relation between CA3 and CA1 in the hippocampal slice. One subarray was designed to stimulate the Schaffer collaterals emanating from CA3, and the other subarray recorded activity from the CA1 pyramidal and dendritic areas. By sequentially stimulating pairs of electrodes, we were able to determine the electrode pair that yielded the maximal response, and immediately identify the location of the latter. Across slices, the single peak in the stimulus optimization graph suggested that a narrow bundle of axons was a dominant afferent input, consistently with the Schaffer collateral fibers being the major excitatory inputs to CA1 (Andersen et al., 1971). Although the sequential approach is reminiscent of microadvancing single electrodes, it is different in that it does not induce incremental damage to the tissue, and in being automatable in software without the need for manual repositioning. Automation of this otherwise tedious input/output optimization process would increase efficiency and productivity in pharmacological drug-screening protocols.

Beyond the gain in speed and efficiency, the ability to stimulate different sites in close proximity to each other without disturbing the tissue by repeated electrode insertions will enable more accurate and concise stimulations. The advantage of such fine spatial control over the stimulation site was illustrated with the selective excitation of PP subpathways. The PP–DG–CA3 high-density pMEA was designed to match the region's anatomy in order to stimulate the two adjacent afferent PP subpathways selectively and differentiate between their responses. The design is intended to study the subpathway's differential effects on the CA3 output response; however, for this experiment we focused just on the DG's granule cell's postsynaptic potentials in order to illustrate the pMEA's ability to selectively target each subdivision.

Although the subdivisions can be experimentally distinguished by their responses to pp stimulation, electrophysiological studies often do not specify which of the subdivisions is being stimulated due to the difficulty in selectively targeting them. This pathway differentiation was easily achieved with this pMEA (Figure 6.4), where stimulating the lateral PP with one pair of electrodes exhibited pp facilitation mostly in the lateral molecular layer, whereas stimulating the medial PP produced pp depression in the medial molecular layer, which reversed into current sources in the lateral molecular layer. These responses were consistent with the electrode's anatomical location. This experiment demonstrated the advantage of high-density pMEAs in effortlessly stimulating specific brain subdivisions and distinguishing their responses in the same slice. The versatility of custom designing such high-density pMEAs will enable and speed up new electrophysiological experiments in various tissues where there is a need for selective stimulation of adjacent pathways. Research on brain slices with sharp cytoarchitectural boundaries such as the rat barrel cortex (Wirth and Luscher, 2004), or defined fiber tracts such as the cerebellum (Heck, 1995, Egert et al., 2002), would benefit significantly from high-density pMEAs that conform to their geometries.

pMEAs are, however, more than just a sum of many electrodes that allow rapid accumulation of many channels of electrophysiological data from a single slice. pMEAs allow the creation of spatiotemporal activity maps that enable many new applications not previously possible with single-electrode recordings. Spatially, activity maps allow tracing of the spread of a response across a tissue. Figure 6.5 illustrated how this information can be useful in an LTP experiment in the hippocampal slice. We can trace the area responding to a stimulation by color-mapping the EPSP amplitude on the slice image (A,B), and we can similarly map out the amount of potentiation (C) in order to visually determine the extent and spread of these parameters over the entire slice. Mapping various activity parameters spatially onto the slice is especially useful for studying topographically organized brain structures such as cortical areas (visual, auditory, barrel, etc.), which have functionally distinct subdivisions.

Information about the activity in different cellular layers (*strata*) can only be mapped with higher-density arrays that sample closely enough to enable CSD analysis. This mathematical transformation is necessary to compensate for the field potentials intermixing in conductive tissue, and to resolve current sources and sinks. The electrode spacing requirement depends on the distance from the source and the size of different spatial domains in the slice, and as such has been estimated at 50 μm (Novak and Wheeler, 1989). Our high-density pMEAs have electrodes spaced 50 μm apart (center to center) specifically for this purpose. Figure 6.6 illustrated the ability of CSD analysis to reduce the spatial spread of field potentials and accurately localize current sources and sinks and even unmask some that may be overpowered by larger potentials. CSD analysis produced a more accurate laminar profile of the CA1 pyramidal cell EPSP and population spike. The CSD profiles could thus be used to map activity along axo-dendritic trees that densely orient parallel to each other. Kim et al. (submitted) used our rectangular high-density pMEA to localize estradiol's effect on CA3 pyramidal cells and compare the activity maps to the distribution of estrogen receptors. High-density pMEAs can thus be used for generating accurate high-resolution spatial activity maps of a tissue, thereby complementing larger-scale maps generated by more widely spaced pMEAs.

There are many applications to purely temporal or long-term recordings from pMEAs as illustrated in Chapters 16 and 17 of this book, which investigate circadian rhythmicity and slice cultures, or the temporal dynamic investigations of Beggs et al. (2003) and Jimbo et al. (1999). However, we did not present examples of such approaches here, as our high-density conformal pMEAs do not confer any significant advantages for solely temporal experiments, and in the case of long-term cultures, the tissue changes its position relative to the electrodes during the span of the culture in ways that are not controllable, and thus makes conformal pMEA design obsolete. On the other hand, in Chapter 7, the authors presented methods of patterning adhesion molecules onto arrays in order to control the development of networks according to set geometries. These patterned networks can be aligned to electrode locations and thereby allow long-term monitoring and investigations of their activity.

Spatiotemporal information is, however, the most distinguishing advantage of MEAs, as they permit the simultaneous recording of activity at many points across a tissue, thereby revealing correlations and interactions between different regions. The activity in a slice spreads from one area to another along excitatory pathways. This propagation can be traced in space over time with pMEAs. We traced the EPSP produced by Schaffer collateral activation at CA3. First, cells in CA3 itself responded through recurrent afferents, the response then propagated to CA2 pyramidal cells, and proceeded to CA1. A movie of the activity could be generated by color-coding the potentials and interpolating between them at every time frame. Four frames of such a movie are shown in Figure 6.7 to illustrate the spread of the activity across the slice over time.

The interaction between hippocampal areas was more specifically explored in our TMPP investigations. TMPP, a $GABA_A$ inhibitor, reduces inhibition in hippocampal slices. In CA3, large number of inhibitory interneurons usually prevents repetitive bursting due to activation of recurrent fibers. In the presence of TMPP, this inhibition was blocked, and CA3 pyramidal cells would burst synchronously when stimulated either directly or upstream, through PP stimulation. The burst of population spikes was easily localized to area CA3 by the pMEA. A facilitated propagation of the response to CA1 was also observed, and the delay in CA1 EPSP indicated the number of synapses traversed by the spreading activity. Monosynaptic responses could be easily compared to di- and trisynaptic ones by overlaying recorded potentials and comparing lag times. The effect of TMPP on the hippocampal slice was thus more than just disinhibiting CA3 pyramidal cells to the point of evoking epileptiform activity: it also facilitated propagation of this activity to CA1. This network effect of TMPP on the hippocampal slice could not have been observed with single-electrode intracellular recordings, although the latter revealed epileptiform discharges in CA1 that we did not detect extracellularly (Lin et al., 2001). MEAs are therefore extremely useful for investigating the propagation of epileptiform activity in vivo and in vitro, spontaneous or drug-induced (Nagao et al., 1996; Harris and Stewart, 2001). pMEAs have been used to localize the origin of epileptic seizures (optical multi-site recordings; Colom and Saggau, 1994), to determine the propagation direction (Harris and Stewart, 2001), and calculate changes in propagation velocity across fibers compared to normal states (Holsheimer and Lopes da Silva, 1989). Beyond drug screening, such monitoring of multi-synaptic activity in a tissue is thus useful for investigations of network connections and dynamics in normal or diseased states. In these investigations, observing the entire slice with pMEAs not only allows localization of the site of activity of the drug, but can also unveil related effects on other areas.

pMEAs have been used for time spreading activity or trace waves in several tissue preparations: retina (Syed et al., 2004), cerebellum (Egert et al., 2002), barrel cortex (Wirth and Luscher, 2004), and so on. pMEAs allow easy quantification of propagation velocity across a slice. In our first activity propagation example, the monosynaptic response spread along the Schaffer collaterals at 0.25 m/sec was consistent with the 0.25 m/sec reported average velocity for this weakly myelinated fiber (Andersen et al., 2000). There are, however, many other measurements of axonal propagation speed that show different values, and our synaptic delay times calculated from the TMPP experiment appear longer than those reported in the literature. The speed of propagation of activity in a tissue preparation is generally thought of as a function of the axon thickness, their myelination, length, and also the number of traversed synapses. There are, however, numerous other factors that can influence the propagation velocity across the multi-synaptic pathways, beyond simple experimental manipulations such as temperature or ionic concentration differences (our unpublished data). In particular for hippocampus, the behavioral state (REM, slow-wave sleep, or alert; Winson and Abzug, 1978) irregular or theta EEG (Herreras et al., 1987), anesthesia (Buzsaki et al., 1986; Pare and Llinas, 1994), pharmacological disinhibition or activation facilitation (Sirvio et al., 1996), stimulation frequency (Herreras et al., 1987; Yeckel and Berger, 1990), stimulation intensity and relative location to recording sites (Pare and Llinas, 1994; our unpublished observations), and animal species (Andersen et al., 1978) have all been shown to affect activity propagation in the trisynpatic pathway. pMEAs are extremely well suited for investigating these variables to elucidate the interrelations between connected regions in multi-synaptic circuits.

Such an understanding of network dynamics is critical for building biomimetic devices or neuroprosthetics. The retinal prosthesis project relies on pMEA recordings from in vitro retinal preparations to develop stimulation algorithms for its implanted MEAs (Meister et al., 1994; Warland et al., 1997; Chichilnisky and Kalmar, 2003; Humayun et al., 2003; Frechette et al., 2004). In our project to replace the CA3 function in vivo, we needed an in vitro model upon which to test our nonlinear mathematical models of the input–output properties CA3. The first step entailed development of a slice-pMEA preparation that allows experimental characterization of the combined nonlinear dynamics of the intrinsic hippocampal trisynaptic circuit. The conformality aspect of the newly designed pMEA was crucial to obtain trisynaptic recordings from hippocampal slices because it is extremely difficult if not impossible to obtain simultaneous dentritic and somatic responses in the three subregions of the hippocampus using the commercially available MEAs with their sparse and symmetrical matrices of electrodes. The novel pMEA has recording electrodes in the DG, CA3, and CA1 regions as well as stimulating electrodes in the DG and the CA1 regions. The DG stimulating electrodes were used to activate the perforant path, and the CA1 stimulating electrodes were used to channel the FPGA/VLSI output to the CA1 pyramidal cells.

Using the designed pMEA, we have recorded trisynaptic responses in the disinhibited hippocampal slice in response to PP stimulation trains consisting of 2000 electrical pulses with Poisson-distributed random intervals. We recorded population spikes in DG and CA3 and dentritic EPSPs in CA1, with delays comparable to those obtained under TMPP disinhibition, but without the CA3 epileptiform activity. The population spike amplitude was extracted to build a nonlinear mathematical model of CA3 (Gholmieh et al., 2004b) that is now implemented in hardware using an FPGA device (Granacki et al., 2004). Currently, we are exploring the next stage of the project, which involves replacing CA3 with a FPGA. First we cut the Schaffer collaterals, and then we send the DG's responses into the FPGA and redirect the FPGA's output (CA3 replacement) to the CA1 stimulating electrodes. We then record the corresponding CA1-evoked responses and compare them to those obtained from intact slices. Our preliminary results indicate that our VLSI model is succeeding at replacing the severed CA3 area (Gholmieh et al., 2004b). This in vitro slice preparation on conformal pMEA approach is allowing us to test several computational and analytical parameters rapidly before any implantation into live animals. Once the FPGA/VLSI model is fine tuned and accurately reproduces CA1 output comparable to the intact, we will connect it to MEAs implanted in rat hippocampus, and repeat the same experiments. Our pMEA work would have thus expedited the development process, from the generation of the nonlinear model to its VLSI testing, and thereby also minimized animal use (because we obtain several hippocampal slices from each rat brain).

In summary, rapid transition of cell types in the brain mandates high spatial resolution sampling whereas the nonsymmetrical organization of the brain in conjuction with the limited number of recording channels requires conformal electrode designs. In this chapter, four pMEAs were introduced that illustrated the advantages of the conformality and high density for in vitro electrophysiological experiments with acute hippocampal slices. The high-density conformal designs confer new abilities to the planar MEA research tool whether it is applied to rapidly and accurately localize target field responses, to increase sampling sites for high-throughput drug investigations, or to trace spatiotemporal network activity propagation and dynamics for biomimetic neuroprosthetic applications. The work presented here lays the ground for the custom design of conformal high-density pMEAs that meet the individual needs of their application.

References

Andersen, P., Bliss, T.V.P., and Skrede, K.K. (1971). Lamellar organization of hippocampal excitatory pathways. *Exper. Brain Res.* 13: 222–238.

Andersen, P., Silfvenius, H., Sundberg, S.H., Sveen, O., and Wigstrom, H. (1978). Functional characteristics of unmyelinated fibres in the hippocampal cortex. *Brain Res.* 144(1): 11–18.

Andersen, P., Soleng, A.F., and Raastad, M. (2000). The hippocampal lamella hypothesis revisited. *Brain Res.* 886(1–2): 165–171.

Beggs, J.M. and Plenz, D. (2003). Neuronal avalanches in neocortical circuits. *J. Neurosci.* 23(35): 11167–11177.

Berger, T.W., Ahuja, A., Courellis, S.H., Erinjippurath, G., Gholmieh, G., Granacki, J.J., Marmarelis, V.Z., Srinivasan, V., Dong, S., Tanguay, A.R., and Wills, J. (2005). Brain-implantable biomimetic electronics as neural prostheses to restore lost cognitive function. Towards Replacement Parts for the Brain: Implantable Biomimetic Electronics as Neural Prostheses. Eds. Berger, T.W. and Glanzman, D.L. MIT Press, Cambridge MA.

Berger, T.W., Baudry, M., Brinton, R.D., Liaw, J.S., Marmarelis, V.Z., Park, AY., Sheu B.J., and Tanguay, A.R.J. (2001). Brain-implantable biomimetic electronics as the next era *Neural Prosthet. Proc. IEEE* 89(7): 993–1012.

Buzsaki, G., Czopf, J., Kondakor, I., and Kellenyi, L. (1986). Laminar distribution of hippocampal rhythmic slow activity (RSA) in the behaving rat: Current-source density analysis, effects of urethane and atropine. *Brain Res.* 365(1): 125–137.

Chapin, J.K., Moxon, K. A., Markowitz, R.S., and Nicolelis, M.A. (1999). Real-time control of a robot arm using simultaneously recorded neurons in the motor cortex. *Nat. Neurosci.* 2(7): 664–670.

Chichilnisky, E.J. and Kalmar, R.S. (2003). Temporal resolution of ensemble visual motion signals in primate retina. *J. Neurosci.* 23(17): 6681–6689.

Colom, L.V. and Saggau, P. (1994). Spontaneous interictal-like activity originates in multiple areas of the CA2–CA3 region of hippocampal slices. *J. Neurophysiol.* 71(4): 1574–1585.

Dahl, D., Burgard, E.C., and Sarvey, J.M. (1990). NMDA receptor antagonists reduce medial, but not lateral, perforant path-evoked EPSPs in dentate gyrus of rat hippocampal slice. *Exp. Brain Res.* 83(1): 172–177.

Duport, S., Millerin, C., Muller, D., and Correges, P. (1999). A metallic multisite recording system designed for continuous long-term monitoring of electrophysiological activity in slice cultures. *Biosens. Bioelectron.* 14(4): 369–376.

Egert, U., Heck, D., and Aertsen, A. (2002). Two-dimensional monitoring of spiking networks in acute brain slices. *Exp. Brain Res.* 142(2): 268–274.

Frechette, E.S., Sher, A., Grivich, M.I., Petrusca, D., Litke, A.M., and Chichilnisky, E.J. (2005). Fidelity of the ensemble code for visual motion in primate retina. *J. Neurophysiol.* 94: 119–135.

Gholmieh, G., Courellis, S., Dimoka, A., Wills, J.D., LaCoss, J., Granacki, J.J., Marmarelis, V., and Berger, T.W. (2004a). An algorithm for real-time extraction of population EPSP and population spike amplitudes from hippocampal field potential recordings. *J. Neurosci. Meth.* 136(2): 111–121.

Gholmieh, G., Courellis, S., Fakheri, S., Cheung, E., Marmarelis, V., Baudry, M., and Berger, T.W. (2003). Detection and classification of neurotoxins using a novel short-term plasticity quantification method. *Biosens. Bioelectron.* 18(12): 1467–1478.

Gholmieh, G., Courellis, S.H., Hsiao, M., Srinivasan, V., Ahuja, A.K., LaCoss, J., Wills, J.D., Tanguay, A.R., Jr., Granacki, J.J., and Berger, T.W. (2004b). A biomimetic electronic prosthetic for hippocampus: Proof-of-concept using the *in vitro* slice. *Soc. Neurosci.* Online Abstract Viewer.

Gholmieh, G., Soussou, W., Courellis, S., Marmarelis, V.Z., Berger, T.W., and Baudry, M. (2001). A biosensor for detecting changes in cognitive processing based on nonlinear systems analysis. *Biosens.Bioelectron.* 16(7–8): 491–501.

Gholmieh, G., Soussou, W., Han, M., Ahuja, A., Hsiao, M.-C., Dong, S., Tanguay, A.R. Jr., and Berger, T.W. (2005). Custom-designed high-density conformal planar multielectrode arrays for brain slice electrophysiology. *J Neurosci Meth.* 10.1016.

Granacki, J.J., Wills, J.D., LaCoss, J., Courellis, S.H., Marmarelis, V.Z., Gholmieh, G., and Berger, T.W. (2004). A biomimetic electronic prosthetic for hippocampus: hardware model of CA3 nonlinear dynamics, *Soci. Neurosci.* Online Abstract Viewer.

Gross, G.W. and Schwalm, F.U. (1994). A closed flow chamber for long-term multichannel recording and optical monitoring. *J. Neurosci. Meth.* 52(1): 73–85.

Gross, G.W., Keefer, E.W., Pancrazio, J.J., and Stenger, D.A. (1999). Rapid determination of toxic effects of trimethylol propane phosphate usina neuronal networks on microelectrode arrays. *Soc. Neurosci. Abstr.* 1(230.13).

Gross, G.W., Rhoades, B.K., Azzazy, H.M., and Wu, M.C. (1995). The use of neuronal networks on multielectrode arrays as biosensors. *Biosens. Bioelectron.* 10(6–7): 553–567.

Harris, E. and Stewart, M. (2001). Propagation of synchronous epileptiform events from subiculum backward into area CA1 of rat brain slices. *Brain Res.* 895(1–2): 41–49.

Heck, D. (1995). Investigating dynamic aspects of brain function in slice preparations: spatiotemporal stimulus patterns generated with an easy-to-build multi-electrode array. *J. Neurosci. Meth.* 58(1–2): 81–87.

Herreras, O., Solis, J.M., Martin del Rio, R., and Lerma, J. (1987). Characteristics of CA1 activation through the hippocampal trisynaptic pathway in the unanaesthetized rat. *Brain Res.* 413(1): 75–86.

Holsheimer, J. and Lopes da Silva, F.H. (1989). Propagation velocity of epileptiform activity in the hippocampus. *Exp. Brain Res.* 77(1): 69–78.

Humayun, M.S., Weiland, J.D., Fujii, G.Y., Greenberg, R., Williamson, R., Little, J., Mech, B., Cimmarusti, V., Van Boemel, G., Dagnelie, G., and de Juan, E. (2003). Visual perception in a blind subject with a chronic microelectronic retinal prosthesis. *Vis. Res.* 43(24): 2573–2581.

Ishizuka, N., Cowan, W.M., and Amaral, D.G. (1995). A quantitative analysis of the dendritic organization of pyramidal cells in the rat hippocampus. *J. Comp. Neurol.* 362(1): 17–45.

Jahnsen, H., Kristensen, B.W., Thiebaud, P., Noraberg, J., Jakobsen, B., Bove, M., Martinoia, S., Koudelka-Hep, M., Grattarola, M., and Zimmer, J. (1999). Coupling of organotypic brain slice cultures to silicon-based arrays of electrodes. *Methods* 18(2): 160–172.

Jimbo, Y. and Robinson, H.P. (2000). Propagation of spontaneous synchronized activity in cortical slice cultures recorded by planar electrode arrays. *Bioelectrochem.* 51(2): 107–115.

Jimbo, Y., Tateno, T., and Robinson, H.P. (1999). Simultaneous induction of pathway-specific potentiation and depression in networks of cortical neurons. *Biophys. J.* 76(2): 670–678.

Kawaguchi, H., Tokioka, R., Murai, N., and Fukunishi, K. (1996). Multichannel optical recording of neuronal network activity and synaptic potentiation in dissociated cultures from rat hippocampus. *Neurosci. Lett.* 205(3): 177–180.

Keefer, E.W., Gramowski, A., Stenger, D.A., Pancrazio, J.J., and Gross, G.W. (2001). Characterization of acute neurotoxic effects of trimethylolpropane phosphate via neuronal network biosensors. *Biosens. Bioelectron.* 16(7–8): 513–525.

Kim, M.T., Gholmieh, G., Soussou, W., Ahuja, A., Tanguay, Jr., A.R., Berger, T.W., and Brinton, R.D. (submitted). 17-beta estradiol potentiates fEPSP within each subfield of the hippocampus with greatest potentiation of the associational/commissural afferents of CA3.

Lin, J., Ritchie, G.D., Stenger, D.A., Nordholm, A.F., Pancrazio, J.J., and Rossi, J., 3rd (2001). Trimethylolpropane phosphate induces epileptiform discharges in the CA1 region of the rat hippocampus. *Toxicol. Appl. Pharmacol.* 171(2): 126–134.

McNaughton, N. and Miller, J.J. (1984). Medial septal projections to the dentate gyrus of the rat: electrophysiological analysis of distribution and plasticity. *Exp. Brain Res.* 56(2): 243–56.

Meister, M., Pine, J., and Baylor, D.A. (1994). Multi-neuronal signals from the retina: acquisition and analysis. *J. Neurosci. Meth.* 51(1): 95–106.

Miyakawa, H. and Kato, H. (1986). Active properties of dendritic membrane examined by current source density analysis in hippocampal CA1 pyramidal neurons. *Brain Res.* 399(2): 303–309.

Morin, F.O., Takamura, Y., and Tamiya E. (2005). Investigating neuronal activity with planar microelectrode arrays: achievements and new perspectives. *J Biosci Bioeng.* 100(2): 131–143.

Nagao, T., Alonso, A., and Avoli, M. (1996). Epileptiform activity induced by pilocarpine in the rat hippocampal-entorhinal slice preparation. *Neuroscience* 72(2): 399–408.

Nakagami, Y., Saito, H., and Matsuki, N. (1997). Optical recording of trisynaptic pathway in rat hippocampal slices with a voltage-sensitive dye. *Neuroscience* 81(1): 1–8.

Nicholson, C. and Freeman, J.A. (1975). Theory of current source-density analysis and determination of conductivity tensor for anuran cerebellum. *J. Neurophysiol.* 38(2): 356–368.

Novak, J.L. and Wheeler, B.C. (1989). Two-dimensional current source density analysis of propagation delays for components of epileptiform bursts in rat hippocampal slices. *Brain Res.* 497(2): 223–230.

Pare, D. and Llinas, R. (1994). Non-lamellar propagation of entorhinal influences in the hippocampal formation: multiple electrode recordings in the isolated guinea pig brain in vitro. *Hippocampus* 4(4): 403–409.

Pine, J. (1980). Recording action potentials from cultured neurons with extracellular microcircuit electrodes. *J. Neurosci. Meth.* 2(1): 19–31.

Sekino, Y., Obata, K., Tanifuji, M., Mizuno, M., and Murayama, J. (1997). Delayed signal propagation via CA2 in rat hippocampal slices revealed by optical recording. *J. Neurophysiol.* 78(3): 1662–1668.

Shimono, K., Baudry, M., Panchenko, V., and Taketani, M. (2002). Chronic multichannel recordings from organotypic hippocampal slice cultures: Protection from excitotoxic effects of NMDA by non-competitive NMDA antagonists. *J. Neurosci. Meth.* 120(2): 193–202.

Shimono, K., Brucher, F., Granger, R., Lynch, G., and Taketani, M. (2000). Origins and distribution of cholinergically induced beta rhythms in hippocampal slices. *J. Neurosci.* 20(22): 8462–8473.

Singer, W. (2000). Why use more than one electrode at a time? *New Technol. Life Sci: Trends Guide* 2000: 12–17.

Sirvio, J., Larson, J., Quach, C.N., Rogers, G.A., and Lynch, G. (1996). Effects of pharmacologically facilitating glutamatergic transmission in the trisynaptic intrahippocampal circuit. *Neuroscience* 74(4): 1025–1035.

Steward, O. (1976). Topographic organization of the projections from the entorhinal area to the hippocampal formation of the rat. *J. Comp. Neurol.* 167(3): 285–314.

Stoppini, L., Duport, S., and Correges, P. (1997). A new extracellular multirecording system for electrophysiological studies: Application to hippocampal organotypic cultures. *J. Neurosci. Meth.* 72(1): 23–33.

Syed, M.M., Lee, S., He, S., and Zhou, Z.J. (2004). Spontaneous waves in the ventricular zone of developing mammalian retina. *J. Neurophysiol.* 91(5): 1999–2009.

Thiebaud, P., Beuret, C., Koudelka-Hep, M., Bove, M., Martinoia, S., Grattarola, M., Jahnsen, H., Rebaudo, R., Balestrino, M., Zimmer, J., and Dupont, Y. (1999). An array of Pt-tip microelectrodes for extracellular monitoring of activity of brain slices. *Biosens. Bioelectron.* 14(1): 61–5.

Thiebaud, P., de Rooij, N.F., Koudelka-Hep, M., and Stoppini, L. (1997). Microelectrode arrays for electrophysiological monitoring of hippocampal organotypic slice cultures. *IEEE Trans. Biomed. Eng.* 44(11): 1159–63.

Warland, D.K., Reinagel, P., and Meister, M. (1997). Decoding visual information from a population of retinal ganglion cells. *J. Neurophysiol.* 78(5): 2336–2350.

Wheeler, B.C. and Novak, J.L. (1986). Current source density estimation using microelectrode array data from the hippocampal slice preparation. *IEEE Trans. Biomed. Eng.* 33(12): 1204–1212.

Winson, J. and Abzug, C. (1978). Dependence upon behavior of neuronal transmission from perforant pathway through entorhinal cortex. *Brain Res.* 147(2): 422–427.

Wirth, C. and Luscher, H.R. (2004). Spatiotemporal evolution of excitation and inhibition in the rat barrel cortex investigated with multielectrode arrays. *J. Neurophysiol.* 91(4): 1635–1647.

Yeckel, M.F. and. Berger, T.W. (1990). Feedforward excitation of the hippocampus by afferents from the entorhinal cortex: Redefinition of the role of the trisynaptic pathway. *Proc. Natl. Acad. Sci.* 87: 5832–5836.

Yeckel, M.F. and Berger, T.W. (1995). Monosynaptic excitation of hippocampal CA1 pyramidal cells by afferents from the entorhinal cortex. *Hippocampus* 5(2): 108–114.

7
Pattern Technologies for Structuring Neuronal Networks on MEAs

John C. Chang and Bruce C. Wheeler

7.1   Introduction

Much progress has been made over the past several centuries regarding the understanding of the human central nervous system. This progress has taken us from knowing the brain as a functionally homogeneous organ to one with interconnected networks of specialized neurons. The initiating study was conducted by Broca who examined a patient with a left-sided posterior frontal lobe lesion which resulted in dysphasia. Broca concluded that the ability to generate coherent speech resided in the dominant posterior frontal lobe and theorized that distinct regions of the brain functioned uniquely. This theory was further developed by Wernicke who speculated that distinct regions of the brain intercommunicated to achieve the observed complex human behavior. Findings from both electrical stimulation and seizure studies support this theory as motor and sensory functions can be localized to specific regions of the brain (Kandel et al., 2000). These experiments provided the foundation for the interconnected network theory, the details of which are being expounded daily by the numerous functional imaging studies reported in the literature (Kandel et al., 2000).

Although the progress in understanding the regional interaction between brain regions has been exceptional, the detailed understanding of the neuronal firing patterns, either in aggregate as in electroencephalography (EEG) or in simultaneous multi-site monitoring as in cortical probes, is desirable. The advantage of understanding the patterns lies in the ability to diagnose diseased states or in cognitive control of exogenous actuators such as robotic arms or input devices to computers (Kandel et al., 2000; Nicolelis, 2001). Diagnosis of seizures is usually confirmed when third-person observation provides suitable symptoms such as absence, tonic or clonic muscular activities, and loss of consciousness, but EEGs can be diagnostic when proper neuronal activities are detected (Kandel et al., 2000). On the other hand, neuronal firing patterns can be harvested to control electronic gadgets as experiments with trained monkeys have shown (Carmena et al., 2003). Thus, knowledge of the patterns of neuronal firing can allow one to decode and predict the state of an organism's neurological function.

With in vivo recording, the successes have come from the easily accessible areas of the brain such as the motor, sensory, and visual cortices. Although it is possible to attain deep brain recording, the difficulty in data interpretation increases dramatically as these structures interact with multiple regions of the brain and are influenced by the behavioral state of the animal (Kandel et al., 2000; Nicolelis and Fanselow, 2002). The multiple factors that influence an intact mammalian nervous system make controlling in vivo networks a monumental task, but properly controlled cell cultures provide an excellent environment for controlling aspects of the experiment that one could not control in the living animal.

Traditionally, cultures of neurons are grown from dissociated cells or slices that preserve the local architecture of the network. In dissociated cultures the innate structure of the brain is destroyed to allow the formation of random new connections. This type of culture grants unprecedented control over the chemical milieu of the culture and access to identified synapses, both of which are difficult to accomplish within the living being (Tao et al., 2000; Ming et al., 2002). To compensate for the loss of network architecture, brain slices have been cultured to maintain the local integrity of the connections, but the tissue architecture is generally lost in the long term (Gaehwiler et al., 1998). Electrical recordings of cultured neurons in both settings with pipette electrodes have demonstrated learning through synaptic potentiation and depression (Cash et al., 1996; Lopez-Garcia et al., 1996; Engert and Bonhoeffer, 1997; Urban and Barrionuevo, 1998); however, extended recordings have been difficult to conduct as the introduction of pipette electrodes can contaminate the culture.

This inability to achieve long-term, multi-site recording has led scientists to pursue recording of cultured neurons on multi-electrode arrays (MEAs). Thomas et al. (1972) first demonstrated the feasibility of recording on MEAs by studying myocardial activity (Thomas et al., 1972), but it was Gross (1979) who demonstrated neuronal recording with MEAs. Since then, numerous neuronal recordings with MEAs have found that dissociated neurons cultured on MEAs retain their dynamic interaction and respond to chemicals in similar fashion as their counterparts in vivo and in vitro (Gross et al., 1997; Jimbo et al., 1998, 1999, 2000; Morefield et al., 2000). Although progress has been made regarding recordings of slice cultures on MEAs, the progress has been limited to the characterization of electrodes and the presence of neuronal activity (Stoppini et al., 1997; Thiebaud et al., 1997; Egert et al., 1998). Although slice cultures may retain the active connections similar to those in vivo, the current MEA technology makes specific site recording difficult in these cultures, as the recording sites are pre-specified by the manufacturer. Thus, the ideal situation would be the maintenance of neurological connectivity with placement of the desired control point over the electrodes.

This goal has been made possible by the recent development of surface modification techniques that have been applied to alter the surface properties of both inorganic (silicon oxide and metals) and organic (polymeric) substrates (Weetall, 1976; Moses et al., 1978; Facci and Murray, 1980; Laibinis et al., 1989; Hook et al., 1991; Vargo et al., 1992; Ferguson et al., 1993; Kumar et al., 1994; Lee and

Viehbeck, 1994; Corey et al., 1996; Garbassi et al., 1998). These techniques have successfully modified commonly used insulator and electrode materials such as silicon oxide, silicon nitride, polyimide, polysiloxane, platinum, and gold. The simplest of all techniques is the physisorption technique which can be applied to all surfaces. Next in ease of applicability is silane grafting which can be applied to any surface presenting hydroxyl groups (silicon oxide, silicon nitride, oxidized polyimide, oxidized polysiloxane, and oxidized platinum). Deposited silane presents a uniform functional group that permits specific chemisorption of growth guidance materials. Alternatively, pure noble metal electrodes (platinum and gold) can be modified using functionalized thiol such as 11-mercaptoundecanoic acid. For polymers, however, chemical functionalization (such as hydrolysis) can also produce specific functional groups for chemisorbing macromolecules (Garbassi et al., 1998). These techniques have produced patterned neuronal cultures on substrates without electrodes and are being applied to MEAs (Hickman and Stenger, 1994; Singhvi et al., 1994; Branch et al., 1998; Ma et al., 1998; Ravenscroft et al., 1998; Branch et al., 2000; Chang et al., 2000; James et al., 2000; Chang et al., 2001a,b; Vogt et al., 2003; Nam et al., 2004; Vogt et al., 2004).

The results from patterning neurons in culture have demonstrated that patterned neurons can survive for the long term, establish interconnections, and yield action potentials when grown on MEAs (Kleinfeld et al., 1988; Jimbo et al., 1993; Ma et al., 1998; Branch et al., 2000; Chang et al., 2000; Liu et al., 2000; Prinz and Fromherz, 2000; Lauer et al., 2002). Branch et al. (2000) have demonstrated that neurons cultured in serum-free media can remain patterned for more than four weeks. Patterned neurons have also been shown to interact even at four weeks old through synaptic transmissions (Ma et al., 1998; Chang et al., 2000; Liu et al., 2000). These results demonstrate the feasibility of reconstructing neuronal networks within culture to model in vivo circuitry.

To successfully reconstruct the in vivo circuits, one needs to understand the combination of materials selection, surface functionalization, and polymer delivery techniques that will work with the MEAs employed for culture. The goals of this review are to focus on the materials available to guide neuronal growth, the patterning techniques used to deliver the growth-inducing polymers, and the functionalization methods for insulator and electrode materials encountered in MEAs. As space is limited, the physicochemical properties of the molecules used are not discussed in this review; instead, we focus on the application and the techniques for patterning neurons. A short review of the known properties of patterned networks of neurons and their potential are given to conclude this chapter.

7.2   Materials Selection

In guidance molecule selection, the right combination of foreground (location with cells) and background (location without cells) materials permits the retention of excellent patterns over the duration of the culture. Both foreground and background materials can be classified into biologically active or inactive materials (Corey and Feldman, 2003). The foreground materials that are biologically inactive include aminated molecules such as poly-lysine, polyethyleneimine (PEI), and aminosilanes which provide an area that is permissive for cellular growth but have not been shown to induce intracellular signaling (Kleinfeld et al., 1988; Hickman and Stenger, 1994; Corey et al., 1996; Rutten et al., 1999; Chang et al., 2000). The active foreground materials include extracellular matrix materials (ECMs) such as laminin, neuronal-cell-adhesive molecules (NCAM), fibronectin, and collagen which induce focal adhesions that may provide intracellular signals (Clark and Brugge, 1995; Schwartz et al., 1995; Ingber, 1997; Banker and Goslin, 1998; Xiao et al., 1999; Kandel et al., 2000). These ECMs can further interact with growth factors such as neurotrophins (NT) or nerve growth factor (NGF) to induce axonal growth (Hari et al., 2004).

For the background, inactive materials include alkylsilanes and polyethyleneglycol which create either a hydrophobic or a protein-resistant surface that repels cells, respectively (Israelachvili, 1997; Sheth and Leckband, 1997; Zhang et al., 1998; Branch et al., 2000; Corey and Feldman, 2003). The active background materials include chondroitin sulfates, myelin-associated glycoprotein, semaphorins, albumin, and serine which range from proteins to amino acids and can resist growth by mechanisms such as cell signaling and protein resistance (Branch et al., 2000; Corey and Feldman, 2003; Jones et al., 2003).

Although numerous guidance molecules exist, not all are useful to the experiments that one has in mind. For example, if one simply desires random growth, one can easily neglect the background molecules. However, if one desires a directed neuronal network, then one will have to select a proper combination of growth-permissive molecule (such as poly-lysine or ECM) plus a growth factor (either NGF or NT) for the foreground and cell-repulsive molecule (such as polyethyleneglycol or chondroitin sulfate) for the background. The complexity of the selection process increases with the directionality within the network.

The complexity of network design using materials selection is demonstrated by the complex intracellular and extracellular signaling that occurs. Reports have shown that substrate adhesivity influences the specificity of the neurites, but that this influence can be molecule and pattern specific (Lochter et al., 1995; Esch et al., 1999). Furthermore, molecules may switch their function based on the intracellular state of the neurons, for example, netrin-1 and semaphorins can alter their attractiveness/repulsiveness based on intracellular concentrations of cyclic nucleotides (Caroni, 1998; Song et al., 1998; Takei et al., 1998; Polleux et al., 2000). As if these complexities were not enough, the developmental history and the chemical gradient further modify the effects of signaling molecules on the growing neurites (Hoepker et al., 1999; Ming et al., 2002). Perhaps the simplest design technique is that proposed by Stenger and Hickman in which speed bumps that slow neurite growth lead to predominant development of dendrites, but even this is technique dependent (Stenger et al., 1998; Vogt et al., 2003, 2004). Thus, directionality design in a neuronal network can be an extremely complicated task (Figure 7.1).

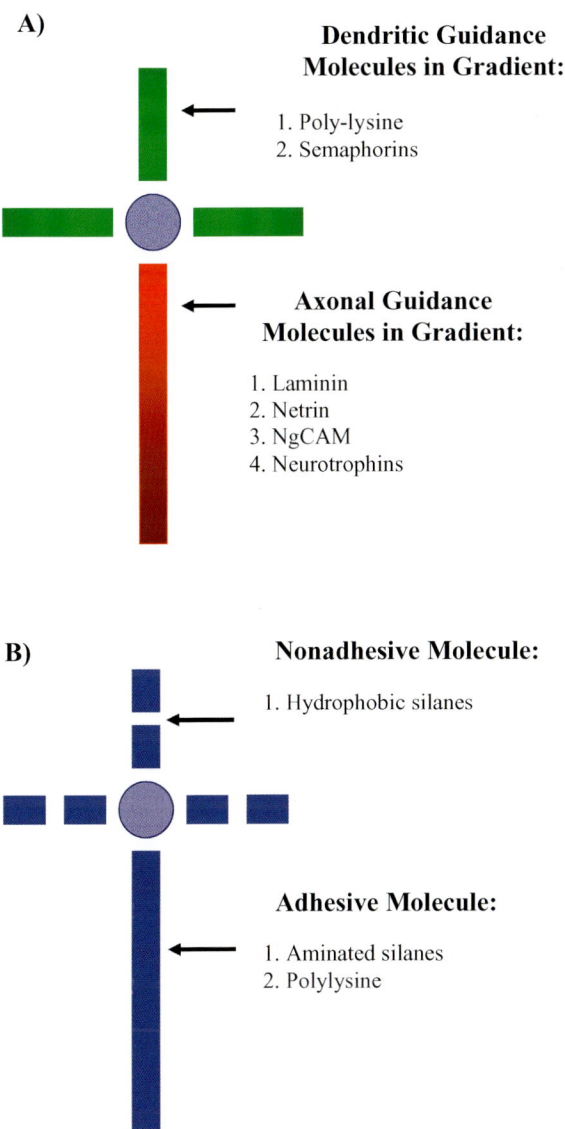

FIGURE 7.1. Proposed directed network designs: two possible designs to achieve directed neuronal networks have been proposed. (A) Signaling technique employs biologically active molecules as the signals to the cells to direct axonal versus dendritic growth. A gradient of these molecules may be incorporated into the design as an additional technique to improve the directionality of the network (Venkateswar et al., 2000). (B) Speed-bump technique employs adhesive and repulsive regions to enhance or retard the growth of the neurites, respectively (Stenger et al., 1998). However, this technique seems to depend on the patterning technique used to create these regions (Vogt et al., 2004).

7.3   Surface Modification Techniques

When culturing neurons on MEAs, the cells encounter a combination of organic and inorganic surfaces that constitute the insulation and the electrodes. The common organic materials used to insulate the MEA are polyimides and polysiloxanes, the structures of which are shown in Figure 7.1 (Gross et al., 1997; Chang et al., 2000). If unmodified, these highly hydrophobic surfaces will not only resist wetting by aqueous solutions but also physisorption and chemisorption of macromolecules conducive for cellular growth (Gross et al., 1997). Thus, it is critical that these surfaces be activated to permit deposition of macromolecules for proper signaling. Alternatively, the electrodes can be insulated using silicon oxides or silicon nitrides (Wise and Weissman, 1971; Shahaf and Marom, 2001). Although these silicon surfaces are more hydrophilic than the organic surfaces, they provide a minimally permissive surface for neuronal survival that is not sufficient for either good cell survival or cell repulsion (a necessary component for maintaining good neuronal patterns in culture). To achieve good patterning and survival on these surfaces, one needs to refunctionalize these surfaces.

Finally, neurons growing on MEA will also encounter electrodes of gold, iridium oxide, platinum, or titanium nitride (Gross et al., 1997; Chang et al., 2000; Weiland and Anderson, 2000; Shahaf and Marom, 2001; Nam et al., 2004). Although metals are permissive for cellular growth, such nonspecific signals to cells may disrupt an experiment designed to study the cellular response to certain extracellular signals (Letourneau, 1975). In addition, electrodes exposed to a proteinaceous solution can accumulate a relatively thick layer of protein that can distort the desired guidance properties of a designed path (Gesteland et al., 1959), and stimulated electrodes can shed platinum into the culture medium (Brummer et al., 1977). Thus, modifying the MEA surface can improve the cell growth, specify desired cell–cell and cell–substrate interactions, and construct specific circuits for neuronal network studies.

The modification techniques can be categorized into three classes, namely, physisorption, functionalization, or grafting. Physisorption simply requires the protein to adhere to the MEA by either the van der Waals or electrostatic forces and involves no chemistry for surface alteration (Lom et al., 1993). Functionalization involves the simple conversion of the original functional group into an alternate functional group that is more amenable for attachment of macromolecules (Garbassi et al., 1998). Grafting, on the other hand, involves attaching a polymeric layer (perhaps more than one molecular layer) that presents a uniform functional group for linking guidance molecules. Under this classification, wet chemical modification of organic surfaces functionalizes the substrates, whereas silane and thiols in general graft the substrates (Garbassi et al., 1998).

7.3.1   Physisorption of Protein

As stated previously, physisorption is the simplest and most generalizable technique for modifying the surfaces of an MEA. Its simplicity and generalizability lie in its independence of specific chemical reaction, because the technique employs either van der Waals or electrostatic forces for binding guidance molecules to the surface of the MEA (Sadana, 1992; Cheng et al., 1994; Bekos et al., 1995; Taborelli et al., 1995). For this technique to succeed, however, one needs to achieve a clean activated surface for protein or peptide deposition. This can be achieved with sonication for reasonably hydrophilic surfaces (e.g., silicon-oxide and silicon-nitride) or sonication plus gaseous plasma/UV exposure for more hydrophobic surfaces (e.g., polysiloxane and polyimide; Figure 7.2; Chang et al., 2001a,b). Our

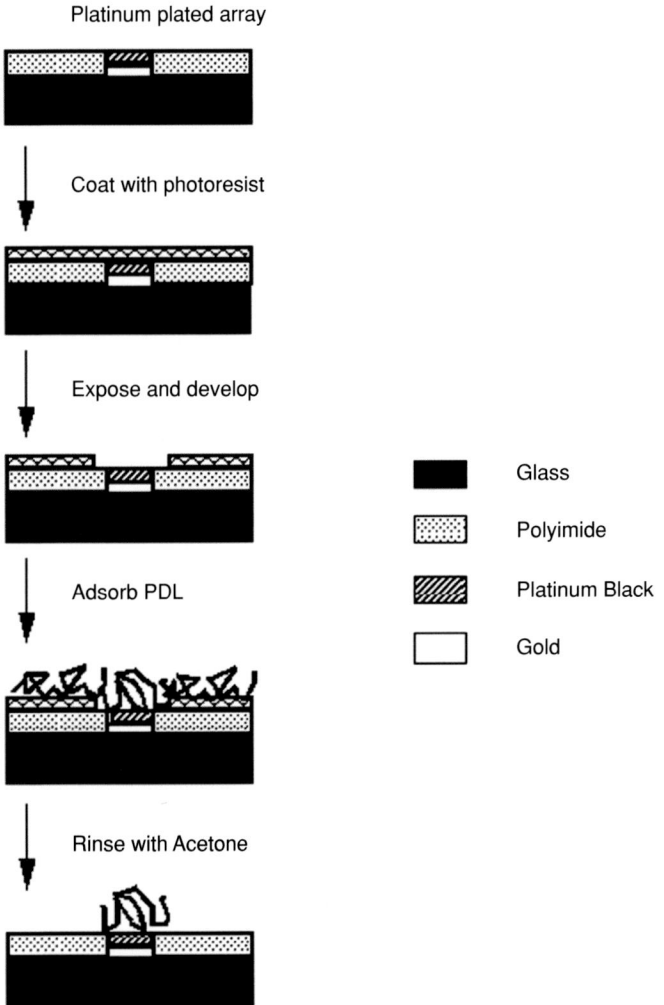

FIGURE 7.2. Lithographic process for patterned adsorption of poly-lysine on MEAs: the process is similar to any lithographic process encountered in a micro-electronic laboratory. The substrate (an MEA in this case) is cleaned via sonication followed by photoresist coating and UV exposure. The developed substrate is then exposed to a protein/macro-molecule solution that deposits the guidance material. Finally, the photoresist is removed via acetone or other solvents that dissolve the photoresist. (Figure reprinted with permission from Chang, et al. 2000; © Kluwer Academic Publishers.)

experience with this technique has been that cells demonstrate robust growth on physisorbed poly-d-lysine surfaces and that patterning can be achieved for short- or long-term depending on the type of background materials used.

Although this technique is simple and generalizable, it does have drawbacks. First, multi-protein deposition in specified patterns is extremely difficult, if not impossible. In order to pattern protein by physisorption, one depends on the successful transfer of pattern to the deposited photoresist using micro-lithography, a process that exposes proteins to denaturants and oxidizing agents, such as acetone, photoresist stripper, and oxygen plasma (Lom et al., 1993; Chang et al., 2000). These harmful agents may destroy the necessary protein conformation that signals the neurons, but the effects may be lessened with dextrose coating (Sorribas et al., 2002). Alternatively, one can ablate the adsorbed protein using a high-energy laser, which may not be easily accessible (Corey et al., 1991). Second, the fidelity of the network structure may depend on the MEA insulation, which may not be the right material for the background. Although it is possible to adsorb a material that inhibits cell growth, long-term patterns have not been achieved with adsorbed background material. In addition, our experience suggests that commonly used silicon nitride does not provide the necessary repulsion that is needed for good, long-term patterns (Chang, 2002), but that oxidized polyimide may provide reasonable background material (Chang et al., 2000). Third, control of the protein conformation is difficult to achieve with physisorption. Although control over the conformation of physisorbed albumin by substrate hydrophicity has been demonstrated (Cheng et al., 1994; Bekos et al., 1995; Taborelli et al., 1995), such a technique may not be applicable to a conformationally sensitive protein which may lose its signaling property with altered structure.

Despite these complexities, we have been quite successful at structuring networks of neurons on MEAs using physisorption and micro-lithography. We have demonstrated long-term patterning of neurons on polyimide, polysiloxane, and silicon nitride MEAs (Figure 7.3A–D; Chang et al., 2000, 2001a,b; Chang, 2002; Chang et al., 2003a). These neurons are electrically active and communicate through synaptic transmissions (Figure 7.3E; Chang et al., 2001a). It is our observation, however, that polyimide in general yields a much better pattern at four weeks than silicon nitride surfaces; we currently do not have sufficient data regarding polysiloxane because these arrays have not survived our culture process on a regular basis. Although silicon nitride can provide good pattern quality at four weeks, this was very sporadic. We speculate that silicon nitride may be less of an ideal surface for patterning because silicon nitride presents an aminated surface which is known to permit cellular growth (Kleinfeld et al., 1988). In order to correct this, one can seek to control the surface groups to regulate culture growth.

7.3.2   Silane Chemistry

Silane modification of substrates is more complicated because of the number (maximum of three) and type of functional groups present on the silicon atom within the silane molecule. The types of functional groups available at the silane terminal are

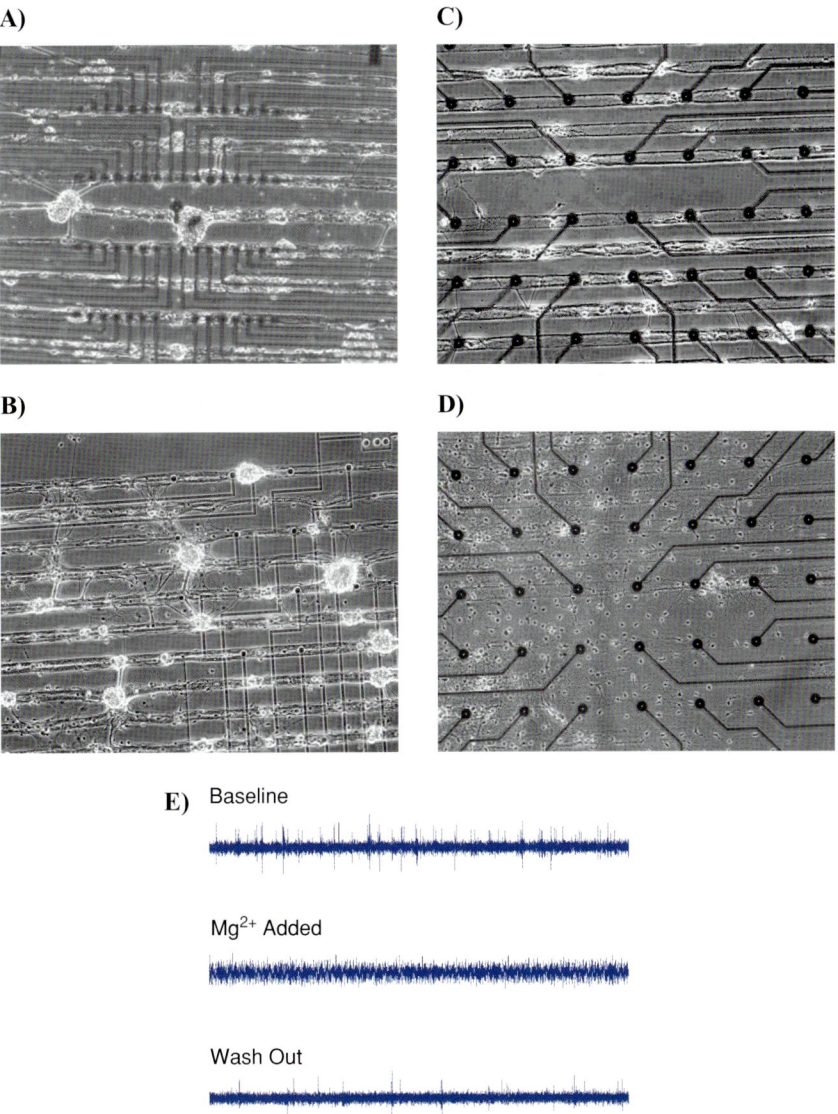

FIGURE 7.3. Neuronal patterning on MEAs with different insulation: all cultures have the same pattern, 40 μm poly-lysine and 60 μm insulation. (A) Eleven days in vitro (DIV) culture on polyimide MEA. (B) Twenty DIV culture on polysiloxane MEA. (C) Thirteen DIV culture on silicon nitride MEA, with good pattern quality. (D) Fourteen DIV culture on silicon nitride MEA, with poor pattern quality. (E) Sample recording of neuronal activity (1.25 sec). Activity can be suppressed with $MgCl_2$, which suggests synaptic communication among the neurons. (A–D are reprinted with permission from Chang, 2002; © 2002 J.C. Chang. E is reprinted with permission from Chang et al. 2001a; © 2001 Elsevier Science.)

methyl (-CH$_3$, inert), methoxy (-OCH$_3$, reactive), ethoxy (-OCH$_2$CH$_3$, reactive), and chloro (-Cl, highly reactive in the presence of water). As methyl groups are inert with regard to the polymerization, increasing the number of methyl groups at the silicon atom decreases the polymerization potential. With chlorosilanes, the high tendency of the chlorine atom to leave the silicon atom makes this a highly reactive terminal, especially when water is present (Figure 7.4A). Thus, chlorosilane

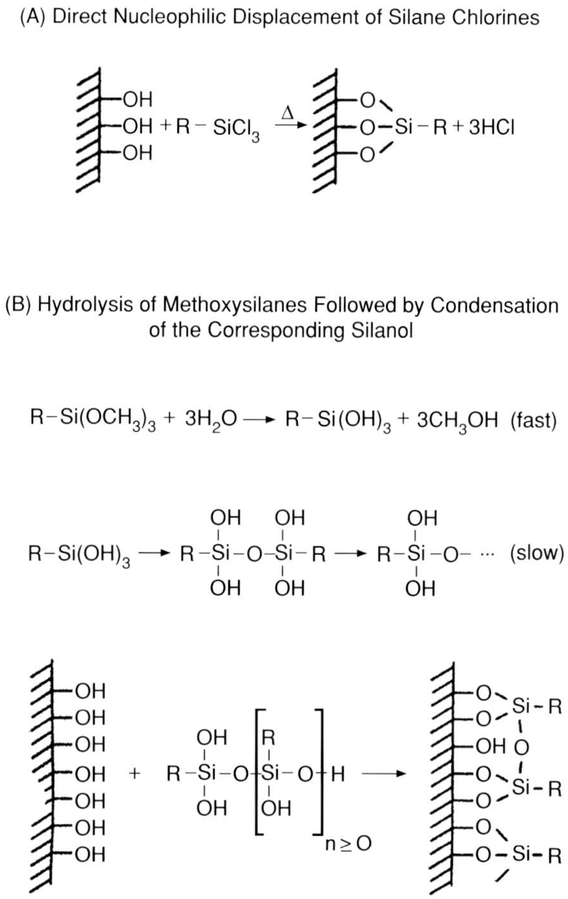

(A) Direct Nucleophilic Displacement of Silane Chlorines

(B) Hydrolysis of Methoxysilanes Followed by Condensation of the Corresponding Silanol

$$R-Si(OCH_3)_3 + 3H_2O \longrightarrow R-Si(OH)_3 + 3CH_3OH \quad (fast)$$

FIGURE 7.4. Silane reaction with hydroxylated surface: (A) chlorosilane reaction with surface-bound hydroxyl groups is a single-step reaction between the chlorine and the hydroxyl groups. Rapid polymerization among the silane molecules can occur with the presence of water. (B) Methoxysilane (and also ethoxysilane) reaction with surface-bound hydroxyl groups requires initial activation with water that converts the methoxysilane to silanols (Si-OH). Silanols then react with surface-bound hydroxyl groups to form a siloxane bond by the condensation reaction (eliminates a water molecule). (Reprinted with permission from Kleinfeld et al., 1988; © 1988 by the Society for Neuroscience.)

generally polymerizes rapidly, sometimes within the solution rather than on the substrate. In contrast to the high reactivity of chlorine, the methoxy and ethoxy groups have a moderate tendency to leave the silicon atom in the presence of water (Figure 7.4B), which makes them ideal candidates for grafting to the hydroxyl presenting substrate. When water is present, the methoxy/ethoxy terminal is converted into a hydroxy terminal that hydrogen bonds with the substrate hydroxy groups and with other silane molecules. When baked, the hydrogen bond is converted into a siloxane bond by the elimination of water molecules. The hydrogen bonds among the silane molecules lead to polymerization, which can be quite extensive. For more details on the silane reaction, one can refer to the book by Pluedemman and the review by Larson (Larson, 1991; Plueddemann, 1991). The readers are also encouraged to contact vendors' technical support to discover more suitable silanes.

In addition to the functional groups present at the silane terminal, the functional groups present at the alkane terminal are equally critical to the success of neuronal patterning. This terminal in combination with the deposited protein or macromolecule determines the type of reaction available for cross-linking to the substrate surface. It typically involves the amine group ($-NH_3$), mercapto group ($-SH$), carboxyl group ($-COOH$), or hydroxyl group ($-OH$), all of which can be linked to the amine group of proteins or macromolecules through a cross-linker (Hermanson, 1996). Although it is possible to radically polymerize this end to proteins using the vinyl group ($-CHCHCH_3$), this type of chemistry is destructive to the conformationally sensitive protein. Thus, the success of patterning depends on the proper selection of the silane, the protein, and the cross-linker. In addition to silane chemistry, the thiol chemistry can also play an important role in structuring neuronal networks in culture.

7.3.3   Thiol Chemistry

Alkanethiols have been shown to interact strongly with metals such as gold, silver, copper, and gallium-arsenide through the formation of the thiolate-metal bond in which sulfur loses its accompanying hydrogen (Dubois and Nuzzo, 1992). This reaction exists naturally when gold is used as the electrode; however, in order to pattern neurons using this technique, one would have to coat the surface of the MEA with a layer of gold, which is the only biocompatible metal named in the list (Nam et al., 2004). Although thiols interact with metals through a single thiolate bond (Figure 7.5), the resulting molecular layer presents a stable uniform layer with minimal defect density for further chemistry and culture (Mrksich et al., 1996; Zhao et al., 1996; Kane et al., 1999). However, the selection of thiols is limited as

$$Au + HS(CH_2)_{10}COOH \xrightarrow{\ -H\ } Au-S^-(CH_2)_{10}COOH$$

FIGURE 7.5. Thiolate formation with gold substrates: thiols react with gold surfaces by the formation of thiolates, which occurs with hydrogen loss by the sulfur atom.

11-mercaptoundecanoic acid is the only commercially available thiol for surface modification. On the other hand, if one's lab has excellent chemical expertise, one can easily synthesize a variety of thiols with different functional groups (Kane et al., 1999).

7.3.4   Inorganic Surface Modification

7.3.4.1   Overview of MEA Materials and Their Reactivity to Silanes

As discussed previously, cells cultured on MEAs can encounter several different inorganic surfaces as either insulation or electrode that may have been modified. Silicon-oxide and silicon-nitride are commonly used as insulation in MEAs and present numerous hydroxyl groups for silane deposition (Corey et al., 1996; Egert et al., 1998; Jung et al., 1998). Platinum, being a noble metal, is generally inert to oxidation, but hydroxyl groups can be created to react with silanes (Moses et al., 1978; Proctor et al., 1985). The details of our experience with these two materials are fully addressed in the following sections, but we confine the discussion on iridium oxide and titanium nitride to the following paragraph because we have limited experience with them.

Although silane linkage to iridium oxide (IrO) and titanium nitride (TiN) has not been demonstrated, silane linkage to these substrates should occur. Reported research has shown that when iridium is oxidized electrochemically to form a thin oxide layer (about 8 molecular layers), the bound oxygen is in the -OH state; however, when the oxide layer becomes thick (about 100 molecular layers), the oxygen is completely bound to the iridium in the form of IrO_2 (Augustynski et al., 1984). Thus, a thin iridium oxide formed from electrochemical activation should be amenable for reaction with silanes (Weiland and Anderson, 2000). However, limiting the activation process to a thin layer may also limit the impedance enhancement gained with a fully activated iridium electrode. This contrasts with TiN electrodes, which permit co-existence of highly reduced electrode impedance and silane modification. Research has shown that sputtered TiN electrode actually consists of titanium nitride, titanium oxynitride (TiN_xO_y), and titanium oxide (TiO_2) at the surface of the electrode (Cyster et al., 2002). When the newly sputtered TiN electrode is exposed to the environment, the surface layer (up to 100 nm) oxidizes to form TiO_2, which can react with silane (Untereker et al., 1977; Nanci et al., 1998). Thus, the columnar structures formed during sputtering improve electrode impedance, whereas oxidation during ambient exposure creates the necessary oxides for silane reaction. However, all this discussion regarding hydroxyl groups on TiN and IrO may be moot as silane deposition may not require the presence of -OH, as demonstrated by silane deposition on gold (Allara et al., 1995).

7.3.4.2   Silicon-Based Insulation Modification by Silane

Although modification of MEAs has considerable similarity to coverslip modification, key differences also exist. The foremost difference is that a MEA is a composite of materials that behave differently with regard to cleaning agents. Although the silicon-based insulation is stable when exposed to acids such as nitric and sulfuric acid, the electrodes generally are very unstable and will require a gentler cleaning procedure. In addition, the presence of multiple materials also requires one to consider the electrode modification process when designing the insulation modification procedure. Next, the MEA surface can be discontinuous which contrasts with the smooth surface of the coverslips. Although this is a minor point, one should still consider whether the discontinuity would disrupt the integrity of the modification layer, such as a Langmuir–Blodgett layer.

The majority of the goals in modifying MEAs are exactly the same as those for modifying coverslips. With MEA modification, one prefers the cross-linking process to be optimal in the sense that the maximal proportion of desired molecule is delivered to the surface. For example, if one desires to link amine-terminated silanes and lysine on proteins, using glutaraldehyde as a cross-linker, one will cross-link many surface sites even before proteins are linked. Thus, it is desirable to cross-link different functional groups on proteins from those present on the silane. Second, it is preferable that the protein be deposited with specific orientation so that the active sites are exposed to the cells to direct cellular actions. This is especially important when one employs biologically active materials such as laminin or neurotrophins. Third, it is important that the total linkage remain intact for as long as possible so that the neurons retain the pattern for the duration of the experiment. Last, the background should be as free of cells as possible.

In our pursuit of these goals, we have tested a few silanes on coverslips regarding their ability to maintain neuronal patterns. The inspiration for our work includes the reports of Kleinfeld et al. (1988) and Lom et al. (1993). In Kleinfeld's work, a two-step deposition process using micro-lithography was employed to deposit alkylsilane and aminosilane separately to guide the growth of neurons (Figure 7.6A; Kleinfeld et al., 1988). This process first deposits the alkylsilanes on the developed area followed by full-coverslip deposition of the aminosilane. Using the same technology, Lom et al. (1993) tested neuronal preference of substrates and found that neurons preferred ECM proteins, for example, laminin, fibronectin, and collagen IV, over aminosilanes and glass. The least preferred surfaces were alkylsilane or albumin covered (Lom et al., 1993). To improve on pattern quality, Corey et al. (1996) reversed the deposition process (i.e., first patterning the aminosilane and then uniform deposition of the alkylsilane, Figure 7.6B) and obtained excellent neuronal compliance to patterns. In addition, it was found that the pattern geometry also influenced the fidelity of the patterns (Corey et al., 1991, 1996). Although silanes alone are capable of achieving patterned networks of neurons, they do not regulate cellular interactions. To achieve this, one can attach signaling molecules to the silanes.

With protein attachment, however, one desires to maximize the surface concentration of the protein on the substrates. Our initial attempt at delivering proteins to aminosilane modified substrate surface succeeded in patterning neurons (Branch et al., 1998, 2000), but the amount was not optimal because glutaraldehyde

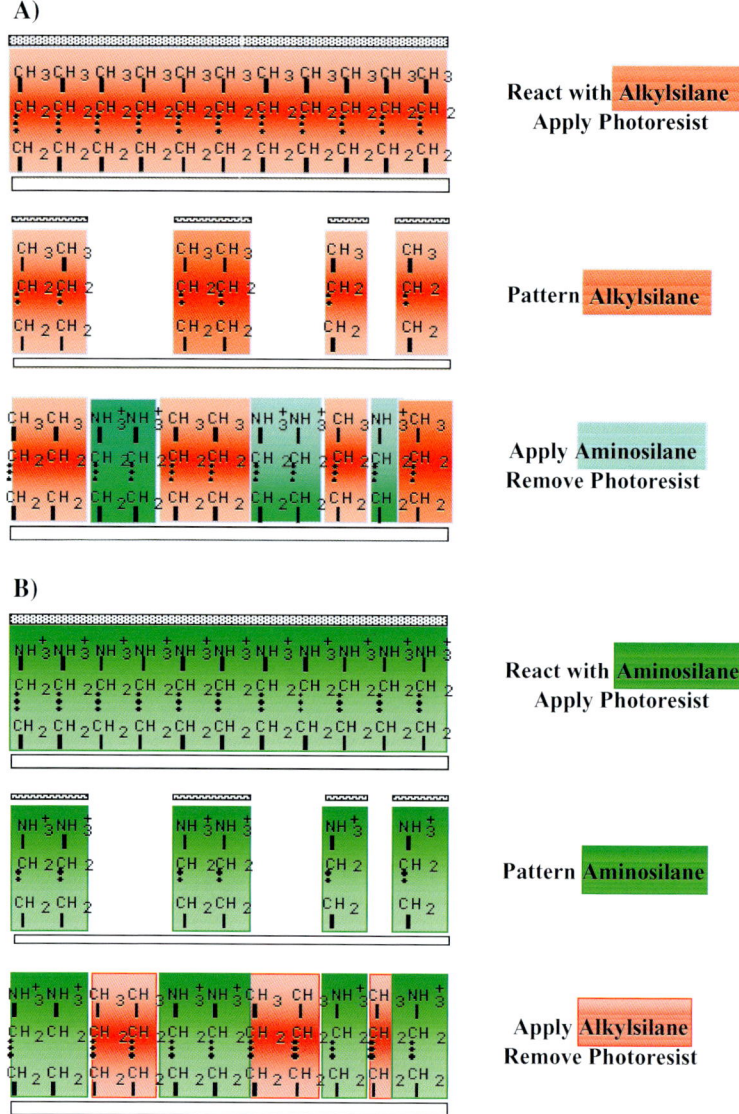

FIGURE 7.6. Silane deposition for neuronal patterning: two types of silane deposition for neuronal patterning have been employed in the literature. (A) The background-first technique was popularized by Kleinfeld et al. (1988) which deposits alkylsilanes first and aminosilanes last. This leads to deposition of aminosilanes throughout the foreground and defects within the background. (B) The selective silane removal technique was popularized by Corey et al. (1996) which deposits aminosilanes first and alkylsilanes last. This technique improves the quality of the patterns because aminosilane is only deposited at the desired location.

cross-linked the amines present on the substrate surface (Figure 7.7A). The linking scheme was improved by using the mercaptosilane to amine linkage that Bhatia et al. (1989) employed to link protein onto glass substrates which markedly improved surface protein density over that of aminosilane to lysine linkage (Figure 7.7B; Bhatia et al., 1989; Branch, 2000).

In addition to optimizing the surface protein density, one also desires longevity in the deposited layers. It is known that silane–glass linkage is sufficiently strong that the film does not break down for nearly four weeks when soaked in 70°C solution (Plueddemann, 1991). This should translate into a silane stability of well over four weeks in 37°C. However, the deposited layer also includes the linkage between the silane and the protein. Our lab had tested the stability of both the silane and the protein layers using patterned neuronal cultures (as an indirect test) and ellipsometry (direct measurement). Branch et al. (2000) found that the neurons retain their patterns, with some breakdown, for well over four weeks. The breakdowns can be attributed to pattern geometry in part (Fromherz and Schaden, 1994; Corey et al., 1996), but also the breakdown of the silane–protein linkage which occurred around the fourth week of soaking (Figure 7.8; Branch et al., 2001). Therefore, the silane–protein linkage should be the target of future improvements as it is currently one of the limiting factors in long-term pattern quality.

With the testing of pattern technology being successful on coverslips, we applied this technology to culturing neurons on MEAs (Figure 7.9; Chang, 2002). In doing so, we had to make some adjustments due to the sensitivity of the electrodes to strongly oxidizing agents. Instead of cleaning the surfaces with pirhana etch or sulfuric acid which can strip the contact pads and the electrodes, we cleaned the arrays by sonicating in acetone and etching in oxygen plasma for ten minutes. Silane deposition on silicon nitride MEAs (purchased from Multi Channel Systems) yielded a silane layer similar to that on coverslips when judged by the static water contact angle. Patterned deposition of poly-d-lysine and polyethyleneglycol resulted in high-fidelity patterns at four weeks; however, pattern breakdown through cell detachment does occur. We speculate that in addition to the silane–protein interface breakdown the adsorbed hydrocarbons on the MEA surface may detach leading to corruption of the neuronal patterns; perhaps a longer oxygen-plasma cleaning will correct this breakdown.

In our work, we have not addressed the protein orientation problem because we consistently used poly-d-lysine as the growth-permitting molecule. However, as we continue to pursue a designed network of neurons, we will inevitably use protein molecules for guidance. Work on inducing axonal growth with laminin has begun in our lab (Corey, 1997), and preliminary work suggests that laminin retains some conformation when linked to the coverslip (Esch et al., 1999). Whether these conformations are sufficient for guidance is a matter that is currently being tested in our lab. If it is the case that laminin is deposited in the wrong orientation, one can alter the amino acid sequences of laminin to aid the proper deposition, a technique that has been applied to orient cytochrome b (Firestone et al., 1996).

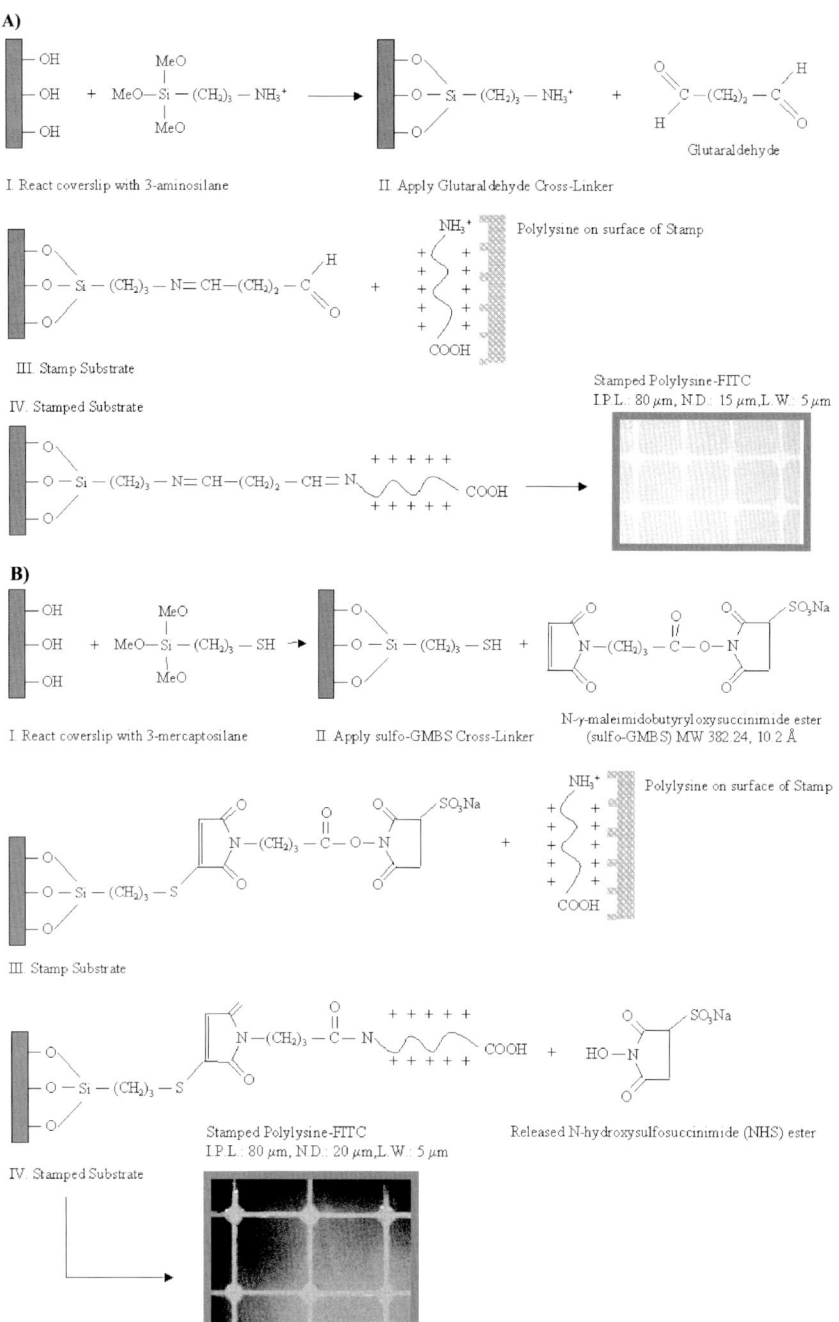

FIGURE 7.7. Reactions of two types of cross-linkers for protein linking to surface bound silanes: (A) aminosilane-coated substrates react with glutaraldehyde to link proteins, but some amounts of the surface amines also cross-link to each other. (B) Mercaptosilane-coated substrate reacts with N-γ-maleimidobutyryloxysulfosuccinimide ester (sulfo-GMBS) to link proteins. Because the functional groups being linked are different, no surface cross-linking is expected.

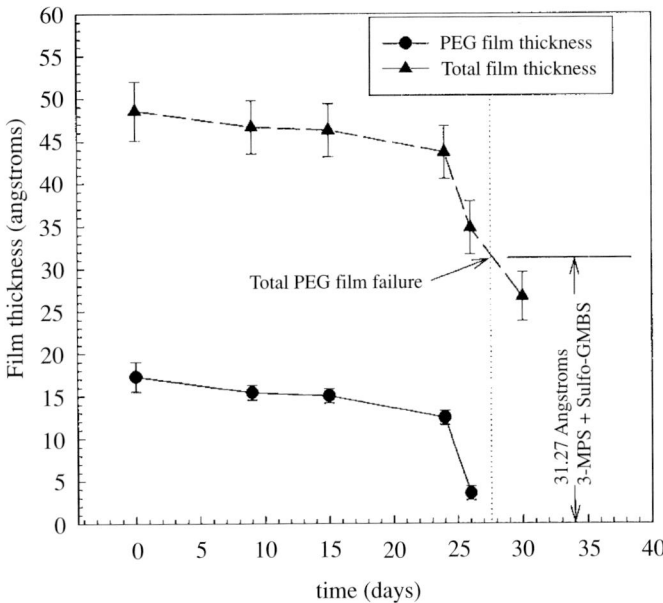

FIGURE 7.8. Degradation of silane–protein linkage in culture condition: the layers tested are the silane and the protein/polyethylene glycol (PEG) layers under culture condition (37°C, PBS, 10% oxygen, and 5% CO_2). The interface between silane and PEG breaks down after three weeks in culture. (Reprinted with permission from Branch et al., 2001; © 2001 Elsevier Science.)

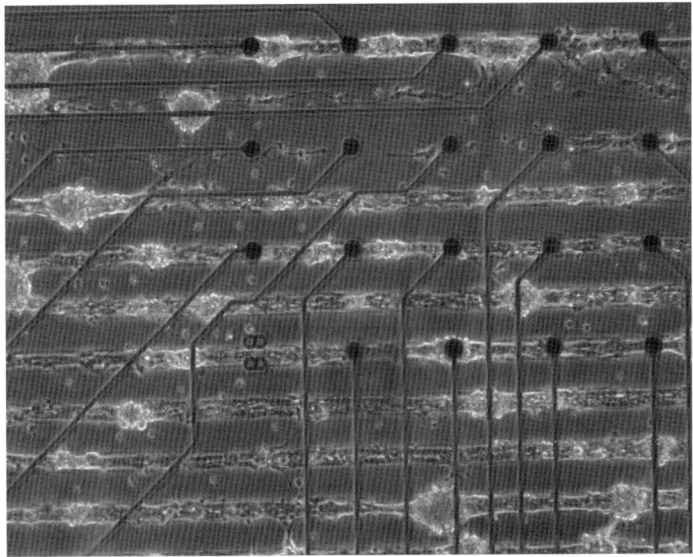

FIGURE 7.9. Patterned culture using silane-coated MEA: the culture is 25 DIV on silicon nitride MEA. The culture remains well patterned in general at three weeks age, but areas of the pattern do suffer breakdown after three weeks.

7.3.4.3   Platinum Modification by Silane

In addition to silicon oxide and silicon nitride, platinum can also be encountered by the neurons on MEAs. Typically, platinum is deposited as platinum black rather than crystalline platinum in order to lower the electrode impedance. For our work, we studied our silane deposition technique on platinum foils to expedite the experiments and applied the technique to platinum black electrodes. The success of the deposition on electrodes was tested by electrode impedance. Successful deposition would lead to an increase in electrode impedance, whereas failed deposition would not cause any change in impedance.

Although platinum is considered a noble metal with resistance to oxidation by oxygen atoms, research has shown that oxygen adsorption results in the formation of hydroxyl groups that are capable of reacting with silanes (Moses et al., 1978). Several techniques have been reported to deposit oxygen onto platinum surfaces. The first and perhaps the simplest is the flame technique in which platinum is heated to incandescence to cause the migration of silicon contaminants to the surface of the platinum (Proctor et al., 1985). This leads to potential silicon oxides on the platinum surface that can react with silanes. Although flaming has been used to activate inert polymer surfaces (Gross, 1979; Gross et al., 1997), it is infeasible to actually heat an MEA to incandescence. Alternatively, electrochemical oxidation of platinum by anodization also places reactive oxygen onto platinum surfaces that can react with silanes (Lenhard and Murray, 1977; Facci and Murray, 1980; Proctor et al., 1985). The final and perhaps the most feasible technique of oxidizing platinum is by oxygen plasma, a frequently used technique for surface cleaning in micro-fabrication. Platinum electrodes have been modified using oxygen plasma to enhance chemical interaction and to produce oxidized species (Chang, 2002; Ward et al., 2002). The species produced are generally PtO which have been shown to react with silanes (Allen et al., 1974; Hammond and Winograd, 1977; Facci and Murray, 1980; Chang, 2002). These data show that oxygen plasma cleaning of MEA results not only in a clean silicon surface for silanization, but also in activated platinum surface for silanization.

In silanizing the platinum surface, we applied our glass silanization protocol and found that silanes can be stably deposited for up to four weeks and resist desorption by electrical stimulation (Chang, 2002). Our data showed that the water contact angle after platinum silanization is similar to that from glass silanization (Chang, 2002). However, this angle was much lower if the platinum had been pretreated with overnight sulfuric acid to remove adsorbed hydrocarbons. This may be attributed to the reaction between the sulfhydryl group of the silane and the unoxidized platinum surfaces. Despite this low contact angle, XPS studies show that silanes are chemically linked regardless of the acid clean status. The silanes remain stably attached at four weeks when immersed in 1X PBS (phosphate buffered saline) at 37°C and exposed to the culture environment (10% oxygen and 5% carbon dioxide). However, there was initial desorption when the substrates were first exposed to the PBS solution. In addition, the silane resisted desorption by applied bipolar pulses lasting for one minute and up to 1.5 V. Thus, we expect the silanes to remain relatively stable on the electrodes for at least four weeks.

Although the silane may attach to platinum substrates, it is not guaranteed that their presence will make platinum electrodes functional. We compared the impedance of the silane modified and sham-treated platinum electrodes. The results showed an eighteenfold increase in electrode impedance (raising electrode impedance from <100 kΩ to >1.5 MΩ) for silane-modified electrodes versus a twofold increase in sham-treated electrodes. We speculate that silane polymerization may contribute to this increase and can be decreased by decreasing the number of hydroxyl groups at the silicon atom. Alternatively, one may consider a waterless reaction that usually deposits a monolayer of silanes (Stayton et al., 1992; Firestone et al., 1996; Nelson et al., 2001). Our normal reaction process includes the addition of water to catalyze the conversion of methoxysilanes to silanols, a step that enhances the polymerization. This process can be contrasted with alternative protocols in which the reaction takes place in a waterless environment where the silanes deposited form a relatively thin layer (Stayton et al., 1992; Firestone et al., 1996; Nelson et al., 2001). The disadvantage of the waterless process is the time needed to complete the reaction, usually 12 to 48 hours.

7.3.4.4   Silane Deposition on Titanium Nitride Electrodes

In our work, we applied the silanization process to MEAs with TiN electrodes and grew neurons in patterns for recording purposes (Figure 7.9; Chang, 2002; Chang et al., 2003a). Although we did not directly measure the presence of silanes with XPS, our results were consistent with the findings from platinum electrodes. First, the measured electrode impedance increased to that of the shunt impedance, suggesting that the electrode impedance after modification was much greater than the shunt impedance. Second, the number of active electrodes recorded from patterned cultures on silanized MEA decreased dramatically. Third, even previously silanized MEAs tended to lag unsilanized MEAs in the percentage of electrodes being active. Thus, our data indirectly suggest that TiN can be modified with silane and that silane deposition on TiN electrodes corrupts the impedance.

7.3.4.5   Thiol Deposition on Gold

Although thiols cannot modify glass, they are the most commonly used modifier for noble metal surfaces. This technique can combine the modification of the MEA surface and the electrode into one convenient step if the electrodes were of gold. Nam et al. (2004) demonstrated the feasibility of this process by coating the MEA surface with a thin layer of gold that reacts with thiols (Figure 7.10A,B). It was shown that the 11-mercaptoundecanoic acid modified gold can support neuronal growth in patterns when the poly-d-lysine is cross-linked. This technique did not increase the impedance as much as the silane technique (because recording

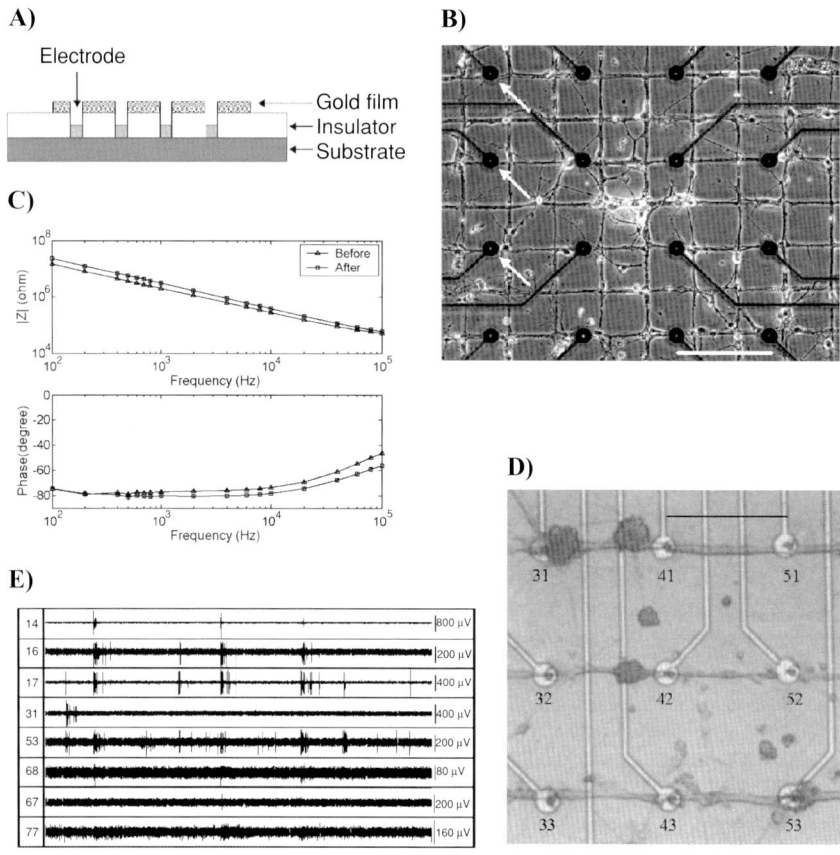

FIGURE 7.10. Patterned culture on thiol-coated MEA: (A) a cross-section view of gold-coated MEA to allow thiol deposition. (B) Twelve DIV culture patterned on thiol-coated MEA. White arrow shows neurites passing by the electrodes. (C) Impedance spectroscopy of electrodes before and after gold coating. The impedance stayed between two to three megohms both before and after the coating. (D) Seventeen DIV culture patterned on thiol-coated MEAs. The electrodes are numbered for reference to the recordings shown in (E). (E) Recordings of neuronal activity from the culture shown in (D). (Reprinted with permission from Nam et al., 2004; © 2004 IEEE.)

is possible, see Section 7.4), most likely due to the presence of defects within the film and to the lack of polymerization between the thiol molecules (Dubois and Nuzzo, 1992). Despite the high impedance of the gold electrodes both before and after thiol modification (Figure 7.10C), the impedance was sufficiently low compared to the shunt impedance that recording was possible, unlike the silane-modified electrodes (Nam et al., 2004). Thus, the combination of gold electrodes and gold-coated MEA surface permits patterned growth and recording of neurons.

7.3.5   Organic Surface Modification

7.3.5.1   Overview of Polymeric MEA Insulation and Its Modification Potential

In addition to silicon-based and platinum substances, the MEAs may incorporate polymeric substances as insulation. Two commonly used polymeric insulators are polyimide and polysiloxane. Polyimide has been used as insulation for nearly 20 years and gives excellent shunt impedance values (Novak and Wheeler, 1986; Boppart et al., 1992; Chang et al., 2000). The procedures for obtaining a stable and biocompatible insulation have also been described in the literature for use in serum-free neuronal cultures (Chang et al., 2000; Nam et al., 2004). In addition to these advantages, polyimide modification procedures by oxygen plasma oxidation and base hydrolysis are well characterized in the literature (Chou et al., 1987; Dunn et al., 1989; Lee and Viehbeck, 1994). As an alternative, one may choose polysiloxane as insulation, a highly hydrophobic material that resists water penetration, because of the ease of modification by oxygen plasma oxidation (Ferguson et al., 1993; Chaudhury, 1995). However, our experience with this insulation in terms of stability has not been as good as polyimide. This insulation tends to detach from the MEA after several uses or during soaks to ensure biocompatibility for cell culture. In contrast, Gross has successfully recorded from polysiloxane coated arrays for nearly a year (Gross, 1994). This may involve a suboptimal interfacial condition between the MEA surface and the polysiloxane that could be improved upon with further experiments.

7.3.5.2   Polyimide Modification

As previously stated, polyimide can be modified by either plasma oxidation or by base hydrolysis. Plasma oxidation relies on the exposure of polyimide to gaseous plasma, usually oxygen when hydroxyls are desired. This process produces several species of oxygen on the polymer surface, such as -OH, -COOH, and -OO (Figure 7.11). Of these, the useful specie is the hydroxyl group when one desires to modify the polyimide with silane (Chou et al., 1987; Vargo et al., 1992). However, if one wishes for direct linkage of protein to the modified polymer, one could choose to use the carboxyl group which reacts with 1-ethyl-3-(3-dimethylaminopropyl)carbodiimide hydrochloride (EDC) to link protein onto the polymer surface (Hermanson, 1996). Of these two options, we experimented with linking oxidized polyimide with silane molecules.

In our experience, oxidized polyimide can be modified with silane just as clean coverslips can. The deposited silane layer yields a static water contact angle similar to that deposited on glass coverslip (Table 7.1). After a soaking period of two weeks in PBS under culture environment, the water contact angle decreased slightly (from 56.5 to 50.7), suggesting perhaps film degradation or polymer reorientation. We believe the latter to have stronger influence on static water contact angle because the sulfur-to-nitrogen ratio, which measures the amount of mercaptosilane present on polyimide, was not significantly different between freshly silanized and soaked silanized samples. The silanized polyimide substrates were also used to culture

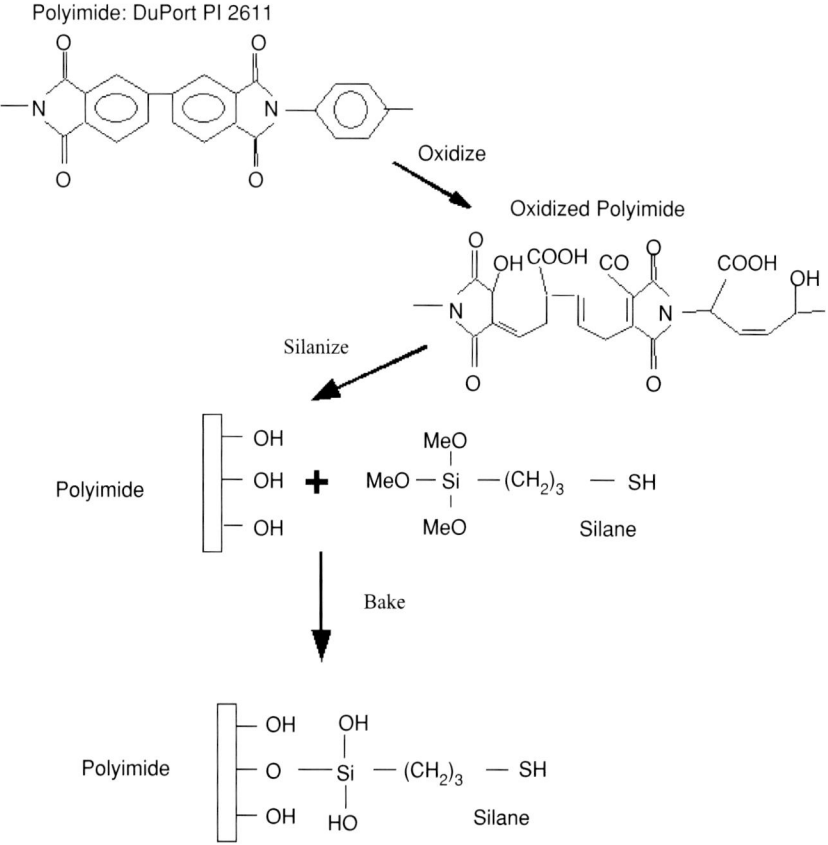

FIGURE 7.11. Polyimide oxidation with gaseous oxygen plasma: oxygen plasma treatment of polyimide produces -COOH, -OH, and -COO·, of which the -OH groups can react with silanes. In addition, the -COOH can react with EDC to cross-link lysines on proteins. (Reprinted with permission from Chang, 2002; © 2002 J.C. Chang.)

patterned neurons, which remained in patterns after two weeks. This was an improvement in pattern quality over unsilanized polyimide substrate that has been stamped with poly-lysine patterns (Figure 7.12A,B).

In addition to oxidizing polyimide with oxygen plasma, one can also generate functional groups with base hydrolysis. Lee and Viehbeck (1994) have shown

TABLE 7.1. Water contact angle of silanized polyimide substrates.

DIV	Contact angle (Mean $\pm$ SEM)
0	56.5 $\pm$ 2.66
7	48.42 $\pm$ 1.12
14	50.67 $\pm$ 0.94

A)   **B)**

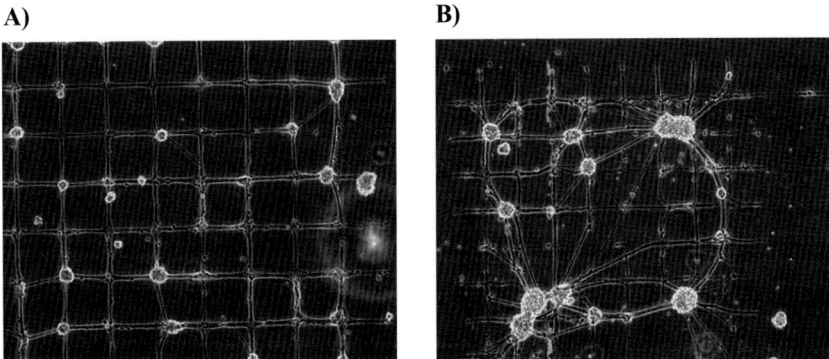

FIGURE 7.12. Comparison of pattern quality on polyimide substrates at eight DIV: (A) silane-linked polyimide yielded high quality patterns at eight DIV. (B) Oxidized polyimide yielded poor quality patterns at eight DIV. ((A) is reprinted with permission from Chang et al., 2003a; © 2003 Elsevier Science Ltd. (B) is reprinted with permission from Chang, 2002; © 2002 J.C. Chang.)

that polyimide exposed to potassium-hydroxide (KOH) produces –COOK and a secondary amine (-NH-). The potassium carboxylate can be exchanged to carboxyl with exposure to an acid solution (Figure 7.13), making the functional group reactive to the previously stated EDC molecule. This is indirectly supported by the water contact angle which becomes very low after exposure to KOH. However, because we desired a simple modification scheme for both the electrode and the insulation, we chose not to pursue this route of modification. Thus, this technique can work in theory, but has not been demonstrated to work in the literature.

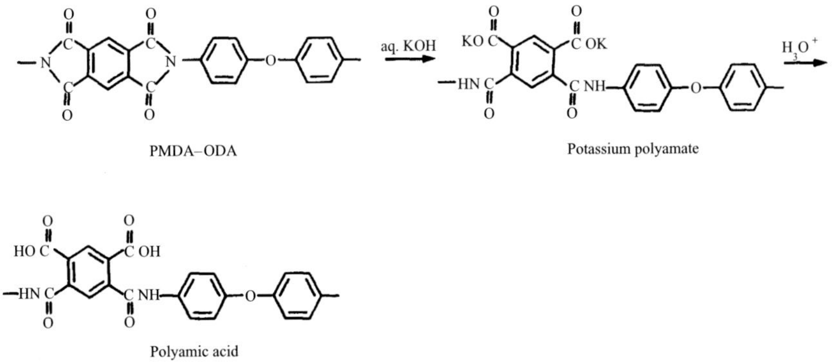

PMDA–ODA   Potassium polyamate

Polyamic acid

FIGURE 7.13. Chemical modification of polyimide: reaction of polyimide with a base (KOH) yields carboxylate (-COOK) groups that can be converted to carboxyl groups (-COOH) through exchange with an acidic solution. The carboxyl groups can be linked to lysines by using EDC. (Reprinted with permission from Lee and Viehbeck, 1994; © 1994 International Business Machine Corp.)

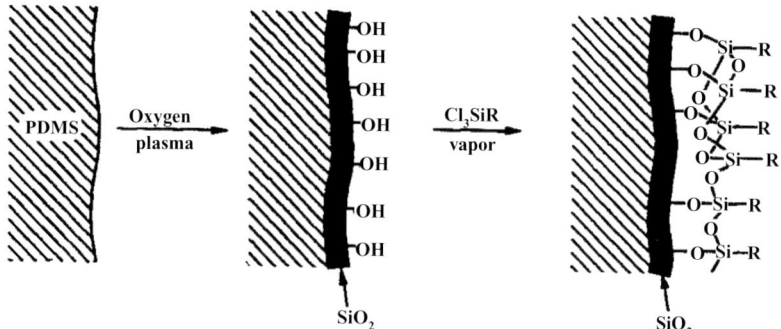

FIGURE 7.14. Oxygen plasma oxidation of poly-dimethyl-siloxane (PDMS): oxygen plasma oxidation of PDMS yields a surface with many hydroxyl groups that can react with silanes, chlorosilanes in this case. (Reprinted with permission from Ferguson et al., 1993; © 1993 American Chemical Society.)

7.3.5.3   Polysiloxane Modification

Another commonly used polymer for MEA insulation is polysiloxane, which is easily modified by oxygen plasma (Gross et al., 1997). The exposure leads to the production of hydroxyl surface groups that permit silane chemisorption (Figure 7.14; Ferguson et al., 1993; Chaudhury, 1995). Published results show that silane films deposited on oxidized polysiloxane are similar to those on silicon oxide surfaces (Ferguson et al., 1993). Although this technique is feasible, we do not use siloxane-insulated MEAs frequently and have not pursued this technique to pattern neurons on MEAs.

7.4   Protein Patterning Techniques

With the surfaces modified, the next step in the patterning of neurons on MEAs is the protein patterning techniques that guide the growth of neurons. These techniques can be classified into two categories, namely, lithography-based and contact-printing-based categories. As the names suggest, lithography-based techniques deliver protein patterns by using micro-lithography, whereas contact-printing-based techniques rely on a molded polymeric structure to transfer protein patterns. Lithographic techniques are simple, especially to the many engineers involved in MEA work. However, this technique requires that the protein be deposited by physisorption and is not compatible with the deposition of multiple proteins. On the other hand, contact-printing uses molded PDMS either as a stamp or as a flow channel to guide the deposition of multiple guidance molecules (Chiu et al., 2000; Jo et al., 2000; Thiebaud et al., 2002). This is a critical issue because patterns with multiple proteins will certainly be used to design directed neuronal networks (Hickman and Stenger, 1994; Wheeler and Brewer, 1994; Stenger et al., 1998; Vogt et al., 2004).

7.4.1 Micro-Lithography

Micro-lithographically based protein patterning has been applied to patterning neurons since the early 1990s and employs micro-lithography to guide protein delivery. The general process is demonstrated in Figure 7.2. The initial step in this class of patterning technique involves the deposition of photoresist which is patterned by exposure to UV light. Depending on which type of photoresist is used (positive or negative), the exposed areas can be foreground or background. After development, the exposed or unexposed areas can adsorb the guidance molecules that remain on the surface during the step to remove the photoresist, assuming a sufficiently gentle process is used to preserve the important protein conformations.

In our experience, this process is the easiest method to achieve at least minimal patterning with excellent survival of the network of neurons because this process does not require complex structures or complex chemistry. The quality of the patterning depends strongly on the quality of the background materials. With polyimide-insulated MEAs, we obtained excellent patterning up to four weeks (Figure 7.3A) for a simple pattern of alternating lines with wide background (60 μm). However, low-fidelity patterning was the norm (18 out of 23 cultures) for silicon-nitride MEAs (Figure 7.3C,D), most likely as a result of the nitrogen within the silicon nitride. Thus, when using micro-lithography plus physisorption, one must consider the type of MEA because the insulation material becomes a critical factor in the quality of the patterns.

Although the background material is crucial, it is possible to present a background different from the insulator; two solutions come to mind. The first is the flood adsorption of background materials, similar to the flood silanization technique that Corey et al. (1996) presented to improve silane patterning. In this method, a background material can be adsorbed to the whole array, which as Corey demonstrated should minimally alter the quality of the foreground pattern. It is possible for one to adsorb high molecular weight PEG or albumin to enhance cell repulsivity of the background (Hoffman, 1996). Alternatively, one could also deposit a hydrophobic silane layer to the nonelectrode areas first followed by micro-lithographic patterning of the foreground. This ought to provide the necessary background while preserving the electrode impedance.

7.4.2 Contact-Printing Techniques

Contact-printing techniques involve either stamps or fluid channels made from poly-dimethyl-siloxane (PDMS), an elastomer that conforms to uneven surfaces, by pouring the monomers into a master mold produced from either silicon or polymeric substrates (Figure 7.15A,B; Branch et al., 1998; Xia and Whitesides, 1998; Corey and Feldman, 2003). When one produces the master from polymeric substrate, reactive-ion etching (RIE) is used to create the grooves within the pattern, producing a valley with right angles. In contrast, silicon masters produced from anisotropic etching have V-shaped valleys. In the direct stamping technique, protein is adsorbed onto the stamp and transferred to the MEA through contact with the

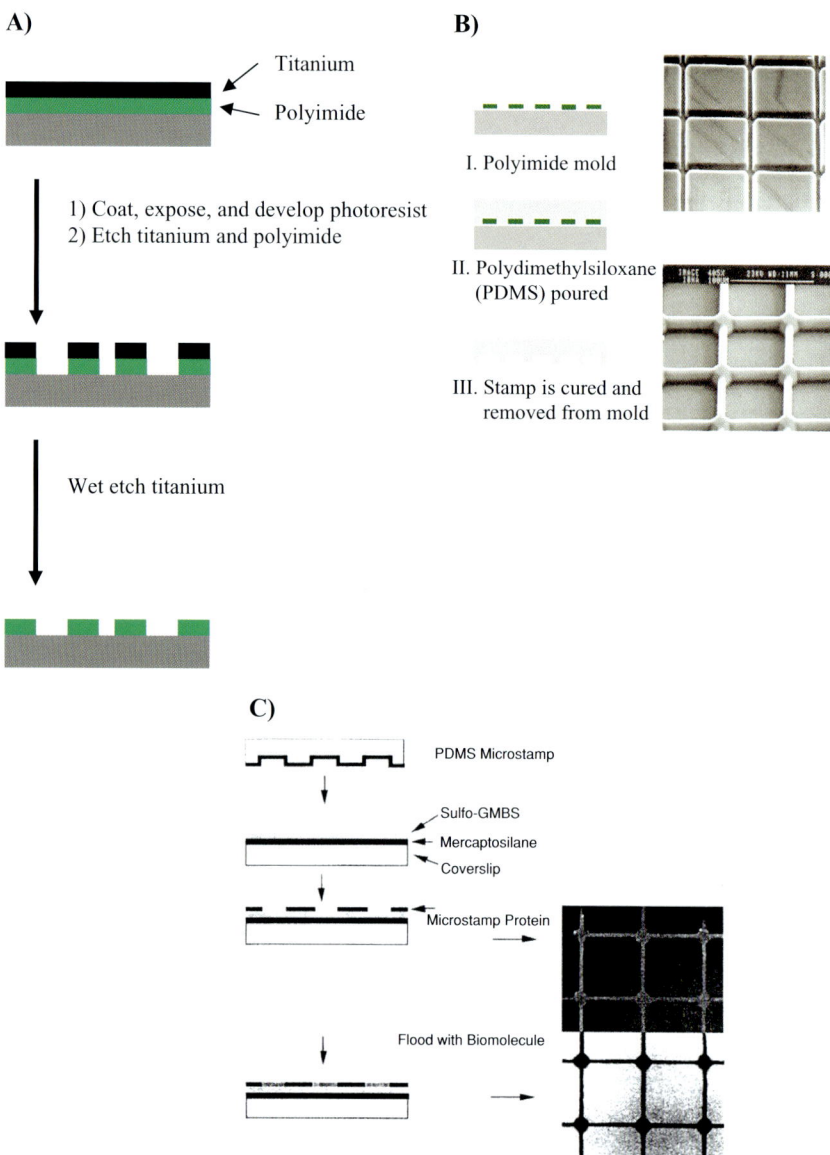

FIGURE 7.15. Fabrication of PDMS micro-stamps: (A) master mold is fabricated on polyimide-coated glass substrate by anisotropic etching of polyimide to create deep relief for stamp fabrication. (B) PDMS stamp is fabricated from the mold by pouring PDMS resin into the mold and curing. (C) Micro-stamping process in which the PDMS stamp is inked with protein and stamped onto activated substrates. The background is flooded with a cell-repulsive material such as PEG. ((C) is reprinted with permission from Branch et al., 2000; © 2000 IEEE.)

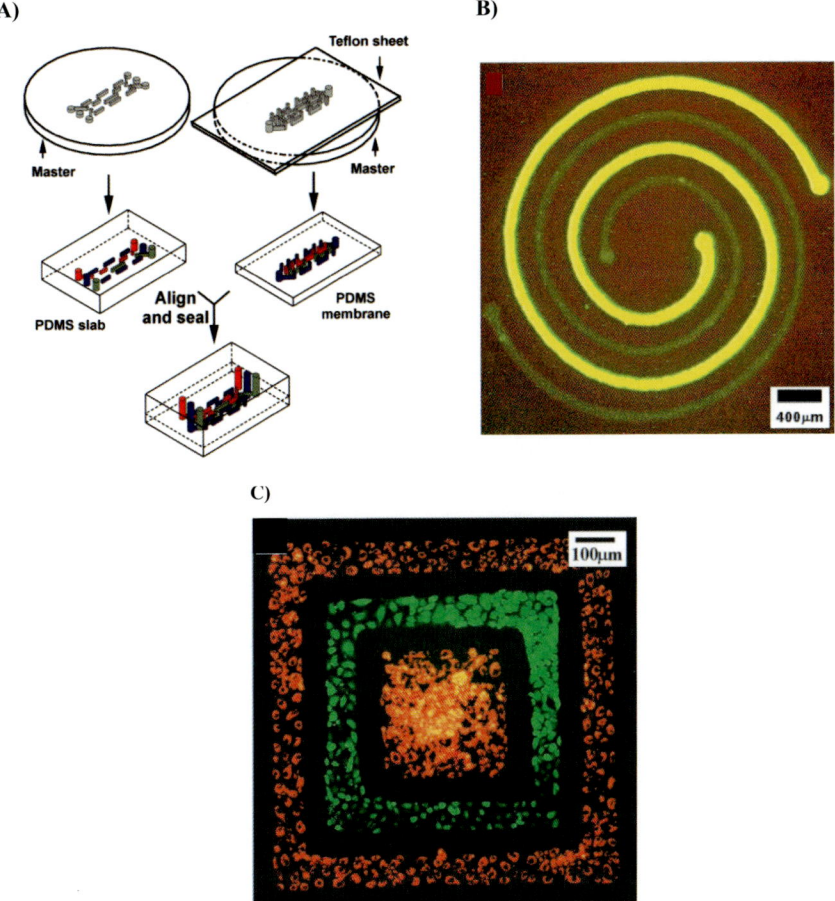

FIGURE 7.16. Micro-fluidic patterning of proteins and cells: (A) the design and assembly of the micro-fluidic channels from two-dimensional sheets of PDMS. The scheme demonstrates a two-layer design. (B) Simultaneous deposition of two proteins both of which are labeled with fluorescein isothiocyanate (FITC). Bovine serum albumin (BSA) is shown in bright green whereas fibrinogen is shown in light green. (C) Simultaneous deposition of two types of cells. Human bladder cancer cells (ECV304 cell line) are shown in green whereas bovine adrenal capillary endothelial cells are shown in red. (Reprinted with permission from Chiu et al, 2000; © 2000 National Academy of Science, U.S.A.)

PDMS stamp (Figure 7.15C). In the fluid-channel technique, micro-channels are designed into the mold to allow in- and out-flow of the protein solution (Figure 7.16A). When the mold contacts the substrates, protein solution is driven into the channels to allow protein adsorption or linking and then washed out with saline. In both techniques, multiple proteins have been delivered to the substrates to guide the growth of cells, though not necessarily neurons (Branch et al., 1998; Wheeler et al., 1999; Branch, 2000; Chiu et al., 2000).

In our experience with patterning neurons using the micro-stamping technique, several factors influence the reliability of the process and the quality of the protein patterns. The first factor is the importance of a stiff backing for the stamp. As PDMS stamps are quite flexible, the backing provides the stiffness to stabilize the patterns during the transfer process (Corey, 1997; Branch, 2000). The second factor is the aspect-ratio between the height and the width of the patterns. Our experience has been that high aspect-ratios are important in obtaining faithful patterns after transfer (Branch et al., 1998). However, others using stamps produced from silicon masters have shown that shallow ridges can work as well (Bernard et al., 1998). The difference may be the different forces transferred onto the patterns due to the shape of the contact areas. The third is the release layer that improves the detachment of the protein from the stamp. In our experience, more proteins are delivered onto the substrate when a release layer (sodium laurylsulfate, in our lab) is used than when no release layer is used (Chang et al., 2003a). This release layer also allows a stamp to be used for several months, versus needing a new stamp after several uses due to stamp exhaustion (Bernard et al., 2001).

We have used this technique to pattern hippocampal neurons on MEAs and cultured the patterned networks for long term. When the MEAs are silanized or thiolated, we can cross-link the protein and a background material (amine-terminated polyethylene glycol) onto the MEA surface (Branch et al., 2000). The patterns on these arrays can be maintained for up to four weeks, but as stated previously, some patterns may detach from the MEA (Figure 7.9; Chang, 2002; Chang et al., 2003a; Nam et al., 2004). Although we apply a background material, this may not be necessary depending on the pattern geometry because the silane and the thiol both present cell-repulsive surfaces (silane is hydrophobic whereas thiol is negatively charged). However, recordings on silanized arrays have been rare due to dramatically increased electrode impedance, but we have recorded electrical activities from patterned networks on thiolated MEAs (Figure 7.10D,E; Chang, 2002; Nam et al., 2004). We have not done multiple protein patterning on MEAs, but this is not only possible but also an important step in preparing for future experiments.

Although direct stamping is simple, the process can potentially alter the conformation of proteins and is limited to delivering macro-molecules. These deficits can be corrected with the micro-channel technique which delivers either protein or cell solutions directly to the substrate (Chiu et al., 2000; Thiebaud et al., 2002). The tradeoff is that the design complexity of the initial master mold increases dramatically as one must design the in- and out-flow channels and solve the assembly problem in order to create a three-dimensional structure with embedded fluid channels. In addition, the channel surfaces need to be altered so that the fluid flows when pressure is applied, as hydrophobic channels require high pressure to drive aqueous fluid. However, the advantage is that channels can easily deliver multiple proteins to the substrate with one contact versus multiple contacts involved in the direct stamping technique.

Using channel-embedded PDMS structures, several groups have demonstrated the ability to pattern proteins onto substrates to guide cellular growth (Anderson et al., 2000; Chiu et al., 2000; Jo et al., 2000; McDonald et al., 2000). The three-dimensional structures are formed from two-dimensional sheets of PDMS carrying various parts of the final assembly which are assembled after activation by gaseous plasma that makes PDMS sheets extremely reactive towards each other (Anderson et al., 2000; Jo et al., 2000; McDonald et al., 2000). To pattern the substrates, the final assembly is aligned relative to a fiduciary and pressed against the substrate, followed by solution delivery (Chiu et al., 2000; Jo et al., 2000; Thiebaud et al., 2002). The channels guide the delivery of either the cells or protein solution to the designated spots on the substrates (Chiu et al., 2000). Bovine adrenal capillary endothelial and human bladder cancer cells have been patterned and cultured with these channels for approximately one day, and bovine serum albumin and fibrinogen patterning have been demonstrated (Figure 7.16 B,C; Chiu et al., 2000). Similarly, neurons have also been patterned by this technique (Thiebaud et al., 2002). These data strongly suggest that MEA patterning with micro-fluidics should be feasible and should simplify the delivery of multiple protein patterns onto MEAs.

7.5   Effects of Structuring Network: Early Results

These patterning techniques have been applied to the structure networks of neurons on both substrates and MEAs (Ma et al., 1998; Ravenscroft et al., 1998; Chang et al., 2000; James et al., 2000; Liu et al., 2000; Chang et al., 2001a,b; Chang, 2002; Chang et al., 2003a,b). These studies have demonstrated synaptic communication among interconnected cells on both substrates (Ma et al., 1998; Ravenscroft et al., 1998; Liu et al., 2000) and on MEA (Chang et al., 2001). In addition, the patterned networks have been shown to develop robust electrical activity when compared to random networks (Chang et al., 2001a,b, 2003b). When tested on substrates, it was shown that the patterned networks achieved greater glial and synaptic density for the first three weeks of the culture (Chang et al., 2003b; Chang, 2002). However, our results suggest that a random network may catch up in terms of glial and synaptic density after three weeks, which will have to be tested with additional experiments. Thus, the early data suggest that giving neuronal networks a defined structure may influence the developmental process of the network (i.e., neuron and glia interaction) in addition to regulating the interconnectivity between neurons. This finding may have an impact on our understanding mental retardation and genetic diseases that influence the brain as stunted neuron–glia interaction may slow nervous system development and alter information processing.

7.6   Conclusion

Current patterning technology has not only demonstrated feasibility with patterned neuronal culture on MEAs, but also the possibility of altering the developmental process of the network on the culture. These abilities will help us discover principles of network dynamics and information processing in regard to the functioning of neuronal networks. In addition, the structured development may provide an alternative paradigm for studying neuron–glia interaction in nervous system development. However, the current technology will need improvement in order to reach these important goals. First, stronger linkage between the guidance molecule and the substrate is necessary to prevent degradation of the designed structure. Second, rules of connectivity specification need to be developed, a task that is currently under intense research by several groups but for which practical findings remain elusive.

As we achieve better understanding of internetwork information processing by the central nervous system, we may be able to replace lost neural function regardless of the location of the lesions. As stated in the introduction, robotic arms can be controlled by the firing patterns of prefrontal neurons, implying that we will soon replace lost motor function. However, the challenge of replacing lost dopaminergic neurons (Parkinson's) will be much more formidable as simple cell replacements have induced severe side effects (Freed et al., 2001). Perhaps the solution is to regulate the connectivity of the replacement neurons because the original dopaminergic neurons had very specific connections. Achieving this will make structured network replacement a standard therapy in many central nervous system illnesses as well!

Acknowledgement. The authors would like to thank the National Science Foundation and the National Institute of Health for their funding of this project (NSF EIA 013828, NIH R01 EB000786, NIMH R55 RR13220, and NIMH R21 NS38617). JC thanks NIMH for a MD/PhD fellowship (F30 MH12897) and Beckman Foundation for a predoctoral fellowship. The authors would also like to thank Gregory Brewer for his insightful discussions regarding the projects and John Torricelli for his assistance with cell cultures. The authors would also like to thank Joe Corey, Darren Branch, Kumar Venkateswar, Yoonkey Nam, and Rudi Scharnweber for their contribution to the research projects. Any opinions, findings, and conclusions or recommendations expressed in this publication are those of the author(s) and do not necessarily reflect the views of the NSF, NIH, or the Beckman Foundation.

References

Allara, D.L., Parikh, A.N., and Rondelez, F. (1995). Evidence for a unique chain organization in long chain silane monolayers deposted on two widely different solid substrates. *Langmuir* 11: 2357–2360.

Allen, G.C., Tucker, P.M., Capon, A., and Parsons, R. (1974). X-ray photoelectron spectroscopy of adsorbed oxygen and carbonaceous species on platinum electrodes. *J. Electroanal. Chem.* 50: 335–343.

Anderson, J.R., Chiu, D.T., Jackman, R.J., Cherniavskaya, O., McDonald, J.C., Wu, H., Whitesides, S.H., and Whitesides, G.M. (2000). Fabrication of topologically complex three-dimensional microfluidic systems in PDMS by rapid prototyping. *Anal. Chem.* 72: 3158–3164.

Augustynski, J., Koudelka, M., and Sanchez, J. (1984). ESCA study of the state of iridium and oxygen in electrochemically and thermally formed iridium oxide films. *J. Electroanal. Chem.* 160: 233–248.

Banker, G., and Goslin, K. (eds.). (1998). *Culturing Nerve Cells.* MIT Press, Cambridge.

Bekos, E.J., Ranieri, J.P., Aebischer, P., Gardella, J.A., Jr., and Bright, F.V. (1995). Structural changes of bovine serum albumin upon adsorption to modified fluoropolymer substrates used for neural cell attachment studies. *Langmuir* 11: 984–989.

Bernard, A., Delamarche, E., Schmid, H., Michel, B., Bosshard, H.R., and Biebuyck, H. (1998). Printing patterns of proteins. *Langmuir* 14: 2225–2229.

Bernard, A., Fitzli, D., Sonderegger, P., Delamarche, E., Michel, B., Bosshard, H.R., and Biebuyck, H. (2001). Affinity capture of proteins from solution and their dissociation by contact printing. *Nat. Biotech.* 19: 866–869.

Bhatia, S.K., Shriver-Lake, L.C., Prior, K.J., Georger, J.H., Calvert, J.M., Bredehorst, R., and Ligler, F.S. (1989). Use of Thiol-Terminal silanes and heterobifunctional crosslinkers for immobilization of antibodies on silica surfaces. *Anal. Biochem.* 178: 408–413.

Boppart, S.A., Wheeler, B.C., and Wallace, C.S. (1992). A flexible perforated microelectrode array for extended neural recordings. *IEEE Trans. Biomed. Eng.* 39: 37–42.

Branch, D. (2000). Elastomeric stamping: Design considerations for long-term maintenance of neuronal networks. Doctoral Dissertation. Department of Biophysics and Computational Biology. University of Illinois, Urbana, IL.

Branch, D.W., Corey, J.M., Weyhenmeyer, J.A., Brewer, G.J., and Wheeler, B.C. (1998). Microstamp patterns of biomolecules for high-resolution neuronal networks. *Med. Biol. Eng. Comp.* 36: 135–141.

Branch, D.W., Wheeler, B.C., Brewer, G.J., and Leckband, D.E. (2000). Long-term maintenance of patterns of hippocampal pyramidal cells on substrates of polyethylene glycol and microstamped polylysine. *IEEE Trans. Biomed. Eng.* 47: 290–300.

Branch, D.W., Wheeler, B.C., Brewer, G.J., and Leckband, D.E. (2001). Long-term stability of grafted polyethylene glycol surfaces for use with microstamped substrates in neuronal cell culture. *Biomaterials* 22: 1035–1047.

Brummer, S.B., McHardy, J., and Turner, M.J. (1977). Electrical stimulation with pt electrodes: Trace analysis for dissolved platinum and other dissolved electrochemical products. *Brain Behav. Evol.* 14: 10–22.

Carmena, J.M., Lebedev, M.A., Crist, R.E., O'Doherty, J.E., Santucci, D.M., Dimitrov, D., Patil, P.G., Henriquez, C.S., and Nicolelis, M.A. (2003). Learning to control a brain-machine interface for reaching and grasping by primates. *PLos Biol.* 1: E42.

Caroni, P. (1998). Driving the growth cone. *Science* 281: 1465–1466.

Cash, S., Zucker, R.S., and Poo, M.-M. (1996). Spread of synaptic depression mediated by presynaptic cytoplasmic signaling. *Science* 272: 998–1001.

Chang, J. (2002). Technologies for and electrophysiological studies of structured, living, neuronal networks on microelectrode arrays. Doctoral Dissertation. Department of Electrical Engineering. University of Illinois, Urbana.

Chang, J., Brewer, G.J., and Wheeler, B.C. (2000). Microelectrode array recordings of patterned hippocampal neurons for four weeks. *Biomed. Microdev.* 2: 245–253.

Chang, J.C., Brewer, G.J., and Wheeler, B.C. (2001a). Modulation of neural network activity by patterning. *Biosens. Bioelectron.* 17: 527–533.

Chang, J.C., Brewer, G.J., and Wheeler, B.C. (2001b). Spatial patterning of neuronal cultures increases apparent network activity. *31st Annual Meeting. Society for Neuroscience*, San Diego, CA.

Chang, J.C., Brewer, G.J., and Wheeler, B.C. (2003a). A modified microstamping technique enhances polylysine transfer and neuronal cell patterning. *Biomaterials* 24: 2863–2870.

Chang, J.C., Brewer, G.J., and Wheeler, B.C. (2003b). A patterned substrate modulates glial development and synaptic density in a serum-free hippocampal culture. *33rd Annual Meeting. Society for Neuroscience*, New Orleans.

Chaudhury, M. (1995). Self-assembled monolayers on polymer surfaces. *Biosens. Bioelectron.*10: 785–788.

Cheng, S.-S., Chittur, K.K., Sukenik, C. N., Culp, L.A., and Lewandowska, K. (1994). The conformation of fibronectin on self-assembled monolayers with different surface composition: An FTIR/ATR study. *J. Coll. Int. Sci.* 162: 135–143.

Chiu, D.T., Jeon, N.L., Huang, S., Kane, R.S., Wargo, C.H., Choi, I.S., Ingber, D.E., and Whitesides, G.M. (2000). Patterned deposition of cells and proteins onto surfaces by using three-dimentional microfluidic systems. *Proc. Nat. Acad. Sci.* 97: 2408–2413.

Chou, N.J., Paraszczak, J., Babich, E., Heidenreich, J., Chaug, Y.S., and Goldblatt, R.D. (1987). X-ray photoelectron and infrared spectroscopy of microwave plasma etched polyimide surfaces. *J. Vac. Sci. Technol. A* 5: 1321–1326.

Clark, E.A. and Brugge, J.S. (1995). Integrins and signal transduction pathways: The road taken. *Science* 268: 233–239.

Corey, J.M. (1997). Cell patterning: Building living neural networks. Doctoral Dissertation, Neuroscience Program. University of Illinois, Urbana.

Corey, J.M. and Feldman, E.L. (2003). Substrate patterning: An emerging technology for the study of neuronal behavior. *Exp. Neurol.* 184: S89–S96.

Corey, J.M., Wheeler, B.C., and Brewer, G.J. (1991). Compliance of hippocampal neurons to patterned substrate networks. *J. Neurosci. Res.* 30: 300–307.

Corey, J.M., Wheeler, B.C., and Brewer, G.J. (1996). Micrometer resolution silane-based patterning of hippocampal neurons: Critical variables in photoresist ad laser ablation processes for substrate fabrication. *IEEE Trans. Biomed. Eng.* 43: 944–955.

Cyster, L. A., Grant, D.M., Parker, K.G., and Parker, T.L. (2002). The effect of surface chemistry and structure of titanium nitride (TiN) films on primary hippocampal cells. *Biomol. Eng.* 19: 171–175.

Dubois, L. and Nuzzo, R. (1992). Synthesis, structure, and properties of model organic surfaces. *Annu. Rev. Phys. Chem.* 43: 437–463.

Dunn, D.S., Grant, J.L., and McClure, D.J. (1989). Texturing of polyimide films during O_2/CF_4 sputter etching. *J. Vac. Sci. Technol. A* 7: 712–1718.

Egert, U., Schlosshauer, B., Fennrich, S., Nisch, W., Fejtl, M., Knott, T., Mueller, T., and Haemmerle, H. (1998). A novel organotypic long-term culture of the rat hippocampus on substrate-integrated multielectrode arrays. *Brain Res. Prot.* 2: 229–242.

Engert, F. and Bonhoeffer, T. (1997). Synapse specificity of long-term potentiation breaks down at short distances. *Nature* 388: 279–284.

Esch, T., Lemmon, V., and Banker, G. (1999). Local presentation of substrate molecules directs axon specification by cultured hippocampal neurons. *J. Neurosci.* 19: 6417–6426.

Facci, J. and Murray, R.W. (1980). Silanization and non-aqueous electrochemistry of two oxide states on platinum electrodes. *J. Electroanal. Chem.* 112: 221–229.

Ferguson, G.S., Chaudhury, M.K., Biebuyck, H.A., and Whitesides, G.M. (1993). Monolayers on disordered substrates: Self-assembly of alkyltrichlorosilanes on surface-modified polyethylene and poly(dimethylsiloxane). *Macromolecules* 26: 5870–5875.

Firestone, M.A., Shank, M.L., Sligar, S.G., and Bohn, P.W. (1996). Film architecture in biomolecular assemblies. Effect of linker on the orientation of genetically engineered surface-bound proteins. *J. Am. Chem. Soc.* 118: 9033–9041.

Freed, C.R., Greene, P.E., Breeze, R.E., Tsai, W.-Y., DuMouchel, W., Kao, R., Dillon, S., Winfield, H., Culver, S., Trojanowski, J.Q., Eidelberg, D., and Fahn, S. (2001). Transplantation of embryonic dopamine neurons for severe Parkinson's disease. *N. Engl. J. Med.* 344: 710–719.

Fromherz, P. and Schaden, H. (1994). Defined neuronal arborizations by guided outgrowth of leech neurons in culture. *Eur. J. Neurosci.* 6: 1500–1504.

Gaehwiler, B.H., Thompson, S.M., McKinney, R.A., Debanne, D., and Robertson, R.T. (1998). Organotypic slice cultures of neural tissue. In: Banker, G. and Goslin, K., eds., *Culturing Nerve Cells*. The MIT Press, Cambridge.

Garbassi, F., Morra, M., and Occhiello, E. (1998). *Polymer Surfaces: From Physics to Technology*. John Wiley & Sons, New York.

Gesteland, R.C., Howland, B., Lettvin, J.Y., and Pitts, W.H. (1959). Comments on microelectrodes. *Proc. IRE* 47: 1856–1862.

Gross, G.W. (1979). Simultaneous single unit recording in vitro with a photoetched laser deinsulated gold multimicroelectrode surface. *IEEE Trans. Biomed. Eng.* 26: 273–279.

Gross, G.W. (1994). Internal dynamics of randomized mammalian neuronal networks in culture. In: Stenger, D.A. and McKenna, T.M., eds., *Enabling Technologies for Cultured Neural Networks*. Academic Press, San Diego, pp. 277–317.

Gross, G.W., Harsch, A., Rhoades, B.K., and Gopel, W. (1997). Odor, drug and toxin analysis with neuronal networks in vitro: Extracellular array recording of network responses. *Biosens. Bioelectron.* 12: 373–393.

Hammond, J.S. and Winograd, N. (1977). XPS spectroscopic study of potentiostatic and galvanostatic oxidation of Pt electrodes in H_2SO_4 and $HClO_4$. *J. Electroanal. Chem.* 78: 55–69.

Hari, A., Djohar, B., Skutella, T., and Montazeri, S. (2004). Neurotrophins and extracellular matrix molecules modulate sensory axon outgrowth. *Int. J. Devl. Neurosci.* 22:113–117.

Hermanson, G.T. (1996). *Bioconjugate Techniques*. Academic Press, Inc., San Diego.

Hickman, J.J. and Stenger, D.A. (1994). Interactions of cultured neurons with defined surfaces. In: Stenger, D.A. and McKenna, T.M., eds., *Enabling Technologies for Cultured Neural Networks*. Academic Press, San Diego, pp. 51–76.

Hoepker, V., Shewan, D., Tessier-Lavigne, M., Poo, M.-M., and Holt, C. (1999). Growth-cone attraction to netrin-1 is converted to repulsion by laminin-1. *Nature* 401: 69–73.

Hoffman, A.S. (1996). Surface modification of polymers: Physical, chemical, mechanical and biological methods. *Macromolec. Symp.* 101: 443–454.

Hook, D.J., Vargo, T.G., Gardella, J.A., Jr., Litwiler, K.S., and Bright, F.V. (1991). Silanization of radio frequency glow discharge modified expanded poly(tetrafluoroethylene) using (aminopropyl)triethoxysilane. *Langmuir* 7: 142–151.

Ingber, D.E. (1997). Tensegrity: The architectural basis of cellular mechanotransduction. *Ann. Rev. Physiol.* 59: 575–599.

Israelachvili, J. (1997). The different faces of poly(ethylene glycol). *Proc. Nat. Acad. Sci.* 94: 8378–8379.

James, C.D., Davis, R., Meyer, M., Turner, A., Turner, S., Withers, G., Kam, L., Banker, G., Craighead, H., Isaacson, M., Turner, J., and Shain, W. (2000). Aligned microcontact printing of micrometer-scale poly-L-lysine for controlled growth of cultured neurons on planar microelectrode arrays. *IEEE Trans. Biomed. Eng.* 47: 17–21.

Jimbo, Y., Kawana, A., Parodi, P., and Torre, V. (2000). The dynamics of a neuronal culture of dissociated cortical neurons of neonatal rats. *Biol. Cybern.* 83: 1–20.

Jimbo, Y., Robinson, H.P.C., and Kawana, A. (1993). Simultaneous measurement of intracellular calcium and electrical activity from patterned neural networks in culture. *IEEE Trans. Biomed. Eng.* 40: 804–810.

Jimbo, Y., Robinson, H.P.C., and Kawana, A. (1998). Strengthening of synchronized activity by tetanic stimulation in cortical cultures: Application of planar electrode arrays. *IEEE Trans. Biomed. Eng.* 45: 1297–1304.

Jimbo, Y., Tateno, T., and Robinson, H.P.C. (1999). Simultaneous induction of pathway-specific potentiation and depression in networks of cortical neurons. *Biophys. J.* 76: 670–678.

Jo, B.-H., Van Lerberghe, L.M, Motsegood, K.M., and Beebe, D.J. (2000). Three-dimensional micro-channel fabrication in polydimethylsiloxane (PDMS) elastomer. *J. Microelectromech. Syst.* 9: 76–81.

Jones, L.L., Sajed, D., and Tuszynski, MH. (2003). Axonal regeneration through regions of chondroitin sulfate proteoglycan deposition after spinal cord injury: A balance of permissiveness and inhibition. *J. Neurosci.* 23: 9276–9288.

Jung, D.R., Cuttino, D.S., Pancrazio, J.J., Manos, P., Cluster, T., Sathanoori, R.S., Aloi, L.E., Coulombe, M.G., Czarnaski, M.A., Borkholder, D.A, Kovacs, G.T.A., Bey, P., Stenger, D.A., and Hickman, J.J. (1998). Cell-based sensor microelectrode array characterized by imaging x-ray photoelectron spectroscopy, scanning electron microscopy, impedance measurements, and extracellular recordings. *J. Vac. Sci. Technol. A* 16: 1183–1188.

Kandel, E.R., Schwartz, J.H., and Jessell, T.M. (eds.) (2000). *Principles of Neural Science.* McGraw-Hill, New York.

Kane, R.S., Takayama, S., Ostuni, E., Ingber, D.E., and Whitesides, G.M. (1999). Patterning proteins and cells using soft lithography. *Biomaterials* 20: 2363–2376.

Kleinfeld, D., Kahler, K.H., and Hockberger, P.E. (1988). Controlled outgrowth of dissociated neurons on patterned substrates. *J. Neurosci.* 8: 4098–4120.

Kumar, A., Biebuyck, H.A., and Whitesides, G.M. (1994). Patterning self-assembled monolayers: Applications in materials science. *Langmuir* 10: 1498–1511.

Laibinis, P.E., Hickman, J.J., Wrighton, M.S., and Whitesides, G.M. (1989). Orthogonal self-assembled monolayers: Alkanethiols on gold and alkane carboxylic acids on alumina. *Science* 245: 845–847.

Larson, G. (1991). An introduction to organosilicon chemistry, *Hul's Silicon Compounds— Register And Review* (1991), pp. 8–23.

Lauer, L., Vogt, A., Yeung, C.K., Knoll, W., and Offenhaeusser, A. (2002). Electrophysiological recordings of patterned rat brain stem slice neurons. *Biomaterials* 23:3123–3130.

Lee, K.-W. and Viehbeck, A. (1994). Wet-process surface modification of dielectric polymers: adhesion enhancement and metallization. *IBM J. Res. Dev.* 38: 457–474.

Lenhard, J. and Murray, R. (1977). Chemically modified electrodes part vii. Covalent bonding of a reversible electrode reactant to pt electrodes using an organosilane reagent. *J. Electroanal. Chem.* 78: 195–201.

Letourneau, P.C. (1975). Cell-to-substratum adhesion and guidance of axonal elongation. *Dev. Biol.* 44: 92–101.

Liu, Q.-Y., Coulombe, M., Dumm, J., Shaffer, K.M., Schaffner, A.E., Barker, J.L., Pancrazio, J. J., Stenger, D.A., and Ma, W. (2000). Synaptic connectivity in hippocampal neuronal networks cultured on micropatterned surfaces. *Dev. Brain Res.* 120: 223–231.

Lochter, A., Taylor, J., Braunewell, K.-H., Holm, J., and Schachner, M. (1995). Control of neuronal morphology in vitro: Interplay between adhesive substrate forces and molecular instruction. *J. Neurosci. Res.* 42: 145–158.

Lom, B., Healy, K.E., and Hockberger, P.E. (1993). A versatile technique for patterning biomolecules onto glass coverslips. *J. Neurosci. Meth.* 50: 385–397.

Lopez-Garcia, J.C., Arancio, O., Kandel, E.R., and Baranes, D. (1996). A presynaptic locus for long-term potentiation of elementary synaptic transmission at mossy fiber synapse in culture. *Proc. Nat. Acad. Sci.* 93: 4712–4717.

Ma, W., Liu, Q.-Y., Jung, D., Manos, P., Pancrazio, J.J., Schaffner, A.E., Barker, J.L., and Stenger, D.A. (1998). Central neuronal synapse formation on micropatterned surfaces. *Dev. Brain Res.* 111: 231–243.

McDonald, J.C., Duffy, D.C., Anderson, J R., Chiu, D.T., Wu, H., Schueller, O.J.A., and Whitesides, G.M. (2000). Fabrication of microfluidic systems in poly(dimethylsiloxane). *Electrophoresis* 21: 27–40.

Ming, G.-I., Wong, S.T., Henley, J., Yuan, X.-B., Song, H.-J., Spitzer, N.C., and Poo, M.-M. (2002). Adaptation in the chemotactic guidance of nerve growth cones. *Nature* 417: 411–418.

Morefield, S.I., Keefer, E.W., Chapman, K.D., and Gross, G.W. (2000). Drug evaluations using neuronal networks cultured on microelectrode arrays. *Biosens. Bioelectron.* 15: 383–396.

Moses, P.R., Wier, L.M., Lennox, J.C., Finklea, H.O., Lenhard, J.R., and Murray, R.W. (1978). X-ray photoelectron spectroscopy of alkylamine-silanes bound to metal oxide electrodes. *Anal. Chem.* 50: 576–585.

Mrksich, M., Chen, C.S., Xia, Y., Dike, L.E., Ingber, D.E., and Whitesides, G.M. (1996). Controlling cell attachment on contoured surfaces with self-assembled monolayers of alkanethiolates on gold. *Proc. Nat. Acad. Sci.* 93: 10775–10778.

Nam, Y., Chang, J.C., Wheeler, B.C., and Brewer, G.J. (2004). Gold-coated microelectrode array with thiol linked self-assembled monolayers for engineering neuronal cultures. *IEEE Trans. Biomed. Eng.* 51: 158–165.

Nanci, A., Wuest, J.D., Peru, L., Brunet, P., Sharma, V., Zalzal, S., and McKee, M.D. (1998). Chemical modification of titanium surfaces for covalent attachment of biological molecules. *J. Biomed. Mater. Res.* 40: 324–335.

Nelson, K.E., Gamble, L., Jung, L.S., Boeckl, M.S., Naeemi, E., Golledge, S.L., Sasaki, T., Castner, D.G., Campbell, C.T., and Stayton, P.S. (2001). Surface characterization of mixed self-assembled monolayers designed for streptavidin immobilization. *Langmuir* 17: 2807–2816.

Nicolelis, M.A. (2001). Actions from thoughts. *Nature* 409: 403–407.

Nicolelis, M.A. and Fanselow, E.E. (2002). Dynamic shifting in thalamocortical processing during different behavioural states. *Philos. Trans. R. Soc. Lond. B Biol. Sci.* 357: 1753–1758.

Novak, J.L. and Wheeler, B.C. (1986). Recording from the aplysia abdominal ganglion with a planar microelectrode array. *IEEE Trans. Biomed. Eng.* 33: 196–202.

Plueddemann, E.P. (1991). *Silane Coupling Agents.* Plenum Press, New York.

Polleux, F., Morrow, T., and Ghosh, A. (2000). Semaphorin 3A is a chemoattractant for cortical apical dendrites. *Nature* 404: 567–573.

Prinz, A. and Fromherz, P. (2000). Electrical synapses by guided growth of cultured neurons from the snail Lymnaea stagnalis. *Biol. Cyb.* 82: L1–L5.

Proctor, A., Castner, J.F., Wingard, L.B., Jr., and Hercules, D.M. (1985). Electron spectroscopic (ESCA) studies of platinum surfaces used for enzyme electrodes. *Anal. Chem.* 57: 1644–1649.

Ravenscroft, M.S., Bateman, K.E., Shaffer, K.M., Schessler, H.M., Jung, D.R., Schneider, T.W., Montgomery, C.B., Custer, T.L., Schaffner, A.E., Liu, Q.Y., Li, Y.X., Barker, J.L., and Hickman, J.J. (1998). Developmental neurobiology implications from fabrication and analysis of hippocampal neuronal networks on patterned silane-modified surfaces. *J. Am. Chem. Soc.* 120: 12169–12177.

Rutten, W.L.C., Smit, J.P.A., Frieswijk, T.A., Bielen, J.A., Brouwer, A.L.H., Buitenweg, J.R., and Heida, C. (1999). Neuro-electronic interfacing with multielectrode arrays. *IEEE Eng. Med. Biol.*1999: 47–55.

Sadana, A. (1992). Protein adsorption and inactivation on surfaces. influence of heterogeneities. *Chem. Rev.* 92: 1799–1818.

Schwartz, M.A., Schaller, M.D., and Ginsberg, M.H. (1995). Integrins: Emerging paradigms of signal transduction. *Ann. Rev. Cell Dev. Biol.* 11: 549–599.

Shahaf, G. and Marom, S. (2001). Learning in networks of cortical neurons. *J. Neurosci.* 21: 8782–8788.

Sheth, S.R. and Leckband, D. (1997). Measurements of attractive forces between proteins and end-grafted poly(ethylene glycol) chains. *Proc. Nat. Acad. Sci.* 94: 8399–8404.

Singhvi, R., Kumar, A., Lopez, G.P., Stephanopoulos, G.N., Wang, D.I.C., Whitesides, G.M., and Ingber, D.E. (1994). Engineering cell shape and function. *Science* 264:696–698.

Song, H.-J., Ming, G.-L., He, Z., Lehmann, M., McKerracher, L., Tessier-Lavigne, M., and Poo, M.-M. (1998). Conversion of neuronal growth cone responses from repulsion to attraction by cyclic nucleotides. *Science* 281: 1515–1518.

Sorribas, H., Padeste, C., Tiefenauer, L. (2002). Photolithographic generation of protein micropatterns for neuron culture applications. *Biomaterials* 23: 893–900.

Stayton, P.S., Olinger, J.M., Bohn, P.W., and Sligar, S.G. (1992). Genetic engineering of surface attachment sites yields oriented protein monolayers. *J. Am. Chem. Soc.* 114: 9298–9299.

Stenger, D.A., Hickman, J.J., Bateman, K.E., Ravenscroft, M.S., Ma, W., Pancrazio, J.J., Shaffer, K., Schaffner, A.E., Cribbs, D.H., and Cotman, C.W. (1998). Microlithographic determination of axonal/dendritic polarity in cultured hippocampal neurons. *J. Neurosci. Meth.* 82: 167–173.

Stoppini, L., Duport, S., and Correges, P. (1997). A new extracellular multirecording system for electrophysiological studies: Application to hippocampal organotypic cultures. *J. Neurosci. Meth.* 72: 23–33.

Taborelli, M., Eng, L., Descouts, P., Ranieri, J.P., Bellamkonda, R., and Aebischer, P. (1995). Bovine serum albumin conformation on methyl and amine functionalized surfaces compared by scanning force microscopy. *J. Biomed. Mat. Res.* 29: 707–714.

Takei, K., Shin, R.-M., Inoue, T., Kato, K., and Mikoshiba, K. (1998). Regulation of nerve growth mediated by inositol 1,4,5-Triphosphate receptors in growth cones. *Science* 282: 1705–1708.

Tao, H.W., Zhang, L.I., Bi, G.-Q., and Poo, M.-M. (2000). Selective presynaptic propagation of long-term potentiation in defined neural networks. *J. Neurosci.* 20: 3233–3243.

Thiebaud, P., de Rooij, N.F., Koudelka-Hep, M., and Stoppini, L. (1997). Microelectrode arrays for electrophysiological monitoring of hippocampal organotypic slice cultures. *IEEE Trans. Biomed. Eng.* 44: 1159–1163.

Thiebaud, P., Lauer, L., Knoll, W., and Offenhaeusser, A. (2002). PDMS device for patterned application of microfluids to neuronal cells arranged by microcontact printing. *Biosens. Bioelectron.*17: 87–93.

Thomas, C.A., Jr., Springer, P.A., Loeb, G.E., Berwald-Netter, Y., and Okun, L.M. (1972). A miniature microelectrode array to monitor the bioelectric activity of cultured cells. *Exp. Cell Res.* 74: 61–66.

Untereker, D.F., Lennox, J.C., Wier, L.M., Moses, P.R., and Murray, R.W. (1977). Chemically modified electrodes Part IV. Evidence for formation of monolayers of bonded organosilane reagents. *J. Electroanal. Chem.* 81: 309–318.

Urban, N. and Barrionuevo, G. (1998). Active summation of excitatory postsynaptic potentials in hippocampal CA3 pyramidal neurons. *Proc. Nat. Acad. Sci.* 95: 11450–11455.

Vargo, T.G., Thompson, P.M., Gerenser, L.J., Valentini, R.F., Aebischer, P., Hook, D.J., and Gardella, J.A., Jr. (1992). Monolayer chemical lithography and characterization of fluoropolymer films. *Langmuir* 8: 130–134.

Venkateswar, R.A., Branch, D.W., and Wheeler, B.C. (2000). An electrophoretic method for microstamping biomolecule gradients. *Biomed. Microdev.* 2: 255–264.

Vogt, A.K., Lauer, L., Knoll, W., and Offenhausser, A. (2003). Micropatterned substrates for the growth of functional neuronal networks of defined geometry. *Biotechnol. Prog.* 19: 1562–1568.

Vogt, A.K., Stefani, F.D., Best, A., Nelles, G., Yasuda, A., Knoll, W., and Offenhausser, A. (2004). Impact of micropatterned surfaces on neuronal polarity. *J. Neurosci. Meth.* 134: 191–198.

Ward, W.K., Jansen, L.B., Anderson, E., Reach, G., Klain, J.-C., and Wilson, G.S. (2002). A new amperometric glucose microsensor: In vitro and short-term in vivo evaluation. *Biosens. Bioelectron.* 17: 181–189.

Weetall, H.H. (1976). Covalent coupling methods for inorganic support materials. *Meth. Enzymol.* 44: 134–148.

Weiland, J. and Anderson, D. (2000). Chronic neural stimulation with thin-film, iridium oxide electrodes. *IEEE Trans. Biomed. Eng.* 47: 911–918.

Wheeler, B.C., and Brewer, G.J. (1994). Multineuron patterning and recording. In: Stenger, D.A. and McKenna, T.M., eds., *Enabling Technologies for Cultured Neural Networks*. Academic Press, San Diego, pp. 167–185.

Wheeler, B.C., Corey, J.M., Brewer, G.J., and Branch, D.W. (1999). Microcontact printing for precise control of nerve cell growth in culture. *J. Biomech. Eng.* 121: 73–78.

Wise, K. D. and Weissman, R. H. (1971). Thin films of glass and their application to biomedical sensors. *Med. Biol. Eng.* 9: 339–350.

Xia, Y. and Whitesides, G.M. (1998). Soft lithography. *Angew. Chem. Int. Ed.* 37: 550–575.

Xiao, Z.-C., Ragsdale, D.S., Malhotra, J.D., Mattei, L.N., Braun, P.E., Schachner, M., and Isom, L.L. (1999). Tenascin-R is a functional modulator of sodium channel β subunits. *J. Biol. Chem.* 274: 26511–26517.

Zhang, M., Desai, T., and Ferrari, M. (1998). Proteins and cells on PEG immobilized silicon surfaces. *Biomaterials* 19: 953–960.

Zhao, X.-M., Wilbur, J.L., and Whitesides, G.M. (1996). Using two-stage chemical amplification to determine the density of defects in self-assembled monolayers of alkanethiolates on gold. *Langmuir* 12: 3257–3264.

II
MEA Applications:
Dissociated Cell Cultures

8
Emerging Histiotypic Properties of Cultured Neuronal Networks

Guenter W. Gross and Kamakshi V. Gopal

8.1 Introduction and Background

In the past decade, substantial progress has been made with substrate-integrated micro-electrode arrays and the growth of viable networks on such arrays. However, such progress has not been accompanied by the acceptance of dissociated (primary) neuronal cultures as reliable systems for pharmacological and toxicological studies. Given that such quasi-monolayer cultures provide strong long-term adhesion stability, optical access to major components of the network circuitry, and high signal-to-noise ratios for multi-site recording of spatiotemporal action potential (spike) patterns, it is essential that such systems be validated as representative of the parent tissue. In addition, action potential waveshapes can also be monitored quantitatively over long periods of time for assessment of channel pharmacology and statistical surveys of discharges from different neuronal compartments. This chapter summarizes characteristic tissue specificities of native activity and histiotypic pharmacological and toxicological responses in order to demonstrate that appropriate dissociation and culture maintenance techniques can generate spontaneously active networks with remarkable similarity to parent tissue responses in vivo.

Work with neuronal tissue in culture can be classified as part of two major mechanistic domains: (1) receptor-dependent studies and (2) circuit-dependent studies. The second domain has received limited attention, however, the first domain is less complex, has been explored by several laboratories, and is supported by a rapidly growing database attesting to the histiotypic nature of network responses in culture. In this chapter, we show that different tissues from the murine CNS have different native activity states and may also differ quantitatively in their pharmacological responses. Nevertheless, the overall pharmacological responses agree well with in vivo data. These results imply that, compared to the parent tissue in situ, primary cultures retain the same general ratio of cells (neurons and glia), receptor properties, synaptic mechanisms, and inherent cellular spike generation characteristics. The lower synaptic density and shallow three-dimensional tissue layer do not seem to impair or alter the type and character of pharmacological responses.

The complexities seen in behavioral phenomena such as tinnitus caused by quinine and salicylate, or metal excitotoxicity due to exposure to mercury, or intoxication induced by ethanol can be studied in vitro in a much more controlled and simpler environment. This provides an opportunity to not only analyze global activity changes brought about by various chemicals, but also offers a platform for careful and objective scrutiny of changes in complex spatiotemporal patterns, and enables quantitative investigations of cell culture correlates of various behavioral phenomena.

8.2   Methods

8.2.1   MEA Fabrication

The techniques used to fabricate and prepare micro-electrode arrays (MEAs) have been described in several publications (Gross, 1979; Gross et al., 1985; Gross and Kowalski, 1991; Gross, 1994; Kovacs, 1994). Because of a lack of commercial MEA supply in the early 1980s, the Center for Network Neuroscience (CNNS) was forced to manufacture MEAs "in-house" using students. This approach necessitated simple procedures: (1) use of commercially sputtered glass plates, (2) spin insulation with resins, and (3) deinsulation with single laser shots. This was followed by electrolytic gold-plating of the exposed metal to adjust the interface impedance to approximately 1 MΩ at 1 KHz (Gross et al., 1985). In 1984, thin film indium-tin oxide (ITO) on soda–lime glass was explored and accepted as the preferred electrode material because it was transparent and commercially produced as highly uniform sputtered glass plates for the liquid crystal display industry (Gross et al., 1985).

Silicone resins such as the dimethyl polysiloxanes (Dow Corning DC 648) and methyltrimethoxysilane (PS233, United Chemical Technologies) have been preferred as insulation materials because of ease of application and curing, and excellent stability under warm saline. Because these resins are hydrophobic, surface activation via flaming is required (Lucas et al., 1986). This provided an additional advantage: the generation, by flaming through masks, of 1 to 4 mm diameter adhesion islands centered on the recording matrix. Subsequent steps involve addition of poly-D-lysine and laminin before cell seeding. Various electrode conductor patterns are presently in use of which three patterns are shown in Figure 8.1.

8.2.2   Culturing Primary Neuronal Networks on Micro-Electrode Arrays

Dissociated tissue cultures were prepared according to the basic method established by Ransom et al. (1977). Various CNS tissues were obtained from embryonic Balb-C and (more recently) from HSD:IRC mice (Charles River Laboratories, Wilmington, MA). Spinal cord tissue was dissected at 14 to 15 days gestation

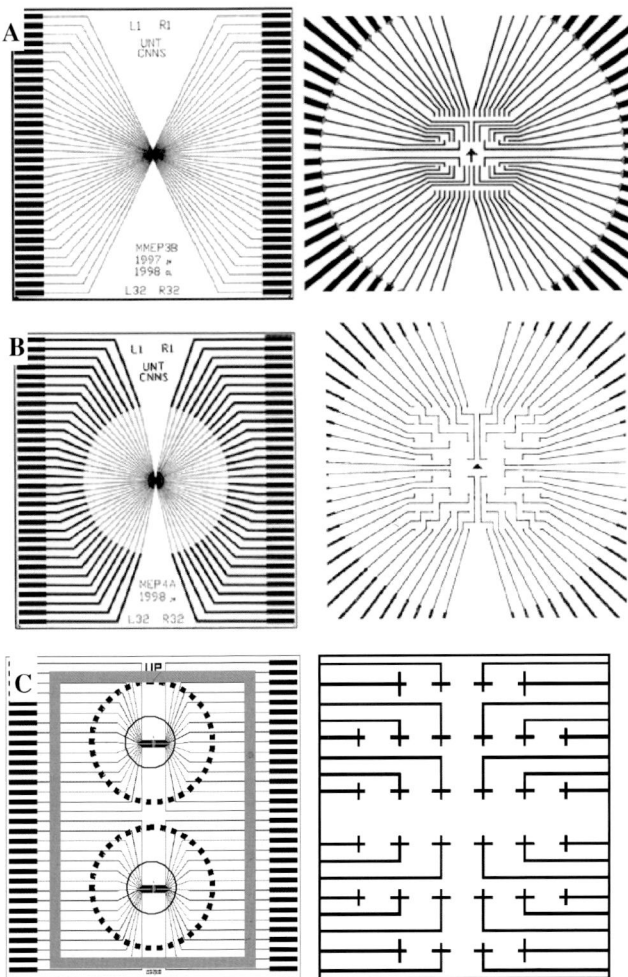

FIGURE 8.1. Micro-electrode array plates in use by the CNNS. All plates measure 5 × 5 cm and are 1.1 mm thick (left column). The right column depicts the recording areas in greater detail. Amplifier edge contacts are the same for all arrays. (A) MMEP 3C with 64 conductors terminating in a 0.8 mm² recording area in 4 rows of 16 columns. Electrode spacing: 40 μm between electrodes and 200 μm between rows. Electrode area: 120 μm². Impedance at 1 kHz: 1 to 2 megohms. (B) MMEP 4A with 64 conductors terminating in a 1.2 mm² recording matrix of 8 rows by 8 columns. Electrode spacing is equidistant at 150 μm. Electrode area is roughly 800 μm², impedance: 0.8 megohms. (C) MMEP 5. Electrode array plate featuring two separate recording islands with 32 micro-electrodes each. The center-to-center distance between the recording areas is 2.24 cm. Electrodes are spaced equidistant at 200 μm. Cruciform electrodes can increase the recording probability in low-density cultures. Dashed circles indicate location of O-ring after chamber assembly.

and cortical tissue at 16 to 17 days. The tissue was dissociated enzymatically and mechanically (via trituration), and seeded at a concentration of 0.2 to 0.5 $\times$ 10^6 cells/ml onto flamed adhesion areas 3 mm in diameter (7 mm^2). The cell pool was seeded as 0.1 ml droplets (approximately 20,000 to 50,000 cells or 2800 to 7100 cells per mm^2). Assuming a neuron to nonneuronal cell ratio of 1:10, this procedure provides a neuronal cell density of approximately 300 to 700 neurons per mm^2. After cell adhesion (2 h), 2 ml medium confined to a 4 cm^2 area by a silicone gasket was added. Spinal cord and cortical cultures were incubated in minimal essential medium (MEM) or Dulbecco's modified minimal essential medium (DMEM), respectively, each supplemented with 5% horse serum and 5% fetal bovine serum for a week. Thereafter, the cells were fed twice weekly using half media changes with the respective media containing 5% horse serum.

The networks were maintained at 37°C in an atmosphere of 10% CO_2 and 90% air until ready for use, usually three weeks after cell seeding (Gross, 1994). Spontaneous activity generally appeared after approximately 5 to 7 days in the form of random spiking and stabilized in terms of coordinated spike and burst pattern by 15 days in vitro. Networks can remain active and pharmacologically responsive for more than six months (Gross, 1994; Potter and DeMarse, 2001). Figure 8.2 shows a three-month-old culture on the 64-electrode MMEP-3.

8.2.3   MEA Assembly and Recording

MEAs were assembled into open or closed flow chambers (Gross and Schwalm, 1994, Gross, 1994). The former provides a constant medium volume recording chamber that is simpler to use for pharmacological experiments and was generally employed in about 80% of experiments. The latter allows long-term recording for over one month. The chambers were kept at 37°C on an inverted microscope stage. The pH was stabilized at 7.4 by passing a stream of humidified 10% CO_2 in air (flow rate of 8 to 12 ml/min) through a cap fitted with a heated ITO window to prevent condensation. Long-term experiments also employed an infusion pump to compensate for slow evaporation via addition of water at a rate of approximately 30 µl/h.

Neuronal activity was recorded with a two-stage, 64-channel amplification and signal processing system (Plexon Inc., Dallas), served by a Dell 800 MHz computer workstation. Total system gain used was normally 10 K. The neuronal activity per channel was usually discriminated with a template-matching algorithm (Plexon Inc.) in real-time to provide single unit spike rate data as well as continual single unit wave shape information. Data were generally displayed in real-time as mean network activity per minute, allowing a visualization of long periods of the temporal evolution of such activity.

After channel and active unit selection, spontaneous activity was generally recorded for 30 min. In most experiments, this episode (termed "native activity") was followed by a medium change using a replacement medium pool sufficient

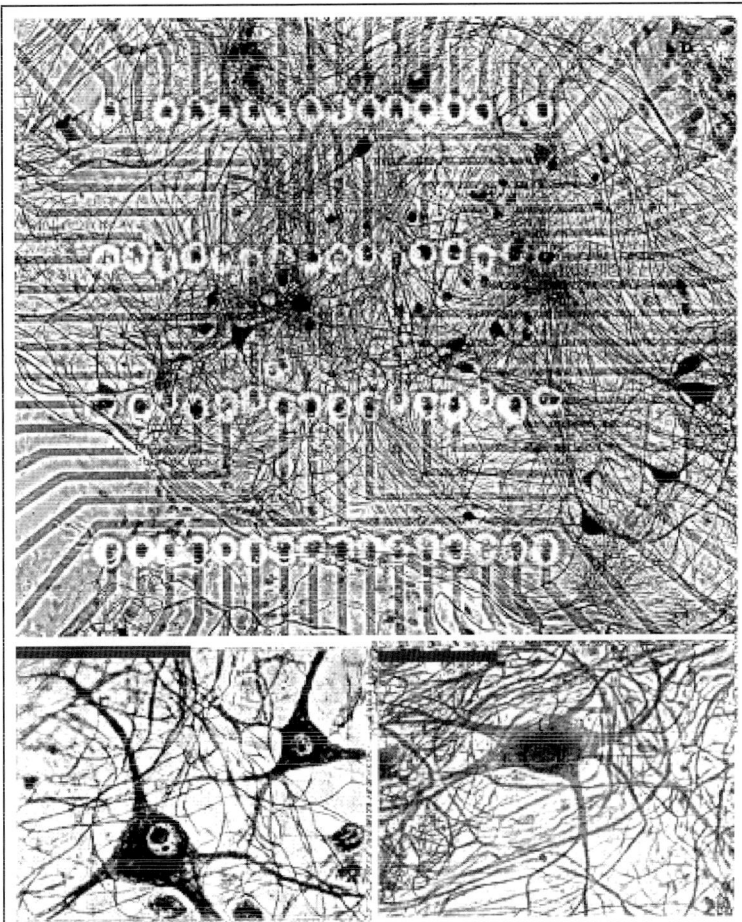

FIGURE 8.2. Top panel: neuronal network on micro-electrode array 96 days after seeding (spinal cord). Conductors in recording matrix are 10 μm wide. Lower panels: enlarged neurons outside the recording matrix. Bars represent 40 μm. Loots-modified Bodian stain.

to allow several medium changes during the course of an experiment. This new network state was termed "reference activity". After 30 min of reference activity recording, aliquots of test compounds were added sequentially to the 2 ml static bath to raise the compound concentration in a stepwise fashion to levels that affected the network activity. Before proceeding to the next higher concentration, sufficient time was given to allow the network to achieve a new stable activity state defined as a temporal evolution of minute means that formed a level plateau with individual minute mean fluctuations not exceeding ±10% of the plateau mean.

8.2.4   Data Acquisition, Display, and Analysis

Recording workstations follow the general arrangement shown in Figure 8.3. Amplifiers, A/D conversion, unit identification and selection, and storage of time stamps and waveshapes were accomplished with hardware and software from Plexon Inc. Custom programs for the real-time display of averaged network spike and burst rates (NACTAN) and some offline automatic plotting of activity variables were added by the CNNS. Although tape records are sometimes kept for detailed analysis of waveshapes, such records are no longer used routinely, primarily because of time limitations. Chart recordings are still considered useful, especially for the determination of burst pattern regularization, as integrated amplitude and period can be determined visually in real-time. However, translation into quantitative data is difficult and such approaches are yielding to new methods of digital processing.

Although the digitized data clearly dominate, it should not be assumed that the digital domain is problem free. For example, waveshape separation in real-time has not yet reached a level of sophistication required by high-density networks where overlapping action potential shapes cannot be recognized. Spike amplitude changes, due to high frequency discharges that enter the relative refractory period of the active neuronal compartment being monitored, also represent a remaining challenging problem. Figure 8.4 depicts the minute-by-minute evolution of network activity as presented by the "Autoplot" program. The same information also can be presented in real-time on the computer screen. Note that the reference activity (REF) stabilizes within 30 min after the start of the recording and is monitored for another 40 min before botulinum toxin is applied. This is a general protocol

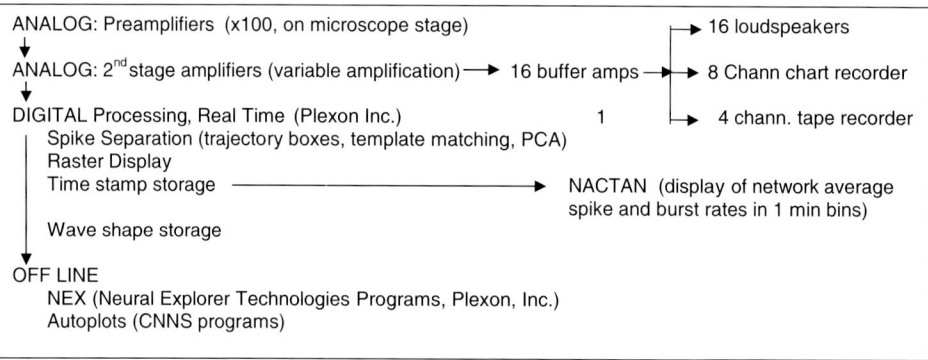

FIGURE 8.3. Data acquisition and processing scheme. Amplification, analog/digital; conversion and storage of time stamps and waveforms were performed by the Plexon MNAP system. Analog signals were also sent to loudspeakers, a chart recorder, and a 14-channel tape recorder. Real-time display of network means (for spike and burst rates) was accomplished with a custom program (NACTAN). Offline analyses included programs offered by NEX (Plexon) and custom burst analysis programs (CNNS Autoplots).

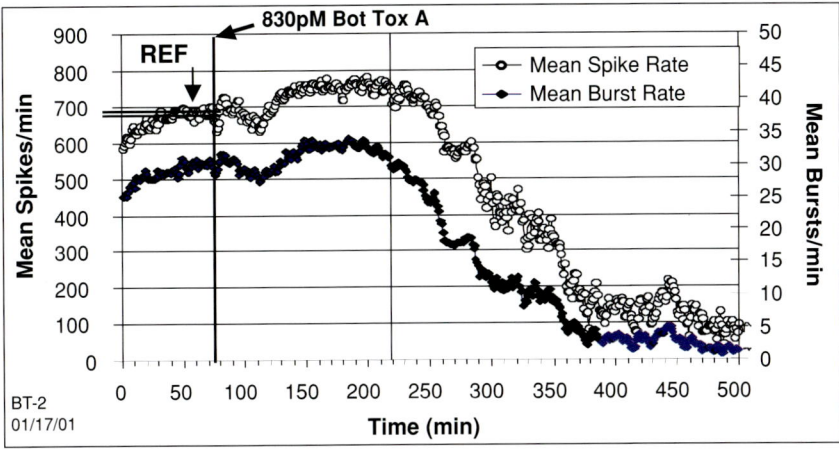

FIGURE 8.4. Evolution of network activity plotted as average values across all channels in bins of one minute. This simple depiction provides mean spike (left ordinate) and burst (right ordinate) data and shows network responses to botulinum toxin A. The first vertical line represents the time of application, the second the first discernable decrease of activity. REF (arrow) refers to the reference episode with the horizontal double line identifying a stable network state before the addition of a test compound.

used in all experiments. Depending on the network behavior, the pre-experiment reference episode may sometimes be extended to 1 h.

8.3 Tissue-Specific Native Activity

A comparison of native network activity from different tissues cultured routinely in the CNNS labs shows interesting differences. Although all tissues are sponta-neously active, many generate unique temporal patterns that are reflected in the spike production and, especially, the burst duration and coordination among chan-nels. Figure 8.5 gives a pictorial summary of the baseline (reference) activity states from networks derived from four different regions of the central nervous system: auditory cortex (AC), frontal cortex (FC), inferior colliculus (IC), and spinal cord (SC). The panels show data from 20 units in terms of spike time stamps for pe-riods of 44 to 70 seconds. IC networks were found to have substantially higher spike and burst rates compared to the cortical networks (auditory and frontal) with little overt coordination among the channels. Both FC and AC networks displayed patterns with coordinated bursting of short burst durations. Spinal networks gen-erally showed a great variety of burst durations, often with several subpopulations participating in different coordinated burst patterns.

Generally ACNs and FCNs generated more regularized activity compared to ICNs. By "regularization" we refer to positive coordination of bursts with overlap-ping burst durations and reduced variability in burst duration and spike frequencies

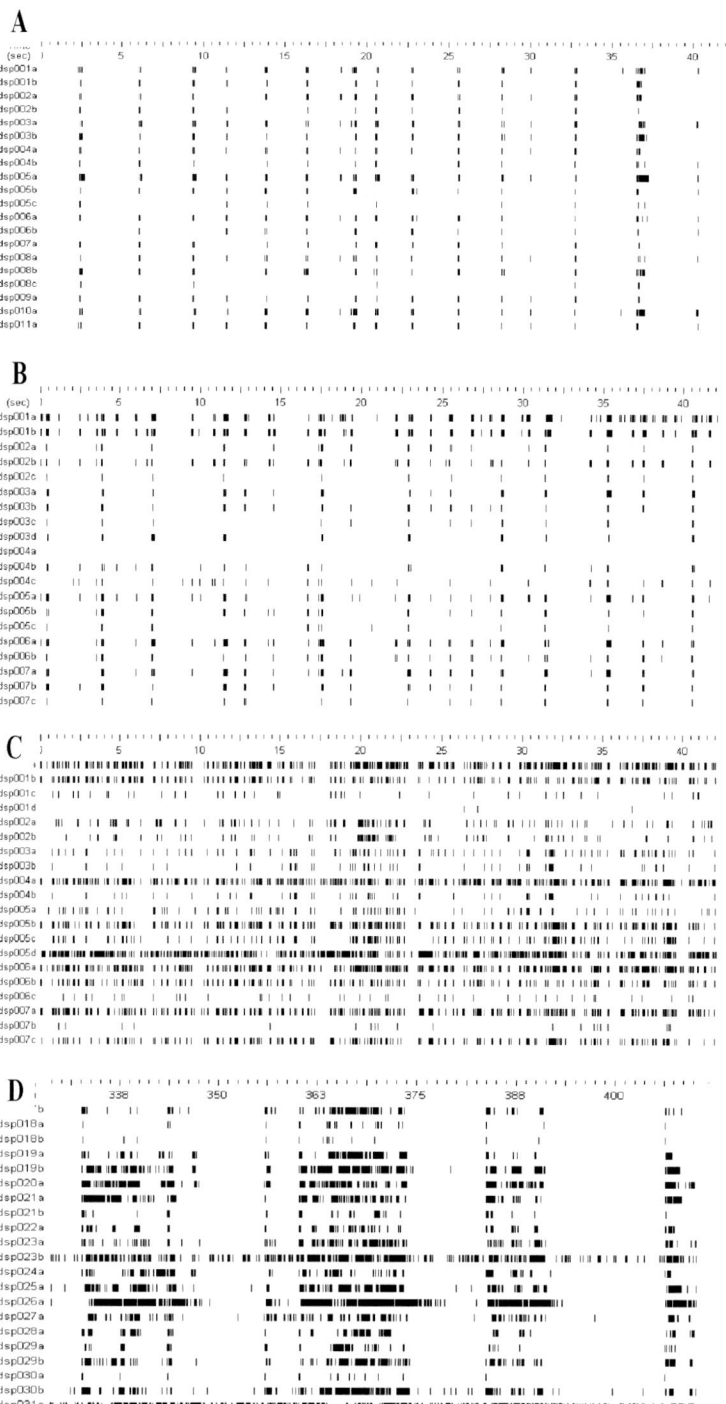

TABLE 8.1. Native activity of networks derived from four
different central nervous system tissues.

Source tissue	NW spike rates (/min) (range)	NW burst rate (/min) (range)	Mean spikes in bursts
Auditory Ctx (N = 8)	330 (165–546)	15 (6–36)	14.8 (4–30)
Frontal Ctx (N = 8)	271 (219–434)	20 (13–39)	9.3 (6–18)
Inferior Collic (N = 5)	1050 (404–1660)	62 (44–94)	11.6 (4–19)
Spinal Cord (N = 33)	532 (150–1500)	15 (3–38)	35* (10–85)

AC, FC, and IC network data from Gopal and Gross 2004. SC data was
randomly selected from the 2003 experimental pool.
* N = 6

in bursts (integrated burst amplitude). In addition, cortical cultures generated little spiking between bursts. ICNs demonstrated a radically different spontaneous
activity with seemingly random bursting, lower coordination of bursting among
channels, and high level of spiking between bursts (Table 8.1).

Table 8.1 represents a simplistic approach to characterizing network patterns.
Clearly, it is not enough to report only spike production, as such spikes can be organized in a great variety of patterns that have neurophysiological significance. Burst
rates add an important dimension. However, bursts can vary greatly in length and
in the spike patterns within the burst. Present methods count only bursts above a
certain threshold and do not yet produce statistical displays of the "types of bursts"
based on distributions of a selected burst variable (e.g., burst duration). Nevertheless, if we look at mean network spike and burst production (averaged across
all channels per minute) and averaged again over a subset of cultures (N), then
one can single out the inferior colliculus (IC) networks as displaying unique native
activity. Spinal cord activity in culture is best identified by the longer and more
intense bursts (more spikes per burst). However, the ranges for all measures are
presently still uncomfortably large. It is therefore necessary to treat each network
as an individual and use only its native activity as the reference for subsequent

←——————————————————————————————————————

FIGURE 8.5. Native raster activity (time stamps) for four different tissues growing on MEAs.
(A) auditory cortex, (B) frontal cortex, (C) inferior colliculus, and (D) spinal cord. Although
temporal patterns can vary widely, the general trend is maintained. Cortical tissues have
short bursts that are coordinated (quasi-synchronized), inferior colliculus tissue generates
highly active cultures with little co-ordination among the channels, and spinal tissue produces strong, relatively long bursts that are often co-ordinated among two or three major
subpopulations.

quantification of changes in activity patterns. Under such normalization protocols, the response repeatability was found more than satisfactory.

8.4   Tissue-Specific and Histiotypic Pharmacological Responses

8.4.1   Ethanol

Ethanol intoxication is usually defined as a blood alcohol level (BAL) of 0.1 g/dL (22 mM). A BAL of 0.05 to 0.10 g/dL (11 to 22 mM) causes slight impairment of balance, speech, judgment, and reasoning. A BAL above 0.5 g/dL (110 mM) leads to coma and death (Little, 1991). Using murine frontal cortex cultures, Xia and Gross (2003) reported excellent agreement of cell culture data with data from human and animal experiments. Repeatable, concentration-dependent sensitivities to ethanol were demonstrated with initial inhibition at 15 to 20 mM and a spike rate EC_{50} of 48.8 ± 5.4 mM. Ethanol concentrations above 100 mM led to a loss of all activity (Figures 8.6 and 8.7, Table 8.2). Complete medium changes after application of 160 mM ethanol returned near normal activity, suggesting no overt cytotoxicity over time periods up to 100 min at this concentration (Figure 8.6B). Longer exposure times were not tested. Optical assessment of neuron and glia appearance after exposure to 160 mM showed no changes detectable with phase contrast microscopy.

Although ethanol did not change the shape of action potentials, it did reveal unit-specific responses. In 14 cultures exposed to ethanol (200 neurons), 40 mM ethanol decreased firing in 71% of the cultures, increased firing in 20%, and generated no effect in 9%. These results also agreed well with the literature (Table 8.3), especially the in vivo experiments reported by Pohorecky and Brick (1977).

TABLE 8.2. Comparison of ethanol effects on mammals and FC cultures.

Mammal	Conc. (mM)	Effect
humans	5–15	slight impairment of attention and judgment
humans	10–22	impairment of speech and balance
humans	15–30	significant sedation
rats	20	sedation
FC cultures	15	first overt change in spike and burst production
humans	30–55	mental confusion
mice	40	loss of righting reflex
FC cultures	48.8	EC_{50} for spike production
humans	100	coma and death
mice	120	coma, hypothermia
FC cultures	100–140	cessation of all spontaneous activity

Human and animal data from Charness et al., 1989; Little, 1991.

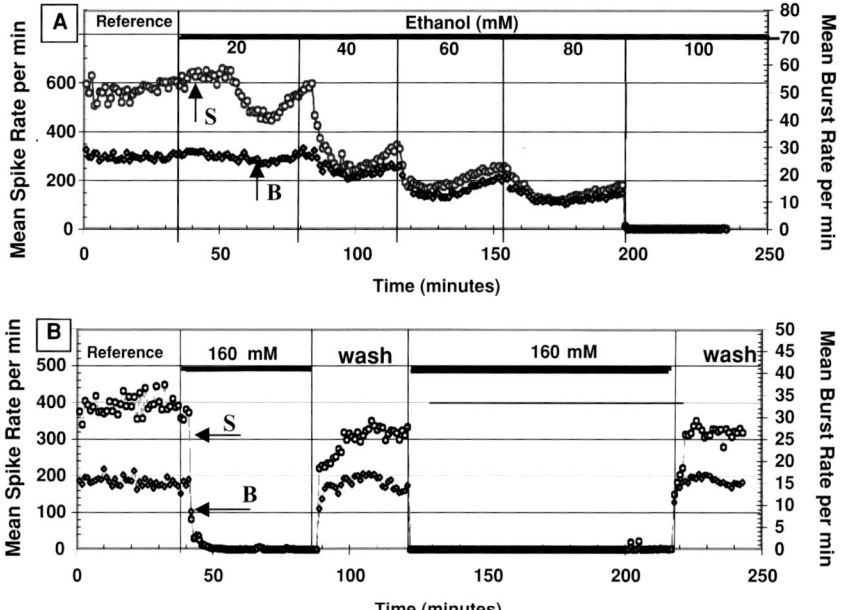

FIGURE 8.6. (A) Typical effect of ethanol on frontal cortex culture spontaneous activity. Data points represent average spike (left ordinate) and burst rate (right ordinate) across all the selected units per minute. Both spike and burst rate decreased under exposure to 20 mM ethanol with progressive reductions associated with increasing concentrations. Ethanol at 100 mM stopped all activity. (B) Reversibility of frontal cortex network activity after exposure to two additions of high concentrations of ethanol. The total cessation of activity at 160 mM was reversed by a complete medium change. The second manipulation displays the same effect. S: spike rate; B: burst rate. (Data from Xia and Gross, 2003.)

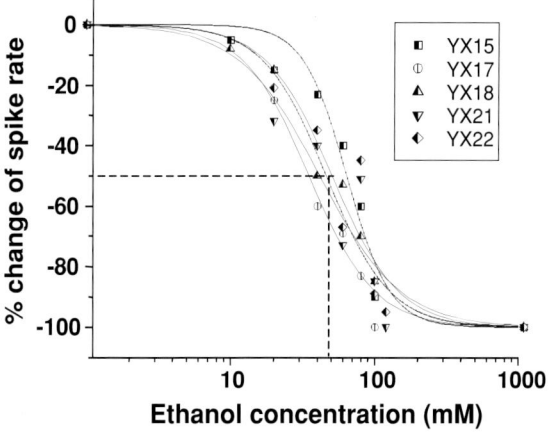

FIGURE 8.7. Concentration–response summary for ethanol using network spike rates. Curves were derived from five different networks, each exposed to sequential ethanol exposures. Spike rate EC$_{50}$: 48.8 ± 5.4 mM (mean ± SE).

TABLE 8.3. Neuron-specific responses to ethanol.

Murine FC cultures	Rat locus coerulius	CA1 pyramidal neurons	Rat hippocampal slices	
Ethanol C	40 mM	50 mM	10–350mM	50–200 mM
Decrease	71%	62%	50%	24%
Increase	20%	22%	9%	—
No effect	9%	16%	29%	—
Biphasic	0	0	12%	—
Reference	Xia & Gross (2003)	Pohorecky & Brick ('77)	Siggins. et al. ('87)	Siggins et al. ('87)

8.4.2 Fluoxetine

In a recent study, Xia et al. (2003) used spontaneously active neuronal networks derived from frontal and auditory cortex tissues to quantify responses to fluoxetine. Although both tissues generated major concentration-dependent suprathreshold electrophysiological changes ending in termination of all spontaneous activity, quantitative results showed significant differences. Whereas frontal cortex networks produced monotonic inhibition with an IC_{50} of 5.4 ± 0.7 (mean $\pm$ SE, $n = 6$), the auditory tissue revealed an excitatory phase at low concentrations and an EC_{50} of 15.9 ± 1.0 (mean $\pm$ SE, $n = 10$).

Although some individual units in frontal cortex networks showed slight excitation, the population response was not biphasic. Auditory cortex population responses, however, were clearly biphasic. The excitatory phase in auditory cortex networks and the higher shut-off concentrations held for both spike and burst production. A repeated-measure ANOVA followed by a t-test indicated a significant difference in spike rate ($p = 0.006$) at 5, 10, 15 µM fluoxetine. The burst rate data also showed significant differences between the two tissues when exposed to 5, 10, 15 µM fluoxetine ($p = 0.02$). See Figure 8.8.

Fluoxetine is lipid soluble with a brain-to-blood ratio ranging from 10 to 20. The concentration in brain tissue can reach 20 µM (Karson et al., 1993) although therapeutic plasma concentration was reported at approximately 1 µM (Altamura et al., 1994). The therapeutic brain concentration of up to 20 µM fluoxetine is within the range at which fluoxetine affected frontal and auditory cortex networks. In our cultures, frontal and auditory cortex activities were completely inhibited at 10 to 16 µM and 20 to 25 µM, respectively.

8.4.3 Quinine

Quinine, a primary anti-malarial drug has several side effects such as headache, tinnitus, hearing loss, and dizziness. To investigate the altered neural activity induced by quinine, Gopal and Gross (2004) examined the acute effects of quinine on spontaneously active AC, FC, and IC networks. In AC networks, low concentrations of quinine (1 to 20 µM) increased mean spike rates, burst rates, and spikes in bursts

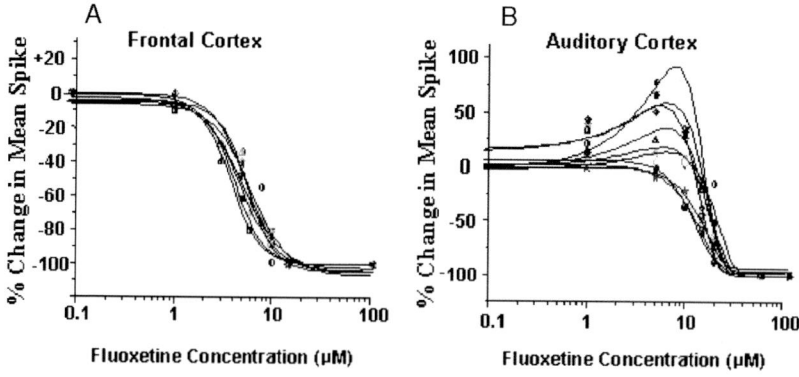

FIGURE 8.8. Concentration–response summary for fluoxetine activity inhibition in networks derived from frontal cortex (A) and auditory cortex (B) tissues. The IC_{50} means $\pm$ S.E. are 5.4 ± 0.7 and 15.9 ± 1.0 μM for frontal and auditory cortex networks, respectively.

compared to the reference state. At concentrations of 30 to 40 μM the mean spike rate decreased although the mean burst rate remained above the baseline level for 1 to 40 μM quinine. Paired samples t-tests comparing the reference state to 1, 10, and 20 μM quinine applications indicated a significant difference between the reference and 1 μM application for spike and burst rates ($p = 0.01$). FC networks, on the other hand, showed little change in both spike and burst rates for 1 μM quinine, but showed decreased mean spike and burst rates at higher concentrations. IC networks also exhibited only inhibitory responses to quinine. EC_{50} values for FC and IC networks were 28.7 ± 4.8 μM, and 23.7 ± 2.1 μM, respectively, but differed greatly for the ACs (42.5 ± 3.9 μM). Please see Figure 8.9.

Another effect of quinine on the networks was the increase in regularization among channels in all three tissues. Regularization was quantified using coefficient

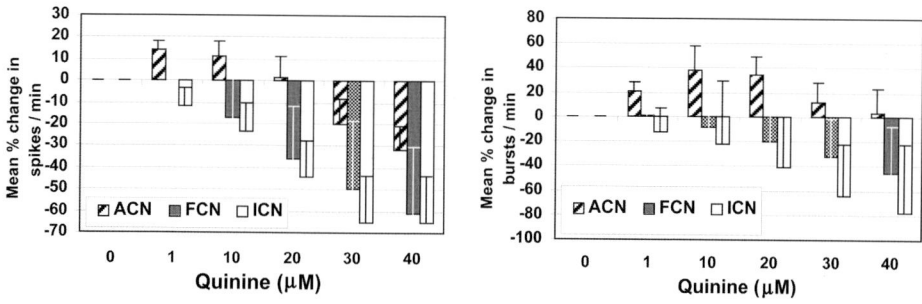

FIGURE 8.9. Summary of acute differential effects of quinine on 8 AC, 8 FC, and 5 IC networks. Averaged network spike and burst production shows a biphasic effect for ACNs and only inhibitory effects for FCNs and ICNs. (Redrawn from Gopal and Gross, 2004.)

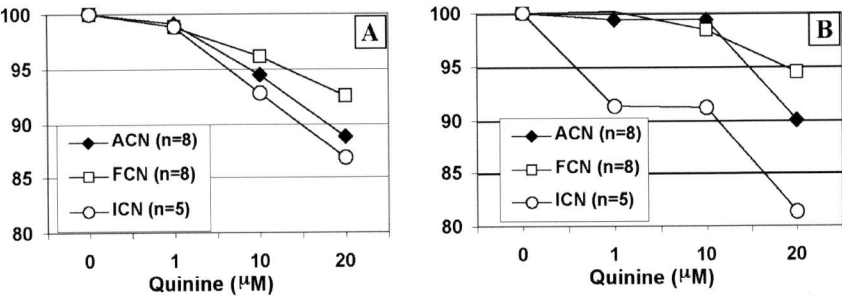

FIGURE 8.10. Percent changes in coefficient of variation (CV) of integrated burst amplitude (A) and burst duration (B) with increasing concentrations of quinine. The greatest change was seen in ICNs, followed by ACNs, and then FCNs.

of variation for burst amplitude and burst duration measures (Figure 8.10). The graph depicts quinine-induced effects at various concentrations with the reference levels normalized to 100. The highest regularization occurred for the ICNs followed by ACNs and then FCNs.

In vivo studies of the cat auditory cortex have shown that quinine increases the spontaneous firing rate in the low-spontaneous rate neurons, and enhances synchrony of firing in all auditory cortical neurons (Ochi and Eggermont, 1997; Kenmochi and Eggermont, 1997). It was suggested that an increase in Ca^{+2} channel conductance leading to increased intracellular calcium concentration leads to increased synchrony in auditory nerve fibers and in auditory cortical neurons, and that this increase in synchronization could be related to the perception of tinnitus (Eggermont, 1990; Ochi and Eggermont, 1997). Using liquid chromatography assays, the concentration of quinine in the cerebrospinal fluid of rats administered 200 mg/kg (a commonly used dosage in quinine investigations) was found to be about 0.5 µg/ml (Chmurzynski, 1997). This translates to approximately 1.4 µM of quinine, which is in the concentration range where an excitatory effect in ACNs was found.

8.4.4   Anandamide

Morefield et al. (2000) showed that the cannabinoid agonist anandamide (AN) decreased spontaneous activity of cultured neuronal networks at concentrations starting below 150 nM. IC_{50} values were approximately 300 nM. Results from in vivo studies correlated well with those obtained from primary cultures. In rats, intravenous AN administration caused immediate, dose-related changes in behavior at concentrations between 3 and 30 mg/kg (approximately 90 to 900 nM) and lasted about 15 minutes (Stein et al., 1996). It is very interesting that the networks in culture also showed spontaneous activity recovery at low concentrations (nanomolar to low micro-molar) in 15 to 20 min after application (Figure 8.11). Morefield concluded that neural tissue in culture, even after total dissociation and

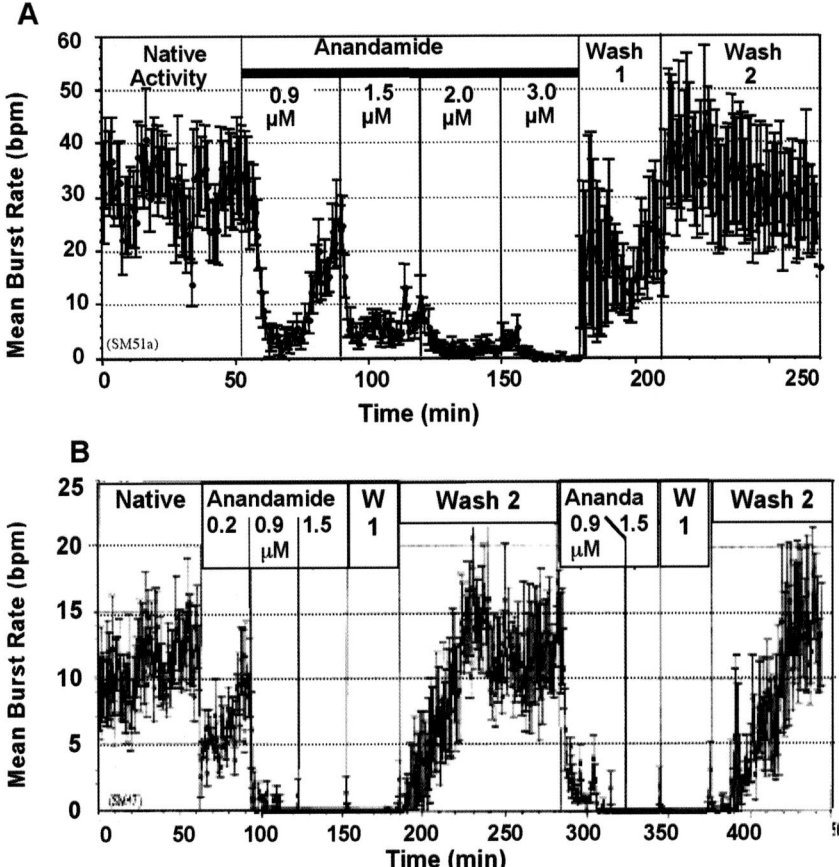

FIGURE 8.11. Responses of an auditory cortex culture (A) and one spinal culture (B) to serial additions of anandamide. Strong responses are already seen at 200 nM in (B) and both networks show a return to near-native activity after two washes (complete medium changes). Note also the spontaneous activity recovery at 900 nM in (A) and at 200 nM in (B), which is also observed in the same time frame in vivo. Data points represent means across all channels in one-minute bins and standard deviations. (Data from Morefield et al., 2000.)

reformation of circuits on the electrode array, remained pharmacologically histio-typic and reflected the properties of the parent tissue.

8.4.5 Toxins

The following short summaries of network responses to various toxins indicate substantial agreement with in vivo data. However, it is difficult to make direct comparisons between IC_{50} in vitro and LD_{50} values in vivo, as the death of an animal is usually not defined electrophysiologically. For example, respiratory insufficiency leading to death is most likely reached before cessation of all bursting activity, but perhaps not at an IC50 level. In addition, compartmentalization, effects of the blood–brain barrier, and detoxification by the liver, also need to be taken into consideration. With these caveats that will require more scrutiny of in vivo experimental protocols, the agreement between these two domains is still remarkable and provides strong evidence on the utility of cultured networks.

Botulinum Toxin A

Preliminary data from botulinum toxin A (Gross et al., 2003) show loss of activity in a time-dependent manner above concentrations of 2 ng/ml (approximately 13 pM). Above this level, the time required for 10%, 50%, and 90% reduction in activity was reduced in a dose-dependent manner. A case for partial reductions with subsequent stabilization could not be made. The only important variable was the time required to reach a particular level of activity reduction. The minimum detectable effect at 2 ng/ml compares well with reports from neuromuscular preparations (1.5 ng/ml; Clark et al., 1987) and mouse LD_{50}s of approximately 3 ng/ml (Pearce et al., 1997) and stoppage of substance P release in cultured dorsal root ganglia at 7.6 ng/ml (Welch et al., 2000).

Cyanide

Dian and Gross (2000) reported that 1 mM sodium cyanide (NaCN) caused irreversible cessation of network spike activity within 2 min whereas 100 μM cyanide resulted in a transient reduction of network spike production by $62.2 \pm 18.6\%$ within 2 min with partial recovery to $21.6 \pm 15.0\%$ of reference after 6 min. This transient was followed by a gradual decline leading to irreversible burst inhibition at 8.3 ± 0.7 h ($n = 4$), and complete spike shutoff at 12.7 ± 1.6 h ($n = 3$).

Dogs treated orally with 20 and 100 mg/kg of potassium cyanide (0.5 mM and 2.5 mM assuming 80% fluid) died in 155 min and 8 min, respectively (Gettler and Baine, 1938). The oral LD_{50} for rats was 6.4 mg/kg of sodium cyanide (ATSDR, 1989) or approximately 160 μM. These values agree well with the observed network responses.

Mercury

Gopal (2002) described acute and chronic effects of mercuric chloride on auditory cortical networks growing on MEAs. Acute application of 100 μM completely and irreversibly inhibited spike and burst activity within 10 to 20 minutes. Networks treated with 10 μM $HgCl_2$ for four weeks from the day of seeding showed no spontaneous activity. These findings are comparable to human and animal reports. In the Minamata Bay exposure, mercury concentration in the brains of patients was estimated to be 1.7 to 26 μM (Friberg and Mottet, 1989). Humans chronically intoxicated with mercury developed symptoms of poisoning when the level in the blood and brain reached 4 to 8 μM (Skerfving, 1972). Chronic mercury exposure of central auditory network activity is within the range known to cause cognitive and behavioral deficiencies in humans.

Tetrodotoxin

Preliminary data indicate IC_{50}s of 1.5 ± 1.1 nM and 6.4 ± 3.1 nM for spinal and cortical networks, respectively, and reversible loss of all spontaneous activity between 20 and 30 nM. These values overlap with LD_{50}s reported for in vivo investigations (mouse) as ranging from 6 to 8 nM after intraperitoneal, intravenous, or subcutaneous injections (Kao and Fuhrman, 1967; Mosher et al., 1964; Lipkind and Fozzard, 1973). However, one report cites an LD_{50} of 0.5 after direct intracerebral injections (Lipkind and Fozzard, 1973).

Trimethyltin Chloride (TMT)

Spinal cord networks respond to TMT at 1 to 2 μM and reliably shut off activity at 4 to 8 μM (Gramowski et al., 2000). The concentrations for a 50% reduction of activity ranged from 1.1 μM to 2.4 μM (spinal cord) and from 3.7 μM to 5.2 μM (auditory cortex) indicating a low interculture variability within one tissue type. The nonoverlapping IC_{50} range for cortical and spinal cord cultures implies tissue specificity for network response to TMT. Brown et al. (1979) determined brain concentration levels after short-term TMT exposure. After a single dose of TMT (10 mg/kg body weight i.v.), a TMT concentration of 3.28 μg/g in wet brain tissue was found. Assuming a tissue water content of 80%, an administration of 10 mg TMT/kg corresponds approximately to a concentration of 20 μM. Our in vitro network responses suggest that brain TMT concentrations of 5 to 10 μM in central nervous system tissue should result in paralysis, respiratory arrest, and death of the experimental animal. Indeed, Wegner et al. (1982) reported 100% death in mice after 48 h under 10 μM.

Trimethyol Propane Phosphate (TMPP)

In murine spinal cultures, a concentration of 2 μM TMPP reorganized network spike activity into almost synchronous, quasi-periodic burst episodes (Keefer et al., 2001). The variability of network burst parameters, quantified as coefficients of variation (CVs) decreased, and the changes in network activity paralleled the effects induced by bicuculline, a known disinhibitory and seizure-inducing drug. Keefer et al. (2001a) demonstrated that nerve cell networks in vitro allowed a rapid classification of TMPP as an epileptogenic compound. In addition, concentration ranges also matched those reported from animal experiments. Acute exposure of rats to TMPP (0.05 to 1.0 mg/kg, intraperitoneal) induces electroencephalographic paroxysms, stereotypic behaviors, subclinical seizures, clinical convulsions, status epilepticus, or lethality in a dose-dependent manner (Bellet and Casida, 1973; Wyman et al., 1993; Lindsey et al., 1998). For a rat of average mass (250 g) and blood volume (64 ml/kg), a concentration of 0.5 mg/kg corresponds to approximately 2 μM TMPP, a level where clear epileptiform burst patterns emerge in the spontaneous activity of spinal networks in vitro.

Chemical Agent Hydrolysis Products

Using neuronal networks on micro-electrode arrays and patch-clamp recording, Pancrazio et al. (2001) showed that the primary hydrolysis product of soman, pinacolyl methylphosphonic acid (PMPA), inhibited network mean burst and spike rates with an EC_{50} of approximately 2 mM. In contrast, the degradation product of sarin, isopropyl methylphosphonic acid (IMPA), and the final common hydrolysis product of soman and sarin, methylphosphonic acid (MPA), failed to affect neuronal network behavior at concentrations reaching 5 mM. Whole-cell patch-clamp records under current-clamp mode also showed that 2 mM PMPA depressed the firing rate of spontaneous action potentials (APs) to 36 ± 6% ($n = 5$ neurons) of control. Previous work in vivo reported that both MPA (Williams et al., 1987) and IMPA (Mecler and Dacre, 1982) possessed very low oral toxicity. Rats that were fed 350 mg/kg IMPA each day exhibited no toxicity after 90 days and lethal doses for each exceeded 5000 mg/kg (Mecler, 1981). In contrast to IMPA and MPA, there are no previous data on the biological activity of PMPA. In this case nontoxic responses in vitro to essentially all three hydrolysis products are mirrored faithfully by in vivo data for MPA and IMPA.

8.5   Discussion and Remaining Problems

Different parent CNS tissues form networks with different characteristic native activities in cell culture. They also yield histiotypic responses to pharmacological agents and toxins to the degree where many EC_{50} values overlap statistically with in vivo data, or are in the same range. Moreover, the data show tissue-specific pharmacological responses, which are beginning to mirror responses in vivo (cf. quinine and auditory cortex). Such results provide hope that improvements in cell culture techniques will allow routine quantitative analyses and studies of great value not just to pharmacology and toxicology, but also to basic neurophysiology, and investigations of information processing.

By improvements in cell culture techniques, we refer primarily to the repeatability of the procedure and consistency of the cultures in terms of cell ratios, cell densities, and synaptic densities. These problems are closely linked to repeatability of the age of the embryonic tissue from which the networks are derived. To routinely obtain timed pregnancies within a 24 h window is not trivial. Yet the enormous speed of developmental changes will most likely force us into using more narrow windows. For example, the longer migratory path of inhibitory neurons in the developing frontal cortex (Parnavelas, 2000) makes the timing of tissue selection critical. If the tissue is too young, the number of differentiated GABA-ergic neurons may be quite low. If we wait too long, the tissue is more difficult to dissociate with a greater loss of the larger neurons. These problems affect cell ratios and, most likely, native activity patterns as well as the accuracy of pharmacological responses.

So, the main problems do not lie in the domain of micro-electrode design or fabrication. Although improvements are always welcome, they will make major contributions to our understanding and application of network dynamics only after the biological variabilities and data interpretation problems are solved. Technical problems that must be faced lie more in the minimization of pH and osmolarity shocks during feeding and in the maintenance of stable environmental conditions during recording, than in the improvement of amplifiers. We must know more about (1) the effects small differences in embryonic age at dissociation can have on spontaneous activity patterns and pharmacological responses, (2) medium biochemistry and effects of changes on activity, (3) effects of osmolarity and pH changes, (4) identification of cells and their ratios in culture, (5) reproducibility of cell ratios by a culture staff, and (6) determination of synaptic densities as a function of time in vitro.

Data processing, display, graphing, and interpretation will also be a major challenge in the foreseeable future. Quantifying network spike and burst rates is only a beginning. Long-term extracellular recording from many neurons in a network provides (1) action potential pattern data and information on different neuronal compartments that fire, (2) individual neuronal spike production, (3) global network spike production, and (4) myriad individual, subpopulation, and global temporal patterns that we have just begun to explore. The secrets of information processing lie in those temporal-spatial patterns and we need more sophisticated, presumably nonlinear, methods to identify and quantify subtle pattern changes. It may be anticipated that we will not be able to understand information processing in the brain until we understand pattern processing in networks. At this juncture in time, the primary method that will provide data on network pattern processing is the micro-electrode array platform that can report massive amounts of spike pattern data for long periods of time from neuronal networks in culture.

References

Altamura, A.C., Moro, A.R., and Percudani, M. (1994). Clinical pharmacokinetics of fluoxetine. *Clin. Pharmacokinetics* 26: 201–214.

ATSDR (Agency for Toxic Substances and Disease Registry) 1989. Toxicological Profile for Cyanide. *ATSDR/TP–88/12; PB90–162058.* Prepared by Syracuse Research Corporation for ATSDR, U.S. Public Health Service, under Contract No. 68–C8–0004.

Bellet, E.M. and Casida, J.E. (1973). Bicyclic phosphorus esters: High toxicity without cholinesterase inhibition. *Science* 182: 1135–1136.

Brown, A.W., Aldrige, W.N., Street, B.W., and Verschoyle, R.D. (1979). The behavioral and neuropathologic sequaela of intoxication by trimethyltin compounds in the rat. *Am. J. Path.* 97: 59–81.

Charness, M.E., Simon, R.P., and Greenberg, D.A. (1989). Ethanol and the nervous system. *N. Engl. J. Med.* 32: 442–454.

Chmurzynski, L. (1997). High-performance liquid chromatographic determination of quinine in rat biological fluids. *J. Chromatogr. B. Biomed. Sci. Appl.* 693: 423–429.

Clark, A.W., Bandyopathyay, S., and Das Gupta, B.R. (1987). The plantar nerve-lumbrical muscles: A useful nerve-muscle preparation for assaying the effects of botulinum neurotoxin. *J. Neurosci. Meth.* 19: 285–295.

Dian, E.E. and Gross, G.W. (2001). Temporal dynamics of cyanide toxicity using cultured neuronal networks: Involvement of NMDA receptor. *Soc. Neurosci. Abst.* 606.6.

Eggermont, J.J. (1990). On the pathophysiology of tinnitus: A review and a peripheral model. *Hear. Res.* 48: 111–124.

Friberg, L. and Mottet, N.K. (1989). Accumulation of methylmercury and inorganic mercury in the brain. *Biol. Trace Elem. Res.* 21: 201–206.

Gettler, A.O. and Baine, J.O. (1938). The toxicology of cyanide. *Am. J. Med. Sci.* 195: 182–198.

Gopal, K.V. (2002). Neurotoxic effect of mercury on auditory cortex networks growing on micro-electrode arrays: A preliminary analysis. *Neurotoxicol. Teratol.* 25: 69–76.

Gopal, K.V. and Gross, G.W. (2004). Unique responses of auditory cortex networks *in vitro* to low concentrations of quinine. *Hear. Res.* 192(1–2): 10–22.

Gramowski, A., Schiffmann, D., and Gross, G.W. (2000). Quantification of acute neurotoxic effects of trimethyltin using neuronal networks cultured on microelectrode arrays. *Neurotoxicol.* 21: 331–342.

Gross, G.W. (1979). Simultaneous single unit recording in vitro with a photoetched laser deinsulated gold multi-microelectrode surface. *IEEEE Trans. Biomed. Eng.* BME–26: 273–279.

Gross, G.W., Wen, W., and Lin, J. (1985). Transparent indium-tin oxide patterns for extracellular, multisite recording in neuronal cultures. *J. Neurosci. Meth.* 15: 243–252.

Gross, G.W. and Kowalski, J.M. (1991). Experimental and theoretical analyses of random network dynamics. In: Antognetti and Milutinovic, eds., *Neural Networks, Concepts, Application and Implementation,* Vol. 4, Prentice-Hall, Englewood Cliffs, NJ, pp. 47–110.

Gross, G.W. (1994). Internal dynamics of randomized mammalian neuronal networks in culture. In: Stenger, D.A. and McKenna, T.M. eds., *Enabling Technologies for Cultured Neural Networks,* Academic Press, New York, pp. 277–317.

Gross, G.W. and Schwalm, F.U. (1994). A closed chamber for long-term electrophysiological and microscopical monitoring of monolayer neuronal networks. *J. Neurosci. Meth.* 52: 73–85.

Gross, G.W., Pancrazio, J.J., and Steele, D. (2003). An *in vitro* assay for botulinum toxin serotypes and their antibodies. *Soc. Neurosci. Abstr.* 122.9.

Kao, C.Y. and Fuhrman, F.A. (1967). Differentiation of the actions of tetrodotoxin and saxitoxin. *Toxicon* 5: 25–34.

Karson, C.N., Newton, J.E.O., Livingston, R., Jolly J.B., Cooper, T.B., Sprigg, J., and Komoroski, R.A. (1993). Human brain fluoxetine concentrations.*J. Neuropsychiatry Clin. Neurosci.* 5: 322–329.

Keefer, E.K., Gramowski, A., Stenger, D.A., Pancrazio, J.J., and Gross, G.W. (2001a). Characterization of acute neurotoxic effects of trimethylolpropane phosphate via neuronal network biosensors. *Biosens. Bioelectron.* 16: 513–525.

Keefer, E.W., Norton, S.J., Boyle, N.A.J., Talesa, V., and Gross, G.W. (2001b). Acute toxicity screening of novel AChE inhibitors using neuronal networks on microelectrode arrays. *Neurotoxicol.* 22 (1): 3–12.

Kenmochi, K. and Eggermont, J.J. (1997). Salicylate and quinine affect the central nervous system. *Hear. Res.* 113: 110–116.

Kovacs, G.T.A. (1994). Introduction to the theory, design, and modeling of thin-film miroelectrodes for neural interfaces. In: Stenger, D.A. and McKenna, T.M., eds., *Enabling Technologies for Cultured Neural Networks.* Academic Press, New York, pp. 121–165.

Lindsey, J.W., Prues, S.L., Alva, C., Ritchie, G.D., and Rossi, J., III (1998). Trimethylolpropane phosphate (TMPP) perfusion into nucleus accumbens of the rat: electroencephalographic, behavioral and neurochemical correlates. *Neurotoxicol.* 19: 215–226.

Lipkind, G.M. and Fozzard, H.A. (1973). An experience on the biological assay of the toxicity of imported fugu (Tetrodon). Shokuhin Eiseigaku Zasshi. *J. Food Hyg. Soc. Japan* 14: 186–190.

Little, H.J. (1991). Mechanisms that may underlie the behavioral effects of ethanol. *Progress Neurobiol.* 36: 171–194.

Lucas, J.H., Czisny, L.E., and Gross, G.W. (1986). Adhesion of cultured mammalian CNS neurons to flame-modified hydrophobic surfaces. *In Vitro* 22: 37–43.

Mecler, F.J. (1981). Mammalian toxicological evaluation of DIMP and DCPD (Phase-3 IMPA). AD-A107574. Fort Detrick, MD, U.S. Army Medical Research and Development Command.

Mecler, F.J. and Dacre, J.C. (1982). Toxicological assessment of isopropyl methylphosphonic acid. *Toxicologist* 2: 35.

Morefield, S.I., Keefer, E.W., Chapman, K.D., and Gross, G.W. (2000). Drug evaluations using neuronal networks cultured on microelectrode arrays. *Biosens. Bioelectron.* 15: 383–396.

Mosher H.S., Fuhrman F.A., Buchwald H.D., and Fischer, H.G. (1964). Tarichatoxin–Tetrodotoxin: A potent neurotoxin. *Science.* 144: 1100–1101.

Ochi, K. and Eggermont, J.J. (1997). Effects of quinine on neural activity in cat primary auditory cortex. *Hear. Res.* 105: 105–118.

Pancrazio, J.J., Keefer, E.W., Ma W., Stenger, D.A., and Gross, G.W. (2001). Neurophysiologic effects of chemical agent hydrolysis products on cortical neurons in vitro. *Neurotoxicol.* 22: 393–400.

Parnavelas, J.G. (2000). The origin and migration of cortical neurones: New vistas. *Trends Neurosci.* 23(3): 126–131.

Pearce, B.L., First, E.R., Maccallum, R.D., and Gupta, A. (1997) Pharmacologic characterization of botulinum toxin for basic science and medicine. *Toxicon* 35: 1373–1412.

Pohorecky, L.A. and Brick, J. (1977). Activity of neurons in the locus coeruleus of the rat: Inhibition by ethanol. *Brain Res.* 131: 174–179.

Potter, S.M. and DeMarse, T.B. (2001). A new approach to neural cell culture for long-term studies. *J. Neurosci. Meth.* 110: 17–24.

Ransom, B.R., Neale, E., Henkart, M., Bullock, P.N, and Nelson, P.G. (1977). Mouse spinal cord in cell culture. I. Morphology and intrinsic neuronal electrophysiologic properties, *J. Neurophysiol.* 40: 1132–1150.

Siggins, G.R., Pittman, Q.J., and French, E.D. (1987). Effects of ethanol on CA1 and CA3 pyramidal cells in the hippocampal slice preparation: An intracellular study. *Brain Res.* 414: 22–34.

Skerfving, S. (1972). Organic mercury compounds. Relation between exposure and effects. In: Friberg, L. and Vostal, J., eds., *Mercury in the Environment,* CRC Press, Cleveland, OH, pp. 141–168.

Stein, E.A., Fuller, S.A., Edgemond, W.S., and Campbell, W.B. (1996). Physiological and behavioural effects of the endogenous cannabinoid, arachidonylethanolamide (anandamide), in the rat. *Brit. J. Pharmacol.* 119: 107–114.

Welch, M.J., Purkiss, J.R., and Foster, K.A. (2000). Sensitivity of embryonic rat dorsal root ganglia neurons to *Clostridium botulinum* neurotoxins. *Toxicon* 38: 245–258.

Wegner, G.R., McMillan, D.E., and Chang, L.W. (1982). Behavioral toxicology of acute trimethyltin exposure in the mouse. *Neurobehav. Toxicol. Teratol.* 4: 157–161.

Williams, R.T., Miller, W.R., III, and MacGillivray, A.R. (1987). Environmental fate and effects of tributyl phosphate and methyl phosphonic acid. CRDEC-CR-87103: NTIS AD-A184 959/5. Aberdeen Proving Ground, MD, U.S. Army Armament Munitions Chemical Command, Chemical Research, Development and Engineering Center.

Wyman, J., Pitzer, E., Williams, F., Rivera, J., Durkin, A., Gehringer, J., Serve, P., Von Minden, D., and Macys, D. (1993). Evaluation of shipboard formation of neurotoxicant (trimethylolpropane phosphate) from thermal decomposition of synthetic aircraft engine lubricant. *Am. Ind. Hyg. Assoc. J.* 54: 584–592.

Xia, Y., Gopal, K.V., and Gross, G.W. (2003). Differential acute effects of fluoxetine on frontal and auditory cortex networks in vitro. Brain Res. 973: 151–160.

Xia, Y. and Gross, G.W. (2003). Histiotypic electrophysiological responses of cultured neuronal networks to ethanol. *Alcohol* 30: 167–174.

9
Closing the Loop: Stimulation Feedback Systems for Embodied MEA Cultures

Steve M. Potter, Daniel A. Wagenaar, and Thomas B. DeMarse

9.1 Introduction

In the vertebrate nervous system, all key functions are performed by many nerve cells working in concert. By contrast, single neurons are sometimes important for specific functions in invertebrates, such as the locust's giant motion detector neuron (LGMD), which integrates visual information and triggers jumping (Gabbiani et al., 1999). To understand more complex processing in vertebrates, we need to know how single-unit activity combines to form the network-level dynamics that take in sensory input, stores memories, and controls behavior. Cultured neuronal networks have provided us with much of our present understanding of ion channels, receptor molecules, and synaptic plasticity that may form the basis of learning and memory (Bi and Poo, 1998; Latham et al., 2000; Misgeld et al., 1998; Muller et al., 1992; Ramakers et al., 1991). To study the nervous system in vitro offers many advantages over in vivo approaches. In vitro systems are much more accessible to microscopic imaging and pharmacological manipulations than are intact animals. Recent developments in multi-electrode array (MEA) technology, including those described below, will enable researchers to answer questions not just at the single-neuron level, but at the network level. Most MEA research has involved recording activity that cultured networks produce spontaneously, via up to 64 extracellular electrodes. Although some studies also included electrical stimulation via the substrate electrodes, it was applied to only one or two of them at a time (Connolly et al., 1990; Fromherz and Stett, 1995; Gross et al., 1993; Jimbo and Kawana, 1992; Jimbo et al., 1998; Maeda et al., 1995; Oka et al., 1999; Regehr et al., 1989; Shahaf and Marom, 2001; Stoppini et al., 1997). We propose that in order to substantially advance our understanding of network dynamics, we need high-bandwidth (many neurons) communication in *both* directions, out of and into the network. This chapter describes technologies that allow recording and stimulation on every electrode of an MEA, and a new closed-loop paradigm that brings in vitro research into the behavioral realm.

9.1.1   The Importance of Embodiment

Nervous systems evolved to aid the survival of motile organisms, by directing their interactions with their surroundings. In the natural environment, sensory input to an animal's nervous system is largely a function of its recent output: as the animal moves and interacts within its environment, its sensory systems are actively oriented so as to provide the information that will be most useful in controlling subsequent behaviors (Nolfi and Parisi, 1999). Animals sense the consequences of an action for a few tens of milliseconds after the motor command is sent to muscles. Neural output is expressed continuously, while it is being modulated by a continuous stream of sensory inputs. This tight sensory–motor loop is likely to be important for learning to predict the consequences of actions and to create and fine-tune adaptive behaviors.

Contrast this with traditional in vivo neurophysiology in the lab: the animal is often anaesthetized and immobilized, unable to behave. Rarified, unimodal sensory input is provided in brief exposures, or "trials", and neural responses are measured directly with electrodes. This open-loop approach has been helpful in understanding cortical maps, and neural receptive fields, among other things. But we must interpret such findings with care; it is likely that the dynamics of the nervous system under these unnatural circumstances are substantially different than in a freely behaving animal (Hartmann and Bower, 2001). We wish to bring in vitro networks closer to the kind of neural processing that nervous systems evolved to do, that is, to take in sensory information continuously, process it continuously, and express adaptive behavior continuously. Each of these processes interacts with the others, so they should not be studied separately.

9.1.2   A New Closed-Loop Research Paradigm

In vitro networks have always been incapable of expressing behavior, by the very fact of being removed from the donor's body. We have developed systems for re-embodying cultured networks, allowing them once again to express behavior. Our embodiments, or neurally controlled animats, are either simulated creatures on the computer (DeMarse et al., 2001), or actual robots (Bakkum et al., 2004). The whole system of MEA culture plus embodiment we call a "hybrot" because it is a hybrid robot with both living and artificial components. The greatest advantage of these systems is that their "brains" hold perfectly still on the microscope stage, amenable to detailed imaging for months at the submicron level, while controlling behavior and receiving sensory inputs (Figure 9.1). Such detailed, extended imaging is not presently possible in the brains of behaving animals. Two-photon time-lapse fluorescence optical microscopy can be used with rodent cortical networks on MEAs in order to find links between morphological and functional dynamics at the network level. This new approach of embodied cultured networks will help us and others address questions about how activity shapes network development, how neuromodulatory systems influence connectivity, and what the bases are of distributed neural activity for sensory processing, memory formation, and behavioral control.

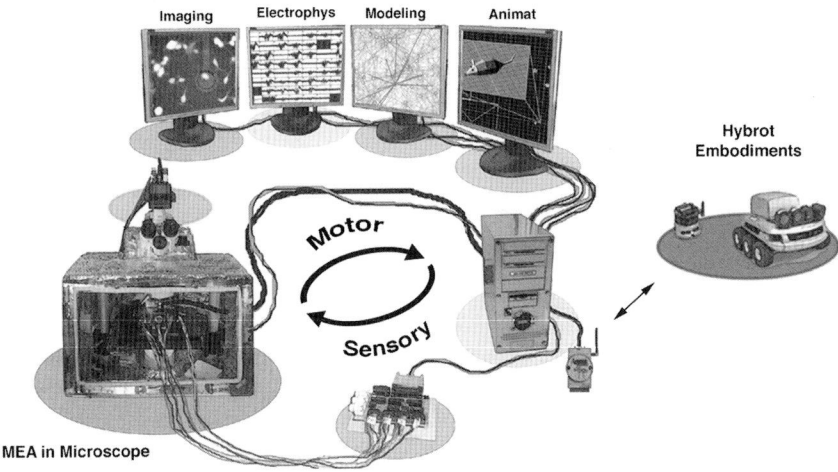

FIGURE 9.1. Embodying cultured neurons. We closed the sensory–motor loop between an MEA, serving as the "brain" of a Hybrot, and a simulated or robotic animat as its embodiment. While controlling behavior and receiving sensory stimulation, the cultured network can be imaged at the micron scale using time-lapse fluorescence microscopy. (From Zenas Chao)

9.2   Long-Term Cell Culture

As with any model system, dissociated primary neuronal cultures have advantages and disadvantages. Unlike cells from other tissues, mammalian neurons are (usually) terminally differentiated when obtained from late-term embryos or neonates, and thus cannot be multiplied by passaging. This means that a ready supply of brain tissue donors must be available, and that the cultures must be pampered more than those prepared from dividing cells, if long-term studies are to be carried out. This includes rigorous adherence to sterile techniques, careful choice and replenishment of media, and maintenance of pH, temperature, and osmolarity (Banker and Goslin, 1998). The reward for this extra effort is the ease of both observation and manipulation of neural circuits, allowing numerous types of inquiries not feasible in humans or even lab animals.

Both the blessing and the curse of neural cultures is that they are much simpler than living brains. Neurons, especially when cultured with glia (Ullian et al., 2001), spontaneously form functional synapses in vitro, and develop complex patterns of activity that closely resemble those recorded from developing brains of animals (Ben-Ari, 2001). Neurons retain their morphological and pharmacological identities in culture, but there are likely to be numerous subtle changes in their properties due to the unnatural environment in which they have been placed. A model system is only helpful if it is simpler than the thing being modeled. But in vitro researchers must always be wary of the limitations of their model systems. We are improving neural cell culture in an effort to remove some of those limitations, as described below.

9.2.1   Dense Monolayer Cultures

Of in vitro model systems, dissociated monolayer cultures provide the best access
by electrodes (whether micro-pipettes or substrate-integrated electrodes), drugs,
and microscopic imaging. If we are to make full use of multi-electrode array substrates, each extracellular electrode should be able to stimulate and record from at
least one neuron. Efforts to trap neurons next to electrodes with micro-engineered
structures such as silicon wells or pillars have been successful, at least for short-
term cultures (Maher et al., 1999; Merz and Fromherz, 2002). Mammalian neurons
survive longer and develop more synapses when in direct contact with glial cells
(Ullian et al., 2001; personal observations), but these often provide the neurons
with the tensile forces needed for escape. In working with silicon neurochips and
neuroprobes in the Pine lab at Caltech, Potter and colleagues Mike Maher and
Hannah Dvorak-Carbone even observed neurons escaping from their cages at the
expense of leaving their nuclei behind! Needless to say, they died soon after. Keeping neurons near electrodes by treating the electrode regions with "neurophilic"
chemicals or the surrounding regions with "neurophobic" chemicals, has also been
reasonably successful. Branch et al. (1998) have maintained adherence to stamped
patterns of poly-D-lysine for about one month in vitro. This approach shows much
promise, not only for keeping neurons in close apposition to electrodes, but for
creating well-defined simple neural circuits in vitro (Nam et al., 2003).

   We chose a different, far simpler approach to ensure that each of the 60 electrodes
in the MEAs we use (Multi Channel Systems, GmbH) is functionally interfaced to
one or more neurons: plate them very densely (Figure 9.2). We usually plate 20,000
to 50,000 mouse cortical cells in a three millimeter diameter region, resulting in
densities of 5,000 to 10,000 cells per square millimeter in the electrode region.

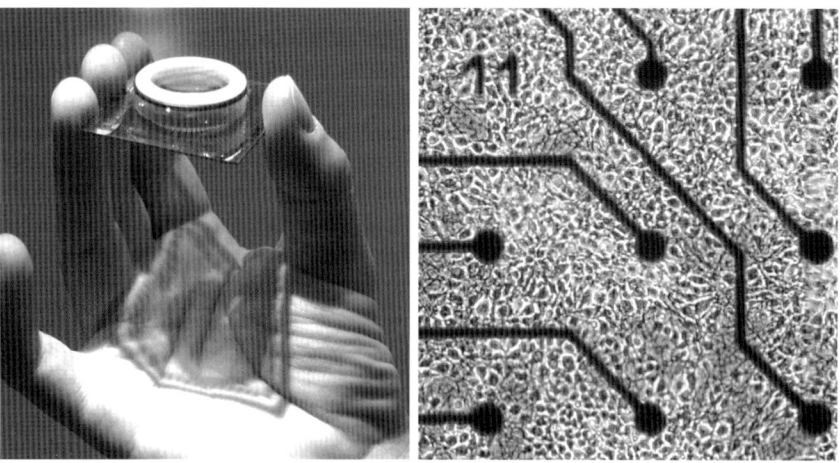

FIGURE 9.2. FEP-membrane sealed MEA (left) and dense monolayer culture of neurons
and glia (right), after several months in vitro.

Thus, each 10 μm diameter electrode will have at least one and usually several neurons within recording and stimulation range.

Considering the three-dimensional structure of intact brains, to grow in a monolayer is clearly not what neurons and glia evolved to do. We and others who tried to produce dense monolayers often observed that cells would tend to clump up and form many-cell balls, following their tendencies to adhere to and migrate along each other during development in vivo. Although such clusters of neurons and glia (which Gross and Kowalski (1999) call "nacelles") are interesting to study since they also form large fascicles of neurites between them that synchronize their activity (Segev et al., 2003), their formation necessarily causes most of the MEA's electrodes to be wasted because the neurons move away from them as they form clusters.

To take full advantage of the multi-unit philosophy behind MEAs, we would like to have as many neurons in contact with electrodes as possible. In order to prevent clumping of neurons and glia in dense cultures, it is necessary to provide the cells with a very adherent substrate, which they prefer over adhering to each other. We found that treating the MEAs with polyethyleneimine (Lelong et al., 1992) and then laminin allows the cells to grow in a monolayer for months (Potter and DeMarse, 2001). Substituting polylysine for polyethyleneimine was not as effective on our MEAs, which have silicon nitride as the insulating layer. It is likely that each type of MEA insulation material will exhibit different cell adhesion properties, and a variety of treatments should be tried, such as exposure to polycations and extracellular matrix proteins, oxidation by flame or plasma, or covalent modification (Bohanon et al., 1996; Lucas et al., 1986; Stenger et al., 1993; Zeck and Fromherz, 2003).

Our dense monolayer cultures are not strictly monolayers: the glial cells, allowed to multiply and eventually become contact-inhibited, often form a very thin layer under, and sometimes over, the neurons and their processes. By scanning labeled cells with a two-photon fluorescence microscope, we have observed that our cultures are 15 to 20 μm thick, and the neuron somata form a monolayer. We make no effort to inhibit the growth or division of glial cells, because they contribute to neuron survival and synapse formation as mentioned above, and because they do not seem to impede the electrodes' ability to stimulate or record neurons. Some electrodes on MEAs become damaged after repeated plating of cultures, from deterioration of the contact pads, titanium nitride electrode surface, or silicon nitride insulation. We routinely have neural activity on every MEA electrode that is physically intact.

9.2.2 Sealed Dishes for Long-Term Cultures

9.2.2.1 Why Neuron Cultures Die

Primary neuron cultures typically survive for less than a couple of months. One common cause of death is obvious: infection. Mold spores are ubiquitous. The air gap in most culture dishes that allows exchange of oxygen and carbon dioxide also allows airborne pathogens to contaminate cultures. The warm humid environment of the incubator is itself often the source of mold, and when one dish gets infected, others are likely to suffer a similar fate. To bring cultures out of the incubator for imaging and manipulations in a typical nonsterile laboratory environment puts them at further risk of infection.

There is another common, but much less obvious, reason why primary neuron cultures die: changes in osmolarity. The humid environment in the incubator is supposed to prevent evaporation of cell culture media. However, unless the incubator door is never opened, the mean humidity inside is substantially less than 100%. In a busy lab where the incubator is opened often, it may be in the 80% range, and cultures suffer from hyperosmolarity due to evaporation. This problem is made worse by the tradition of feeding neural cultures by replacing only half the medium with fresh medium; hyperosmolarity persists even after feeding. This causes a gradual deterioration and death of neurons that is usually considered to be "normal" for neural cultures because checking medium osmolarity is not commonly done.

9.2.2.2   Gas-Permeable Membrane Dishes

We developed a new culturing method that greatly reduces the occurrence of both of these problems, and has allowed us to maintain several neural cultures for well over a year (Potter and DeMarse, 2001), and for over two years in one case. The culture dishes are hermetically sealed with a Teflon membrane (Figure 9.2), fluorinated ethylene-propylene, 12.5 μm (Dupont). Although this membrane has no pores (thus preventing infection), it is quite permeable to some small molecules, notably oxygen and carbon dioxide. It is hydrophobic, and thus relatively impermeable to water and water vapor. (Note the difference from Gore-Tex Teflon material used on rain gear, which has pores, and *is* permeable to water vapor.) This allows us to culture our cells in an incubator maintained at 65% relative humidity. A dry incubator full of sealed dishes never needs to be cleaned or sterilized. Of great interest to us and others using MEAs, the low humidity allows putting expensive electronics inside the incubator without the risk of damage by water condensation (Figure 4.6, Section 4.3). Sealed dishes can be repeatedly removed from the incubator for imaging or MEA electrophysiology without fear of contamination.

The membrane slows the shift in pH of carbonate-buffered media caused by removal from an incubator with 5% CO_2 atmosphere, by about a factor of two compared to a standard culture dish with an air gap (Potter and DeMarse, 2001). In practical terms, this allows 30 min to 1 h of experimentation outside the incubator before the medium must be exchanged or re-equilibrated with 5% CO_2. The membrane is transparent and amenable to imaging on an upright microscope. The sealed MEA cultures are fed by removing the special Teflon lids (available from ALA Scientific) in the laminar flow hood, and replacing all (not half) of the medium, approximately once per week. Of course, proper sterile technique must be utilized during feeding. If one culture does become infected, the use of sealed dishes prevents mold from spreading to others in the incubator.

The mean concentration of oxygen in the brain is far less than the atmospheric 20%, and usually between 1 and 5% depending on brain region (Studer et al., 2000). Presumably neurons in a 20% O_2 incubator are suffering from an unnatural level of oxidative damage. Therefore, we also routinely use an atmosphere brought to 9% oxygen by injecting pure nitrogen, because this has been shown to enhance survival of primary hippocampal cultures (Brewer and Cotman, 1989).

9.2.2.3   Transporting Live MEA Cultures

It was necessary to ship a number of young and elderly MEA cultures from Caltech to Georgia Tech when the Potter group moved to Atlanta in 2002. We created a system that resulted in viable firing cultures after a transcontinental FedEx journey. The most important consideration is that the cultures not be exposed to turbulence of the medium, which might tear them from the substrate. We made lids consisting of a 2 mm layer of Sylgard (silastic rubber, Dow Corning) on a glass microscope slide, and pressed the Sylgard layer against the glass ring of the MEA after over-filling the dish with degassed Hibernate medium (Brewer and Price, 1996), being very careful that no air remained in the dish. As long as there are no bubbles in this rigid vessel, neural cultures are extremely resistant to damage by shock (g-forces). The cultures were placed in styrofoam boxes with 4°C cold packs, to reduce metabolism. Hibernate medium is buffered for ambient carbon dioxide levels, so no special consideration for maintenance of pH was made. Upon arrival, the medium was replaced with standard serum-containing medium (Potter and DeMarse, 2001), and the glass/Sylgard lids were replaced with Teflon membrane lids. Neural activity was recorded as soon as the cultures warmed up and equilibrated to the culture medium.

9.3   Real-Time Data Processing

Most experimental scientists would prefer to have the results of their experiment as soon as possible, ideally, with intermediate results appearing on the computer screen even before the experiment is over. This allows the experimenter to make the best of unexpected contingencies that might warrant a redirection of efforts. For experiments lasting days, as ours often do, online data processing becomes crucial. Most electrophysiology and imaging systems incorporate this "online" philosophy. But online is not necessarily real-time, although the terms are often used interchangeably. By "real-time," we mean systems in which results are available in milliseconds, and in which maximum delays are known and guaranteed. Real-time systems are necessary for closed-loop electrophysiology. Because real-time systems are not part of commercially available MEA setups, and we need real-time feedback for our closed-loop paradigm, we developed our own system.

9.3.1   Preventing Data Glut

Multi-electrode arrays are capable of generating large amounts of data in a short period; sampling each of 60 channels at 25 kHz creates a data stream of several megabytes per second, or tens of gigabytes in one afternoon. Even with the plummeting cost of disk storage, the largest affordable current storage systems are taxed by continuous recording for days. Clearly, some data reduction strategy is necessary, and this usually takes the form of extraction of spikes from the raw data stream.

It is assumed by most MEA users that neural signals smaller than action potentials, such as postsynaptic potentials, are hidden in the noise of an extracellular recording, so it makes sense only to record action potentials. This is not strictly true. We have observed that noise levels are higher when recording from a living culture than from a clean MEA with just medium in it. By blocking all sodium-channel-dependent activity in an MEA culture with tetrodotoxin, we verified that indeed, many of the tiny voltage peaks often called "noise" are actually biogenic, and may include both subthreshold depolarizations as well as spikes recorded from neurons at some distance from the electrode. This exercise is helpful in setting an appropriate threshold for detection of action potentials, which usually ranges from four to six times the standard deviation of the background signal, depending on our desire to include or exclude questionable peaks.

9.3.1.1   MeaBench Software

Wagenaar developed a suite of software modules to allow us to do MEA recording and stimulation in real-time. It was necessary to move from Windows to the Linux operating system, because Windows does not allow tight enough control of low-level interrupts that may disrupt real-time processing. Our philosophy is to make MeaBench open-source, extensible, and scriptable with a standard UNIX command-line interface. Data streams or files from MeaBench modules can also be sent to other software packages, such as MatLab (Table 9.1).

Spike-detection thresholds may change during the course of an experiment, especially long experiments where the physical relations between neurons and electrodes are changing due to cell growth. MeaBench has a module that adaptively adjusts the spike-detection threshold to track the RMS background signal, per electrode, on a time-constant of one second. To reduce the number of unusually large noise peaks detected as spikes, and to avoid counting a multiphasic spike as two or more spikes, a simple spike shape criterion (suggested by P. P. Mitra) is applied to all candidate spikes. MeaBench saves 3 ms of "context" centered on the peak of each identified spike, that is, a small segment of raw data, which may be used later for spike sorting. By saving just the spike time, electrode number, and context for each spike detected, and discarding the rest of the data, we usually reduce data storage requirements by a factor of a hundred, compared to saving raw data. Each spike saved with context is 164 bytes. Another order of magnitude of data reduction can be realized by forfeiting spike waveforms and only saving spike

TABLE 9.1. MeaBench features.

- Modular, open-source design
- Direct streaming of raw electrode voltage traces to disk
- Real-time artifact suppression
- Real-time spike detection, with easily replaceable detection algorithms
- Playback of raw electrode traces and recorded spikes at any speed
- Online visualization of electrode traces, with:
  - Markers for spikes
  - Variable window size
  - Option to trigger off auxiliary channel pulses
  - Unique "scrollback" buffer, to allow closer inspection of interesting recent events
- Online generation of raster plots, in three ways:
  - One raster showing all activity (accumulated over electrodes)
  - 8 × 8 geometric display of recording electrodes
  - One raster per stimulus type
- Raster plot can be scrolled back during experiment to view any previous interval
- Online sonification of spikes (stereo and tonal mapping)
- A suite of MatLab functions to import data from MeaBench to MatLab
- An easy interface that allows any program to hook on to any of MeaBench's data streams
- Linux OS, command-line based, modules are easily scripted

times and electrode numbers (16 bytes/event). By contrast, compression of raw data files by lossless algorithms such as the UNIX gzip (LZ77) function typically only reduces file size by a factor of two. Part of dealing with data glut is to decide carefully beforehand which features of the recordings are crucial for the questions being asked, and which are not.

9.3.1.2   Spike Sorting

In our dense cultures, each electrode records signals from three to ten neurons, and typically there are few inactive electrodes. Thus, spike sorting in real-time is still not feasible on a desktop computer without custom digital signal-processing hardware, such as Plexon's Multichannel Acquisition Processor, designed for that purpose. For that reason, we usually combine the activity from several neurons recorded by any given electrode, in defining a relevant motor pattern. Spike sorting is still fraught with uncertainty due to spike overlap, especially during bursts, and variability due to cellular changes and noise. Recent theoretical advances such as noise modeling, clustering, and decomposition show promise of overcoming these problems (Shoham et al., 2003) but will be difficult to implement in real-time (millisecond latency) on desktop computers for more than a few channels.

9.3.1.3   Line-Noise Filter

Even with careful attention to grounding and shielding, it is usually not possible to eliminate all interference from devices running off the mains power. MeaBench has a module that can eliminate interference from the mains, including 60 Hz (or 50 Hz in Europe), and its harmonics. Note that it is a bad idea to use a 60 Hz notch filter for this purpose, because the interference is seldom, if ever, a pure sine wave, and because much of the spectral content of interesting neural signals is near 60 Hz, such as gamma oscillations (Cunningham et al., 2003). We use a simple home-made mains thresholder consisting of two diodes and a voltage-divider, that plugs into the mains and sends out low-voltage transitions in phase with the 60 Hz high voltage. This is recorded on one of the analog input channels along with the neural data. MeaBench creates a template of one period of the mains-related interference on each electrode, by averaging many 60 Hz periods, triggered by the mains thresholder. This template is then subtracted from the incoming data stream from each electrode. This approach could be adapted to other periodic noise sources present in some lab environments, such as motors.

9.4   Stimulation Systems

Speaking in terms of embodied networks, the input or sensory side of MEA technology is not as technically well-developed as the output or motor side. But it is equally important in our closed-loop paradigm. The very same MEA electrodes used for recording can (and should) also be used to stimulate neurons.

9.4.1   Optimization of Stimulation Parameters

A wide variety of pulse shapes have been used by the groups mentioned in Section 9.1 to evoke neuronal activity through MEA electrodes, including voltage-controlled and current-controlled pulses. How does one decide what kind of stimulation to use? We have systematically investigated the efficacy of both types of stimuli as a function of amplitude and pulse width (Wagenaar et al., 2004). We found that in most cases negative current pulses are what excite neurons to fire action potentials. That does not mean that voltage-controlled pulses are not useful. In fact, positive-first biphasic voltage-controlled pulses were the most effective stimuli in our repertoire. This is easily understood if one realizes that the sharp downward voltage transient between the two phases corresponds to a strong negative current pulse.

Researchers have traditionally preferred current-controlled stimuli, because their efficacy appears more amenable to modeling (Buitenweg et al., 2002), as the electric field and potential resulting from stimulation are directly proportional to the current passing through an electrode. There are, however, significant advantages to using voltage control: not only is the circuitry needed to control voltages simpler, but more important, with voltage control it is possible to avoid electrochemical reactions. Current-controlled stimuli can easily exceed voltages that can damage electrodes and harm neurons; this becomes significant when electrode voltages exceed one volt. Although the damage can be reduced by employing charge-balanced pulses, it is still desirable to avoid it altogether. Under current control, this is only possible if the impedance spectra of all the electrodes in the array are known. Moreover, the key advantage of current control—the ability to calculate the resulting electric field in the medium surrounding the

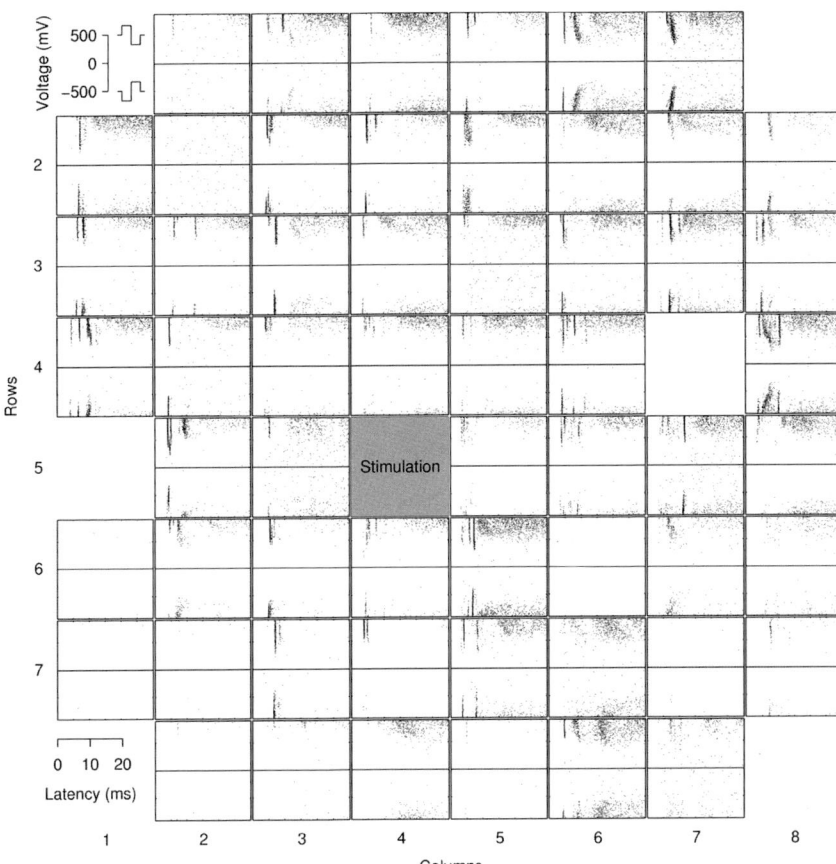

FIGURE 9.3. Short-latency action potentials in response to different stimuli. Each dot in a raster plot is one action potential induced by a biphasic pulse delivered to electrode 45 at time zero. As voltage was increased from zero to 1 volt, more responses were observed across the network. (Reprinted from Wagenaar et al. (2004) with permission from Elsevier.)

electrode—is compromised in MEAs because leakage currents through the insulation layer can reduce the current actually passing through the electrode by as much as 30% depending on the integrity and thickness of the insulation. Consequently, we use voltage-controlled positive-first biphasic pulses of less than 1 volt.

Pulse amplitude is the main determinant of stimulus efficacy (Wagenaar et al., 2004) (Figure 9.3). The number of cells directly stimulated by a given voltage-controlled pulse grows linearly with the amplitude of that pulse. Cultured networks are not as sensitive to the width of voltage pulses. Of course, the pulse must be wide enough to allow the cell membrane and all the parasitic capacitances in the system time to charge, (approximately 400 microseconds in our system). Increasing the width beyond 400 microseconds has little effect on neural response.

In Section 9.4.3 we describe two MEA stimulation systems designed with slightly different philosophies; the RACS by Wagenaar (Wagenaar and Potter, 2004), and the 64-CNS by DeMarse. But first we consider ways to deal with stimulation artifacts.

9.4.2   Stimulation Artifacts and Software Solutions

Anyone who uses MEA electrodes to stimulate and record neural activity knows the two are difficult to combine. Multi-electrode arrays typically record signals of 10 to 100 µV, whereas stimuli are on the order of one volt. The large disparity between these two voltage ranges means that artifacts resulting from stimulation can easily swamp out recorded action potentials, even on electrodes distant from the stimulation site.

Several factors contribute to stimulation artifacts (Grumet et al., 2000). A combination of capacitive crosstalk between electrode traces and conduction through the culture medium couples the stimulated electrode to all of the other recording electrodes. If the resulting transient is larger than the dynamic range of the amplification system—as is often the case—the nonlinear properties of saturated amplifiers and the connected filters greatly increase the size and duration of the artifact. We often observe cross-channel stimulus artifacts of several hundred microvolts lasting tens of milliseconds. Because many neurons respond to stimulation within this period, it is important to be able to record immediately after stimulation, so artifact suppression is essential.

Some hardware approaches based on track-and-hold (active suppression) circuits have been successful (Jimbo et al., 2003; Novak and Wheeler, 1988), but are not commercially available. One recently developed commercially available hardware solution (MEA1060-BC by Multi Channel Systems, which we have not tried) grounds the amplifiers during stimulation, and is claimed by the manufacturer to "completely remove any stimulus artifacts." However, without a sample-and-hold circuit, the DC offset voltage often found on MEA electrodes can itself produce a substantial artifact when the amplifier is switched back into the circuit (Jimbo et al., 2003). For stimulation systems that don't already have active artifact suppression in hardware, software solutions can be easily and cheaply applied.

We developed a software approach to artifact suppression that allows us to record within 1 to 2 msec after stimulation from all but the stimulated electrode itself (Wagenaar and Potter, 2002). Artifacts depend on the stimulation history and on the individual recording electrode, so each artifact tends to be different from any other artifact. This makes template-based algorithms (which subtract a fixed model of the artifact) perform poorly. We found that the shape of the artifacts is not well modeled even by a variable exponential decay. Instead, our SALPA* algorithm models each individual artifact, in real-time, by a curve constructed from locally fitting polynomials to the recorded signals. Subtraction of this model

---

* Suppression of Artifacts by Local Polynomial Approximation.

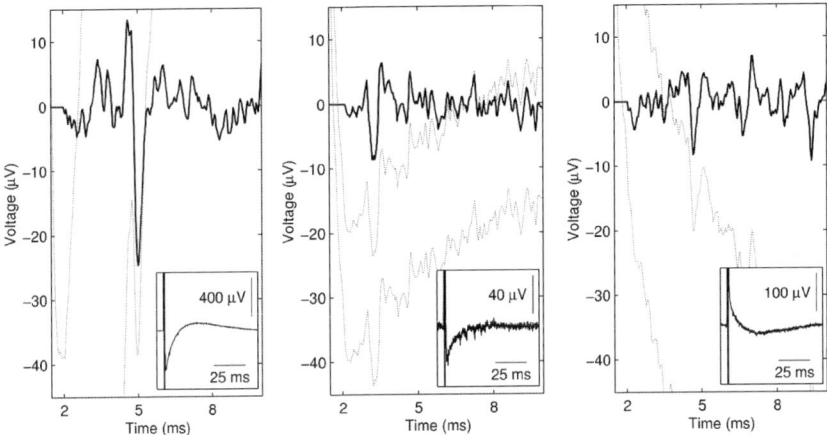

FIGURE 9.4. MEA recordings without (dotted traces and insets) and with SALPA (solid traces) applied, to practically eliminate stimulation artifacts on nonstimulated channels within one millisecond after amplifier becomes unpegged from its supply rails. (Reprinted from Wagenaar and Potter (2002) with permission from Elsevier.)

from the recording leaves an artifact-free signal in which spikes can be detected by voltage thresholding (Figure 9.4).

By leveling large artifacts upon which tiny action potentials ride, SALPA enables their detection an order of magnitude sooner, and can be used in online closed-loop systems in which the spikes detected are used to trigger stimuli. This opens up a new window for studying stimulus responses in culture, at latencies of only a few milliseconds. In particular, spikes from neurons directly stimulated by the electric pulse, without synaptic communication, became detectable for the first time (Figure 9.5). This revealed that a pulse delivered to a single electrode directly stimulates cells with arborizations covering the entire electrode array (Figure 9.3).

9.4.3   Stimulation Hardware

We developed two stimulation systems that can be plugged directly into the MEA1060 preamplifier (Multi Channel Systems) to enable one to stimulate any of the 60 electrodes of the MEA substrate. Commercial stimulation systems to date have been limited to small numbers of electrodes (usually fewer than ten), and required fully specifying a stimulation protocol ahead of time. We enabled short-latency stimulation feedback based on recorded signals. Our systems were designed with the idea of delivering a wide variety of spatiotemporal stimulus patterns to a cultured network, while recording continuously. Rapid switching between stimulation and recording on any electrode is crucial, especially if stimuli are to be triggered by the neural activity itself. Previous systems required manual plugging and unplugging of wires, to stimulate different electrodes. Ours use multiplexing of stimuli to isolation switches. If isolation switches are included in a

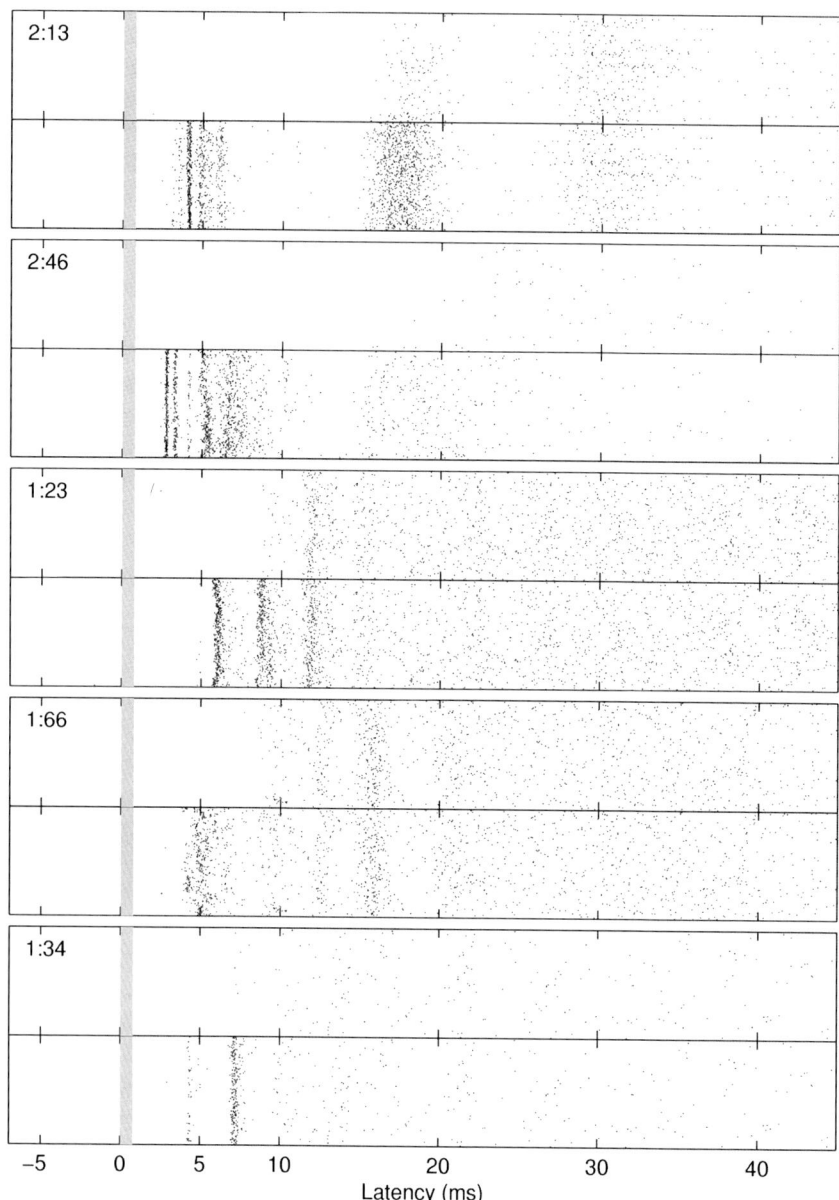

FIGURE 9.5. SALPA allows recording of action potentials (dots in raster plots) at shorter latencies than if no artifact subtraction is used. Two MEAs and five different electrodes' responses are shown (dish:electrode) without (top raster) or with (bottom raster) SALPA applied to raw voltage traces before spike detection by thresholding (at five times RMS noise, without stimuli).

stimulator design, it is important to locate these switches close to the MEA, because any low-impedance path away from the MEA can serve as an antenna, bringing unacceptable noise into recordings. Thus, we put the switches as close as possible to the MEA in both of our simulators, the RACS and the 64-CNS. Switches were carefully selected to have good isolation characteristics, low leakage current, and small charge injection.

9.4.3.1   The RACS: Real-Time All-Channel Stimulator

The philosophy behind the RACS (Wagenaar and Potter, 2004) was to make an inexpensive, flexible real-time stimulation system controlled by a dedicated low-end computer running the RT Linux real-time operating system. RACS is modular, with four banks of 16 stimulator lines that plug directly into each edge of the MEA1060 preamplifier (Figure 9.6, left). Adaptation to other recording systems (including in vivo probes) should be straightforward, provided they allow direct electrical access to the electrodes. The RACS main board (Figure 9.1) connects to the parallel port of a PC, and comprises a digital-to-analog converter and multiplexer. This routes stimuli and switching signals to the four modules that plug into the recording system. There are also auxiliary analog and digital outputs that may be used to trigger other lab equipment or to encode experimental parameters. The PCB layout uses standard 0.1 in. DIP components, and is publicly available (see Wagenaar and Potter (2004)). It can be replicated at an estimated cost of US$ 250 and about one day of work.

Controlling the stimulator is extremely flexible: stimulation sequences can be generated using Perl scripts. By making such scripts read from a MeaBench spike stream, stimuli can be made contingent on the firing pattern of the culture. Because

FIGURE 9.6. All-channel stimulation systems for Multi Channel Systems preamplifiers. The RACS system (left) has a modular design, with one switching circuit plugged in to each side of the preamplifier. It features simple 0.1 in. DIP construction, with complex stimulus patterns being controlled by a separate computer running RT-Linux. The DAC and multiplexers are on a separate board (shown in Figure 9.1). The 64-CNS for "inverted microscope" amplifier (center) and "upright microscope" amplifier (right) has an onboard USB-interfaced microprocessor for control of complex stimulus patterns, and features surface-mount components for ease of access to the MEA by micro-pipettes.

the stimulator can switch between electrodes with microsecond timing, it is possible to stimulate using arbitrarily complex multi-channel patterns. Thus, for the first time, stimulating neuronal ensembles in culture with naturalistic patterns across all electrodes of the MEA is possible. Spatiotemporal analysis of multi-unit recordings is now common, thanks to MEAs and multi-wire probes. But spatiotemporal distributed stimulation is virtually nonexistent in the literature. One notable exception is from Heck (1995), who used distributed patterns of stimulation (across 11 wire electrodes) to test circuit hypotheses in cerebellar slices. This may be a far better way to probe the information-processing capacity of neuronal networks than sending in stimuli on single electrodes.

9.4.3.2   The 64-CNS: 64-Channel Neural Stimulator

Like the RACS, the 64-CNS was also designed to allow a host computer to dynamically stimulate electrodes on a MEA in patterns that can be modified in real-time in response to the ongoing neural activity. Unlike the RACS, it has everything on one low-profile PC board with surface-mount components (Figure 9.6, center and right). The system features an onboard 8 MHz micro-controller, obviating the need for the RT Linux box used by RACS. The micro-controller schedules the timing and delivery of biphasic voltage stimulation pulses to each of the 60 channels on the MEA. This relieves the host computer from the micro-second timing requirements needed to deliver patterned stimulations to the MEA, which is difficult to accomplish on common multi-tasking computers.

The board plugs directly into a version of the MEA1060 preamp built with headers in place of the grounding DIP switches. It has a programmable voltage source and low-noise programmable switches that control the delivery of stimulation pulses to any of the 60 channels on the MEA. A blanking system is used to ground the recording amplifiers during the delivery of stimulation pulses, reducing stimulus artifacts with the limitations described above. The micro-controller receives commands from a host computer via a high-speed USB serial interface. These commands are stored on the microprocessor's program stack and executed when triggered by the host. Stimuli of the same amplitude can be delivered to any number of electrodes simultaneously. Stimuli of different amplitudes can follow each other by as little as 500 µs.

Our hardware and software systems allowed neural activity (analogous to motor commands) to be processed rapidly and used to trigger electrical stimuli via substrate electrodes (analogous to sensory inputs). Using MeaBench, and the RACS stimulation system, our 15 ms loop time includes 60-channel recording, spike extraction, pattern detection, and the triggering and delivery of stimuli to multiple electrodes (Wagenaar and Potter, 2004). This is at the fast end of sensory motor loops in mammals. Activity patterns to be used as triggers may include bursts on certain channels, a vector sum of activity across the array (Lukashin et al., 1996), or the occurrence of precisely timed spatiotemporal sequences of action potentials (Nadasdy, 2000). To help decide which neural activity patterns might serve as effective motor commands, we use a number of online visualization tools included in the MeaBench suite, such as raw voltage traces, raster plots, burst analysis, and stereo sonification of spike data.

9.4.4   The Future of MEA Stimulation

The language that neurons use to communicate with each other is chemical, that is, neurotransmitters. Extracellular electrodes are able to stimulate neurons by the lucky fact that part of the communication process within a neuron is electrical, that is, the opening of voltage-sensitive ion channels. Although it is possible to induce action potentials by electrically depolarizing the excitable membrane of neurons, it would be more natural to induce them via the very chemicals that neurons use to induce them. Some advances have been made in this direction. One approach is to include into the MEA a micro-fluidic system for the localized delivery of neurotransmitters or other neuroactive compounds (Heuschkel et al., 1998). Another approach is to apply neuroactive compounds via a micro-manipulated puffer pipette (Liu and Tsien, 1995). One could potentially address more locations in a network with a scanning laser beam than a micro-pipette. A pulsed infrared laser can be used to "uncage" neurotransmitters (or agonists) that are only pharmacologically active after photolysis of an attached caging group. Denk et al. (1994) used this method to map acetylcholine receptor distribution on a single neuron, by recording the resulting whole-cell current when the agonist carbachol was photo-uncaged next to it. For any of these techniques to find routine applicability to many neurons in a MEA culture, we need advances in development of micro-fluidic structures and in photo-uncagable neuroactive molecules (Furuta et al., 1999). Our collaborators, Ari Glezer and Bruno Frazier at Georgia Tech, have begun building MEA systems with incorporated micro-fluidics, as part of an NIH Bioengineering Research Partnership. Similar projects are underway in Europe (Ziegler, 2000).

9.5   Closed-Loop Multi-Unit Electrophysiology

Closed-loop electrophysiology, where stimulation is contingent on what is recorded, is well established as a useful tool for studying single-neuron and small-circuit properties, using artificial conductance injection with a dynamic clamp (Nowotny et al., 2003; Raikov et al., 2004; Sharp et al., 1992, 1993; Suter and Jaeger, 2004). Glass micro-electrodes are used for both recording and injecting of currents in one or more nearby cells, under the control of a model for some dynamic property of the cell or network. This model may be implemented in software (Kullmann et al., 2004) or hardware (Raikov et al., 2004). This technique was a natural extension of the idea of voltage- (or current-) clamp recording, which dates back to 1948 (Huxley, 2002), except that instead of keeping some cellular parameter constant, it is varied dynamically according to the cell's behavior and the model. The speed of the feedback for these experiments is in the tens of kHz range, corresponding to loop times in the tens of μs range. The need for very fast feedback, and the fact that the technique requires delicate micro-manipulation of electrodes onto neurons, makes it difficult to use for more than a few cells at a time.

Closed-loop multi-unit electrophysiology, by contrast, is only just now coming of age (Ananthaswami, 2002). Because MEAs have extracellular electrodes, it is not feasible to do the sort of detailed conductance-model feedback experiments for which the dynamic clamp has been used. Instead, the recordings and stimuli focus on action potentials, the presumed currency of information transfer in the brain. One type of closed-loop multi-unit electrophysiology is to record from an electrode array implanted in an animal's brain, and to feed back perceptual stimuli through the animal's natural senses. In this approach being pursued by several groups, a monkey or rat is implanted with cortical electrodes (Carmena et al., 2003; Chapin et al., 1999; Taylor et al., 2002). Recorded neural signals control a robotic arm while the animal moves its own arm, and information about the prosthetic arm's movement is fed back to the animal through its eyes: either by watching the prosthetic arm, or by watching moving shapes on a video screen. The animal eventually learns to move the shapes (and the robot arm) without even moving its own arm. In the first closed-loop study to use what we call hybrots (which they call "neurobots") (Kositsky et al., 2003; Reger et al., 2000), Mussa-Ivaldi and Miller (2003) used an acute slice from a lamprey brainstem to control a Khepera wheeled robot. They mapped a circuit that normally processes vestibular information to a phototropism task: the robot moved towards or away from a light. Sensory input (from the Khepera's light sensors) was delivered to the slice via two tungsten wire electrodes, and motor commands were recorded by two glass extracellular micro-pipette electrodes. All of these closed-loop electrophysiology experiments are important steps toward studying distributed processing in embodied, situated neural systems.

9.5.1   Neurally Controlled Animats

Animals are situated and embodied. We want to enable the study of sensory–motor learning in cultured networks. If learning is defined as a process by which experience or practice results in a relatively permanent change in behavior (Morris, 1973), then to learn, a system must have a body to behave with, and an environment in which to behave. We have used the software and hardware systems described above to re-embody cultured cortical networks. We created a virtual environment and a very simple embodiment on the computer, as the first neurally controlled animat (DeMarse et al., 2001; Potter et al., 1997). (An animat is any simulated or robotic animal (Meyer and Guillot, 1994).) It was a neuroethology experiment, in the sense that we did not set the animat to any particular task, but merely observed the effect of the feedback stimulation on its behavior. We used an artificial neural network to cluster firing rate data in a high-dimensional space, and classify recurring patterns (DeMarse et al., 2001). We found that the diversity of activity patterns expressed by cultured cortical networks was enhanced by real-time feedback stimulation, at least while the sensory–motor loop was closed.

We did not observe any evidence of lasting (more than 30 min) changes in the open-loop behavior (driven by spontaneous activity) in the animat, as a result of closed-loop sessions. In a more disembodied closed-loop study by Shahaf and Marom (2001), cortical networks cultured on MEAs were stimulated until they satisfied a "learning" criterion of increased firing at 50 ± 10 ms latency after a probe stimulus. In this case, the turning off of a periodic stimulus (delivered via two substrate electrodes) was the only stimulus parameter contingent on the multi-unit recordings. They propose that the stimuli serve as an exploratory driving force, which is supported by our observation that with feedback stimuli, the cultures expressed more differentiable activity patterns. They hypothesize that by turning off the stimuli upon achievement of the learning criterion, the most recent pattern is "selected" by the network, this being an adaptive or desired response (Shahaf and Marom, 2001).

9.5.2   Hybrots

There are a number of reasons to use physical embodiments (robots) for animats (Holland and McFarland, 2001; Webb, 2002). Simulating the mechanics of the real world to any degree of realism is computationally difficult, but with robots, you get the physics "for free." It is becoming more and more clear that in animals, the physics of their bodies and interactions with the environment do a large amount of the sensory–motor processing that might previously have been attributed to neural systems alone. This is the premise of the interdisciplinary field of embodied cognition (Clark, 1997). Using robots also forces researchers to apply themselves to real-world problems, even if at a simplistic level.

We have used MEA cultures to control several hybrots in a closed-loop paradigm (Bakkum et al., 2004). One of these was the Koala 6-wheeled rover (K-Team, Figure 9.1). Under control of the neuronal network (the hybrot's "brain"), the Koala (the hybrot's body) was commanded to approach and then follow another robot being controlled randomly by the computer. For the control, Alec Shkolnik used a reproducible network property, the reduction or enhancement in dishwide response to the second of two stimulus pulses, depending on the interpulse interval (IPI). At short IPIs (± 20 msec), the response is maximal, roughly the sum of both responses when each stimulus is delivered alone. But at 100 to 300 msec IPI, the network is still in a refractory state from the first stimulus, and the response is minimal (Darbon et al., 2002). Beyond 500 msec, the response is intermediate. The magnitude of the Koala's movement in one feedback cycle is in proportion to this response. The distance to target is encoded in the subsequent IPI. This mapping enabled the Koala to approach a stationary robot, and to follow it at a certain distance when it began to move. Other closed-loop hybrot experiments using cultured networks are under way in Europe, as part of the multi-national EU-funded NeuroBIT project (Martinoia et al., 2004)).

9.6   Combining Imaging With MEA Electrophysiology

One of the most important advantages of in vitro neural models is that they are readily imaged with the light microscope. MEAs are usually transparent, and can be fabricated with clear leads of indium-tin oxide (Gross et al., 1985; Jimbo and Kawana, 1992) on thin glass for imaging on inverted microscopes. The "body" of re-embodied cultured networks (Bakkum et al., 2004) can move and behave, whereas the "brain" holds still on the microscope stage. Although it is possible to do microscopic imaging of living neurons in vivo (Gan et al., 2003; Helmchen et al., 2001; Levene et al., 2004; Majewska and Sur, 2003; Trachtenberg et al., 2002), the animal must be immobilized by physical restraint, paralytics, and/or anaesthesia. In those cases, the animal is incapable of expressing normal behavior, and likely not processing sensory input normally either. Hints at the morphological correlates of learning and memory in vivo must be gleaned by imaging before and after the animal learns, or in pairs of animals undergoing different experiences. Historically, this imaging has been done in slices of aldehyde-fixed brain tissue (Burgess and Coss, 1983; Rollenhagen and Bischof, 1994; Weiler et al., 1995), where inferences about activity-dependent morphological dynamics are difficult to make.

By using MEAs in conjunction with microscopic imaging, it is now possible to observe activity-dependent morphological dynamics, at a variety of time scales, while they are happening. With Scott Fraser of the Biological Imaging Center at Caltech, Potter developed new imaging techniques for maintaining mammalian neurons on the microscope stage for days, to allow time-lapse movies of morphological dynamics. These include advances in two-photon microscopy and specimen life support (Potter, 2000; Potter et al., 1996a, 2001a, 1996b, c, 2001b).

9.6.1   Why Two Photons Are Better Than One

Two-photon microscopy (Denk et al., 1990) allows repeated imaging of fluorescently labeled neurons with little or no light-induced damage (Potter, 1996; Williams et al., 2001). One of the benefits of two-photon microscopy is wasted on monolayer cultures, namely, the ability to image deeper into thick specimens by using infrared illumination. It is important to realize, however, that the two-photon effect limits excitation of the label to a plane about one micron in thickness (Potter et al., 1996c), and that monolayer cultures of neurons and glia are 15 to 20 microns thick. Thus, even for monolayer cultures, the total light dose is greatly reduced compared to one-photon confocal or wide-field fluorescence microscopy.

Commercially available multi-photon microscopes that are also visible-light confocal microscopes are not as efficient as they could be, because the requirements for confocal imaging are different than for multi-photon imaging (Pawley, 1995; Potter, 2000). Notably, with two-photon excitation it is not necessary to focus the light emitted by the specimen to create an in-focus image. This is because, at any moment, light is only emitted by a diffraction-limited focal volume of the scanning infrared laser beam. If emitted light is scattered on its way out of the specimen, those photons can be collected by a photomultiplier tube and referred back to the point of excitation to create the image. This makes two-photon imaging inherently more photon-efficient than one-photon imaging, especially when a direct (nondescanned) detector is used to collect as many scattered photons as possible (Wokosin et al., 1998). To overcome the shortcomings of commercially available multi-photon microscopes, we have built a custom two-photon microscope from the ground up, according to the design of Tsai et al. (2001). This design includes direct detection, and a flexible open architecture that can accommodate new equipment as it becomes available.

9.6.2   Keeping Cells Happy on the Microscope

Because it is difficult to fit a microscope inside a cell incubator, for long-term imaging it is better to bring the incubator to the microscope. We built inexpensive microscope enclosures out of Reflectix insulation (Figure 9.1, http://www.reflectixinc.com/) to warm the culture to rodent body temperature, to block ambient light, and to maintain an atmosphere with 5% carbon dioxide, for pH homeostasis (Potter, 2000). By using Teflon-sealed dishes with "baggy" lids, the objective of an upright microscope can be submerged into the culture medium without compromising sterility or osmolarity. The microscope enclosure serves the additional purpose of preventing focus drift due to changes in room temperature that cause expansion and contraction of microscope components. This fact alone makes it superior to systems that merely warm the MEA itself, when doing time-lapse imaging.

9.6.3   Variegated Neurons

Thanks to the use of fluorescent proteins in more and more transgenic constructs (Hadjantonakis et al., 2003), it is easy to prepare MEA cultures from animals that come prelabeled for fluorescence microscopy. There are now several color variants (xFPs) of the green fluorescent protein from jellyfish (Heim and Tsien, 1996), and red emitting fluorescent proteins from coral (Campbell et al., 2002; Okita et al., 2004). We prepared dissociated cultures from the combined cortices of two or more transgenic and wild-type mice (Feng et al., 2000) to produce cultures in which a small random subset of neurons is fluorescently labeled (Figure 9.7). For dense cultures, a ratio of 1:20 labeled:unlabeled cells allows detailed imaging of neurites without much overlap of labeled cells. Unlike dye labeling, or transient transfection with viruses or plasmids, xFP labeling in neurons from transgenic animals is self-renewing and harmless, allowing individual neurons to be followed in vitro for months.

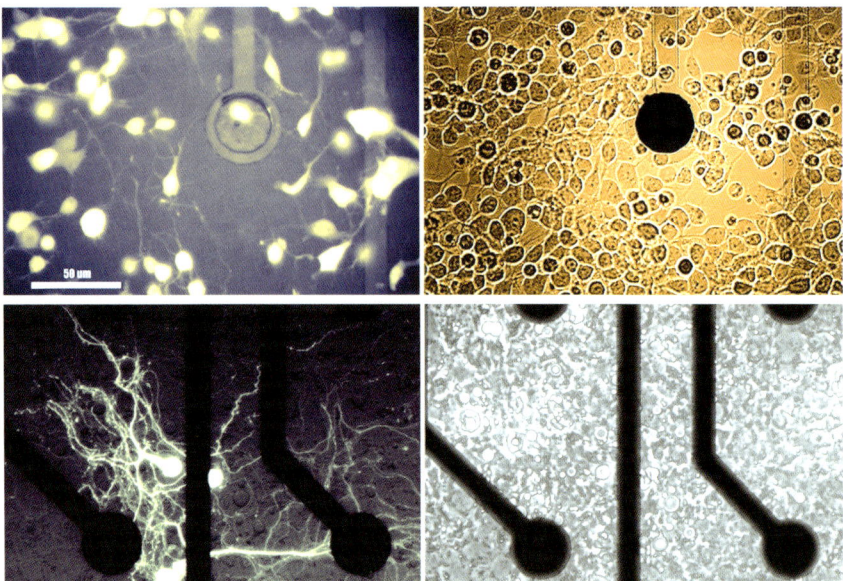

FIGURE 9.7. YFP-labeled neurons on MEAs. Fluorescence (left) and phase-contrast (right) microscopic images from mixed neuron/glia cultures after 3 days in vitro (top) and 50 days in vitro (bottom). In some cases, we mixed cells from transgenic mice in which every cell expressed YFP under an actin promoter with an excess of unlabeled wild-type cortex cells (top). In other cases, we used mice (e.g., Thy1-YFP-H from Feng et al., 2000) that only express YFP in layer 5 pyramidal neurons (bottom). Although the older cultures show vacuoles from cells that had died earlier, many neurons show healthy morphology including complex neurites with dendritic spines and axonal varicosities, as well as robust electrical activity many months after plating.

9.7   Conclusion

By combining MEA electrophysiology with long-term time-lapse imaging, it is possible to make correlations between changes in network function and changes in neuronal morphology. By re-embodying dissociated cultured networks, network function can be mapped onto behavior, and in vitro research can now make use of a new kind of behavioral studies that include detailed (submicron) imaging not possible in vivo. By closing the sensory–motor loop around MEA cultures, they are more likely to shed light on the mechanisms of learning, memory, and information processing in animals.

Acknowledgments. The authors thank the following for their significant contributions to this work: Funding: Whitaker Foundation, Georgia Research Alliance, NSF Center for Behavioral Neuroscience, NIH National Institute of Neurological Disorders and Stroke, NIH National Institute of Biomedical Imaging and

Bioengineering. Postdocs: Axel Blau and Mark Booth. Graduate students: Radhika Madhavan, Peter Passaro, Zenas Chao, Douglas Bakkum, and Komal Rambani. Undergrads: Mike Atkin, Gray Rybka, Sam Thompson, William Findley, Matthew MacDougall, Miro Dudik, John Brumfield, Blythe Towal, and Alec Shkolnik. Collaborators: Guy Ben-Ary, Jerry Pine, Scott Fraser, Josh Sanes, and David Kleinfeld. Technicians: Pooja Bhatia, Eno Ekong, Sheri McKinney, Dustin Romine, and Yan Wu. Lab managers: Mary Flowers and Bryan Williams. Manufacturer: Multi Channel Systems for help with data acquisition hardware.

References

Ananthaswami, A. (2002). Mind over metal. *New Sci.* Feb 23: 26–29.

Bakkum, D.J., Shkolnik, A.C., Ben-Ary, G., Gamblen, P., DeMarse, T.B., and Potter, S.M. (2004). Removing some 'A' from AI: Embodied cultured networks. In: Iida, Pfeifer, R., Steels, L., and Kuniyoshi, Y. (Eds.) pp. 130–45. Springer-Verlag, New York.

Banker, G. and Goslin, K. (1998). *Culturing Nerve Cells, 2nd Edition.* MIT Press, Cambridge, MA.

Ben-Ari, Y. (2001). Developing networks play a similar melody. *Trends Neurosci.* 24: 353-360.

Bi, G.Q. and Poo, M.M. (1998). Synaptic modifications in cultured hippocampal neurons: Dependence on spike timing, synaptic strength, and postsynaptic cell type. *J. Neurosci.* 18: 10464–10472.

Bohanon, T., Elender, G., Knoll, W., Koberle, P., Lee, J.S., Offenhausser, A., Ringsdorf, H., Sackmann, E., Simon, J., Tovar, G., and Winnik, F. M. (1996). Neural cell pattern formation on glass and oxidized silicon surfaces modified with poly(N-isopropylacrylamide). *J. Biomat. Sci. Polymer Edition* 8: 19–39.

Branch, D.W., Corey, J.M., Weyhenmeyer, J.A., Brewer, G. J., and Wheeler, B.C. (1998). Microstamp patterns of biomolecules for high-resolution neuronal networks. *Med. Biol. Eng. Comput.* 36: 135–141.

Brewer, G.J. and Cotman, C.W. (1989). Survival and growth of hippocampal-neurons in defined medium at low-density—Advantages of a sandwich culture technique or low oxygen. *Brain Res.* 494: 65–74.

Brewer, G.J. and Price, P.J. (1996). Viable cultured neurons in ambient carbon dioxide and hibernation storage for a month. *Neuroreport* 7: 1509–1512.

Buitenweg, J.R., Rutten, W.L.C., and Marani, E. (2002). Extracellular stimulation window explained by a geometry-based model of the neuron-electrode contact. *IEEE Trans. Biomed. Eng.* 49: 1591–1599.

Burgess, J.W. and Coss, R.G. (1983). Rapid effect of biologically relevant stimulation on tectal neurons—changes in dendritic spine morphology after 9 minutes are retained for 24 hours. *Brain Res.* 266: 217–223.

Campbell, R.E., Tour, O., Palmer, A.E., Steinbach, P.A., Baird, G.S., Zacharias, D.A.,and Tsien, R.Y. (2002). A monomeric red fluorescent protein. *Proc. Nat. Acad. Sci. U. S. A.* 99: 7877–7882.

Carmena, J.M., Lebedev, M.A., Crist, R.E., O'Doherty, J.E., Santucci, D.M., Dimitrov, D.F., Patil, P.G., Henriquez, C.S., and Nicolelis, M.A.L. (2003). Learning to control a brain–machine interface for reaching and grasping by primates. *PLoS Biol.* 1: 193–208.

Chapin, J.K., Moxon, K.A., Markowitz, R.S., and Nicolelis, M.A.L. (1999). Real-time control of a robot arm using simultaneously recorded neurons in the motor cortex. *Nature Neurosci.* 2: 664–670.

Clark, A. (1997). *Being There: Putting Brain, Body, and the World Together Again.* MIT Press, Cambridge, MA.

Connolly, P., Clark, P., Curtis, A.S., Dow, J.A., and Wilkinson, C.D. (1990). An extracellular microelectrode array for monitoring electrogenic cells in culture. *Biosens. Bioelectron.* 5: 223–234.

Cunningham, M.O., Davies, C.H., Buhl, E.H., Kopell, N., and Whittington, M.A. (2003). Gamma oscillations induced by kainate receptor activation in the entorhinal cortex in vitro. *J. Neurosci.* 23: 9761–9769.

Darbon, P., Scicluna, L., Tscherter, A., and Streit, J. (2002). Mechanisms controlling bursting activity induced by disinhibition in spinal cord networks. *Euro. J. Neurosci.* 15: 671–683.

DeMarse, T.B., Wagenaar, D.A., Blau, A.W., and Potter, S.M. (2001). The neurally controlled animat: Biological brains acting with simulated bodies. *Auton. Robots* 11: 305–310.

Denk, W., Delaney, K.R., Gelperin, A., Kleinfeld, D., Strowbridge, D.W., Tank, D.W., and Yuste, R. (1994). Anatomical and functional imaging of neurons using 2-photon laser scanning microscopy. *J. Neurosci. Meth.* 54: 151–162.

Denk, W., Strickler, J.H., and Webb, W. W. (1990). 2-photon laser scanning fluorescence microscopy. *Science* 248: 73–76.

Feng, G.P., Mellor, R.H., Bernstein, M., Keller-Peck, C., Nguyen, Q.T., Wallace, M., Nerbonne, J.M., Lichtman, J.W., and Sanes, J.R. (2000). Imaging neuronal subsets in transgenic mice expressing multiple spectral variants of GFP. *Neuron* 28: 41–51.

Fromherz, P. and Stett, A. (1995). Silicon-neuron junction—Capacitive stimulation of an individual neuron on a silicon chip. *Phys. Rev. Lett.* 75: 1670–1673.

Furuta, T., Wang, S.S.H., Dantzker, J.L., Dore, T.M., Bybee, W.J., Callaway, E.M., Denk, W., and Tsien, R.Y. (1999). Brominated 7-hydroxycoumarin-4-ylmethyls: Photolabile protecting groups with biologically useful cross-sections for two photon photolysis. *Proc. Nat. Acad. Sci. U. S. A.* 96: 1193–1200.

Gabbiani, F., Krapp, H.G., and Laurent, G. (1999). Computation of object approach by a wide-field, motion-sensitive neuron. *J. Neurosci.* 19: 1122–1141.

Gan, W.B., Kwon, E., Feng, G.P., Sanes, J.R., and Lichtman, J.W. (2003). Synaptic dynamism measured over minutes to months: Age-dependent decline in an autonomic ganglion. *Nature Neurosci.* 6: 956–960.

Gross, G.W. and Kowalski, J.M. (1999). Origins of activity patterns in self-organizing neuronal networks in vitro. *J. Intell. Mater. Syst. Struct.* 10: 558–564.

Gross, G.W., Rhoades, B.K., Reust, D.L., and Schwalm, F.U. (1993). Stimulation of monolayer networks in culture through thin-film indium-tin oxide recording electrodes. *J. Neurosci. Meth.* 50: 131–143.

Gross, G.W., Wen, W.Y., and Lin, J.W. (1985). Transparent indium-tin oxide electrode patterns for extracellular, multisite recording in neuronal cultures. *J. Neurosci. Meth.* 15: 243–252.

Grumet, A.E., Wyatt, J.L., and Rizzo, J.F. (2000). Multi-electrode stimulation and recording in the isolated retina. *J. Neurosci. Meth.* 101: 31–42.

Hadjantonakis, A.K., Dickinson, M.E., Fraser, S.E., and Papaioannou, V.E. (2003). Technicolour transgenics: Imaging tools for functional genomics in the mouse. *Nature Rev. Genet.* 4: 613–625.

Hartmann, M.J. and Bower, J.M. (2001). Tactile responses in the granule cell layer of cerebellar folium Crus IIa of freely behaving rats. *J. Neurosci.* 21: 3549–3563.

Heck, D. (1995). Investigating dynamic aspects of brain-function in slice preparations—spatiotemporal stimulus patterns generated with an easy-to-build multielectrode array. *J. Neurosci. Meth.* 58: 81–87.

Heim, R. and Tsien, R.Y. (1996). Engineering green fluorescent protein for improved brightness, longer wavelengths and fluorescence resonance energy transfer. *Curr. Biol.* 6: 178–182.

Helmchen, F., Fee, M.S., Tank, D.W., and Denk, W. (2001). A miniature head-mounted two-photon microscope: High-resolution brain imaging in freely moving animals. *Neuron* 31: 903–912.

Heuschkel, M.O., Guerin, l., Buisson, B., Bertrand, D., and Renaud, P. (1998). Buried microchannels in photopolymer for delivering of solutions to neurons in a network. *Sens. Actuat. B-Chem.* 48: 356–361.

Holland, O. and McFarland, D. (2001). *Artificial Ethology*. Oxford University Press, Oxford.

Huxley, A. (2002). From overshoot to voltage clamp. *Trends Neurosci.* 25: 553–558.

Jimbo, Y. and Kawana, A. (1992). Electrical stimulation and recording from cultured neurons using a planar electrode array. *Bioelectrochem. Bioenerget.* 29: 193–204.

Jimbo, Y., Kasai, N., Torimitsu, K., Tateno, T., and Robinson, H.P.C. (2003). A system for MEA-based multisite stimulation. *IEEE Trans. Biomed. Eng.* 50: 241–248.

Jimbo, Y., Robinson, H.P.C., and Kawana, A. (1998). Strengthening of synchronized activity by tetanic stimulation in cortical cultures: Application of planar electrode arrays. *IEEE Trans. Biomed. Eng.* 45: 1297–1304.

Kositsky, M., Karniel, A., Alford, S., Fleming, K.M., and Mussa-Valdi, F.A. (2003). Dynamical dimension of a hybrid neurorobotic system. *Trans. Neural Syst. Rehab. Eng.* 11: 155–159.

Kullmann, P.H.M., Wheeler, D.W., Beacom, J., and Horn, J.P. (2004). Implementation of a fast 16-bit dynamic clamp using LabVIEW-RT. *J. Neurophysiol.* 91: 542–554.

Latham, P.E., Richmond, B.J., Nirenberg, S., and Nelson, P.G. (2000). Intrinsic dynamics in neuronal networks. II. Experiment. *J. Neurophysiol.* 83: 828–835.

Lelong, I.H., Petegnief, V., and Rebel, G. (1992). Neuronal cells mature faster on polyethyleneimine coated plates than on polylysine coated plates. *J. Neurosci. Res.* 32: 562–568.

Levene, M.J., Dombeck, D.A., Kasischke, K.A., Molloy, R.P., and Webb, W.W. (2004). In vivo multiphoton microscopy of deep brain tissue. *J. Neurophys.* 91: 1908–1912.

Liu, G and Tsien, R.W. (1995). Synaptic transmission at single visualized hippocampal boutons. *Neuropharmacol.* 34: 1407–1421.

Lucas, J.H., Czisny, L.E., and Gross, G.W. (1986). Adhesion of cultured mammalian central nervous system neurons to flame-modified hydrophobic surfaces. *In Vitro Cell Dev. Biol.* 22: 37–43.

Lukashin, A.V., Amirikian, B.R., and Georgopoulos, A.P. (1996). A simulated actuator driven by motor cortical signals. *Neuroreport* 7: 2597–2601.

Maeda, E., Robinson, H.P., and Kawana, C.A. (1995). The mechanisms of generation and propagation of synchronized bursting in developing networks of cortical-neurons. *J. Neurosci.* 15: 6834–6845.

Maher, M.P., Pine, J., Wright, J., and Tai, Y.C. (1999). The neurochip: A new multielectrode device for stimulating and recording from cultured neurons. *J. Neurosci. Meth.* 87: 45–56.

Majewska, A. and Sur, M. (2003). Motility of dendritic spines in visual cortex in vivo: Changes during the critical period and effects of visual deprivation. *Proc. Nat. Acad. Sci. U. S. A.* 100: 16024–16029.

Martinoia, S., Sanguineti, V., Cozzi, L., Berdondini, L., van Pelt, J., Tomas, J., Le Masson, G., and Davide, F. (2004). Towards an embodied in vitro electrophysiology: The Neuro-BIT Project. *Neurocomput.* 58–60: 1065–1072.

Merz, M. and Fromherz, P. (2002). Polyester microstructures for topographical control of outgrowth and synapse formation of snail neurons. *Adv. Mater.* 14: 141ff.

Meyer, J.-A. and Guillot, A. (1994). From SAB90 to SAB94: Four years of animat research. In: Cliff, D.H., Meyer, P., Wilson, J.-A., Cambridge, S.W., eds., *From Animals to Animats 3. Proceedings of the Third International Conference on Simulation of Adaptive Behavior.* MIT Press, Cambridge, MA, pp. 2–11.

Misgeld, U., Zeilhofer, H.U., and Swandulla, D. (1998). Synaptic modulation of oscillatory activity of hypothalamic neuronal networks in vitro. *Cell. Molec. Neurobiol.* 18: 29–43.

Morris, C.G. (1973). *Psychology: An Introduction.* Appleton-Century-Crofts, New York.

Muller, T.H., Misgeld, U., and Swandulla, D. (1992). Ionic currents in cultured rat hypothalamic neurones. *J. Physiol.* 450: 341–62.

Mussa-Ivaldi, F.A. and Miller, L.E. (2003). Brain-machine interfaces: Computational demands and clinical needs meet basic neuroscience. *Trends Neurosci.* 26: 329–334.

Nadasdy, Z. (2000). Spike sequences and their consequences. *J. Physiol.-Paris* 94: 505–524.

Nam, Y., Khatami, D., Wheeler, B.C., and Brewer, G.J. (2003). Electrical stimulation of patterned neuronal networks in vitro. *IEEE Eng.Med. Biol.*, Cancun, Mexico.

Nolfi, S. and Parisi, D. (1999). Exploiting the power of sensory-motor coordination. *Advances in Artificial Life, Proceedings.* 1674: 173–182.

Novak, J.L. and Wheeler, B.C. (1988). Multisite hippocampal slice recording and stimulation using a 32 element microelectrode array. *J. Neurosci. Meth.* 23: 149–59.

Nowotny, T., Zhigulin, V.P., Selverston, A.I., Abarbanel, H.D.I., and Rabinovich, M.I. (2003). Enhancement of synchronization in a hybrid neural circuit by spike-timing dependent plasticity. *J. Neurosci.* 23: 9776–9785.

Oka, H., Shimono, K., Ogawa, R., Sugihara, H., and Taketani, M. (1999). A new planar multielectrode array for extracellular recording: application to hippocampal acute slice. *J. Neurosci. Meth.* 93: 61–67.

Okita, C., Sato, M., and Schroeder, T. (2004). Generation of optimized yellow and red fluorescent proteins with distinct subcellular localization. *Biotechniques* 36: 418ff.

Pawley, J.B. (Ed.) (1995). *Handbook of Biological Confocal Microscopy,* 2nd Edition. Plenum Press, New York.

Potter, S.M. (1996). Vital imaging: Two photons are better than one. *Curr. Biol.* 6: 1595–1598.

Potter, S.M. (2000). Two-photon microscopy for 4D imaging of living neurons. In: Yuste, R., Lanni, F., and Konnerth, A., eds., *Imaging Neurons: A Laboratory Manual.* CSHL Press, Cold Spring Harbor, NY, 20.1–20.16.

Potter, S.M. and DeMarse, T.B. (2001). A new approach to neural cell culture for long-term studies. *J. Neurosci. Meth.* 110: 17–24.

Potter, S.M., Fraser, S.E., and Pine, J. (1996a). The greatly reduced photodamage of 2-photon microscopy enables extended 3-dimensional time-lapse imaging of living neurons. *Scanning* 18: 147.

Potter, S.M., Fraser, S.E., and Pine, J. (1997). Animat in a petri dish: Cultured neural networks for studying neural computation. *Proceedings of the 4th Joint Symposium on Neural Computation, UCSD*: 167–174.

Potter, S.M., Lukina, N., Longmuir, K.J., and Wu, Y. (2001a). Multi-site two-photon imaging of neurons on multi-electrode arrays. *SPIE Proceedings* 4262: 104–110.

Potter, S.M., Pine, J., and Fraser, S.E. (1996b). Neural transplant staining with DiI and vital imaging by 2-photon laser-scanning microscopy. *Scan. Microscopy* Supplement 10: 189–199.

Potter, S.M., Wang, C.M., Garrity, P.A., and Fraser, S.E. (1996c). Intravital imaging of green fluorescent protein using 2-photon laser-scanning microscopy. *Gene* 173: 25–31.

Potter, S.M., Zheng, C., Koos, D.S., Feinstein, P., Fraser, S.E., and Mombaerts, P. (2001b). Structure and emergence of specific olfactory glomeruli in the mouse. *J. Neurosci.* 21: 9713–9723.

Raikov, I., Preyer, A., and Butera, R.J. (2004). MRCI: A flexible real-time dynamic clamp system for electrophysiology experiments. *J. Neurosci. Meth.* 132: 109–123.

Ramakers, G.J.A., Raadsheer, F.C., Corner, M.A., Ramaekers, F.C.S., and Vanleeuwen, F.W. (1991). Development of neurons and glial-cells in cerebral-cortex, cultured in the presence or absence of bioelectric activity—morphological observations. *Euro. J. Neurosci.* 3: 140–153.

Rambani, K., Booth, M.C., Brown, E.A., Raikov, I., and Potter, S.M. (2005). Custom-made multiphoton microscope for long-term imaging off neuronal cultures to explore structural and functional plasticity. *Proc. SPIE* 5700: 102–108.

Regehr, W.G., Pine, J., Cohan, C.S., Mischke, M.D., and Tank, D.W. (1989). Sealing cultured invertebrate neurons to embedded dish electrodes facilitates long-term stimulation and recording. *J. Neurosci. Meth.* 30: 91–106.

Reger, B.D., Fleming, K.M., Sanguineti, V., Alford, S., and Mussa-Ivaldi, F.A. (2000). Connecting brains to robots: An artificial body for studying the computational properties of neural tissues. *Artif. Life* 6: 307–324.

Rollenhagen, A. and Bischof, H.J. (1994). Spine morphology of neurons in the avian forebrain is affected by rearing conditions. *Behav. Neural Biol.* 62: 83–89.

Segev, R., Benveniste, M., Shapira, Y., and Ben-Jacob, E. (2003). Formation of electrically active clusterized neural networks. *Phys. Rev. Lett.* 90: Art. No. 168101.

Shahaf, G. and Marom, S. (2001). Learning in networks of cortical neurons. *J. Neurosci.* 21: 8782–8788.

Sharp, A.A., Abbott, L.F., and Marder, E. (1992). Artificial electrical synapses in oscillatory networks. *J. Neurophysiol.* 67: 1691–1694.

Sharp, A.A., O'Neil, M.B., Abbott, L.F., and Marder, E. (1993). The dynamic clamp - Artificial conductances in biological neurons. *Trends Neurosci.* 16: 389–394.

Shoham, S., Fellows, M.R., and Normann, R.A. (2003). Robust, automatic spike sorting using mixtures of multivariate t-distributions. *J. Neurosci. Meth.* 127: 111–122.

Stenger, D.A., Pike, C.J., Hickman, J.J., and Cotman, C.W. (1993). Surface determinants of neuronal survival and growth on self-assembled monolayers in culture. *Brain Res.* 630: 136–147.

Stoppini, I., Duport, S., and Correges, P. (1997). A new extracellular multirecording system for electrophysiological studies: Application to hippocampal organotypic cultures. *J. Neurosci. Meth.* 72: 23–33.

Studer, L., Csete, M., Lee, S. H., Kabbani, N., Walikonis, J., Wold, B., and McKay, R. (2000). Enhanced proliferation, survival, and dopaminergic differentiation of CNS precursors in lowered oxygen. *J. Neurosci.* 20: 7377–7383.

Suter, K.J. and Jaeger, D. (2004). Reliable control of spike rate and spike timing by rapid input transients in cerebellar stellate cells. *Neurosci.* 124: 305–317.

Taylor, D.M., Tillery, S.I.H., and Schwartz, A.B. (2002). Direct cortical control of 3D neuroprosthetic devices. *Science* 296: 1829–1832.

Trachtenberg, J., Chen, B.E., Knott, G.W., Feng, G.P., Sanes, J.R., Welker, E., and Svoboda, K. (2002). Long-term in vivo imaging of experience-dependent synaptic plasticity in adult cortex. *Nature* 420: 788–794.

Tsai, P.S., Nishimura, N., Yoder, E.J., Dolnick, E.M., White, G.A., and Kleinfeld, D. (2001). Principles, design, and construction of a two photon laser scanning microscope for in vitro and in vivo brain imaging. In: Frostig, R., ed., *Methods for In Vivo Optical Imaging.* CRC Press, Boca Raton, FL.

Ullian, E.M., Sapperstein, S.K., Christopherson, K.S., and Barres, B.A. (2001). Control of synapse number by glia. *Science* 291: 657–661.

Wagenaar, D.A. and Potter, S.M. (2002). Real-time multi-channel stimulus artifact suppression by local curve fitting. *J. Neurosci. Meth.* 120: 113–120.

Wagenaar, D.A. and Potter, S.M. (2004). A versatile all-channel stimulator for electrode arrays, with real-time control. *J. Neural Eng.* 1: 39–45.

Wagenaar, D.A., Pine, J., and Potter, S.M. (2004). Effective parameters for stimulation of dissociated cultures using multi-electrode arrays. *J. Neurosci. Meth.* 138: 27–37.

Webb, B. (2002). Robots in invertebrate neuroscience. *Nature* 417: 359–363.

Weiler, I.J., Hawrylak, N., and Greenough, W.T. (1995). Morphogenesis in memory formation—Synaptic and cellular mechanisms. *Behav. Brain Res.* 66: 1–6.

Williams, R.M., Zipfel, W.R., and Webb, W.W. (2001). Multiphoton microscopy in biological research. *Curr. Opin. Chem. Biol.* 5: 603–8.

Wokosin, D.L., Amos, B.G., and White, J.G. (1998). Detection sensitivity enhancements for fluorescence imaging with multi-photon excitation microscopy. *IEEE EMBS* 20: 1707–1714.

Zeck, G. and Fromherz, P. (2003). Repulsion and attraction by extracellular matrix protein in cell adhesion studied with nerve cells and lipid vesicles on silicon chips. *Langmuir* 19: 1580–1585.

Ziegler, C. (2000). Information processing by natural neural networks: INPRO project IST-2000–26463.

10

Emerging Network Activity in Dissociated Cultures of Neocortex: Novel Electrophysiological Protocols and Mathematical Modeling

MICHELE GIUGLIANO, MAURA ARSIERO, PASCAL DARBON, JÜRG STREIT, AND HANS-RUDOLF LÜSCHER

Introduction

10.1   Outline

Since the early attempts at combining micro-fabricated transducers with in vitro neurobiological systems (Gross, 1979; Gross et al., 1985), cultures of neurons dissociated from the vertebrate nervous system have represented a convenient choice for several reasons (Stengler and McKenna, 1994). Neurons can be easily cultured over biocompatible substrates, grown in an incubator and maintained under healthy conditions for several weeks or more (Potter and DeMarse, 2001). This fulfills the requirements of the unconventional approach followed by MEA investigators: instead of invasively probing a single neuron by means of (e.g.) an intracellular glass-pipette electrode, let a population of neurons develop ex vivo and grow around multiple probes, for extended periods of time, under noninvasive conditions. On the other hand, the choice of cultured neurons is also related to the in vitro development of functional synaptic contacts (Nakanishi and Kukita, 1998) and to the emergence of spontaneous patterned electrical activity (Kamioka et al., 1996; Van den Pol et al., 1996). Thus, along a tradition of investigation that is common to physics (Amit, 1989), the possibility of accessing a reduced version of an active nervous system (Bulloch and Syed, 1992) constitutes a unique opportunity for the investigation of network electrophysiology. Indeed, such an approach makes it possible to dissect the interactions among individual neurons of a network and to look for collective mechanisms at the cellular and subcellular levels, through manipulation of the physicochemical conditions.

Under such perspectives, we review in this chapter electrophysiological data obtained from networks of neurons dissociated from the rat neocortex and cultured over arrays of substrate micro-electrodes (MEAs). In particular, we discuss a recent

---

This chapter is dedicated to the memory of the late Professor Massimo Grattarola, who pioneered the MEA technique in Italy.

experimental approach for the study of cortical network electrophysiology, and introduce a simple theory accounting for the emergence of the in vitro spontaneous collective activity. We conclude with some remarks on the perspectives of novel experimental protocols and mathematical modeling, as complementary tools to MEAs and traditional electrophysiological techniques.

10.1.1   Relevance of the Study of In Vitro Neocortical Networks

In the field of life sciences, we are assisting in an increase of interest for the function of biological networks of elements and for the complexity emerging from the interactions and combinations of such elements (e.g., the dynamics of motifs in neuronal/genetic/metabolic/biochemical networks) (Milo et al., 2002). The underlying inspirations of such a trend suggested interesting analogies with the physics of semiconductors and with the development of modern digital electronics (Grattarola and Massobrio, 1998). In fact, the design of semiconductor electronics proceeded first from very simple devices interconnected in complex manners (i.e., transistors and Boolean logic-gate networks), then evolved into a relatively simple interfacing of highly sophisticated units (e.g., cluster computing, parallel architectures, etc.).

In the case of the design of biological systems, evolution followed an opposite path. It started from the simple combination of complex (bio)molecular compounds (e.g., the assembly of a lipidic bilayer) and went on, assembling the nervous system of mammals, which functionally appears as a highly intricate map of networks, each composed of complex (sub)cellular elements. However, the most recent phylogenic outcome of evolution, the neocortex, might reveal simpler principles (Douglas and Martin, 1990), irrespective of its anatomical complexity and heterogeneity. In addition, the very same basic electrophysiological processes might be carried out by each small region of the neocortex, by a kind of general-purpose canonical micro-circuitry. Actually, although there are several differences from layer to layer, with regard to projections, cell density, morphology, and size, a stereotypical organization seems indeed to dominate. For instance, Douglas and Martin (1990) focused their proposal on a canonical building block, underlining and emphasizing the tremendous recurrent excitation, estimated as 90% of the total afferent excitation.

In such a context, and under the perspective of ultimately understanding how the synaptic organization of the neocortex produces the complexity of cortical functions, the convergence of a mathematical theory and experimental results is imperative. Such an approach might be also devised to determine how many of the details underlying single channels, dendrites, neurons, and synapses, play a role at higher levels, and whether these details must be fully retained or largely simplified, at the level of large-scale cortical processing description.

One possibility to challenge existing theories and to develop new ones, is represented by the study of in vitro preparations as reduced and highly simplified neurobiological systems, with regard to the network-level electrical activity.

Nevertheless, in vitro networks are generally not considered to retain enough details and features of the in vivo cortical physiology (Steriade, 2001). This is certainly true with regard to differences in neuronal morphology, firing patterns, and resting properties, as well as in functional aspects and oscillatory organized activities. In the case of cultured networks of dissociated cells, the bidimensional arrangement of random synaptic contacts might be a reason for further skepticism.

However, keeping in mind such limitations and avoiding the tempting attitude of incriminating isolated neuronal properties, channels, and molecules as responsible for complex physiological or behavioral processes, in vitro preparations might still be invaluable research tools. Although cellular diversity and differentiation are retained in cultures (Huettner and Baughman, 1986), it is very unlikely that the precise "signature" of any canonical micro-circuitry is reproduced, as neuronal connections randomly re-organize. However, strong excitatory recurrent coupling characterize in vitro cultures (Nakanishi et al., 1999), and forms of collective electrical activities arise spontaneously, sustained by recurrent synaptic connections (Maeda et al., 1995). These constitute intermediate steps of investigation, crucial to the understanding of in vivo cortical phenomena. Therefore, the inaccuracy of in vitro networks in re-creating faithful replicas of in vivo functions may not represent an obstacle and, on the contrary, it is an ideal framework to develop novel electrophysiological protocols and test theoretical interpretative frameworks.

In such a context, the MEA technique, by increasing the spatial resolution and making it possible to chronically and noninvasively track collective network activity over time, is playing an instrumental role. Referring again to the analogy with digital electronics, the availability of MEAs is providing the conditions for accessing the simultaneous electrical activity of several components, interconnected in a functional circuit. We believe that a step in the direction of understanding the general principles underlying the design of "digital computers" (the design of a neocortex) may come from the understanding of how a subset of "transistors" and "diodes" (neurons and synapses) performs concerted and not isolated functions.

10.2   MEA Experiments

10.2.1   Culture Technique, MEA Recordings, and Single-Neuron Patch-Clamp

According to standard methods, cortical neurons can be enzymatically dissociated from the cortices of early postnatal rats, by exposing brain tissue slices to a trypsin solution. After such a treatment, cells completely lose neurites and synaptic connections and are plated on a MEA surface (Tscherter et al., 2001), previously coated with substances promoting cell adhesion (Stengler and McKenna, 1994). After just a few hours from plating, neurons start to elongate new neurites, establish synaptic contacts, and, after a few weeks in culture, reach a mature stage (Kamioka et al., 1996), and self-organize into a densely interconnected cellular monolayer. Axonal and dendritic branches at this stage extend over 1 mm, resulting in a large number of functional synapses with neighboring cells (Marom and Shahaf, 2002; Nakanishi and Kukita, 1998).

As already mentioned, by culturing dissociated neurons over MEAs, neuronal somata as well as axons are allowed to develop close to the individual substrate micro-electrodes. Such proximity makes it possible to extracellularly detect spontaneous spiking activity from one or more neighboring cells (Streit et al., 2001; Bove et al., 1998; Grattarola and Martinoia, 1993). As this kind of signal transduction procedure is noninvasive, electrical activity of in vitro networks can be characterized as it progresses, evolves, and organizes, during long-term culture, while preserving sterile and physiological conditions.

The MEAs employed in our experiments contained 68 platinum planar electrodes, spaced 200 μm and laid out in the form of a rectangle. During each recording session, channels showing neuronal activity were selected and recordings were digitized to be stored on a hard disk, while monitoring multi-channel amplified raw voltage traces with a custom oscilloscope software. The detection of extracellularly recorded action potentials (i.e., fast voltage transients) and further analysis were performed offline as previously described (Giugliano et al., 2004; Tscherter et al., 2001; Streit et al., 2001). In a series of experiments, a patch-clamp technique was employed in the whole-cell configuration (Hamill et al., 1981). The aim of these experiments was to access the intracellular membrane voltage at the soma of cultured neurons, and to characterize single-cell responses to current injection. Therefore, network activity had to be suppressed by a cocktail of blockers of synaptic transmission. This consisted of D-2-amino-5-phosphonovalerate (D-APV) and 6-cyano-7-nitroquinoxaline-2,3-dione (CNQX), which are competitive antagonists of the NMDA and non-NMDA glutamatergic receptors, respectively. Once these substances were bath-applied, incoming synaptic activity and spontaneous spiking were completely suppressed (see next sections and Figure 10.1). Such a pharmacological manipulation makes it possible to study electrical properties of individual neurons, under isolated conditions, as neurons receive no inputs from the surrounding network, other than what is artificially injected through the patch-pipette by the experimenter.

10.2.2   Development of In Vitro Electrical Activity

Before about seven days in vitro (DIV), most neurites do not reach neighboring neurons (Nakanishi et al., 2001) and the distributed electrical activity consists in the emission of rare action potentials, as detected by individual MEA electrodes (Kamioka et al., 1996). No spatial propagation of signals is observed at this stage and no correlation characterizes neuronal activity at spatially distinct locations. Such spontaneous activity is completely abolished by tetrodotoxin (TTX), a blocker of intrinsic neuronal spiking mechanisms, or by selective antagonists of glutamatergic synaptic receptors or by activation of GABAergic receptors (Kamioka et al., 1996). This evidence suggests that at this stage, the network can be regarded as an arrangement of uncoupled neurons, whose membrane voltage rarely fluctuates, reaching an excitability threshold as a result of spontaneous synaptic receptor activation.

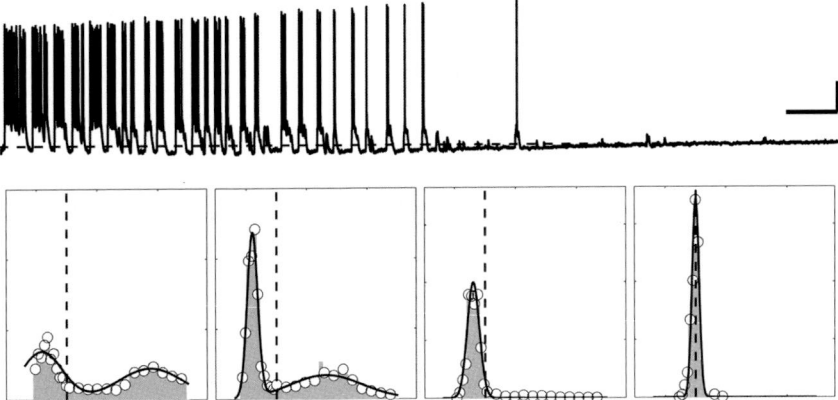

FIGURE 10.1. Synaptic bases of network-driven population bursts. A cocktail of pharmacological blockers of fast excitatory synaptic transmission was bath-applied while recording intracellularly the membrane voltage of a neuron participating to population bursting (upper trace, over 80 sec). As time passes and the number of blocked synaptic receptors increases to saturation, the coupling between neurons of the network becomes weak and finally breaks, as no PB occurs anymore. As the network synaptic drive decreases, a hyperpolarizing contribution fades out, releasing resting membrane voltage to slightly more depolarized values (dashed line). Finally, the cumulative inactivation (see Figure 10.8) of spike-emission processes recovers, and the over-shoot amplitude of individual action potentials increases as the fraction of inactive sodium channels decreases. Under these conditions, the subthreshold voltage amplitude histogram was computed every 20 sec (lower panels: bars and circles): at the beginning the distribution is well fitted (thick lines) by a double Gauss-function and later by a single Gauss-function, while its mean slightly shifts to more depolarized values (calibration: 5 sec/ 20 mV for the upper trace, 9 mV/0.05 mV^{-1} for the two lower-left panels and 9 mV/0.12 mV^{-1} for the others).

Spontaneous presynaptic leakage of neurotransmitter molecules or glutamate spill-over from neighboring synapses can explain such an activity phase (Maeda et al., 1995). Such random uncorrelated firing evolves, at a later stage, into a more organized firing pattern, consisting of isolated spikes and quite regular sequences of bursts of action potentials. These bursts, referred to as population bursts (PBs), occur almost simultaneously across all the MEA micro-electrodes and are not characterized by reproducible spatial trajectories, while propagating across the network (Maeda et al., 1995). This phase characterizes cultured networks around 5 to 16 DIV and it is associated with structural changes, represented by an increase in neurite elongation and by the formation of functional synapses (Nakanishi et al., 2001; Nakanishi and Kukita, 1998). Individual neurons emit rare and irregular spikes or bursts of action potential, superimposed on spontaneous voltage fluctuations around a resting potential of about −60 mV (Nakanishi and Kukita, 1998), similarly to what is observed in adult neocortex (Sanchez-Vives and McCormick, 2000). As time in cultures further advances, the propagation velocity of PBs across the network increases, indicating the formation of an increasing number of reliable synaptic connections. After about 30 DIV, the network reaches a stable condition

Firing rate homeostasis in cultured networks

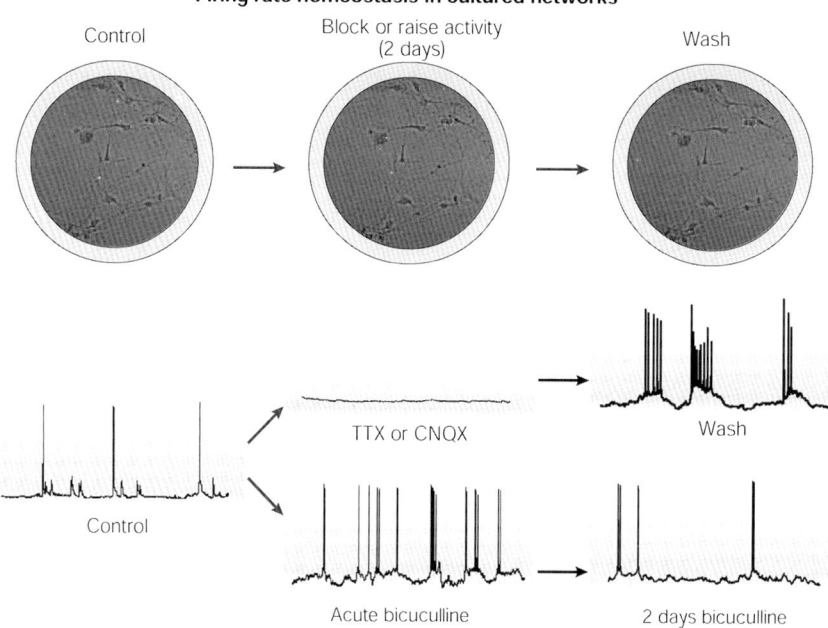

FIGURE 10.2. In vitro chronic pharmacological manipulation of synaptic transmission in cultured networks of neocortical neurons induces rebound phenomena as soon as control conditions are restored. Homeostatic mechanisms of the population firing rate therefore co-exist with heterosynaptic plasticities and participate in adjusting cellular and synaptic properties, to compensate for changes in synaptic drive. As a consequence, single-neuron properties as well as synaptic efficacies are dynamically regulated over time scales ranging from hundreds of milliseconds to minutes and hours, in activity-dependent manners. (Reproduced with permission from *Nature Reviews Neuroscience*; © 2004 Macmillan Magazines Ltd.)

of maturation, exhibiting a richer and more elaborate temporal pattern of irregular, synchronized population bursts (Marom and Shahaf, 2002; Kamioka et al., 1996; Maeda et al., 1995). Maturation of the network is also associated with a transient decline in the number of synapses, which is markedly related to activity-dependent processes (Van Huizen et al., 1987) as well as homeostatic plasticity (Turrigiano and Nelson, 2004; see Figure 10.2).

10.2.3   Population Bursting Is Driven by Network Interactions

The available electrophysiological and pharmacological data convincingly suggest that the spontaneous electrical activity described above emerges as a collective phenomenon, and is sustained by synaptic connectivity (see Figure 10.3). No intrinsic neuronal pacemaker mechanisms have been reported so far, either to account for the random spatial features of the origin and propagation of population bursts (Maeda et al., 1995), or to explain correlations between collective activity and

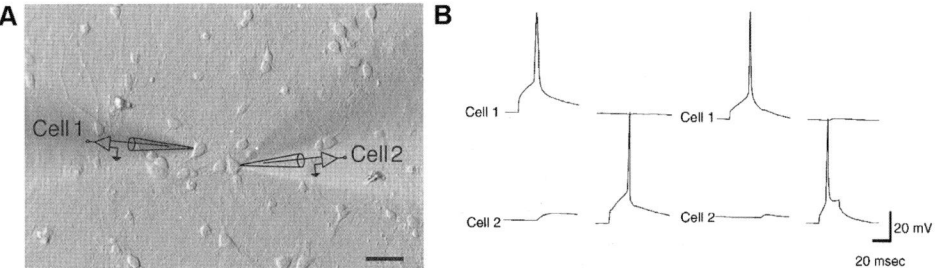

FIGURE 10.3. Cultured neurons, dissociated from neocortex, establish extensive functional chemical synapses, a few weeks after plating. These connections have been studied with simultaneous pair recordings by Nakanishi and collaborators (A) (calibration: 50 μm). Although in vitro connectivity appears to be higher than in intact cortex, emission of a spike by a presynaptic (excitatory) neuron usually evokes (B) weak subthreshold responses in postsynaptic cells, for unidirectional (left) as well as bidirectional coupling (right). (Modified and reproduced with permission, © 1998–1999 Elsevier Ltd.)

plastic changes in synaptic efficacy induced by repetitive electrical stimulation (Jimbo et al., 1999; Maeda et al., 1998). During the mature stage of these cultures, synaptic connectivity of a generic neuron has been estimated to be generally monosynaptic, with propagation delays of a few milliseconds, irrespective of the spatial distance between cells, and to involve 10 to 30% of other neurons (Nakanishi and Kukita, 1998).

Together with the large number of anatomical synaptic contacts (Marom and Shahaf, 2002), and the strong correlation between activity and degree of development of neurite outgrowth (Nakanishi et al., 2001; Kamioka et al., 1996; Muramoto et al., 1993), these considerations support a network architecture able to sustain a reverberating spiking activity through recurrent excitatory connections. Consistent with the in vitro connectivity pattern, relatively restricted to spatially neighboring sites, a mature cultured network could be therefore thought of as a homogeneous chain of synaptically connected subpopulations (Giugliano et al., 2004). Each population would then have some probability to initiate a PB, spreading to the entire culture by means of sparse excitatory connectivity. Physical network sectioning experiments of Maeda et al., (1995), and of Nakanishi and Kukita (1998) are consistent with such an interpretation and with the hypothesis that synchronized bursting is mediated by chemical synapses rather than by way of gap junctions and/or diffusible factors.

10.3   From Single-Neuron Properties to Network Activity

As the electrical activity described in the previous section is an emerging population phenomenon, it is relevant to investigate the quantitative conditions for its occurrence. It is also of particular interest for network neurosciences to characterize the cellular-level features of neurons and synapses, which play a major role at the network level. In fact, it is likely that an effective description of neuronal integration and excitability at the cellular level will provide several benefits for the prediction of the population-level activity, in analogy to the outstanding contributions of the phenomenological (macroscopic) description of the ionic (microscopic) mechanisms proposed by Hodgkin and Huxley (1952) for the generation of an action potential. In particular, the characterization of single-neuron (input–output) response properties, carried out under the appropriate conditions, has been suggested as a fundamental step to predict and quantitatively interpret how a population of neurons interacts. Such a view is strongly supported by several theoretical studies (Salinas, 2003; Mattia and Del Giudice, 2002; Brunel, 2000), where single-neuron response properties have been exploited to derive predictions about collective phenomena, such as the global spontaneous irregular activity (Amit and Brunel, 1997), the emergence of fast network-driven oscillations (Brunel and Wang, 2003; Fuhrmann et al., 2002), and of selective delay-activity states (Wang, 2001; Amit and Brunel, 1997).

10.3.1   How Can We Re-Create a Realistic Input to an Isolated Neuron?

In vivo neocortical neurons continuously produce excitatory and inhibitory postsynaptic currents (EPSCs/IPSCs) (Destexhe and Paré, 1999). Such an intense activity arises from the high degree of connectivity of intracortical afferents, spontaneously active at very low firing rates, making postsynaptic membrane voltage fluctuate as in a random walk (Destexhe et al., 2003; Abeles, 1991; Gerstein and Mandelbrot, 1964). This phenomenon is known to strongly affect intrinsic biophysical properties of neurons, as compared to TTX-induced resting conditions, and to modulate their responsiveness to external inputs (Steriade, 2001; Gerstner, 2000; Destexhe and Paré, 1999). In such perspectives, traditional in vitro electrophysiological protocols, which consist in the evaluation of the transient spiking response to an injection of DC current steps (McCormick et al., 1985), are completely inappropriate to approach the single-neuron level (Holt et al., 1996).

Although it is apparent that the conditions in mature cultures are different from those of an intact cortex, these considerations still hold when considering emerging population activity. In particular, if we are to determine the effective single-neuron response properties, we must employ a protocol that replicates conditions experienced by the same neuron, when participating in the network activity. Thus, we immediately realize that nonstationarity characterizes the inputs to a neuron, during each PB (see Figure 10.1). The amplitude distribution of the subthreshold membrane voltage, recorded in current-clamp, indicates two alternating regimes: a resting condition, with low variance, and an active phase possibly related to a high-conductance state (Destexhe et al., 2003).

In the following, we consider only the active regime. In fact, invoking quasi-stationary conditions, we later recover the nonstationarity and ignore such a simplification (Giugliano et al., 2004). But how do we mimic all possible (stationary) inputs to the network? As described in the previous sections, the whole-cell

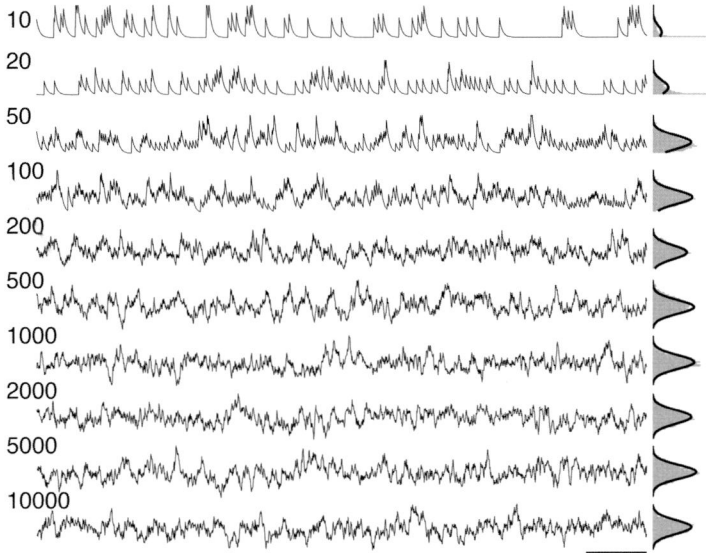

FIGURE 10.4. The overall synaptic input, resulting from stationary activation of a network of presynaptic neurons, can be described as a random walk in time and characterized by a Gauss-distributed stochastic process (i.e., by the central-limit theorem). The figure reports the computer-simulated time course of the total synaptic input to a generic neuron, quantified as conductance or current changes, from a presynaptic neuronal population of increasing size, through identical independent AMPAr-mediated synapses (i.e., 10–10,000, $\tau_I = 10$ msec). The distribution of amplitudes approaches a bell-shaped profile (panels on the right column) for as few as 100 synapses, each spontaneously active at a very low rate of 5 Hz (calibration: 100 msec).

patch-clamp technique can be exploited to inject a desired stimulus waveform into an isolated neuron and to measure the resulting input–output transformation. In the recent literature, there have been several attempts at re-creating in vitro a realistic network input (Silberberg et al., 2004; Chance et al., 2002; Fuhrmann et al., 2002; Protopapas and Bower, 2001; Destexhe and Paré, 2000; Poliakov et al., 1997; Mainen and Sejnowski, 1995). We followed Rauch et al. (2003), who proposed to computer-synthesize a Gauss-distributed noisy waveform and to inject it under current-clamp into the soma. Such nondeterministic stimuli are supposed to re-create realistic fluctuations for a cell embedded in a large cortical network (Figure 10.4; Destexhe et al., 2003).

10.3.2   The Extended Mean-Field Hypothesis

The theoretical framework, which inspires and motivates the approach of Rauch et al. (2003), is related to a powerful mathematical technique that makes it possible to predict the mean firing rate of a population of interacting neurons on the basis of the discharge response of a single cell to a realistic noisy input (Amit and Brunel,

1997). Further validating the hypotheses behind the use of noise injections, we note that there is a higher, and considerably less structured, degree of in vitro synaptic connections, compared to the in vivo cytoarchitecture. Moreover, because the impact of a single EPSC/IPSC on the in vitro postsynaptic membrane voltage is very weak in evoking suprathreshold responses (see Figure 10.3), the overall current experienced by a generic postsynaptic neuron can be indeed approximated by a diffusion stochastic process (see Figure 10.4 and Appendix; Fourcaud and Brunel, 2002; Destexhe and Paré, 2000).

Such considerations are linked to the hypothesis that individual neurons in a homogeneous active network cannot be distinguished in a statistical sense. In fact, because of the very large number of (random) synaptic connections and the presence of nonhomogeneities and noise sources, neurons roughly tend to instantaneously experience the same input current. Of course, each neuron will instantaneously receive a different realization of the same process, but its descriptors (i.e., current mean, variance, and correlation time length) are assumed to be the same. In other words, each neuron experiences the same mean field, extended to the regime of input fluctuations. Under such hypotheses, the characterization of the discharge properties of a single neuron becomes statistically representative of the others, as a whole.

10.3.3   The Noisy Current-Clamp Protocol

Because the nondeterministic stimuli described in the previous section account for different presynaptic network architectures and regimes, let's, for instance, indicate by $N_{e/i}$ the number of excitatory/inhibitory neurons, characterized by stationary mean activation rates $f_{e/i}$ and projecting to a generic neuron with probability $C_{e/i}$. Under the hypothesis that synaptic inputs are approximately independent, the distribution of the resulting postsynaptic somatic current amplitude becomes Gauss-distributed, by the central-limit theorem (see Figure 10.4), with steady-state mean m and variance s^2 given by the expressions reported below (Rauch et al., 2003):

$$m = N_e C_e \langle I_e \rangle f_e \tau_e - N_i C_i \langle I_i \rangle f_i \tau_i$$

$$s^2 = N_e C_e \langle I_e^2 \rangle f_e \tau_e / 2 + N_i C_i \langle I_i^2 \rangle f_i \tau_i / 2$$

where $I_{e/i}$ and $\tau_{e/i}$ are the effective peak-amplitude and decay time constant at the soma, for individual excitatory and inhibitory postsynaptic currents, respectively, and the averaging operator $\langle \rangle$ is intended across the population of synapses. In this example, we can therefore inject a realization of a Gauss-distributed noisy current, characterized by (m, s^2), to recreate in an isolated neuron the input from such an afferent network.

In a more general case, in our experiments we aimed at an exploration of the plane (m, s^2), although some of the combinations might not be consistent with all possible network regimes. For any pair (m, s^2), an iterative expression was thus employed to synthesize a realization $I(t)$ of a current to be injected, under

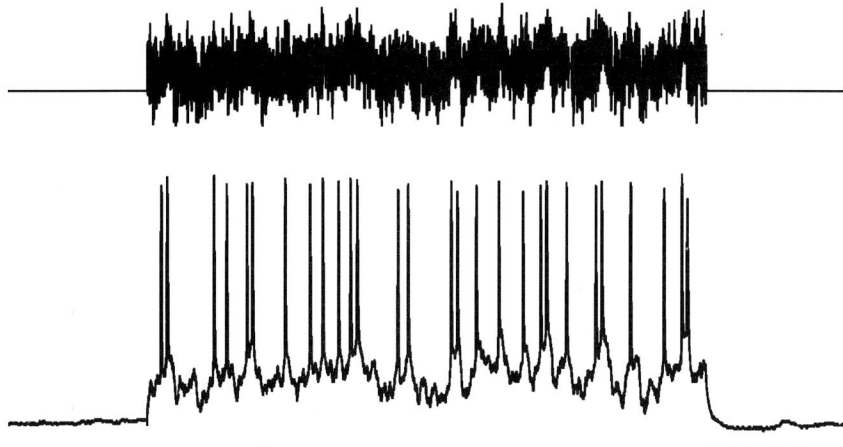

FIGURE 10.5. Electrical response evoked by a noisy stimulus current that mimics a realistic input drive of cortical networks. The response is determined by either mean current or random fluctuations. Each stimulus waveform was generated as an independent realization of the stochastic Ornstein–Uhlenbeck process, fully specified by the steady-state mean m, variance s^2, and autocorrelation time-length τ_I (calibration: 1 sec). (Modified and reproduced with permission, © 2004, The American Physiological Society.)

current-clamp, in the soma of patched neurons (see Figure 10.5):

$$I(t + dt) = I(t) + (m - I(t))\frac{dt}{\tau_I} + s\sqrt{\frac{2dt}{\tau_I}}\,\xi_t, \qquad (10.1)$$

where ξ_t is a unitary Gauss-distributed random variable (Press et al., 1992), updated at a rate of 5 kHz (i.e., $dt = 0.2$ msec). We fixed $\tau_e = \tau_I = \tau_I = 5$ msec, thereby focusing on AMPA- and GABA$_A$-mediated synaptic currents (Destexhe et al., 1994). As a neuronal output, we chose to estimate the steady-state mean firing rate f as a function of m and s^2. Actually, $f(m, s^2)$ is the only relevant descriptor, under the assumption of the mean-field hypothesis, to predict the population firing rate (Amit and Brunel, 1997). Such a characterization allows us to predict, in a self-consistent way, how population activity occurs in a network of connected neurons, because the recurrent synaptic components can be described by recursively considering $m = m(f)$ and $s^2 = s^2(f)$ (Giugliano et al., 2004; Amit and Brunel, 1997).

10.4   Single Neurons Respond as Integrate-and-Fire Units

The results of single-neuron experiments have been summarized in Figures 10.6 and 10.7. As opposed to a DC stimulation, which elicits periodic and regular spike trains, noisy current injection produces neuronal membrane voltage that evolves

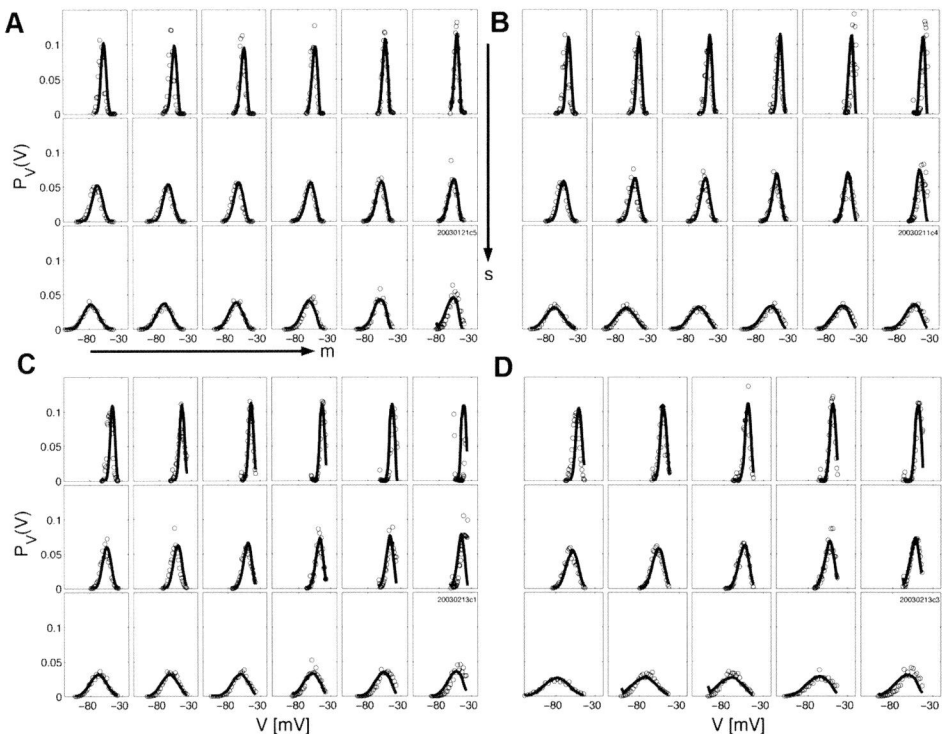

FIGURE 10.6. Steady-state amplitude distribution of subthreshold membrane voltage in single-cell noise injection experiments. The amplitude distribution of membrane voltage in single-cell noise injection experiments (open circles) is in good agreement with the mathematical model prediction (continuous thick line; see Appendix), although its parameters were tuned to match the mean firing rates only (see Figure 10.7 and compare to Figure 10.1). Panels A–D report the results from four different experiments, and in each subplot the voltage amplitude density distribution is plotted for increasing values of m, from left to right, and of s from top to bottom, within the same panel. (Reproduced with permission, © 2004, The American Physiological Society.)

in time as in a random walk, leading to an irregular spike emission (Figure 10.5). Its subthreshold amplitude distribution becomes bell-shaped, similar to what is observed in vivo (but see also Figure 10.1), with mean and standard deviation increasing with m and s^2 (see Figure 10.6). Cultured neurons also showed very similar response properties compared to neurons in acute slices (Rauch et al., 2003), when repetitively stimulated by the same noise realization. This was shown to yield a much higher precision in the timing of individual spikes in slices (Mainen and Sejnowski, 1995), proving that the somatic spike emission mechanism of cultured neurons is intrinsically reliable.

However, the most unexpected and interesting result is the following: as far as the steady-state spiking frequency f is concerned, cultured neocortical neurons

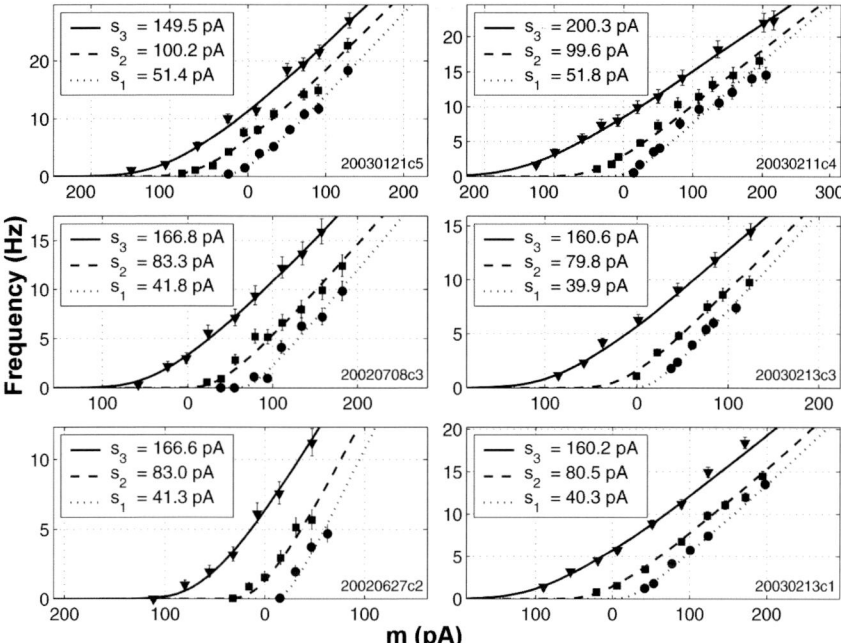

FIGURE 10.7. Results from six different single-cell noise injection experiments (see Figure 10.5). The current-to-discharge rate response of cortical neurons is captured with remarkable accuracy in the plane (m, s^2) by the simplified dynamics of a leaky integrate-and-fire model neuron. Experimental responses were evaluated at the steady-state as a function of the current mean m (markers), for increasing values of the fluctuation amplitude s, and compared to the corresponding model best-fit prediction (lines). (Modified and reproduced with permission, © 2004, The American Physiological Society.)

respond as simple integrate-and-fire (IF) units (Figure 10.7; Giugliano et al., 2004). Actually, the curves $f(m, s^2)$, collected for each cell at the steady-state, are reproduced with high accuracy by the response function $\Phi(m, s^2)$ of an identified IF model (see Equation (10.2) and Appendix). This proves that the biophysical and morphological details of a neuron collapse into an extremely simplified quantitative description of neuronal integration and excitability, when studied under appropriate realistic conditions.

10.4.1   Leaky Integrate-and-Fire Model

A single-compartment description of neuronal excitability was therefore enough to account for the entire experimental data set. This is surprising, as the model relies on a very small number of free parameters (i.e., 5), which were identified in each experiment by numerical optimization techniques. Such a model is known as the Lapicque's or leaky IF (Tuckwell, 1988) and it has been extensively studied (Fourcaud and Brunel, 2002, Mattia and Del Giudice, 2002) and employed in large-scale network simulations (Reutimann et al., 2003). As opposed to biophysically realistic models (Abbott and Dayan, 2001), the leaky IF is characterized just by a single state-variable V, representing the membrane potential, and by a reduced set of effective constant parameters, as follows.

$$C \frac{dV(t)}{dt} = \begin{cases} \bar{g}\,(E - V) + I_m(t) & \text{if } V(t) < \vartheta \\ 0, V(t) = H & \text{if } V(E_0) = \vartheta \text{ and } t \in (t_0; t_0 + \tau_{arp}) \end{cases} \quad (10.2)$$

In the previous definition, the absolute refractory period and hyperpolarization voltage following the emission of an action potential have been indicated by τ_{arp} and H, respectively. Moreover, the integration operated by the neuronal membrane is assumed to be passive and characterized by a voltage-independent ionic conductance (g) and by a capacitance C. All the nonlinearities associated with the emission of an action potential (Hodgkin and Huxley, 1952), have been lumped into a fixed voltage threshold ϑ, therefore describing each spike as a highly stereotyped event that corresponds to a threshold crossing for V (see Appendix). Finally, below threshold, the membrane voltage decays to its resting value E, when the total membrane current $I_m(t) = 0$.

10.4.2   Spike-Frequency Adaptation and Slow/Cumulative Inactivation

In all the noisy current-clamp experiments, the temporal dynamics of the instantaneous output firing rate $f = f(t)$ was characterized by fast and frequency-dependent adaptation components, which occur over a time scale of several hundred milliseconds, and by a slower component, occurring over a time scale of several seconds (Sanchez-Vives et al., 2000; Powers et al., 1999; Sawczuk et al., 1997; Fleidervish et al., 1996; Douglas and Martin, 1990). These mechanisms, together with homosynaptic depression (Tsodyks et al., 2000), have been proposed as candidates for the generation of oscillatory activities in the nervous system, and related to the termination of the PB (Giugliano et al., 2004; van Vreeswijk and Hansel, 2001). In order to account for the single-cell experimental recordings, the IF model further incorporated a simplified adaptation current, describing the contribution of intracellular calcium- and/or sodium-activated outward currents. $I_m(t)$ was therefore identified as the sum of an intrinsic contribution $I_X(t)$, due to adaptation mechanisms, and an external/synaptic current $I(t)$. Each emitted spike was then assumed to cause a sudden increase in the intracellular concentration $X(t)$ of the ions, involved in such an activity-dependent hyperpolarization current, which was formalized as follows.

$$I_X(t) = -\alpha X(t)$$

$$\tau_a \frac{dX(t)}{dt} = -X(t), \ X(t_0^+) \to X(t_0^-) + \tau_a^{-1} \quad (10.3)$$

The stationary effect of fast and slow adaptation processes at steady-state was therefore captured by this simple mechanism, and it reduced the slope of the curve $\Phi(m, s^2)$ by the factor α (see Appendix).

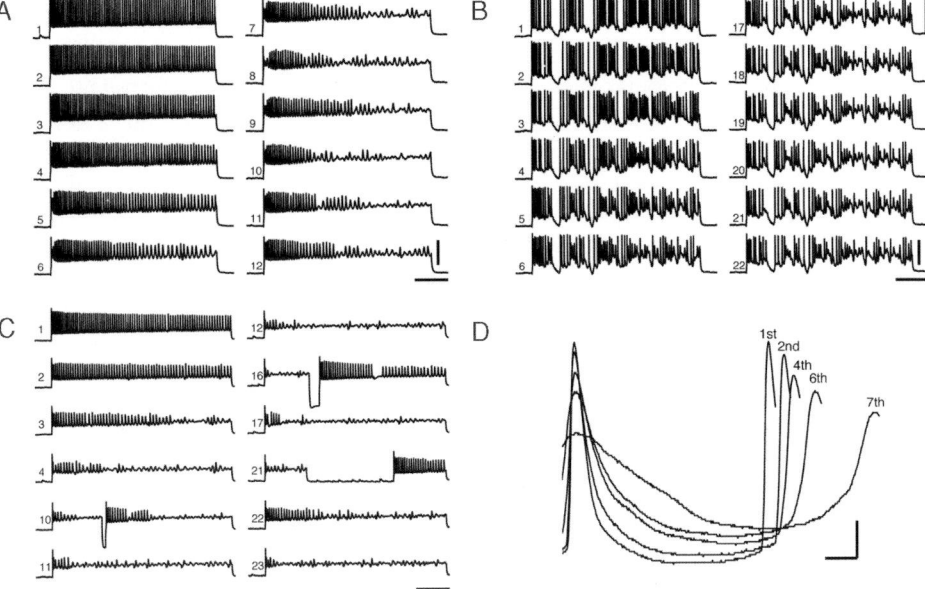

FIGURE 10.8. Slow inactivation of spike-generation mechanisms. Inactivation contributes to determine the maximal steady-state spike frequency that cortical neurons are able to sustain. Similarly to the effect of sodium/calcium-dependent potassium currents, such an inactivation might participate in the termination of each population burst. A repeated intracellular DC current stimulation (A), with very short interstimulus pauses, makes the spike inactivation apparent. Such an inactivation slowly builds in time, and increasingly affects (D) the amplitude, the interspike intervals, and the maximal upstroke velocity of successive action potentials. (B) The same phenomenon affects neuronal responses under noisy current-clamp. However, the voltage fluctuations, induced by the nondeterministic current component, may delay the expression of inactivation, and transiently reverse it. (C) The cumulative character of such a slow inactivation can be revealed by delivering short hyperpolarizing current pulses, interleaved to DC current stimuli: inactivation can be reversed only partially (calibration: 1 sec/50 mV for (A–C); 10 msec/10 mV for (D)).

However, an additional cumulative component, related to the inactivation of the spike generation mechanisms, characterized neuronal responses (see Figure 10.8). The presence of such cumulative inactivation was explicitly tested by extending the protocol described in Fleidervish et al. (1996) and Schwindt et al. (1989), which consists in a repeated pulse-stimulation lasting 1 sec, with a very short recovery time. Under noisy current injection, and employing the same current realization for each repetition, the same phenomenon occurred, although the voltage fluctuations, induced by the nondeterministic stimulus waveform, could delay the onset of inactivation at parity of m, and sometimes transiently reversed the inactivation for a few tens of milliseconds, as compared to DC stimuli. For the sake of simplicity, such an intrinsic inactivation was not considered in the IF model (but see Giugliano et al., 2002).

10.4.3   A Simple Model of Chemical Synaptic Interactions

Although we did not experimentally approach the effective characterization of synaptic physiology so far, there are enough data available from the literature to define a minimal model of chemical synaptic interactions between cortical neurons in vitro (see also Giugliano (2000)). Consistent with the hypotheses underlying the current stimuli injected in single-neuron experiments, synaptic coupling is described by current rather than conductance changes (see La Camera et al. (2003) for a discussion). Interactions between any two connected neurons are triggered by presynaptic emission of an action potential, after an effective delay δ of 1.5 msec, which includes the axonal delay and the synaptic release latency (Nakanishi and Kukita, 1998). Each postsynaptic current consists of an instantaneous rise to J_e (i.e., the synaptic efficacy) and by an exponential decay with a time constant $\tau_e = 5$ msec (see Figure 10.3). In the case of a population of N_e excitatory neurons, the total synaptic current $I_i(t)$ into the neuron i is therefore given by:

$$I_i(t) = \sum_{j=1}^{N_e} \sum_k J_e C_{ij}\, e^{-(t-t_k^j-\delta)/\tau_e}\, \Theta(t - t_k^j - \delta) \qquad (10.4)$$

where $\{t_k^j\}$ are the times of emission of action potentials by the jth presynaptic neuron, C_{ij} is the connectivity matrix, and $\Theta(t)$ is the step function (i.e., $\Theta(t) = 0$, $t < 0$ and $\Theta(t) = 1$, $t \geq 0$). The probability of connection (i.e., $C_{ij} = 1$) was fixed to 0.3 to 0.4 (Nakanishi and Kukita, 1998), and the effect of spontaneous synaptic release and of other sources of randomness (Maeda et al., 1995) was incorporated as an activity-independent additional irregular synaptic drive (Giugliano et al., 2004).

10.5   A Network of IF Model Neurons

The results presented and briefly discussed in previous sections indicate that an extremely simple mathematical model of neuronal electrical activity can be employed to describe the discharge properties of individual dissociated neocortical neurons cultured in vitro. Moreover, the experiments performed by patch-clamp allowed us to identify the numerical parameters of such a description. Thus, such effective values can be immediately incorporated in a model of a network of synaptically connected IF neurons, to be studied and computer simulated with unparalleled realism. We approached the study of such a model network in two ways: by direct numerical simulation and by mean field theoretical analysis. For the sake of simplicity, we focus on the electrical activity of a homogeneous population composed only of excitatory model neurons. With the aim of comparing model performances to real MEA data, we consider a set of experiments obtained in the presence of bicuculline, a blocker of $GABA_A$ inhibitory synaptic receptors (see Figure 10.9).

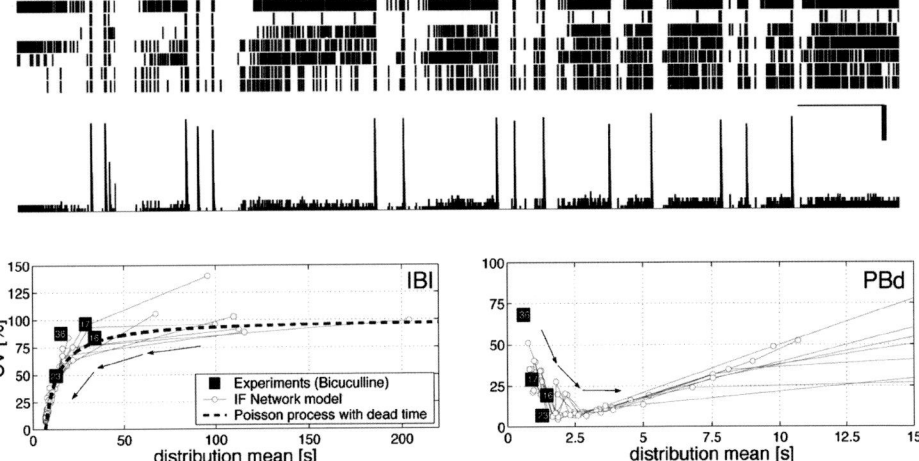

FIGURE 10.9. Network activity emerging in dissociated cultures of neocortical neurons, detected by MEAs. The raster plot (upper panels) indicates the time of occurrence of the events extracellularly detected by 7 substrate electrodes, and below the resulting population mean firing rate is reported (scale bars: 60 sec, 10 Hz). To compare the performances of computer simulations to real network data, results from 4 MEA experiments in the presence of bicuculline were considered, by simultaneously studying (lower panels) the coefficient of variation (CV), and the mean of the interbursts intervals (IBIs) distribution, and of the duration of the population bursts (PBd). Numbers associated with each marker help to identify the same experiment in both plots and indicate the number of PBs, collected over 10 min of continuous recording. Computer simulations of the network of *IF* model neurons reproduced the same features of the spontaneous patterned activity, simultaneously matching the IBIs and the PBd, for increasing values of synaptic coupling. (Indicated by the arrows; modified and reproduced with permission, © 2004, The American Physiological Society.)

10.5.1   Computer Numerical Simulations

The numerical simulation of the network model described above consists in the simultaneous integration of Equations (10.2) through (10.4), for each neuron of the network (Giugliano et al., 2004), by an appropriate discrete-time algorithm implemented on a desktop computer (see, for instance, Reutimann et al. (2003)). To our surprise, as soon as we incorporated the features and parameters estimated from the experiments into the model, spontaneous electrical activity arose and organized into a population bursting. Although individual IF neurons of the simulated network are not intrinsic burster cells (Smith et al., 2000; McCormick et al., 1985), and no pacemaker mechanism had been explicitly introduced and localized in the model, we assisted in the emergence of a repetitive transient networkwide synchronization of electrical activity, strikingly similar to what we measured in a real network of cortical cells cultured over the MEAs (see Figure 10.9, upper panel).

In these simulations, the spatial origin of the PBs varies randomly with each PB, consistent with the lack of spatial structure in the network and with the (experimental) conclusions about the lack of a unique rhythm generator mechanism driving the network. As several investigators noted the regular/irregular character of PBs (Wiedemann and Lüthi, 2003; Nakanishi et al., 2001; Maeda et al., 1995), we took a careful look at the features of activity in our model. This is quite relevant as the highly irregular character of neuronal firing is ubiquitous in in vivo cortical physiology (Destexhe et al., 2003; Shadlen and Newsome, 1998). We found that the frequency and the character of the simulated PBs depended strongly on the synaptic connectivity and on the excitatory coupling J_e. In particular, by increasing J_e, we indeed observed three different regimes: asynchronous irregular firing, periodic synchronized bursting, and nonperiodic synchronized bursting.

First of all, in order to model a condition of very weak excitatory coupling between neurons, or to mimic a very low connectivity, we set J_e very small. Under such conditions, the network of model neurons is characterized by low-rate asynchronous spiking activity. Such a background activity is mainly determined by activity-independent synaptic activation and only weakly by the contribution of recurrent connections. No spontaneous PB occurs and any brief depolarizing stimulus, even delivered to a large fraction of neurons of the network, does not evoke spiking activity that persists more than the duration of the stimulus.

In terms of a theoretical interpretation that is discussed in the following sections, we may conclude that under such conditions, the dynamics of the whole population is dominated by a single low-rate stable state, as the recurrent synaptic feedback is too weak to self-sustain a reverberating activity. We should note that, in our model, the reliability of synaptic transmission remains unchanged and does not "mature" with time (see Equation (10.4)). Anyway, at low firing rates, low values of J_e can also statistically account for low-probability synaptic release, characterizing immature synapses (Tsodyks et al., 2000; Kamioka et al., 1996). As discussed at the beginning of this chapter, such a situation approximates the early developmental stage of cultured networks.

In another set of computer simulations, we substantially increased J_e. A larger value of J_e may correspond to an increase in both the number and efficacy of synaptic contacts, as observed in vitro during maturation (Muramoto et al., 1993). Under such conditions, where the low-rate random spiking is still present and at the same frequency, spontaneous PBs occur very frequently and regularly. Now, even a brief external triggering stimulus successfully recruits neurons to produce evoked PBs, whose duration is inversely determined by α, and independent of stimulus strength.

Actually, in such a regime the global dynamics of the network is transiently bistable: once a PB is started, adaptation slowly redefines the location and existence of the network stable states, until there is suddenly only a single stable state at 0 Hz, similarly to what is observed after each PB in the MEA experiments (see Figure 10.9, upper panel). In details, the adaptation hyperpolarizing currents $I_X(t)$, which start to build up in every neuron recruited by a PB (see Equation (10.3)), decrease the mean input current and therefore lower the output firing rate (see Figure 10.7), until the recurrent synaptic inputs stop. This may be considered as a kind of reset for the collective network activity. Later, the network is refractory as the mechanisms of spike-frequency adaptation recover (i.e., $X(t) \to 0$), for a time that is proportional to τ_a. In particular, we note that although the value of τ_a can be estimated from the transient time course of $f(t)$, upon current stimulation, it is feasible that its actual value changes due to the degree of network maturation.

Finally, for very strong synaptic coupling, the network loses its bistable transient properties, and only a high-frequency firing characterizes the network. We doubt that such a regime can be directly observed in vitro, because of the impact of additional adaptation mechanisms (Figure 10.8), and because of other metabolic constraints on neuronal firing (e.g., the activity of ATP-operated electrogenic pumps). These processes act on a very long time scale, compared to a single spike, and they would switch off network activity by down-regulating intrinsic neuronal excitability (i.e., changing the profile of Φ_{IF}). Such a scenario would probably result in a synchronized bursting with very long burst-recovery intervals, and it would not be always possible to evoke a PB by electrical stimulation, during the spontaneous interburst intervals (Kamioka et al., 1996).

10.5.2 *Theory*

In the case of the network architecture discussed here, the statistics of the total synaptic current received by a generic neuron are described by the following recurrent mean field equations,

$$m(f) = N_e C_{ee} J_e f \tau_e + m_0 \text{ and } s^2(f) = N_e \; C_{ee} \; J_e^2 \; f \tau_e / 2 + s_0^2, \qquad (10.5)$$

where f is the network mean firing rate and J_e the effective peak-amplitude for the individual postsynaptic currents. By m_0 and s_0^2, we indicate two fixed parameters, corresponding to the spontaneous neurotransmitter release and other sources of randomness, independent of f by hypothesis.

As mentioned in the previous sections, the knowledge of the single-neuron response function $f(m,s^2)$, identified in the experiments, lets us derive quantitative predictions and interpretations on the collective network properties. This is possible, under the same extended mean-field hypotheses that underlie the noisy currents (Equation (10.1)), injected into real neurons, and it can be done without running a single numerical simulation.

In fact, an in vitro network of synaptically interacting excitatory neurons may be regarded as a single dynamical system. Its stationary states can be predicted and interpreted, in the limit of an infinite number of neurons $N_e \to \infty$, by employing the mean-field equations (10.5) and studying $\Phi_{IF}(m(f), s(f))$ as a function of f (Amit and Brunel, 1997), in particular looking for its fixed points.

For the sake of clarity, we first consider the collective activity in the absence of spike-frequency dependent adaptation (i.e., $\alpha = 0$). Because of the simultaneous dependence of Φ_{IF} on m and s^2, the collective firing-rate of the network may be characterized by two stable dynamic-equilibrium states, in a range of synaptic coupling J_e (see Figures 10.10A and 10.11A). Such global activity configurations correspond to the solutions f^* of the following self-consistent network equation,

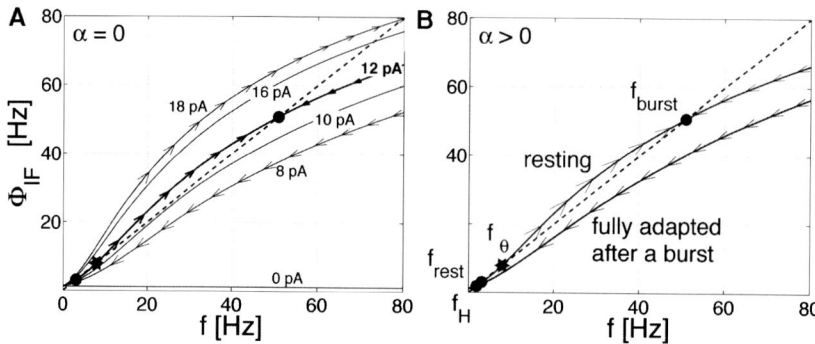

FIGURE 10.10. Prediction of the collective activity emerging in a homogeneous population of excitatory neurons based on single-neuron response properties. (A) The plot represents the profile of $\Phi_{IF}(m(f), s^2(f))$ as a function of network firing rate f, for increasing values of the average synaptic efficacy J_e. In a small range of synaptic coupling, the network may be characterized by two distinct dynamic-equilibrium states for the collective activity ($\alpha = 0$, circles: stable states; star: unstable state). (B) Φ_{IF} was then plotted as a function of f, while comparing the instantaneous network response profile immediately before a PB (resting) and after the complete buildup of adaptation currents I_X, at the end of the same PB (fully adapted). When the network is active at a low frequency regime f_{rest}, the impact of the spike-frequency dependent adaptation is negligible and an additional equilibrium stable state, at a high frequency f_{burst}, may characterize the collective dynamics. However, if a transition $f_{rest} \to f_{burst}$ occurs, the adaptation starts to slowly build up, bending the network response profile until no stable dynamic equilibrium state at high frequency can be sustained anymore ($f(t) \to f_H$). Thus, it can be inferred that the actual network mean firing rate will be flipping between two states, in an alternating and activity-dependent way, as confirmed by computer simulations (see Figures 10.9 and 10.11). (Modified and reproduced with permission, © 2004, The American Physiological Society.)

further satisfying a stability condition:

$$f^* = \Phi_{IF}(m(f^*), s(f^*)) \quad \frac{d}{df}\Phi_{IF}(f^*) < 1.$$

These solutions have been graphically identified as the intersection points of $\Phi_{IF}(f)$ with the unitary-slope line, and marked as circles (stable) and stars (unstable states) (see Figure 10.10A).

When occupying a stable dynamic regime, the activity of the network is the result of the interactions between the neurons, and a transition from one state to the other can occur spontaneously in small networks. Actually, these transitions might be the result of fluctuations induced by finite-size effects (Mattia and Del Giudice, 2002), or they can be triggered by an external brief stimulus (Figure 10.11).

10.5.3   Mechanisms of PB

Considering the full network model, where individual neurons keep adapting their output rate as a function of activity (i.e., $\langle X(t) \rangle \sim f(t)$), it is possible to carry

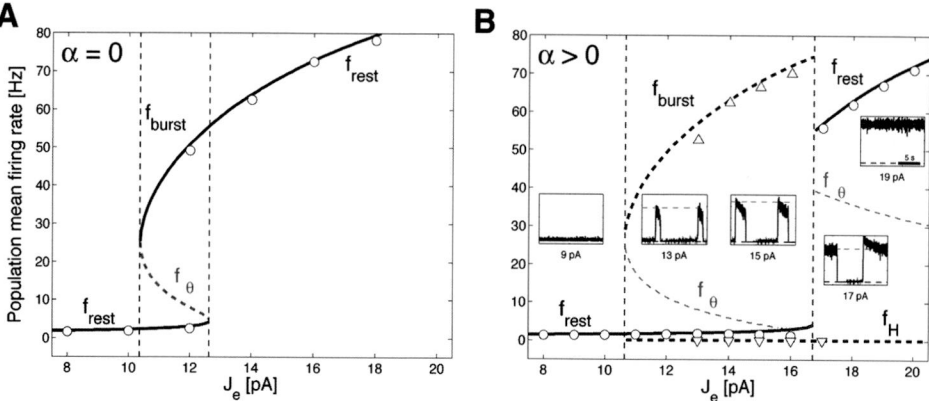

FIGURE 10.11. Network bistability in the absence of adaptation and other sources of non-stationarity in a homogeneous excitatory population, for a small range of synaptic coupling J_e (A). This is a direct consequence of the profile of the network response function, experimentally characterized in Figure 10.7. A similar phase diagram can be considered for the model network, including adaptation (B). By studying the quasi-stationary/instantaneous stable states, it is possible to make quantitative predictions on the mean firing rates during asynchronous regimes as well as during population bursts. The insets report the actual traces of simulated network activity for different synaptic coupling, and compare the actual resting and bursting activity levels with the predictions of the theory (black and gray dotted lines). In both panels, black continuous and dashed traces indicate stationary and quasi-stationary stable states, respectively. Dashed gray lines report the location of unstable stationary and quasi-stationary equilibria, whose distance from f_{rest} determines the regular/irregular character of network oscillations. The markers represent the network mean firing rates, measured in computer simulations under different regimes. Although the simulated network considered was very small, $N_e = 100$, the agreement between theoretical predictions and numerical computer simulations is remarkable. (Reproduced with permission, © 2004, The American Physiological Society.)

out an approximate analysis. Provided that the mechanisms responsible for the reduction of excitability (e.g., adaptation) act on a time scale (τ_a), much longer than the single-neuron dynamics $(c/\bar{g})$, an analysis of the quasi-stationary equilibria may be representative of the collective activity of the network. In other words, by assuming that adaptation is delayed and transiently uncoupled from intrinsic neuronal dynamics, we may consider it as frozen and determined by the previous global conditions.

Let's consider the situation of small J_e: a low-rate stable regime f_{rest} is expected to characterize the network dynamics, as it can be immediately determined from Φ_{IF}. By making explicit the dependence on α (see Equation (10.5) and Appendix), we can write the following self-consistent equation and solve it to find the stable solution f_{rest},

$$f_{rest} = \Phi_{IF}(m(f_{rest}) - \alpha \, f_{rest}, \, s(f_{rest})).$$

As confirmed by computer simulations, the instantaneous network firing rate $f(t)$ fluctuates around f_{rest}, as a result of finite-size network effects. However, each neuron of the network instantaneously responds to its input drive according to a quasi-stationary input–output response function given by

$$\Phi_{IF}(m - \alpha\, f_{rest}, s),$$

with f_{rest} a constant. Neurons are therefore not experiencing immediately the impact of adaptation. Thus, let's assume that the network *was* at "rest". We may now determine the existence of any additional stable dynamical attractor f^*. f^* must satisfy the following equation,

$$f^* = \Phi_{IF}(m(f^*) - \alpha\, f_{rest}, s(f^*)).$$

$f^* = f_{rest}$ is of course a solution of the previous equation, but for increasing values of J_e, $f^* = f_{burst}$ is also a solution ($f_{burst} > f_{rest}$, Figure 10.10B). Under these circumstances there always exists an intermediate unstable equilibrium f'_θ ($f_{rest} < f'_\theta < f_{burst}$), satisfying the previous relationship and separating the two basins of attraction (Figure 10.10B).

Intuitively, when a fluctuation in the global activity of a "resting" network is large enough to overcome the distance $\Delta'_\theta = (f'_\theta - f_{rest})$, the stability of f_{rest} may be (transiently) lost and the entire network synchronously shifts to a new regime where $f(t) = f_{burst}$. The unstable state is therefore acting as a no-return point, similarly to what happens to the membrane voltage of a neuron during the generation of an action-potential.

Anyway, such a regime cannot be sustained indefinitely, because f_{burst} is not a solution of the full, self-consistent equation, including the *updated* effects of adaptation

$$f^* = \Phi_{IF}(m(f^*) - \alpha\, f_{burst}, s(f^*)).$$

Instead, such an equation is satisfied by $f^* = f_H \approx 0$ Hz, with $f_H < f_{burst}$ (see Figure 10.10B). Therefore, although adaptation progressively builds up in individual neurons, network activity decreases and the locations of the f_{burst} and f'_θ tend to approach and finally overlap, until their existence is suddenly lost (see Figure 10.10B). When this occurs, the network dynamics suddenly converges to f_H and most neurons stop firing, similarly to the hyperpolarization experienced by the membrane voltage after an action-potential. This accounts quantitatively for the generation of PBs, as compared to direct network simulations (Figure 10.11B, markers). By analogy, the amount of time spent in PB is therefore related to the distance $\Delta' = (f_{burst} - f'_\theta)$ and to the time requested by adaptation for the full buildup (i.e., α and τ_a). Qualitatively, it can be concluded that for an increasing J_e, Δ'_θ decreases whereas Δ' increases, thus the mean interburst interval (IBI) decreases and the mean duration of PB (PBd) increases.

Finally, as in the previous case, such a new regime cannot be sustained indefinitely, because f_H is not a solution of:

$$f^* = \Phi_{IF}(m(f^*) - \alpha\, f_H, s(f^*)).$$

The network will thus recover its resting activity level $f_{rest} > f_H$, as described at the beginning.

10.5.4   Variability of the IBIs Distribution

From the considerations developed so far, it is possible to interpret the regular/irregular character of the IBIs. In the last paragraphs, an analogy between PBs and action potentials occurring in an excitable membrane was often proposed and it is now exploited. In fact, similarly to the temporal evolution of the membrane potential in the single-neuron IF dynamics, the activity of a network randomly fluctuates as a result of synaptic noise. Occasionally, these fluctuations may be large enough to overcome an excitability threshold (i.e., Δ'_θ). When this happens, a major explosive event occurs ($f \rightarrow f_{burst}$), and later the activity is strongly refractory to any further generation of PBs ($f \approx f_H$). The generation of PBs is somehow similar to the generation of an action potential in a model of integration of a noisy input. By mapping the population mean firing rate f into the membrane voltage of an abstract IF model neuron, we make the previous comparisons explicit, setting the resting membrane voltage to f_{rest}, the spike threshold to f'_θ, the reset potential to f_H, and the absolute refractory period proportional to τ_a.

An increase in network synaptic coupling J_e induces a decrease in the distance $f'_\theta - f_{rest}$ (see Figure 10.11B), so that the rate of threshold crossing is expected to monotonically increase, while preserving an irregular character. Such predictions were confirmed by the simulated network activity, where IBIs statistics are approximated by a Poisson process with a refractory time (see Figure 10.9, lower panel, dotted thick line). This is reminiscent of a widely studied balanced (i.e., drift-free) integration process (see Shadlen and Newsome (1998)), where the threshold crossings are determined by subthreshold fluctuations only, in a noise-dominated regime.

10.5.5   Interpreting MEA Experimental Data

As anticipated, we conclude that the frequency and regular/irregular character of spontaneous population bursting are intimately related to the synaptic strength J_e, and inversely, on the size of the network. The last dependence is associated with the amplitude of random fluctuations in the population firing rate, which are produced by finite-size network effects (Mattia and Del Giudice, 2002). At a parity of network size, we predict and observe a rare and unpredictable occurrence of short and irregular PBs, for intermediate values of J_e, and a frequent and more regular occurrence of longer and more regular PBs, for larger J_e.

Here, we attempt to provide an interpretation of the MEA experiments, consistent with the insights gained from the theory. This brings us to conclude that around 1 to 5 DIV, because the number and effectiveness of in vitro synapses are very low (i.e., J_e is small), no PB can occur because the network is intrinsically incapable of self-sustaining (transient) reverberating activity of a PB. Later, the strong neuritic outgrowth and synaptogenesis turn the network into a bistable device. We speculate that such conditions may correspond to an overshoot phase of in vitro network development, consisting in an over-expression of branches and synaptic boutons, resulting in a large J_e. As described in the previous subsection, the recurrent connections of the network can now transiently sustain a regular population bursting, mainly determined and dominated by the recovery time constants of intrinsic neuronal adaptation. Later, as the network reaches its mature condition, pruning of a large fraction of synaptic contacts, or modulation of synaptic efficacy by homeostatic processes (see Figure 10.2), makes J_e substantially smaller. In such a regime, the statistics of the simulated IBIs and PBd matched the results of the MEA experiments, performed under pharmacological disinhibition (see Figure 10.9).

We propose the diagram of Figure 10.11 to summarize the prediction of the theory, as well as the results of the computer simulation. Any pharmacological and/or ionic manipulation of excitatory synaptic efficacy is expected to alter not only the frequency and the regular/irregular occurrence of PBs, but also the interburst spiking frequency, according to the trajectory indicated by triangles and the dotted thick line of Figure 10.11B.

Lowering extracellular magnesium concentration, by increasing activation of NMDA-receptors, is expected to unveil population bursting in a silent mature culture (Robinson et al., 1993). Similarly, a decrease in the concentration of extracellular calcium ions would tend first to drive a regularly bursting mature network into a rare unpredictable bursting regime, and finally to abolish completely each PB (Canepari et al., 1997). Therefore, the action of bath-applied drugs such as AP-V, CNQX, cyclothiazide, or the modulation of extracellular ions such as Mg^{++} and Ca^{++}, largely involved in excitatory synaptic transmission, can be roughly mapped as modulations of J_e. The outcome of such a modulation depends of course on the intensity and sign of these manipulations, as well as on the previous state of the network (Turrigiano and Nelson, 2004).

From the phase diagram of Figure 10.11, it becomes clear that by a partial blockade of glutamatergic synaptic receptors, PBs disappear whereas asynchronous isolated spiking activity may persist (Kamioka et al., 1996). A complete blockade of synaptic transmission would remove also the background random activity, as neurons become insensitive to the fluctuations induced by spontaneous synaptic release (i.e., $m_0 = s_0 = 0$).

A final prediction of this theory consists in the monotonic dependence on J_e of the interburst firing rate (Nakanishi and Kukita, 1998; Maeda et al., 1998; Canepari et al., 1997; Kamioka et al., 1996). The use of repeated extracellular stimulation known to induce activity-dependent potentiation of synaptic efficacy in a long-term manner, and corresponding to an increase in J_e, is expected to: (i) turn a previously spontaneously silent mature network into an active one by increasing the probability of spontaneous PBs; (ii) turn a silent network, previously nonresponsive to electrical stimulation, into a system that generates a PB upon electrical stimulation; and (iii) increase the number of bursts per minute, while increasing the number of spikes per burst (Maeda et al., 1998). Finally, a graded modulation of synaptic efficacy J_e is expected to induce a graded modification on the network dynamics (Maeda et al., 1998).

10.6 Concluding Remarks

In vivo cortical processing naturally arises from interconnections among large populations of neurons and simultaneous feedback from several brain structures. Such conditions are of course completely lacking in isolated brain slices and cell cultures. Nevertheless, approaching the in vitro network level from the knowledge of single neurons and synapses, understanding how much complexity is the result of the interactions and how much is intrinsic, does represent a powerful tool to challenge our models and predictions on small-scale reduced problems.

In this chapter, single-neuron discharge properties have been investigated in dissociated cultures of neocortex coupled to MEAs, demonstrating that a simplified point-neuron IF model is an adequate description when network mean firing rates are considered. It is significant that spike response properties of cultured neocortical neurons qualitatively resemble those of cells in acute slice preparations. This is of great importance as dissociated neurons might undergo a different ex vivo development of intrinsic biophysical properties.

The same patterned activity, experimentally characterized in vitro by MEA recordings, was reproduced in the simulated networks, and interpreted in terms of the network-response properties, emerging from $f = \Phi(m,s)$. Matching with MEA recordings is satisfactory, indicating that the discharge response properties to noisy current stimuli and the experimentally characterized spike-frequency adaptation are indeed sufficient to account for the emerging collective activity, observed in the experiments. Finally, it is interesting to note that the reduction driven by the experimental data at the single-neuron level did not compromise the richness of phenomena occurring at the level of a population.

Acknowledgments. The authors thank Stefano Fusi, Giancarlo La Camera, and Anne Tscherter for helpful discussions, and Ruth Rubli for assistance. This work was supported by the Swiss National Science Foundation, the Human Frontier Science Program, the EC Thematic Network in Neuroinformatics and Computational Neuroscience, and the University of Genova, Italy.

Appendix A

A.1 The Membrane Potential as a Diffusion Process

In order to characterize the spiking response of an IF model neuron under noisy current inputs, it is convenient to normalize all voltages to the resting membrane voltage E, and to introduce the probability density function $P_V(v, t)$ (Fourcaud and Brunel, 2002; Fusi and Mattia, 1999; Abeles, 1991). Such a function gives at any time t the probability that the membrane voltage $V(t)$ approaches values around v:

$$P_V(v, t)\, dv = \text{Prob}\{v < V(t) \leq v + dv\}. \tag{A.1}$$

Such a characterization accounts for the trajectories of the membrane potential, resulting from any possible current input. In a generalized IF model, with a state- and time-dependent intrinsic membrane current $L(V, t)$ (Fourcaud-Trocmé et al., 2003), the membrane potential evolves subthreshold as

$$C \frac{dV}{dt} = L(V, t) + I(t). \tag{A.2}$$

When $I(t)$ is a delta-correlated and Gauss-distributed process, with infinitesimal mean and variance μ and σ^2, respectively, $V(t)$ is a stochastic diffusion process. As a consequence, the probability density $P_V(v, t)$ satisfies an equation known as the Fokker–Planck equation (Fourcaud and Brunel, 2002; Risken, 1984; Cox and Miller, 1965):

$$\frac{\partial}{\partial t} P_V(v, t) = -\frac{\partial}{\partial v} \phi(v, t). \tag{A.3}$$

In the previous equation, the right-hand side is reminiscent of the gradient of a flux of particles in a diffusion equation, and it can be identified as the sum of a diffusive term and a drift-related term:

$$\phi(v, t) = -\frac{\sigma^2}{2} \frac{\partial}{\partial v} P_V(v, t) + (\mu - L(V, t)) P_V(v, t). \tag{A.4}$$

Readers interested in the derivation of Equation (A.3), may refer to the literature (Risken, 1984; Cox and Miller, 1965).

A.2   The Current-to-Rate Response Function

By analogy with the diffusion of a substance in a medium, it is possible to identify and evaluate the neuron spiking mean rate $f(t)$ at time t, as the number of first-passages $V(t) = \vartheta$, in the time unit. This coincides with the definition of the flux ϕ at the spike-threshold, because of the additional boundary conditions that must satisfy Equation (A.3) (see below):

$$f(t) = \phi(\vartheta, t) \tag{A.5}$$

At steady-state, such a quantity has been defined as the current-to-rate response function (Tuckwell, 1988; Ricciardi, 1977); it was estimated experimentally in rat neocortical neurons (Giugliano et al., 2004; Rauch et al., 2003) and indicated by $\Phi_{IF}(\mu, \sigma^2)$ as a function of the input current statistics.

As anticipated, appropriate boundary conditions must be specified to introduce the nonlinearities of the IF dynamics (see also Fusi and Mattia, 1999). These are related to the excitability threshold ϑ, the reset potential H, and the refractory period τ_{arp}. The analogy with a process of diffusion in a medium can be pursued identifying $P_V(v, t)$ as an instantaneous concentration, and discussing the way it is restricted.

- At $V = \vartheta$, in terms of an absorbing barrier, because the particles (i.e., neurons) that are crossing such a threshold are "absorbed" at any time, and leave the interval $(-\infty; \vartheta)$ to undergo refractoriness:

$$\forall t, \, P_V(\vartheta, t) = 0 \tag{A.6}$$

- At $V = H$, in terms of a discontinuity in the flux $\phi(H, t)$, due to the spike-reset mechanisms and the consequent flow of previously absorbed particles, coming out from refractoriness:

$$\lim_{\varepsilon \to 0} \left(\phi(H + \varepsilon, t) - \phi(H - \varepsilon, t) \right) = f(t - \tau_{\mathrm{arp}}) \tag{A.7}$$

At any time t, $P_V(v, t)$ must further satisfy a normalization condition

$$\int_{-\infty}^{\vartheta} P_V(v, t) \, dv + \int_{t}^{t + \tau_{\mathrm{arp}}} f(t') \, dt' = 1, \tag{A.8}$$

because no neuron is allowed to have a membrane potential outside the range $(-\infty; \vartheta)$, unless it is refractory (i.e., no "particle" is destroyed or generated).

For the leaky IF (Equation (A.2)), we finally report the expression of the current-to-frequency response function at the steady-state (Ricciardi, 1977), which can be derived from Equations (A.3), and (A.6) through (A.8), by setting $L(V, t) = \bar{g}V$ and dropping any temporal dependence:

$$f = \phi(\vartheta) = \Phi = \left[\tau_{\mathrm{arp}} + \tau \sqrt{\pi} \int_{(\hat{H})}^{(\hat{\vartheta})} e^{x^2} (1 + erf(x)) \, dx \right]^{-1} \tag{A.9}$$

where $\tau = C/\bar{g}$ is the neuron membrane time constant, $erf(x)$ indicates the error function (Abramowitz and Stegun, 1994), and

$$\hat{H} = (H - \mu\tau)/(\sigma\sqrt{\tau}), \, \hat{\vartheta} = (\vartheta - \mu\tau)/(\sigma\sqrt{\tau}),$$

$$\mu = m/C, \, \sigma = (s\sqrt{2\tau_1})/C.$$

We note that the current-to-frequency response function, employed for the fit of experimental data curves (see Figure 10.7), incorporated the stationary effect of the adaptation current $I_X(t)$ (see also Equation (A.3)). This is reintroduced by replacing m with $(m - \alpha f)$ in the previous expressions, obtaining an implicit expression to be solved in f. In fact, adaptation currents affect only the steady-state mean current without changing substantially the input variance (La Camera et al., 2002).

References

Abbott, L.F. and Dayan, P. (2001). *Theoretical Neuroscience*. MIT Press, Cambridge.

Abeles, M. (1991). Relations between membrane potential and the synaptic response curve. In *Corticonics: Neural Circuits of the Cerebral Cortex*, Cambridge University Press, Cambridge, UK, Chap. 4, pp. 118–149.

Abramowitz, M. and Stegun, I.A. (1994). *Tables of Mathematical Functions*. Dover, New York.

Amit, D.J. (1989). *Modeling Brain Function*. Cambridge University Press, Cambridge, UK.

Amit, D.J. and Brunel, N. (1997). Model of global spontaneous activity and local structured (learned) delay activity during delay. *Cerebral Cortex* 7: 237–252.

Bove, M., Martinoia, S., Verreschi, G., Giugliano, M., and Grattarola, M. (1998). Analysis of the signals generated by networks of neurons coupled to planar arrays of micro-transducers in simulated experiments. *Biosens. Bioelectron.* 13: 601–612.

Brunel, N. (2000). Persistent activity and the single cell $f–I$ curve in a cortical network model. *Network* 11: 261–80.

Brunel, N. and Wang, X.J. (2003). What determines the frequency of fast network oscillations with irregular neural discharges? I. Synaptic dynamics and excitation-inhibition balance. *J. Neurophysiol.*, 90: 415–30.

Bulloch, A.G.M. and Syed, N.I. (1992). Reconstruction of neuronal networks in culture. *Trends Neurosci.*, 15: 422–27.

Canepari, M., Bove, M., Maeda, E., Cappello, M., and Kawana, A. (1997). Experimental analysis of neuronal dynamics in cultured cortical networks and transitions between different patterns of activity. *Biol. Cybern.*, 77: 153–62.

Chance, F.S., Abbott, L.F., and Reyes, A.D. (2002). Gain modulation from background synaptic input. *Neuron* 35: 773–782.

Cox, D.R. and Miller, H.D. (1965). *The Theory of Stochastic Processes*. Chapman & Hall, London.

Destexhe, A. and Paré, D. (1999). Impact of network activity on the integrative properties of neocortical pyramidal neurons *in vivo*. *J. Neurophysiol.* 81: 1531–1547.

Destexhe, A. and Paré, D. (2000). A combined computational and intracellular study of correlated synaptic bombardment in neocortical pyramidal neurons *in vivo*. *Neurocomput.* 32: 113–119.

Destexhe, A., Mainen, Z.F., and Sejnowski, T.J. (1994). Synaptic transmission and neuromodulation using a common kinetic formalism. *J. Comp. Neurosci.*, 1: 195–230.

Destexhe, A., Rudolph, M., and Paré, D. (2003). The high-conductance state of neocortical neurons *in vivo*. *Nature Rev.* 4: 739–51.

Douglas, R.J. and Martin, K.A. (1990). Neocortex. In: *The Synaptic Organization of the Brain: 3rd edition*. Oxford University Press, New York, Chap. 12, pp. 389–438.

Fleidervish, I., Friedman, A., and Gutnick, M.J. (1996). Slow inactiavation of Na^+ current and slow cumulative spike adaptation in mouse and guinea–pig neocortical neurones in slices. *J. Physiol.* 493.1: 83–97.

Fourcaud, N. and Brunel, N. (2002). Dynamics of the firing probability of noisy integrate-and-fire neurons. *Neural Comp.* 14: 2057–2110.

Fourcaud-Trocmé, N., Hansel, D., van Vreeswijk, C., and Brunel, N. (2003). How spike generation mechanisms determine the neuronal response to fluctuating inputs. *J. Neurosci.*, 23(37): 11628–11640.

Fuhrmann, G., Markram, H., and Tsodyks, M. (2002). Spike frequency adaptation and neocortical rhythms. *J. Neurophysiol.* 88: 761–770.

Fusi, S. and Mattia, M. (1999). Collective behavior of networks with linear (VLSI) integrate and fire neurons. *Neural Comp.* 11: 633–652.

Gerstein, G.L. and Mandelbrot, B. (1964). Random walk models for the spike activity of a single neuron. *Biophys. J.* 4: 41–68.

Gerstner, W. (2000). Population dynamics of spiking neurons: Fast transients, asynchronous states, and locking. *Neural Comp.* 12: 43–89.

Giugliano, M. (2000). Synthesis of generalized algorithms for the fast computation of synaptic conductances with Markov kinetic models in large network simulations. *Neural Comput.* 12(4): 771–799.

Giugliano, M., Darbon, P., Arsiero, M., Lüscher, H.-R., and Streit, J. (2004). Single-neuron discharge properties and network activity in dissociated cultures of neocortex. *J. Neurophysiol.* 92(2): 977–996.

Giugliano, M., La Camera, G., Rauch, A., Lüscher, H.-R., and Fusi, S. (2002). Non-monotonic current-to-rate response function in a novel integrate-and-fire model neuron. In: Dorronsoro, J.R., ed., *Proceedings of ICANN2002*, Springer, New York, pp. 141–146.

Grattarola, M. and Martinoia, S. (1993). Modeling the neuron-microtransducer junction: From extracellular to patch recording. *IEEE Trans. Biomed. Eng.* 40: 35–41.

Grattarola, M. and Massobrio, G. (1998). *Bioelectronics Handbook: MOSFETs, Biosensors, Neurons.* McGraw-Hill, New York.

Gross, G.W. (1979). Simultaneous single-unit recording in vitro with a photoetched laser deinsulated, gold multielectrode surface. *IEEE Trans. Biomed. Eng.* 26: 273–279.

Gross, G.W., Wen, W., and Lin, J. (1985). Transparent indium-tin oxide patterns for extracellular, multisite recordings in neuronal cultures. *J. Neurosci. Meth.* 15: 243–252.

Hamill, O.P., Marty, A., Neher, E., Sakmann, B., and Sigworth, F.J. (1981). Improved patch-clamp techniques for high-resolution current recording from cells and cell-free membrane patches. *Pfluegers Arch.* 391: 85–100.

Hodgkin, A.L. and Huxley, A.F. (1952). A quantitative description of membrane current and its application to conduction and excitation in nerve. *J. Physiol.* 117: 500–544.

Holt, G.R., Softky, W.R., Koch, C., and Douglas, R.J. (1996). Comparison of discharge variability in vitro and in vivo in cat visual cortex neurons. *J. Neurophysiol.* 75(5): 1806–1814.

Huettner, J.E. and Baughman, R.W. (1986). Primary culture of identified neurons from the visual cortex of postnatal rats. *J. Neurosci.* 6: 3044–3060.

Jimbo, Y., Tateno, T., and Robinson, H.P.C. (1999). Simultaneous induction of pathway-specific potentiation and depression in networks of cortical neurons. *Biophys. J.* 76: 670–678.

Kamioka, H., Maeda, E., Jimbo, Y., Robinson, H.P.C., and Kawana, A. (1996). Spontaneous periodic synchronized bursting during formation of mature patterns of connections in cortical cultures. *Neurosci. Lett.* 206: 109–112.

La Camera, G., Rauch, A., Senn, W., Lüscher, H.-R., and Fusi, S. (2002). Firing rate adaptation without losing sensitivity to fluctuations. In: Dorronsoro, J.R., ed., *Proceedings of ICANN2002,* Springer, New York, pp. 180–185.

La Camera, G., Senn, W., and Fusi, S. (2003). Equivalent networks of conductance- and current-driven neurons. In: *Proceedings of ICANN/ICONIP,* Instanbul, Turkey.

Maeda, E., Kuroda, Y., Robinson, H.P.C., and Kawana, A. (1998). Modification of parallel activity elicited by propagating bursts in developing networks of rat neocortical neurones. *Eur. J. Neurosci.* 10: 488–496.

Maeda, E., Robinson, H.P., and Kawana, A. (1995). The mechanisms of generation and propagation of synchronized bursting in developing networks of cortical neurons. *J. Neurosci.* 15(10): 6834–6845.

Mainen, Z.F. and Sejnowski, T. (1995). Reliability of spike timing in neocortical neurons. *Science,* 268: 1503.

Marom, S. and Shahaf, G. (2002). Development, learning and memory in large random networks of cortical neurons: lessons beyond anatomy. *Quart. Rev. Biophys.* 35: 63–87.

Mattia, M. and Del Giudice, P. (2002). Population dynamics of interacting spiking neurons. *Phys. Rev. E* 66(5): 051917.

McCormick, D.A., Connors, B.W., Lightfall, J.W., and Prince, D.A. (1985). Comparative electrophysiology of pyramidal and sparsely spiny stellate neurons of the neocortex. *J. Neurophysiol.* 59: 782–806.

Milo, R., Shen-Orr, S., Itzkovitz, S., Kashtan, N., Chklovskii, D., and Alon, U. (2002). Network motifs: Simple building blocks of complex networks. *Science*, 298(5594): 824–827.

Muramoto, K., Ichikawa, M., Kawahara, M., Kobayashi, K., and Kuroda, Y. (1993). Frequency of synchronous oscillations of neuronal activity increases during development and is correlated to the number of synapses in cultured cortical neuron networks. *Neurosci. Lett.* 163: 163–165.

Nakanishi, K. and Kukita, F. (1998). Functional synapses in synchronized bursting of neocortical neurons in culture. *Brain Res.*, 795(1–2): 137–146.

Nakanishi, K., Kukita, F., Asai, K., and Kato, T. (2001). Recurrent subthreshold electrical activities of rat neocortical neurons progress during long-term culture. *Neurosci. Lett.* 304(1–2): 85–88.

Nakanishi, K., Nakanishi, M., and Kukita, F. (1999). Dual intracellular recording of neocortical neurons in a neuron-glia co-culture system. *Brain Res. Prot.* 4: 105–114.

Poliakov, A.V., Powers, R.K., and Binder, M.D. (1997). Functional identification of the input-output transforms of motoneurones in the rat and cat. *J. Physiol.* 504.2: 401–424.

Potter, S.M. and DeMarse, T.B. (2001). A new approach to neuronal cell culture for long-term studies. *J. Neurosci. Meth.* 59: 782–806.

Powers, R.K., Sawczuk, A., Musick, J.R., and Binder, M.D. (1999). Multiple mechanisms of spike-frequency adaptation in motoneurones. *J.Physiol. (Paris)* 93: 101–114.

Press, W., Teukolsky, S.A., Vetterling, W.T., and Flannery, B.P. (1992). *Numerical Recipes in C: The Art of Scientific Computing.* Cambridge University Press, New York.

Protopapas, A.D. and Bower, J.M. (2001). Spike coding in pyramidal cells of the piriform cortex of rat. *J. Neurophysiol.* 86: 1504–1510.

Rauch, A., La Camera, G., Lüscher, H.-R., Senn, W., and Fusi, S. (2003). Neocortical pyramidal cells respond as integrate-and-fire neurons to *in vivo*-like input currents. *J. Neurophysiol.* 90: 1598–1612.

Reutimann, J., Giugliano, M., and Fusi, S. (2003). Event-driven simulation of spiking neurons with stochastic dynamics. *Neural Comp.* 15: 811–830.

Ricciardi, L.M. (1977). *Diffusion Processes and Related Topics in Biology.* Springer, Berlin.

Risken, H. (1984). *The Fokker–Planck Equation: Methods of Solution and Applications.* Springer, Berlin.

Robinson, H.P.C., Kawahara, M., Jimbo, Y., Torimitsu, K., Kuroda, Y., and Kawana, A. (1993). Periodic synchronized bursting and intracellular calcium transients elicited by low magnesium in cultured cortical neurons. *J. Neurophysiol.* 70: 1606–1616.

Salinas, E. (2003). Background synaptic activity as a switch between dynamical states in a network. *Neural Comp.* 15: 1439–1475.

Sanchez-Vives, M.V. and McCormick, D.A. (2000). Cellular and network mechanisms of rhythmic recurrent activity in neocortex. *Nat. Neurosci.* 3: 1027–1034.

Sanchez-Vives, M.V., Nowak, L.G., and McCormick, D.A. (2000). Cellular mechanisms of long-lasting adaptation in visual cortical neurons *in vitro. J. Neurosci.* 20: 4286–4299.

Sawczuk, A., Powers, R.K., and Binder, M.D. (1997). Contribution of outward currents to spike frequency adaptation in hypoglossal motoneurons of the rat. *J. Physiol.* 78: 2246–2253.

Schwindt, P.C., Spain, W.J., and Crill, W.E. (1989). Long-lasting reduction of excitability by a sodium-dependent potassium current in cat neocortical neurons. *J. Neurophysiol.* 61: 233–244.

Shadlen, M.N. and Newsome, W.T. (1998). The variable discharge of cortical neurons: implications for connectivity, computation, and information coding. *J. Neurosci.* 18(10): 3870–3896.

Silberberg, G., Bethge, M., Markram, H., Pawelzik, K., and Tsodyks, M. (2004). Dynamics of population rate codes in ensembles of neocortical neurons. *J. Neurophysiol.* 91(2): 704–709.

Smith, G.D., Cox, C.L., Sherman, S. M., and Rinzel, J. (2000). Fourier analysis of sinusoidally driven thalamocortical relay neurons and a minimal integrate-and-fire-or-burst model. *J. Neurophysiol.* 83(1): 588–610.

Stengler, D.A. and McKenna, T.M. (1994). *Enabling Technologies for Cultured Neural Networks.* Academic Press, London.

Steriade, M. (2001). Similar and contrasting results from studies in the intact and sliced brain. In: *The Intact and Sliced Brain,* Bradford Books, MIT Press, Cambridge, MA, Chap. 3, pp. 103–190.

Streit, J., Tscherter, A., Heuschkel, M.O., and Renaud, P. (2001). The generation of rhythmic activity in dissociated cultures of rat spinal cord. *Eur. J. Neurosci.* 14: 191–202.

Tscherter, A., Heuschkel, M.O., Renaud, P., and Streit, J. (2001). Spatiotemporal characterization of rhythmic activity in spinal cord slice cultures. *Eur. J. Neurosci.* 14: 179–190.

Tsodyks, M., Uziel, A., and Markram, H. (2000). Synchrony generation in recurrent networks with frequency-dependent synapses. *J. Neurosci.* 20: RC50: 1–5.

Tuckwell, H.C. (1988). *Introduction to Theoretical Neurobiology.* Cambridge University Press, New York.

Turrigiano, G.G. and Nelson, S.B. (2004). Homeostatic plasticity in the developing nervous system. *Nature Rev. Neurosci.* 5(2): 97–107.

Van den Pol, A.N., Obrietan, K., and Belousov, A. (1996). Glutammate hyperexcitability and seizure-like activity throughout the brain and spinal cord upon relief from chronic glutammate receptor blockade in culture. *Neuroscience* 74: 653–674.

Van Huizen, F., Romijn, H.J., and Corner, M.A. (1987). Indications for a critical period for synapse elimination in developing rat cerebral cortex cultures. *Brain Res.,* 428: 1–6.

Van Vreeswijk, C., and Hansel, D. (2001). Patterns of synchrony in neural networks with spike adaptation. *Neural Comp.* 13: 959–992.

Wang, X.J. (2001). Synaptic reverberation underlying mnemonic persistent activity. *Trends Neurosci.* 24(8): 455–463.

Wiedemann, U.A. and Lüthi, A. (2003). Timing of network synchronization by refractory mechanisms. *J. Neurophysiol.* 90: 3902–3911.

11
Analysis of Cardiac Myocyte Activity Dynamics with Micro-Electrode Arrays

ULRICH EGERT, KATHRIN BANACH, AND THOMAS MEYER

Introduction

The analysis of cardiac electrical potentials through electrocardiograms (ECG) in vivo is a well-known technique to reveal information about system properties of the heart, arrhythmia, indications of conduction failures, and so on. Although such recordings are indispensable, the spatial resolution and the opportunities for manipulation in vivo are limited. In contrast, in vitro investigations of isolated organs and tissues (e.g., Purkinje fibers, papillary muscle, Langendorff heart, or patches of cardiac muscle; Yamamoto et al., 1998), are routinely employed to study the mechanisms of the generation and propagation of cardiac potentials at a higher spatial resolution using intracellular, extracellular, or optical recording (Hirota et al., 1987; Mastrototaro et al., 1992; Hofer et al., 1994).

Cultures of cardiac myocytes harvested from cardiac tissue after enzymatic digestion of the tissue offer the opportunity for analyses of single-cell or cell-aggregate properties, for example, for developmental, pharmacological, and biophysical studies (Furshpan et al., 1976; Guevara et al., 1981; Scott et al., 1986; Metzger et al., 1994, 1995; Nag et al., 1996; Egert et al., 2003; Banach et al., 2003). Although these cultures do not maintain the structure of cardiac tissue, the functional properties of action potential (AP) generation and propagation, the contractility, and the ion channel composition of the original cells are conserved or re-established, depending to some extent on the culture system. The facility of production and the simple anatomical structure of these cell and tissue culture systems thus allow the researcher to address questions otherwise not easily approachable in organs or animal preparations. The contractions of the cells in most of these preparations, however, hinder electrophysiological studies with conventional electrodes, in particular intracellular or patch-clamp recording, and in optical recordings with voltage-sensitive dyes, causing various artifacts in the signal.

Although the chemical decoupling of the contractile apparatus can in principle reduce motility, it may have side effects and is thus generally not desirable. Extracellular recording of field potentials (FP) from contracting myocyte cultures is, however, facilitated considerably when the cells are grown directly on the recording electrodes, that is, in culture dishes with integrated microelectrode arrays (see Chapter 2; Thomas et al., 1972; Israel et al., 1984; Rohr, 1990; Igelmund et al., 1999; Kehat et al., 2002; Meiry et al., 2001; Banach et al., 2003; Halbach et al., 2003). Cells grown on MEAs will adhere tightly to the substrate and contract isometrically, avoiding the motion artifacts that usually deteriorate the signal-to-noise ratio (SNR). These devices enable noninvasive simultaneous, multi-site, extracellular recordings from myocytes, excluding the mechanical stimulation of the cell that can hardly be avoided with conventional wire- and micro-pipette recording techniques.

The multiplicity of recording sites also allows us to keep track of slow structural displacements during proliferation of excitable tissue when the tissue grows, shifting propagation pathways during differentiation of the cellular network, and during pharmacological manipulation of, for example, the excitability or the intercellular coupling via gap junctions. The combination of these properties enables studies that were hitherto not possible because they required sterile long-term monitoring of activity patterns, re-identification of relevant regions or measurements that would be confounded by mechanical, chemical, or photochemical stress (as in recordings with voltage-sensitive dyes). In addition, the spatiotemporal structure of activity and of the FP waveform can be monitored in real-time even during irregular activity, with changing pacemaker dominance or propagation pathways.

In this chapter, we introduce the recording of cell cultures prepared from cardiac tissue on MEAs, and the basis for the visualization and interpretation of the data obtained in such extracellular recordings. Our intention is to give the user a general starting point without specializing too much towards a particular application, and to illustrate what can be gained from such data. Detailed protocols appear in Egert and Meyer (2005).

MEAs with 60 to 70 electrodes are produced with thin-film photolithographic techniques and have become commercially available. Dedicated full recording systems are currently available as the "MEA1060" system from Multi Channel Systems (Reutlingen, Germany)[*] and as "MED64" from Alpha MED Sciences (Tokyo, Japan).[†] Other groups have fitted their own setups or use similar MEAs produced elsewhere.[‡]

11.1 Methods

11.1.1 Recording Equipment

The MEAs used in our experiments have 60 microelectrodes, with a diameter of 30 μm, positioned on an 8×8 grid with 200 μm spacing (see Chapter 2). The recording area thus covers 1.4×1.4 mm^2. The electrodes themselves are flat, with a rough surface of titanium nitride (TiN), 80 to 200 MΩ impedance (at 1 kHz) and

---

[*] www.multichannelsystems.com.
[†] www.med64.com.
[‡] For example, Ayanda Biosystems.

a max. 1 µm recess into the substrate. The culture chamber is formed by a 2 cm diameter glass ring of 6 mm height glued to the MEA base plate, resulting in a chamber volume of ca. 1.8 ml. Recordings were carried out with an MEA1060 system (Multi Channel Systems, Reutlingen, Germany), amplifier bandwidth 0.1, 1, or 10 Hz to 3.2 kHz, amplification 1000x or 1200x).

The sampling frequency was 10 to 25 kHz. Further analyses were carried out offline, using tools written for MATLAB (The Mathworks, Natick, USA).

11.1.2   Cell Culture

To promote the adhesion of cells, MEAs were coated with polyethylenimine (PEI, all chemicals were obtained from Sigma-Aldrich, Deisenhofen, Germany, unless stated otherwise), cellulose nitrate (Schleicher & Schuell, Dassel, Germany (Egert et al., 2002)) or fibronectin (Becton Dickinson, Heidelberg, Germany). A suspension of cardiac cells was prepared from embryonic chicken ventricles (E10 to E13). The ventricular myocardium was isolated, minced, and digested with 0.05% trypsine (original activity 10400 U/mg) in phosphate buffered saline (PBS) at 37°C, washed and re-suspended in DMEM with 3% FCS to approximately 10^7 cells/ml. Approximately 2×10^5 cells from this suspension were placed onto the electrode field of a MEA and supplemented with 1 ml medium after 1 to 2 min. The cell culture medium was exchanged every other day. Cultures with spontaneous activity were recorded after 3 to 7 days.

The isolated or weakly coupled islets of contracting cells and cell aggregates each have their individual pacemakers. With increasing incubation time the cells proliferate and after a few days the islets merge into a continuous sheet. The contractions become increasingly dominated by one site, which interferes with and eventually entrains the initiation centers and the surrounding cellular network. The extent and completeness of this entrainment depends on the intercellular resistance across the gap junctions, the Na^+-channel density, the current generated by the pacemakers, and other factors. Under suitable conditions, the contraction cycle eventually becomes stable and regular, and is initiated from one region only. The exact time course of this sequence depends on the initial seeding density, the substrate coating, and the age of the animal from which the cells were prepared. Cardiac myocytes easily proliferate and become spontaneously active if harvested from embryonal or perinatal animals. The number of spontaneously active cells and the contraction rate decrease with increasing age of the animal.

11.1.3   Analysis of MEA Recordings from Cardiac Myocytes

As with conventional recording techniques, FPs correlate in time with the contraction cycle (FP spike rate) and can therefore be analyzed for contraction rate and spatiotemporal patterns of arrhythmia. Several aspects need to be considered in these analyses.

Shortly after seeding of the cardiac myocytes onto the MEA, some cells started to contract spontaneously. During the transition period, when the contraction of the cell sheet was still initiated by several pacemaker regions, the current produced by the eventually dominating pacemaker and other excitable cells was often not large enough to depolarize the cell membrane across larger distances. This led to the interference of additional pacemakers. The effectiveness of this interference often depended on the phase of the beating cycles relative to each other, causing irregular beating with variable FP waveforms (similar to the Wenckebach periodicity in the intact heart (Castellanos et al., 1998a,b; Schwab et al., 2004)). The propagation pathways varied for FP cycles initiated at different pacemaker regions, resulting in different latencies between the peak time of FP_{min} at the MEA electrodes. Although these arrhythmias are an interesting subject themselves, analyses of delay times between electrodes and arrhythmia patterns need a preliminary detection of this situation to correctly assign the originating pacemaker in each contraction cycle and correctly identify the propagation sequence. This is often possible based on the different waveforms of FPs generated by the two pacemakers at a particular electrode. Please see Figure 11.1.

In particular in arrhythmic tissue, it is obvious that the beat-to-beat variation of the peak timing prohibits the reconstruction of the efficacy and path of propagation based on sequential recordings at different locations. The MEA recording technique enables the detailed simultaneous assessment of event times, their variability across beats, the correlation of timing and variability between locations, and thus of the dynamic processes in the tissue at many positions. Besides regularity analyses, the most salient aspect of this is the propagation of the AP wave across the tissue. The propagation pathway, speed, and variability depend on a number of variables, and particularly on the intercellular resistance across gap junctions and the magnitude of the depolarizing sodium current (Spach et al., 1979; Spach, 1983; Spach and Heidlage, 1995). Although the spatial resolution of MEAs is not as good as in video-enhanced microscopy, we can easily identify the sharp peak associated with the Na^+ current, and thus determine the temporal succession of the depolarization across the tissue with high temporal resolution and describe the propagation of excitation (Kleber et al., 1996; Rohr and Kucera, 1997; Rohr et al., 1997a,b).

11.2   Results

Even though the timing of FP peaks can be determined with high temporal resolution, it does not allow a straightforward determination of the conduction velocity. The pathway of excitation propagation can only be projected from the succession of the peak times at the electrodes. With an electrode spacing of 200 μm this may not be unequivocally feasible as the real propagation pathway may be tortuous and narrow. Furthermore, the density of Na^+-channels and the local intercellular resistance contribute to the conduction velocity. Local variations (e.g., delays caused by high-resistance sections, bifurcations, etc.) contribute to the variability of the conduction velocity. Because of these sources of ambiguity, a general "conduction velocity" is often difficult to determine and verify satisfactorily. In most cases, however, the interpretation of experimental findings can be based equally well

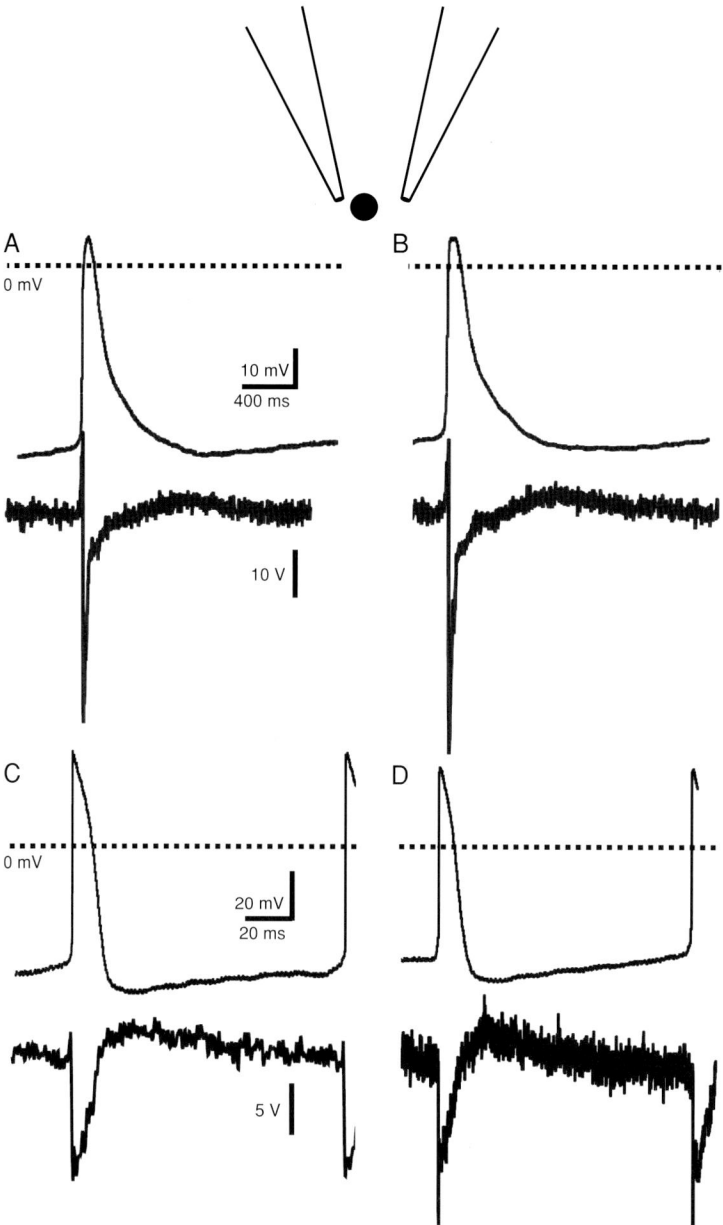

FIGURE 11.1. Comparison of extracellular FP and action potentials recorded intracellularly. Simultaneous paired recordings of APs (upper traces) and FPs (lower traces) in spontaneously active cardiac myocyte cultures. Whereas the FP minima correlate in time with the AP upstroke, the repolarization of the AP coincides with the subsequent FP maximum. APs ((AB) and (CD), respectively) were recorded successively from different cells around the same MEA electrode as illustrated in the sketch. The FP recorded simultaneously with this MEA electrode is shown underneath. The examples shown in (AB) and (CD) were obtained from two different cultures (Adapted with permission from Halbach et al., 2003, Figure 3).

on interelectrode delays directly, avoiding the need to determine the length and homogeneity of the propagation pathway.

11.2.1   Origin of the Signal Recorded

As described by Rall (Rall, 1969; Plonsey, 1977) for neuronal axons, the time course of the potential recorded extracellularly at an axon is related to the time course of the first derivative of the membrane potential of the cell studied. This corresponds to the sum of the depolarizing and hyperpolarizing currents across the membrane. Transient inward currents would be detected as negative voltages, outward currents as positive deflections. The main part of these currents is carried by Na^+, K^+, and Cl^- ions flowing through channels in the membrane and of compensating capacitive currents simultaneously flowing in the respective opposite directions. The spatial distribution of the latter very much depends on the distribution of the internal and leak resistances of the cell, which in turn depend on the morphology of the cells. Because of the complex and ramified dendritic tree of neurons it is essentially impossible to fully reconstruct the transmembrane potential or its time course from an extracellular recording in nervous tissue (Henze et al., 2000). In addition, at any one site in the extracellular space the local potential will be the sum of contributions from all current sources within the recording horizon of the electrode (Egert et al., 2002; Halbach et al., 2003). In native tissue this encompasses a large number of cells.

The situation is simplified, however, if the cell or tissue under study can be considered large with respect to the electrode, homogeneous in its morphology, and if the resistance between the recording electrode and the reference electrode is high. These conditions are largely met when cardiac myocytes with their extensive network of cells coupled through gap junctions are cultured on MEAs. In contrast to recordings from neuronal tissue, the time course of FPs recorded in cardiac tissue should therefore more closely correspond to that expected from theoretical considerations.

The stable mechanical situation during MEA recordings from cardiac myocyte cultures enables intracellular recordings to be performed simultaneously with the extracellular MEA recordings (Halbach et al., 2003). An analysis of the intra- and extracellular time course indicated that the current components mentioned above could in part be identified in the MEA recording. In the region initiating the activity, a sharp negative peak in the FP accompanied the depolarizing upstroke of the intracellular AP that is mainly driven by Na^+. Conversely, a positive peak paralleled the repolarization phase carried by K^+. At recording sites along the propagation pathway a positive peak, reflecting passive outward currents, preceded the negative peak. This TTX-sensitive current likely corresponds to a compensating outward current matching the depolarizing inward Na^+-current (Figure 11.1).

Detailed analyses of the FP shape that can reveal the spatial distribution of features of the FP are therefore possible (Figure 11.2; Sprössler et al., 1999; Halbach et al., 2003). The basis of this interpretation is the correlation between changes of some transmembrane currents and those of certain components of the FP. This

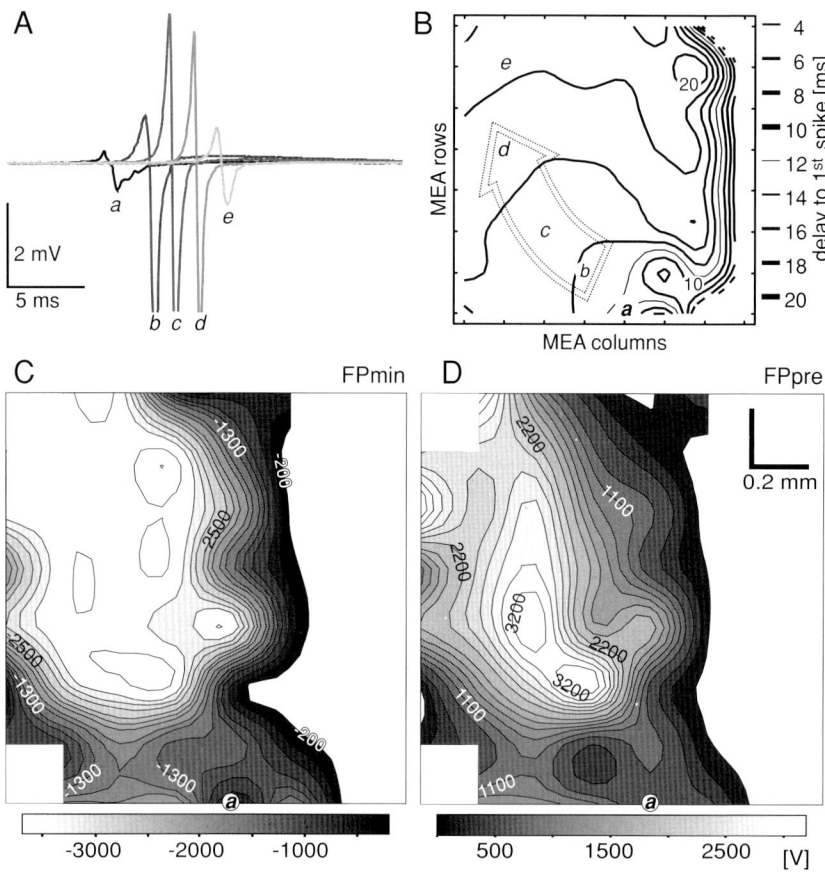

FIGURE 11.2. Two-dimensional analysis of excitation spread and FP parameters. The contracting region of this culture of chicken cardiac myocytes covered the left part of the MEA. (A) FP traces recorded at 5 MEA electrodes in the presumed initiation region (a) with the earliest spike detected and at positions along the propagation pathway (b–e) with increasing delays to (a). The FP recorded in the initiation region is small and has only a minor positive peak (FP_{pre}) preceding the minimum (FP_{min}). This suggests that the electrode does not record from the pacemaker site proper, where we would expect no preceding positive peak. FPs recorded at positions with increasing delays had a prominent FP_{pre} and dominating FP_{min}. (B) Contour plot of the isochrones of the FP_{min} with respect to position (a). The dense lines to the right stem from border artifacts. The arrow indicates a likely path of propagation of the underlying AP. (C) Map of the magnitude of FP_{min}. (D) Map of the magnitude of FP_{max}. These maps indicate that larger peaks coincide with positions along the path of propagation, rather than the initiation region of the spike, which is similar to the situation in native tissue.

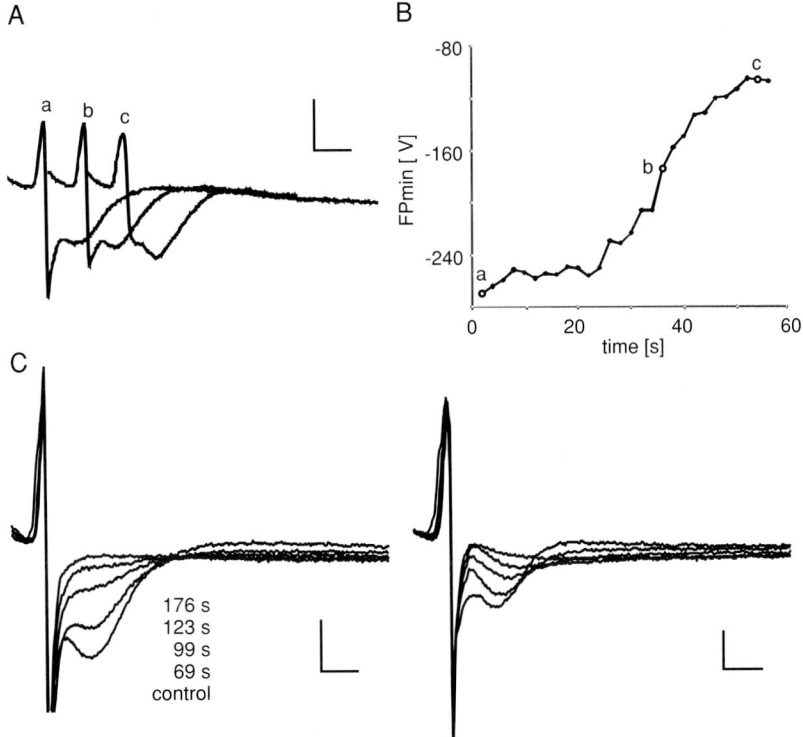

FIGURE 11.3. Origin of the FP: (A) FPs recorded in a cardiac myocyte culture during Na^+ withdrawal by superfusion with nominally Na^+-free buffer at the points in time indicated in (B). (B) FP_{min} magnitude decreased continuously during the washout of Na^+ starting at 20 min. Blocking Na^+-channels with tetrodotoxin produced a similar effect and increased the conduction delay, which is known to correlate with the magnitude of the Na^+-current (Spach and Kootsey, 1983). (C) This experiment illustrates another current component directly influencing the FP shape. Superfusion of the culture with nominally Ca^{2+}-free solution resulted in a magnitude decrease of the delayed second minimum of the FP, indicating its Ca^{2+} dependence (two recording sites of the same culture). This period of the FP coincides with the plateau phase of the AP. (scale bars are 100 $\mu V/20$ msec). (Reproduced with permission from Halbach et al., 2003, Figures 7AB and 9AB.)

allows relating the waveform recorded extracellularly and changes thereof to changes of the shape of the underlying APs and to the known sequence of Na^+-, K^+- and Ca^{2+}-currents (Figures 11.1 and 11.3).

11.2.2   Spatial Resolution

Even though the recording horizon of MEA electrodes was estimated in several simulation studies (Fromherz et al., 1991; Buitenweg et al., 2000; Heuschkel et al., 2002), it is not clear how this translates into the detectability of a signal in native

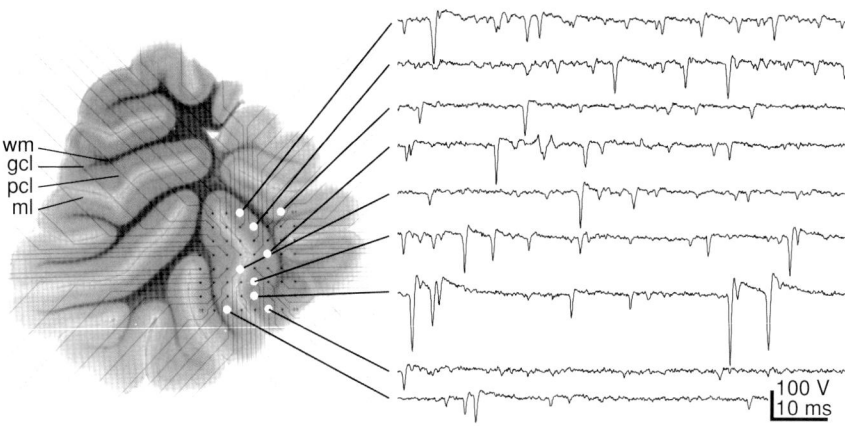

FIGURE 11.4. Spike activity in acute brain slice: (A) cerebellar slice from a juvenile rat in the recordings situation on an MEA. (B) Raw signal recorded at the electrodes marked in (A). Spikes can be readily detected. The signal-to-noise ratio is typical for this preparation.

tissue considering the noise from technical and biological sources, or the geometry of the dendritic tree.

Because of the extensive and overlapping dendritic trees of many neurons it is not possible to identify straightforwardly where the neuronal signal recorded at an electrode was generated. Obviously, neuronal spikes will decrease in amplitude with distance from the neuron. A second scaling factor is the size of a neuron; or rather its membrane area that in part determines the cells input resistance. Small neurons procure only weak currents to depolarize the capacitance formed by their cell membrane. Their extracellular signal is therefore smaller than that of larger cells (large cell bias). This ambiguity impairs the identification of the signal source and its distance to the recording electrode in single-unit recordings.

We further investigated this question and measured the relative change of the amplitude of neuronal spikes in MEA recordings from acute parasagittal slices from the rat cerebellum (Figure 11.4). To assess the decay of the neuronal spike with distance from the origin a compact source is necessary. The structure of the cerebellum is such that in these slices the large, essentially two-dimensional dendritic tree of the Purkinje cells (PC) lies flat on the MEA, extending into the molecular layer only. Their cell bodies are aligned in the Purkinje cell layer (PCL) and the neurons are spontaneously active. PC neurites do not extend into the plane of the PCL and only the axon of the cell passes through the granule cell layer. Within the PCL the spikes detected therefore essentially originate from the somata of the PCs only. Probing the region around an MEA electrode that picked up large spikes with a single micro-pipette electrode, we measured the amplitude distribution of the potential recorded with a micro-pipette. The change of the ratio of the signals recorded simultaneously at both electrodes as a function of the distance between the electrodes yielded an estimate of the detectable FP of the cell studied (Figure 11.5).

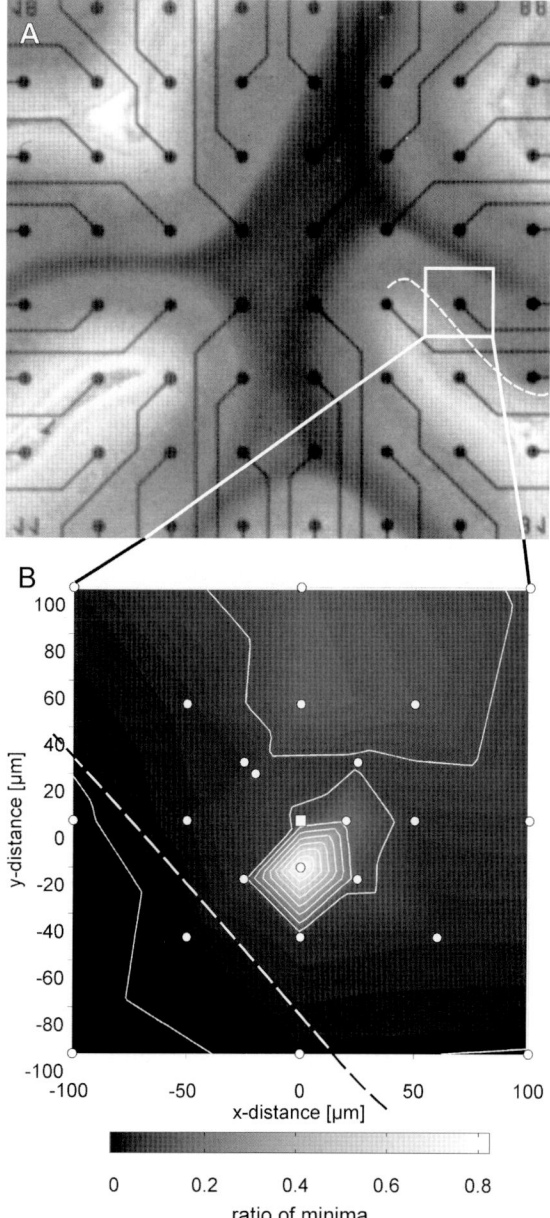

FIGURE 11.5. Recording horizon of an MEA electrode: The extent of the FP corresponding to a neuronal action potential was estimated by comparing the amplitude of a spike presumably recorded from a Purkinje cell soma with an MEA electrode to the signal simultaneously recorded at sites around the MEA electrode with a micro-pipette electrode. The ratio of the peak voltage on these electrodes was used to estimate the range within which an MEA electrode would likely pick up the spike from the noise. For somatic spikes this was 60 μm at best. This value will scale with the current produced by the cell, and hence generally with the cell size. Unitary myocyte FPs, as the basic component of the compound FPs recorded in a culture, are therefore likely detectable at a larger distance. (Reproduced with permission from Egert et al., 2002, Figure 4, © Springer-Verlag.)

The resulting potential surface indicated that the recording horizon of such an electrode has a radius of approximately 60 µm (Egert et al., 2002) at best for PCs; the fields of view of the MEA electrodes spaced on a 200 µm grid therefore do not overlap.

This footprint is obviously valid for unit spikes only and the electrode horizon scales with the size of the cell for reasons explained above. In addition, extensive dendritic trees in some cases will allow the recording of spikes at greater distances. In recordings from the neocortex in vivo with single electrodes spikes are often detected first at approximately 80 µm from the point of maximal SNR which is still within the range of our estimate measured above. Cardiac myocytes, on the other hand, form a special case in that neighboring myocytes are coupled by gap junctions, effectively forming a large cell in which membrane depolarizations spread across the cell, similar to the situation in long axons and dendrites. The frequently large amplitude of spikes found in MEA recordings from cardiac myocytes can be attributed to this extensive sheet of membrane, de- and hyperpolarized by comparatively large currents.

11.2.3   Evaluation of Cardioactive Drugs

The interpretation of the FP shape given above suggests that an analysis of changes of the FP in cultures of cardiac myocytes might be used to assess the effect of cardioactive drugs under conditions similar to those in intact tissue. In contrast to isolated cells, the cells in culture interact and differentiate to form an excitable network, generating and propagating excitation with pacemaker regions and propagation pathways. A comparison of the FP amplitude in the pacemaker region with those along the propagation pathways shows increased SNRs along the pathway, indicating larger currents generated there. The interaction between the cardiomyocytes in culture also induces a differential expression of ion channels. Compared to single-channel studies revealing detailed information on the site of action of a drug, studies in cell cultures with analyses of the FP shape could yield information on the response of cardiomyocytes as a whole and of the cellular network.

11.2.3.1   Extracting Action Potential Properties from Extracellular Recordings

An important parameter for the functional interplay in the heart is, for instance, the duration of the action potential (AP). In the ECG this is determined from the delay between the Q and the T waves (QT-interval) that marks the depolarization and repolarization phases of the AP in the ventricular myocardium. A lengthening of the QT-interval has been associated with lethal side effects of certain drugs (Vieweg, 2002; Redfern et al., 2003; Fermini and Fossa, 2003). Tests for QT prolongation have therefore become obligatory in safety pharmacology (Fermini and Fossa, 2003). On the basis of data and considerations on cell cultures we analyzed the changes of the FP duration in response to the application of several drugs known to prolong the cardiac AP during the recording. We observed a

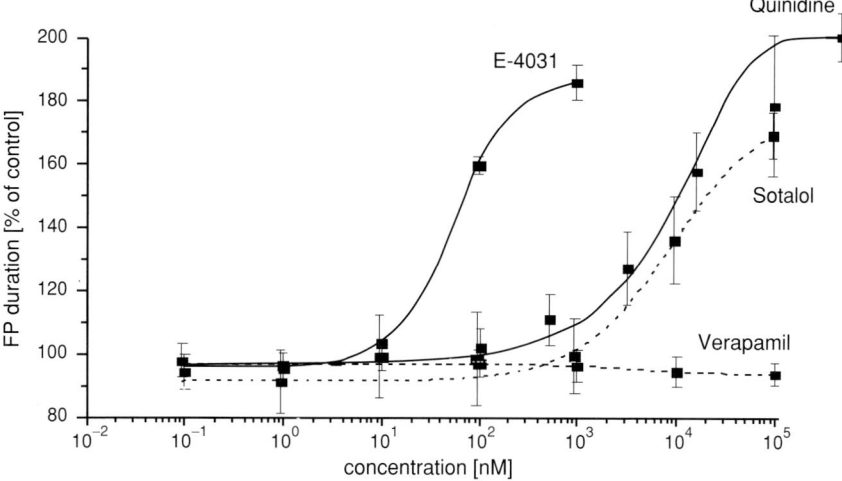

FIGURE 11.6. Estimating QT duration changes by cardioactive drugs with extracellular recordings: MEA recordings from cardiac myocytes facilitate in vitro prescreening for QT prolongation in cardiac myocytes. These dose–response curves were determined from cardiac myocyte cultures on 96-well arrays (see Figure 11.7). All drugs tested thus far changed the field potential duration as expected from ECG recordings (Lazzara, 1993; Redfern et al., 2003).

concentration-dependent prolongation of the FP duration, which is closely related to the QT-interval with E4031, a specific blocker of the HERG channel, and the anti-arrhythmic drugs Quinidine, and Sotalol. Verapamil, a Ca^{2+}-ion influx inhibitor, and Amiodarone, both substances with anti-arrhythmic properties, did not change the QT-interval (Figure 11.6). These effects, and the concentration ranges in which they occur, are comparable to those reported in the literature (Lazzara, 1993; De Ponti et al., 2002; Redfern et al., 2003). In particular the results with Verapamil match the in vivo profile, which is not predictable from HERG-channel data alone (De Ponti et al., 2002).

11.3   Discussion

11.3.1   Other Approaches to the Analysis of Spatio-Temporal Cardiac Excitation Patterns

In addition to the approaches described thus far, other techniques have been developed to investigate the structure of cardiac excitation patterns in different preparations and on various time scales. Cultures of dissociated cardiomyocytes from neonatal rats on MEAs have been used, for example, to investigate structure-dependent impulse conduction and beat rate variability (Rohr et al., 1997b; Kucera et al., 2000).

In cardiology, the spread of activity, its embryonic development, and its pathological conditions, such as arrhythmias, are highly important issues. The combination of MEA recording with cultures of cardiac myocytes thus marks a significant progress, supported by the fortuitous situation that the contraction is essentially isometric at the tissue surface adhering to the substrate. The signal amplitude can be very high and motion artifacts are not observed. This allows the observation and analysis of physiological processes in active cardiac tissue over periods of several days, enabling a new class of experiments on cardiac development (Rohr et al., 1997a; Igelmund et al., 1999). Lu et al. (2003) explanted whole mouse hearts, and even whole mouse embryos on MEAs to study the differentiation of cardiac activity (Reppel et al., 2004). Myocyte cultures have also been recorded with transistor-based electrode arrays (Sprössler et al., 1999).

The requirement for QT-evaluation in safety pharmacology necessitates tests on a large number of drugs, either under development or already on the market. Using in vivo tests is inconvenient because of the associated high costs and low throughput. Because several ion channels and receptors are involved in the dynamics determining the AP duration, with the HERG channel being the most important one, data obtained from cell lines expressing a single ion-channel type can be misleading, resulting in false negatives or false positives (De Ponti et al., 2002). The interplay of the various cellular mechanisms is better captured in native cells. MEAs provide a simple tool to implement semi-automatic arrays on such cultures.

11.4   Outlook

The applications and techniques presented above illustrate the wide range of questions that can be approached with substrate-integrated microelectrode arrays.

To increase the throughput and cut the costs of corresponding experiments in drug research it is desirable to run several preparations simultaneously. This requires that such systems can be run largely unsupervised during the experiment, necessitating highly stable recording, stimulation, perfusion, and drug application configurations. MEAs enable this approach, in particular with the perspective for new, multi-well plate arrays (Figure 11.7). Perforation of MEAs to optimize perfusion will further improve this tool.

The option to combine MEAs with conventional electrophysiological techniques, and transparent indium-tin oxide (ITO) electrodes facilitating optical measurements, for example, with Ca^{2+}-sensitive dyes (Banach et al., 2002), promises further uses for this technique.

We expect that the availability of these techniques, of the corresponding data acquisition and data analysis tools, and an increasing number of experimental protocols will contribute to our understanding of signal propagation and information processing in neuronal and cardiac networks, as well as pathological conditions thereof and approaches for their treatment.

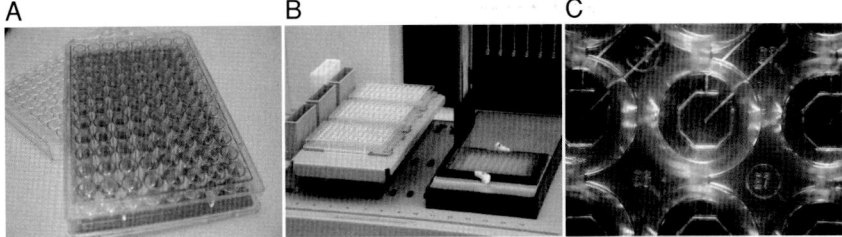

FIGURE 11.7. A new recording system with substrate-integrated microelectrodes produced on printed circuit boards with standard 96-well plate format. (A) 96-well plate with integrated electrodes. Cardiac myocytes were successfully cultured on these arrays for tests of QT-prolongation. (B) The standard 96-well layout and pipetting robot facilitate the adaptation to common laboratory procedures and increase the throughput of the assay. The data acquisition electronics is integrated into the base plate underneath the 96-well array. (C) Close-up of a well with a central electrode and ring-shaped reference electrode. The actual electrode at the tip of the lead is 100 μm in diameter. (Images courtesy of Multi Channel Systems, Reutlingen, Germany.)

Acknowledgments. This work was funded by grants of the German BMBF (FKZ 0310964D and 16SV1743), the Land Baden-Württemberg (U.E.), and the American Heart Association (K.B., AHA-0330393Z). We would also like to thank Multi Channel Systems for providing some of the data and images used in this manuscript.

References

Banach, K., Halbach, M., and Blatter, L.A. (2002). Spatio-temporal organization of calcium signaling and electrical activity in multicellular preparations of neonatal rat heart. *Biophys. J.*, 82: 3188.

Banach, K., Halbach, M., Hu, P., Hescheler, J., and Egert, U. (2003). Development of electrical activity in cardiac myocyte aggregates derived from mouse embryonic stem cells. *Am. J. Physiol.* 284: H2114–H2123.

Buitenweg, J.R., Rutten, W.L., and Marani, E. (2000). Finite element modeling of the neuron-electrode interface. *IEEE Eng. Med. Biol. Mag.* 19: 46–52.

Castellanos, A., Moleiro, F., Acosta, H., Ferreira, A., Cox, M.M., Interian, A., Jr., and Myerburg, R.J. (1998a). Sudden Wenckebach periods and their relationship to neurocardiogenic syncope. *Pacing Clin. Electrophysiol.* 21: 1580–1588.

Castellanos, A., Moleiro, F., Pastor, J.A., Interian, A., Jr., and Myerburg, R.J. (1998b). Reverse alternating Wenckebach periods and other modes of regression of > or = 8:1 to 2:1 atrioventricular block. *Am. J. Cardiol.* 82: 528–531.

De Ponti, F., Poluzzi, E., Cavalli, A., Recanatini, M., and Montanaro, N. (2002). Safety of non-antiarrhythmic drugs that prolong the QT interval or induce torsade de pointes: an overview. *Drug Saf.* 25: 263–86.

Egert, U., Heck, D., and Aertsen, A. (2002). 2-Dimensional monitoring of spiking networks in acute brain slices. *Exp. Brain Res.* 142: 268–274.

Egert, U. and Meyer, T. (2005). Heart on a chip—Extracellular multielectrode recordings from cardiac myocytes in vitro, In: Dhein, S., Mohr, F.W., and Delmar, M., eds., *Practical Methods in Cardiovascular Research*. Springer, Stuttgart.

Egert, U., Meyer, T., and Banach, K. (2003). Heart on a chip, In: Dhein, S. and Delmar, M., eds., *Methods in Cardiovascular Research*. Springer, Stuttgart.

Fermini, B. and Fossa, A.A. (2003). The impact of drug-induced QT interval prolongation on drug discovery and development. *Nat. Rev. Drug Discov.* 2: 439–447.

Fromherz, P., Offenhäusser, A., Vetter, T., and Weis, J. (1991). A neuron-silicon junction: A Retzius cell of the leech on an insulated-gate field-effect transistor. *Science* 252: 1290.

Furshpan, E.J., Macleish, P.R., O'Lague, P.H., and Potter, D.D. (1976). Chemical transmission between rat sympathetic neurons and cardiac myocytes developing in microcultures: Evidence for cholinergic, adrenergic, and dual-function neurons. *Proc. Nat. Acad. Sci. U. S. A.* 73: 4225–4229.

Guevara, M.R., Glass, L., and Shrier, A. (1981). Phase locking, period-doubling bifurcations, and irregular dynamics in periodically stimulated cardiac cells. *Science* 214: 1350–1353.

Halbach, M.D., Egert, U., Hescheler, J., and Banach, K. (2003). Estimation of action potential changes from field potential recordings in multicellular mouse cardiac myocyte cultures. *Cell. Physiol. Biochem.* 13: 271–284.

Henze, D.A., Borhegyi, Z., Csicsvari, J., Mamiya, A., Harris, K.D., and Buzsáki, G. (2000). Intracellular features predicted by extracellular recordings in the hippocampus in vivo. *J. Neurophysiol.* 84: 390–400.

Heuschkel, M.O., Fejtl, M., Raggenbass, M., Bertrand, D., and Renaud, P. (2002). A three-dimensional multi-electrode array for multi-site stimulation and recording in acute brain slices. *J. Neurosci. Meth.* 114: 135–148.

Hirota, A., Kamino, K., Komuro, H., and Sakai, T. (1987). Mapping of early development of electrical activity in the embryonic chick heart using multiple-site optical recording. *J. Physiol. (Lond.)* 383: 711–728.

Hofer, E., Urban, G., Spach, M.S., Schafferhofer, I., Mohr, G., and Platzer, D. (1994). Measuring activation patterns of the heart at a microscopic size scale with thin-film sensors. *Am. J. Physiol.* 266: H2136–H2145.

Igelmund, P., Fleischmann, B.K., Fischer, I.V., Soest, J., Gryshchenko, O., Sauer, H., Liu, Q., and Hescheler, J. (1999). Action potential propagation failures in long-term recordings from embryonic stem cell-derived cardiomyocytes in tissue-culture. *Pflug. Arch. Eur. J. Phys.* 437: 669–679.

Israel, D.A., Barry, W.H., Edell, D.J., and Mark, R.G. (1984). An array of microelectrodes to stimulate and record from cardiac cells in culture. *Am. J. Physiol.* 247: H669–H674.

Kehat, I., Gepstein, A., Spira, A., Itskovitz-Eldor, J., and Gepstein, L. (2002). High-resolution electrophysiological assessment of human embryonic stem cell-derived cardiomyocytes: a novel in vitro model for the study of conduction. *Circ. Res.* 91: 659–661.

Kleber, A.G., Fast, V.G., Kucera, J., and Rohr, S. (1996). Physiology and pathophysiology of cardiac impulse conduction. *Zeitschrift für Kardiologie* 85: 25–33.

Kucera, J.P., Heuschkel, M.O., Renaud, P., and Rohr, S. (2000). Power-law behavior of beat-rate variability in monolayer cultures of neonatal rat ventricular myocytes. *Circ. Res.* 86: 1140–1145.

Lazzara, R. (1993). Antiarrhythmic drugs and torsade de pointes. *Eur. Heart J.* 44: 88–92.

Mastrototaro, J.J., Massoud, H.Z., Pilkington, T.C., and Ideker, R.E. (1992). Rigid and flexible thin-film multielectrode arrays for transmural cardiac recording. *IEEE Trans. Biomed. Eng.* 39: 271–379.

Meiry, G., Reisner, Y., Feld, Y., Goldberg, S., Rosen, M., Ziv, N., and Binah, O. (2001). Evolution of action potential propagation and repolarization in cultured neonatal rat ventricular myocytes. *J. Cardiovasc. Electrophysiol.* 12: 1269–1277.

Metzger, J. M., Lin, W.I., and Samuelson, L.C. (1994). Transition in cardiac contractile sensitivity to calcium during the in vitro differentiation of mouse embryonic stem cells. *J. Cell Biol.* 126: 701–711.

Metzger, J.M., Lin, W.I., Johnston, R.A., Westfall, M.V., and Samuelson, L.C. (1995). Myosin heavy chain expression in contracting myocytes isolated during embryonic stem cell cardiogenesis. *Circ. Res.* 76: 710–719.

Nag, A.C., Lee, M.L., and Sarkar, F.H. (1996). Remodelling of adult cardiac muscle cells in culture: Dynamic process of disorganization and reorganization of myofibrils. *J Muscle Res. Cell. Motil.* 17: 313–34.

Plonsey, R. (1977). Action potential sources and their volume conductor fields. *Proc. IEEE* 65: 601–611.

Rall, W. (1969). Distribution of potential in cylindrical coordinates and time constants for a membrane cylinder. *Biophys. J.* 9: 1509–1541.

Redfern, W.S., Carlsson, L., Davis, A.S., Lynch, W.G., MacKenzie, I., Palethorpe, S., Siegl, P.K., Strang, I., Sullivan, A.T., Wallis, R., Camm, A.J., and Hammond, T.G. (2003). Relationships between preclinical cardiac electrophysiology, clinical QT interval prolongation and torsade de pointes for a broad range of drugs: Evidence for a provisional safety margin in drug development. *Cardiovasc. Res.* 58: 32–45.

Reppel, M., Pillekamp, F., Lu, Z.J., Halbach, M., Brockmeier, K., Fleischmann, B.K., Hascheler, J. (2004). Microelectrode arrays: a new tool to measure embryonic heart activity. *J. Electrocardiol.* 37 Suppl: 104–109.

Rohr, S. (1990). A computerized device for long-term measurements of the contraction frequency of cultured rat heart cells under stable incubating conditions. *Pflug. Arch. Eur. J. Phys.* 416: 201–206.

Rohr, S. and Kucera, J. P. (1997). Involvement of the calcium inward current in cardiac impulse propagation: Induction of unidirectional conduction block by nifedipine and reversal by Bay K 8644. *Biophys. J.* 72: 754–766.

Rohr, S., Kucera, J.P., and Kleber, A.G. (1997b). Form and function: Impulse propagation in designer cultures of cardiomyocytes. *News Physiol. Sci.* 12: 171–177.

Rohr, S., Kucera, J.P., Fast, V.G., and Kleber, A.G. (1997a). Paradoxical improvement of impulse conduction in cardiac tissue by partial cellular uncoupling. *Science* 275: 841–844.

Schwab, J.O., Eichner, G., Schmitt, H., Schrickel, J., Yang, A., Balta, O., Luderitz, B., and Lewalter, T. (2004). Heart rate variability in patients suffering from structural heart disease and decreased AV-nodal conduction capacity. Insights into the formation of heart rate variability. *Z. Kardiol.* 93: 229–233.

Scott, J.A., Khaw, B.A., Fallon, J.T., Locke, E., Rabito, C.A., Peto, C.A., and Homcy, C.J. (1986). The effect of phenothiazines upon maintenance of membrane integrity in the cultured myocardial cell. *J. Mol. Cell Cardiol.* 18: 1243–1254.

Spach, M.S. (1983). The role of cell-to-cell coupling in cardiac conduction disturbances. *Adv. Exper. Med. Biol.* 161: 61–77.

Spach, M.S. and Heidlage, J.F. (1995). The stochastic nature of cardiac propagation at a microscopic level: Electrical description of myocardial architecture and its application to conduction. *Circ. Res.* 76: 366–380.

Spach, M.S. and Kootsey, J.M. (1983). The nature of electrical propagation in cardiac muscle. *Am. J. Physiol.* 244: H3–22.

Spach, M.S., Miller, W.T., Miller–Jones, E., Warren, R.B., and Barr, R.C. (1979). Extracellular potentials related to intracellular action potentials during impulse conduction in anisotropic canine cardiac muscle. *Circ. Res.* 45: 188–204.

Sprössler, C., Denyer, M., Britland, S., Knoll, W., and Offenhäusser, A. (1999). Electrical recordings from rat cardiac muscle cells using field-effect transistors. *Phys. Rev. E* 60: 2171–2176.

Thomas, C.A., Springer, P.A., Loeb, G.W., Berwald-Netter, Y., and Okun, L.M. (1972). A miniature microelectrode array to monitor the bioelectric activity of cultured cells. *Exp. Cell Res.* 74: 61–66.

Vieweg, W.V.R. (2002). Mechanisms and risks of electrocardiographic QT interval prolongation when using antipsychotic drugs. *J. Clin. Psychiatry* 63: 18–24.

Yamamoto, M., Honjo, H., Niwa, R., and Kodama, I. (1998). Low-frequency extracellular potentials recorded from the sinoatrial node. *Cardiovasc. Res.* 39: 360–372.

III
MEA Applications:
Acute/Cultured Slices

12
A Hippocampal-Based Biosensor for Neurotoxins Detection and Classification Using a Novel Short-Term Plasticity Quantification Method

GHASSAN GHOLMIEH, SPIROS COURELLIS, VASILIS MARMARELIS, MICHEL BAUDRY, AND THEODORE W. BERGER

12.1   Introduction

Current sensors such as the handheld chemical agent monitor (CAM by Graseby Dynamics) and the M90-D1-C (Environics, Finland) are capable of detecting and identifying nerve agents but come short of identifying unknown chemical compounds. Hence, there is a need to develop a new type of sensor that can detect newly uncharacterized chemical compounds. In this chapter, we present a newly developed biosensor for rapid neurotoxins screening based on the quantification of short-term plasticity (STP) of CA1 hippocampal system in vitro using a newly developed analytical approach.

The hippocampal slice with its preserved trisynaptic pathway (Andersen et al., 1971) has been a very popular experimental preparation for investigating memory mechanisms and drug effects using long-term potentiation (LTP) and short-term plasticity (Bliss and Lomo, 1973; Fountain and Teyler, 1995; Xie et al., 1997; Buonomano, 1999, 2000). Short-term plasticity (STP)-based methods are typically chosen to assess drug effects of rapidly acting chemical compounds inasmuch as they require less experimental time.

Current experimental methods for measuring STP are mostly based on the analysis of paired pulse (Dobrunz et al., 1997; Leung et al., 1994; Creager et al., 1980) or fixed frequency train stimulation (Papatheodoropoulos et al., 2000; Buonomano 1999; Pananceau et al., 1998; Castro-Alamancos et al., 1997). These methods lack the ability to test all the possible time intervals in a time-efficient manner. Based on a variant of the Volterra approach, we have developed and implemented an efficient method to characterize and quantify STP using Poisson distributed random train (RIT) stimulation (Gholmieh et al., 2002). It can be viewed as a hybrid between the paired pulse and the fixed frequency train approaches. It is, however, more time efficient than both of them. Responses to a large variety of interimpulse intervals can be evaluated in three minutes in contrast to several hours when using the paired pulse approach (Gholmieh et al., 2001, 2002).

The computed STP descriptors of the underlying nonlinear dynamics are the first- and second-order Volterra kernels that describe the state of the system.

The second-order kernel can be further linearly decomposed into nine Laguerre functions. The first-order kernel and the Laguerre coefficients of the second-order kernel expansion can be viewed as the characteristics of the system. These characteristics are speculated to change under different chemical environments, giving each drug a unique signature.

The proposed biosensor is comprised of a multi-electrode array, a hippocampal slice, and MATLAB software for data acquisition and online nonlinear STP analysis. It was tested using chemical compounds belonging to six different classes: DNQX and CNQX (AMPA receptor antagonists), DAP5 (an NMDA receptor antagonist), carbachol (a cholinergic agonist), trimethylopropane phosphate (TMPP) and picrotoxin ($GABA_A$ antagonists), tetraethylammonium (a potassium channel blocker), and valproate (an antiepileptic medication).

The current chapter summarizes our successful efforts for the last four years to build a hippocampal-based biosensor for neurotoxin detection and classification. Section 12.2 describes the novel modeling approach in detail. Section 12.3 presents the nonlinear characteristics of the CA1 region obtained through the described nonlinear modeling approach. Section 12.4 proves that the novel method can detect changes in STP. Section 12.5 tests our hypothesis that each drug can have a unique signature, and Section 12.6 presents a proof of principle solution for long-term monitoring.

12.2   The Novel Modeling Approach

Experimental studies investigating short-term plasticity use impulse sequence stimuli, as this type of signal comprises the most common form of input in biological neural systems. In particular, paired pulse stimulation of variable interimpulse intervals (Creager et al., 1980; Leung et al., 1994; Dobrunz et al., 1997) and short impulse trains at a fixed frequency (Yamamoto et al., 1980; Turner and Miller, 1981; Landfield et al., 1986; Fueta et al., 1998; Pananceau et al.,1998), varying from one impulse train to another, have been used very widely in experimental STP research. Each method identifies areas of facilitation and inhibition within the STP mechanisms only for a limited number of interimpulse intervals or impulse train frequencies. Using these methods, extension of the experimental protocol to cover a wide range of intervals and frequencies may prolong the experimental time considerably, to the point that the viability of the experimental preparation may be affected.

A new method to quantitatively describe short-term plasticity in biological neural systems that uses impulse sequence stimuli (electrical pulses) of variable interimpulse interval (Poisson distribution) has been introduced (Gholmieh et al., 2001). The new method innovates by combining impulse sequence stimuli with quantitative STP descriptors derived using the Volterra modeling approach, adapted for impulse sequences of varying interimpulse intervals at the input and variable population spike amplitude sequences at the output.

Our choice of stimuli and responses can be viewed as a hybrid of the variable interimpulse intervals used by the paired pulse approach and the pattern of repeated impulses at fixed intervals utilized by the fixed frequency approach. Although the electrical cell measure used in establishing the novel method was the amplitude of the population spikes, other measures could have been used instead including dendritic and somatic excitatory postsynaptic potentials (EPSP) slope and amplitude.

The adapted Volterra approach considers the interimpulse intervals $(n_i - n_j)$ as input and a measure of the cell response to evoked stimulation (population spike amplitude) as output ($y(n_i)$). In our analysis, we employed the first- and second-order Volterra kernels, enabling us to construct the model shown in Equation (12.1):

$$y(n_i) = k_1 + \sum_{n_i - \mu < n_j < n_i} k_2(n_i - n_j), \tag{12.1}$$

where n_i is the time of occurrence of the ith impulse, n_j is the time of occurrence of jth impulse before the ith impulse, $y(n_i)$ is the amplitude of the population spike in response to the impulse that occurred at time n_i, μ is the memory of the biological system, k_1 is the first-order Volterra kernel, and k_2 is the second-order Volterra kernel. Equation (12.1) describes the amplitude of the population spike at time (n_i) in terms of the first-order kernel (k_1) and the second-order kernel (k_2), the impulse evoking the population spike, and the effect of past impulses. The Volterra kernels k_1 and k_2 capture the nonlinear dynamics of the underlying STP mechanisms and comprise the quantitative STP descriptors.

The Volterra–Poisson kernels were estimated using the Laguerre–Volterra method (Marmarelis, 1993), an approach that reduced the kernel estimation effort to computing the coefficients of a set of Laguerre functions. The Laguerre basis functions form an orthonormal set and are defined as follows.

$$L_l(n) = \alpha^{(n-l)/2}(1 - \alpha)^{1/2} \times \sum_{k=0}^{l}(-1)^k \binom{n}{k}\binom{l}{k}\alpha^{(l-k)}(1 - \alpha)^k \tag{12.2}$$

where $L_l()$ is the Laguerre base function of lth order, and alpha (α) is a parameter that varies between 0 and 1 and affects the time extent of the basis functions. The order of a Laguerre function corresponds to the number of the times the function crosses the "zero line". Figure 12.1A shows the zeroth-, second-, and fourth-order Laguerre functions for a fixed value of alpha ($\alpha = 0.94$). Figure 12.1B shows the second-order Laguerre functions for $\alpha = 0.90, 0.94$, and 0.95.

Using L Laguerre basis functions, the second-order kernel can be expressed as

$$k_2(n_i - n_j) = \sum_{l}^{L} c_l L_l(n_i - n_j) \tag{12.3}$$

where $L_l()$ is the lth order Laguerre function and c_l is the corresponding coefficient. Combining Equations (12.1) and (12.3), we obtain:

$$\left\{ y(n_i) = k_1 + \sum_{n_i - \mu < n_j < n_i} \sum_{l}^{L} c_l L_l(n_i - n_j) \right\}_{i=1,\ldots,N} \tag{12.4}$$

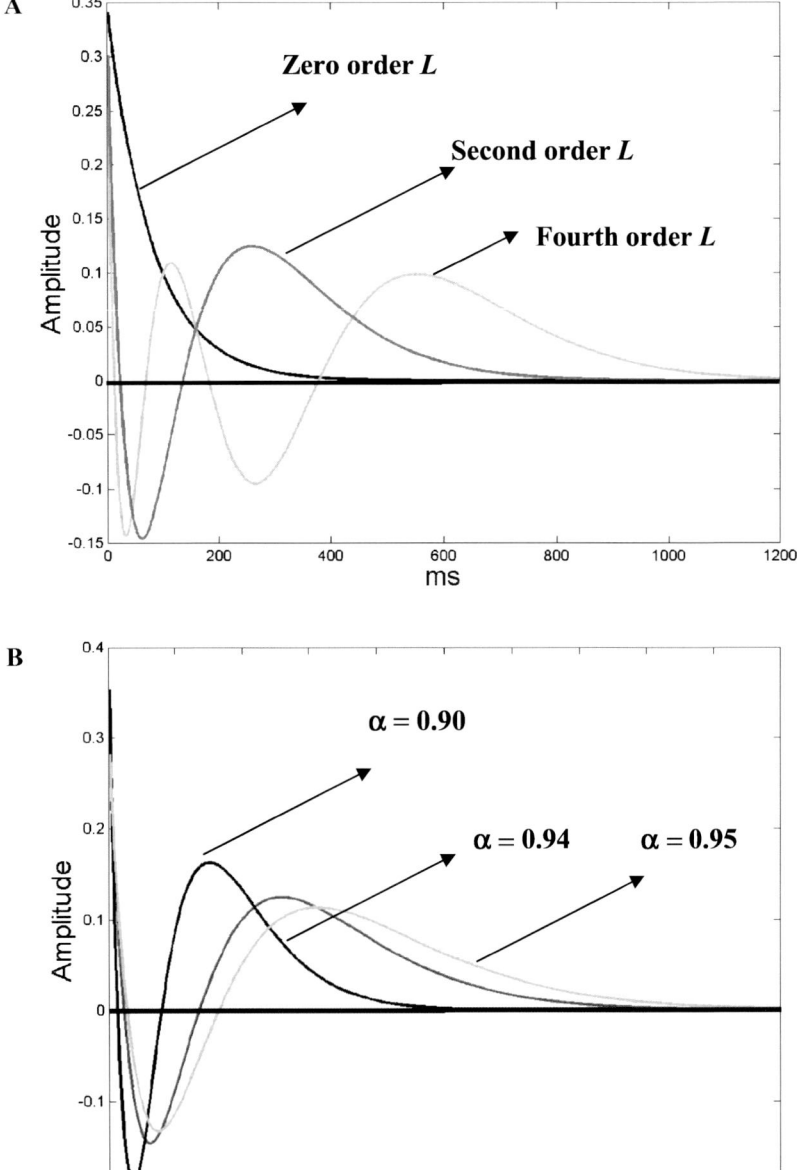

FIGURE 12.1. Example of Laguerre functions: A: zeroth-, second-, and fourth-order Laguerre functions for $\alpha = 0.94$. The order of the function correlates with the number it crosses the zero line. B: The second-order Laguerre function for different values of alpha ($\alpha = 0.90, 0.94, 95$). As alpha increases, the time extent of the Laguerre functions increases (Gholmieh et al., 2003).

Because c_l are constant, Equation (12.3) can be rearranged as follows:

$$\left\{ y(n_i) = k_1 + \sum_l c_l \sum_{n_i - \mu < n_j < n_i} L_l(n_i - n_j) \right\}_{i=1,\dots,N} \tag{12.5}$$

Equation (12.5) applies to every population spike amplitude ($y(n_i)$) in the dataset. For example, 400 equations are formed using 400 consecutive stimulation impulses in order to calculate the value of k_1 and the Laguerre coefficients (c_l) that lead to the estimation of k_2. The first-order kernel (k_1) and the Laguerre coefficients (c_l) were calculated using least squares on the set of N equations expressed by (12.5). Substituting the estimated Laguerre coefficients in Equation (12.3), we obtained the estimate for the second-order kernel (k_2). Appendix A presents a detailed numerical example.

The Volterra–Poisson kernel estimates can be interpreted as follows. The first-order kernel can be interpreted as the mean of the population spike amplitude, and the second-order kernel represents the effect of the interaction between the current impulse and past impulses (within a memory window μ) on the amplitude of the current population spike. The role of the first- and second-order kernel in the model shown in Equation (12.1) is illustrated in Figure 12.2. The first-order kernel contributes to the model response with a constant value and the second-order kernel provides the requisite amplitude adjustment, based on the time interval between the current impulse and past impulses that occurred within a memory window μ. We have normalized second-order kernels by dividing them with the corresponding first-order kernels. Thus, second-order kernels represent percentage adjustments with respect to the first-order kernel.

The results obtained through the Volterra approach can be compared to those obtained in the literature using the paired pulse approach. We have previously concluded (Gholmieh et al., 2002) that the amplitude of the conditioning response corresponds to k_1, the amplitude of the test response corresponds to $k_1 + k_2$, and the paired pulse ratio corresponds to $1 + k_2/k_1$. Because the interimpulse intervals chosen for testing chemical compounds in most paired pulse studies usually correspond to the maximum possible facilitation, the amplitude of the test response and the paired pulse ratio correspond to $k_1 + k_{2\,max}$ and $1 + k_{2\,max}/k_1$, respectively. Therefore, if both k_1 and k_2 increased or decreased, the conditioned response amplitude would follow.

The ability of the Volterra–Poisson approach to accurately model and track STP changes was evaluated using the normalized mean square error (NMSE). The NMSE is defined as the ratio of the sum-of-squares of the model residuals over the sum-of-squares of the recorded response amplitudes.

$$NMSE_{pr} = \frac{\sum_i (Y_{pr_i} - Y_{data_i})^2}{\sum_i Y_{data_i}^2} \tag{12.6}$$

where Y_{pr} is the predicted amplitude of the population spike using the estimated kernels and Y_{data} is the amplitude extracted from the recorded data. The NMSE

A

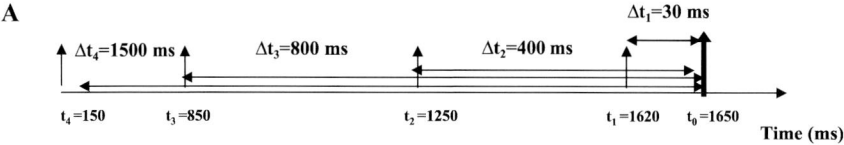

Population Spike

B

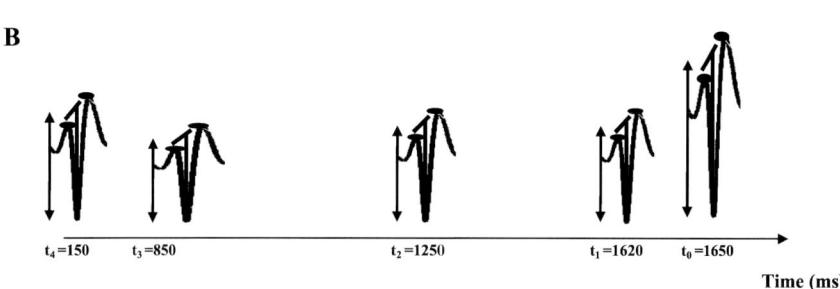

C

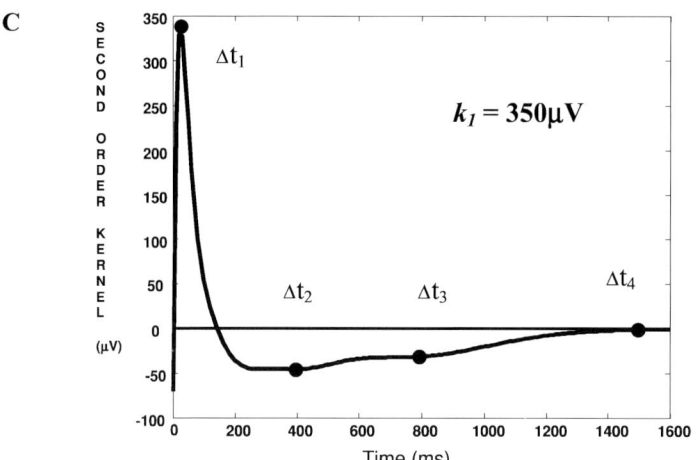

FIGURE 12.2. Predictive power of kernels. A: A series of input electrical stimuli applied through a stimulating electrode to the Schaffer collaterals where Δt indicate the time difference between the present impulse and the past impulses B: The corresponding recorded population spikes. The amplitude of each population spike was measured as the difference of the distance between the population spike minimum and the midpoint of the line that joins the two positive peaks. C: The first- and second-order kernel. The amplitude of the response for the last impulse (bold arrow, at 1650 msec) can be estimated using the first-order kernel, the second-order kernel, and Equation (12.1) as follows.

$$y(1650) = k_1 + k_2(1650 - 1620) + k_2(1650 - 1250) + k_2(1650 - 850) + k_2(1650 - 150)$$
$$= k_1 + k_2(\Delta t_1) + k_2(\Delta t_2) + k_2(\Delta t_3) + k_2(\Delta t_4)$$
$$= k_1 + k_2(30) + k_2(400) + k_2(800) + k_2(1500)$$
$$= 350 + 340 - 30 - 30 + 0$$
$$= 630\,\mu V.$$

is a measure of how well kernels can capture the system nonlinear dynamics. Small NMSE values indicate that the STP descriptors sufficiently captured the underlying nonlinear STP dynamics and that the associated STP model is reliable. Large NMSE values suggest that higher-order terms (third, fourth, etc.) need to be included or that the data is noisy.

12.3   Nonlinear Characteristics of CA1 In Vitro

Extracellular recordings were achieved using acute hippocampal slices and a microelectrode array. The acute hippocampal slices were prepared as follows: adult rats were decapitated after being fully anesthetized with Halothane. The hippocampus was extracted in a chilled aCSF bath and transverse slices (500 μm thick) were obtained using a Leika vibrotome (VT 100S). Each slice was positioned over a multi-electrode array (Figure 12.3) with the guidance of an inverted microscope (Leica DML 4x). A bipolar stimulation electrode (twisted Nichrome wires) was placed in the Schaffer collaterals region.

The setup consisted of a multi-micro-electrode plate (MMEP4; Gross et al., 1993), pre-amplifiers, two data acquisition boards (Microstar, DAP 3200/214e series) at 7.35 kHz sampling rate per channel, and custom-developed software. A MATLAB-based custom user interface was employed to control the data acquisition boards, perform data extraction, and conduct the data analysis.

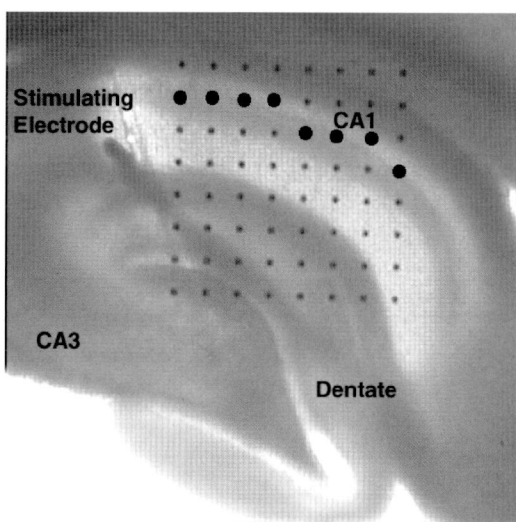

FIGURE 12.3. Picture of a hippocampal slice positioned over the multi-electrode (8 × 8) array. Each black dot represents an electrode. The blackened electrodes show (potential) recording sites from CA1 (Gholmieh et al., 2002).

The stimulation intensity was adjusted to evoke a response with amplitude 15% of the maximum population spike amplitude recorded in the I/O curve. Two stimulation sequences were then applied. Each stimulation sequence consisted of a random impulse train (RIT) of 200 pulses (Poisson distributed interimpulse intervals with a mean of 2 Hz).

The extracted data, consisting of interimpulse intervals (input) and the corresponding evoked population spike amplitude (output), was analyzed using the Volterra–Poisson approach. The calculated CA1 second-order STP descriptor (kernel) was characterized by a fast rising facilitatory phase (0 to 25 msec), a peak between 25 and 50 msec, a fast declining facilitatory phase (50 to 200 msec), and a slow depressive phase (200 to 2000 msec; Figure 12.3C). The extent (i.e., memory, μ in Equation (12.1)) of k_2 was in the range of 1600 to 2000 msec. The estimation of the second-order kernel employed nine ($L = 9$) Laguerre functions.

In establishing a measure of comparison with existing research, we employed the cross-correlation method for kernel estimation as well (Krausz and Friesen, 1975; Berger et al., 1988a; Scalabassi et al., 1988). The resulting cross-correlation-based kernels were compared to kernels obtained using the adapted Volterra method. The comparison was used to establish consistency between the methods and continuity with research results reported earlier in the literature. Figure 12.4A shows an example of the second-order kernel obtained using the cross-correlation method (wiggly line) and the adapted Volterra method (smooth line) for the same recording. Smoothing the cross-correlation estimate with a triangular moving average window, we obtained the result shown in Figure 12.4B. A visual comparison verifies the similarity between the two kernels. Furthermore, the square root of the normalized mean square error (NMSE) for each case (3.82% for cross-correlation versus 3.60% for adapted Volterra method), the first-order kernel (345 μV for cross-correlation versus 353 μV for adapted Volterra method), and the correlation coefficient (0.983) confirm that both methods report comparable results.

The main contribution of the novel method, however, is the tenfold decrease in the experimental time because the required input is only around 400 electrical stimulation impulses in comparison to 4000 impulses required by the cross-correlation method. Hence, the characterization of STP required significantly reduced data collection time and employed computationally efficient data analysis methods, allowing for the data analysis potentially to be done online.

12.4   Detecting Changes in STP

We examined the ability of the developed modeling approach to track STP changes in the CA1 region of the hippocampus under different conditions. Again, we applied random interimpulse intervals stimuli to the Schaffer collaterals and recorded

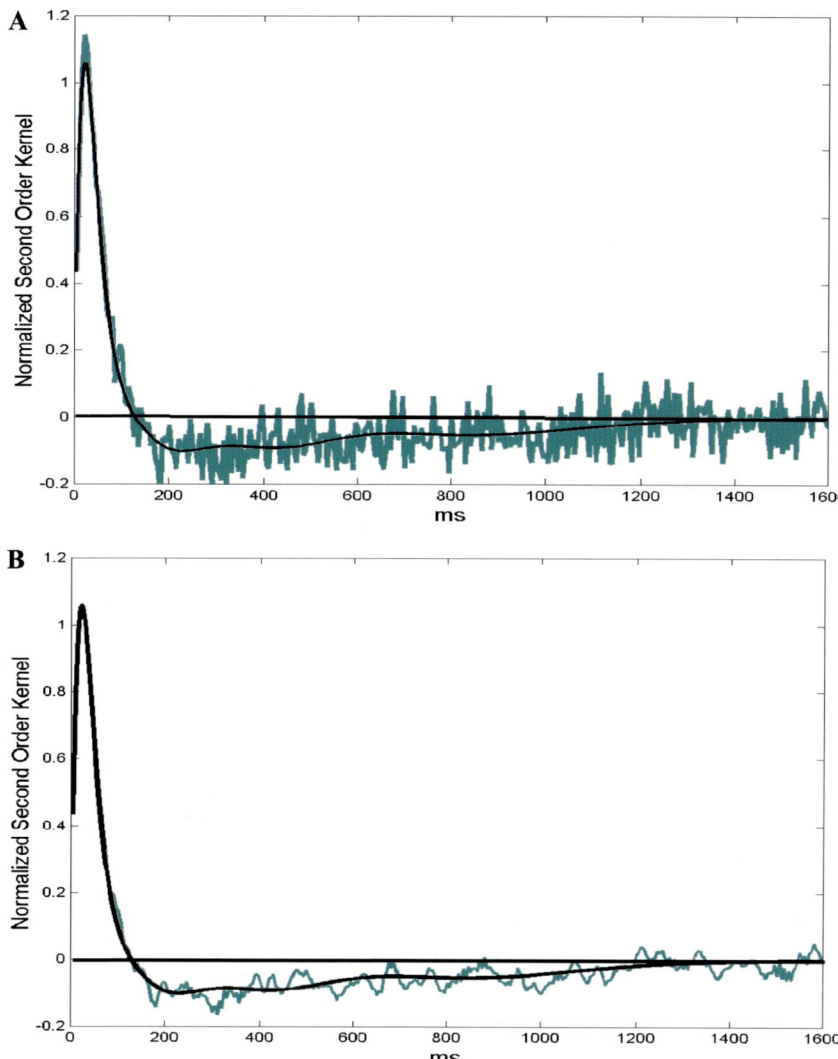

FIGURE 12.4. Comparison of the second-order kernels obtained through cross-correlation method and the adapted Volterra approach. A: Second-order kernels obtained using the cross-correlation method (wiggly line) and the adapted Volterra method (smooth line) for the same recording. B: Smoothing the cross-correlation estimate with a triangular moving average window we obtained a relatively smoother curve. Visual comparison verifies the similarity of the second-order kernels from the two methods (Gholmieh et al., 2002).

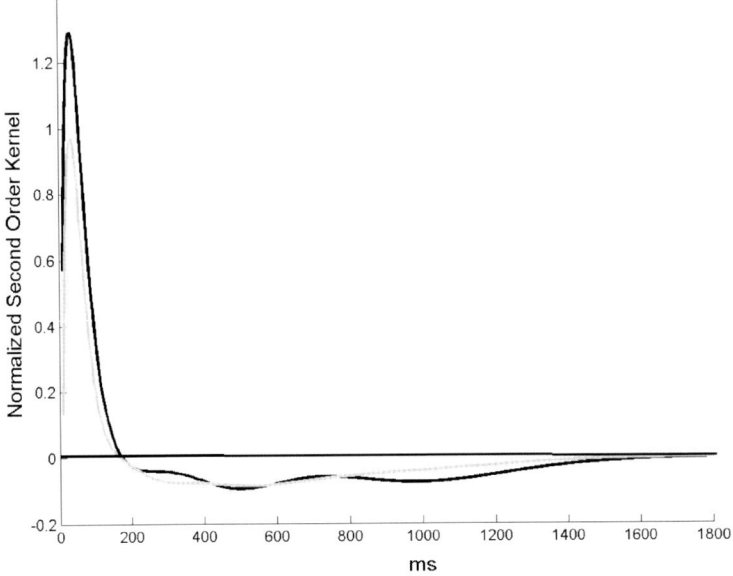

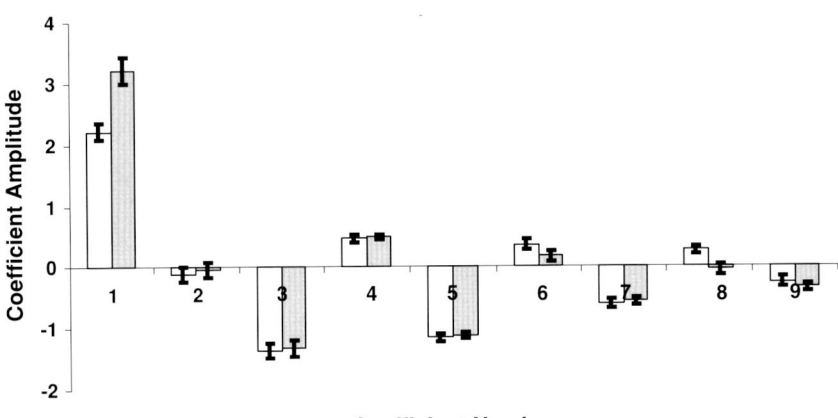

FIGURE 12.5. A: Averages of the computed second-order STP descriptors (k_2) before (gray) and after (black) adding the 0.5 mM calcium medium. B: Low calcium concentrations effect on the Laguerre coefficients. White columns: baseline. Gray columns: Low (Ca) effect. Error bars represent one standard deviation (Gholmieh et al., 2005.)

the amplitude of the corresponding population spike responses at the CA1 cell body layer. The nonlinear dynamic behavior of STP in CA1 was investigated by varying the stimulus intensity (i.e., the impulse magnitude) and by modifying the chemical consistency of the in vitro preparation (Gholmieh et al., 2005). The first- and second-order kernels (k_1 and k_2) were computed for two different stimulus intensity levels and two different chemical agents (picrotoxin and low calcium concentration). The current section reexamines the data corresponding to the calcium ion concentration change.

Calcium ion is thought to play an important role in the presynaptic neurotransmitter release (Katz and Miledi, 1968; Augustine et al., 1985; Augustine and Charlton, 1986). Decreasing the extracellular calcium concentration caused a decrease in the value of the first-order kernel and increased the peak facilitation value of the second-order kernel (Figure 12.5). Mirroring the changes in the second-order kernels, the associated Laguerre coefficient values showed a significant change in the value of the first coefficient only.

Previous I/O and paired pulse studies showed that lowering calcium concentration decreased the unconditioned response (Stringer et al., 1988) and increased the conditioned response amplitude (Sagratella et al., 1991; Iglemund et al., 1995). Our results are in agreement with the literature inasmuch as k_1 decreased and k_2 peak increased. In addition, we found no changes in STP properties at long interimpulse intervals.

One of the biggest strengths of the proposed Volterra–Poisson models of STP is that they provide model prediction to arbitrary impulse sequence stimuli. We demonstrated this property by employing the out-of-sample prediction paradigm that involves the following steps. (1) STP kernels were computed using data from a subset of recordings (estimation subset), (2) the stimulus from a recording outside the estimation subset was used as the input to the model, and (3) the resulting predicted model response was compared to the corresponding recorded response. The adequacy of the in-sample and out-of-sample predictions was judged by the associated NMSE.

Figure 12.6A shows a segment before changing the medium to low calcium concentration with the out-of-sample predicted response (circles) and the corresponding datapoints of the recorded response (diamonds). The NMSE for the out-of-sample prediction was 9.25% and the NMSE for the in-sample prediction was 7.54% for the analyzed dataset. Figure 12.6B shows a segment of an out-of-sample predicted response (circles) and the corresponding datapoints of the recorded response (diamonds) in the case of low calcium concentrations. The NMSE for the out-of-sample prediction was 11.48% whereas the NMSE for the in-sample prediction was 7.71% for the analyzed dataset. The average NMSE values over all recordings between in-sample and out-of-sample predictions were within 5%.

These results suggest that the computed kernel STP descriptors are adequately capturing the changes in the nonlinear dynamics of the underlying STP mechanisms and that the proposed STP models are reliable.

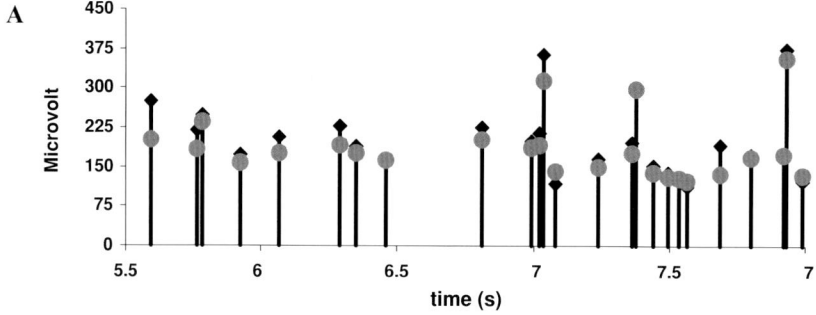

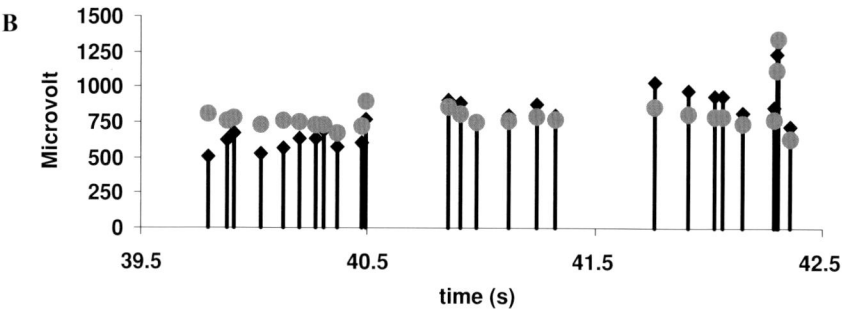

FIGURE 12.6. STP model out-of-sample prediction: Black diamonds represent recorded values and dark circles represent out-of-sample predicted values. A: Segment of recorded and predicted responses in the control case. B: Segment of recorded and predicted responses in low calcium concentration medium (Gholmieh et al., 2005).

12.5   Developing a Tissue-Based Biosensor Using the Novel Method

12.5.1   Drug Effects on the First- and Second-Order Kernels

Armed with the developed methodology and the characterization of the CA1 nonlinear dynamics, a new tissue-based biosensor was developed for screening chemical compounds that rapidly affect the nervous system (Gholmieh et al., 2003). Describing the functional state of the biosensor, the Volterra–Poisson kernels changed accordingly as specific chemical compounds were added. The second-order kernel was decomposed into nine Laguerre functions whose coefficients along with the first-order kernel were used as features for classification purposes using an artificial neural network (ANN).

The biosensor was tested using chemical compounds consisting of six different classes: DNQX and CNQX (AMPA receptor antagonits), DAP5 (an NMDA receptor antagonist), carbachol (a cholinergic agonist), trimethylopropane phosphate

(TMPP) and picrotoxin ($GABA_A$ antagonists), tetraethylammonium (a potassium channel blocker), and valproate (an antiepileptic medication). The concentrations used were as follows: picrotoxin (100 µM); trimethylopropane phosphate (10 µM); tetraethylammonium (4 mM); valproate (5 mM); carbachol (1 mM); DAP5 (25 µM); CNQX (3 µM); and DNQX (0.15, 1.5, 3, 5, and 10 µM). Each chemical compound gave, as expected, a different feature profile corresponding to its pharmacological class.

The effect of various chemical compounds on the first-order kernel (k_1) was quantified as the ratio of post-drug k_1 value to pre-drug k_1 value (Figure 12.7A). A chemical compound caused an increase or a decrease in k_1 if the ratio was greater or less than one, respectively. No change in the k_1 ratio was seen in the negative control set. Picrotoxin, TMPP, and TEA caused an increase in the k_1 ratio, and DNQX, CNQX, DAP5, carbachol, and valproate caused a decrease in the k_1 ratio.

The negative control study is summarized in Figure 12.7B. The six conducted experiments showed very low variability in the first- and second-order kernels (Figure 12.4B) because there was no statistically significant change in the value of any of the Laguerre coefficients (inset) or in the value of the first-order kernel ($p < 0.01$). Moreover, the second-order kernel exhibited the same characteristics as previously described.

$GABA_A$ inhibitors such as picrotoxin and bicuculline are well known for their ability to inhibit early postsynaptic inhibitory potential (Davies et al., 1990). Picrotoxin ($n = 5$) caused an increase in the value of the first-order kernel by 27% (sd ±15%), and an increase in peak facilitation value of k_2 by 75%. The picrotoxin effect on STP (Figure 12.8A) caused a significant increase in the value of the first ($p < 0.01$), third ($p < 0.01$), fourth ($p < 0.01$), fifth ($p < 0.05$), and ninth ($p < 0.05$) Laguerre coefficient (inset). These results are consistent with previous findings that $GABA_A$ antagonist increased the PS amplitude of both the conditioning and test response in paired pulse studies (Leung et al., 1994).

Several recent studies have shown that TMPP induces epileptiform activity in CA1 (Lin et al., 1998; Lin et al., 2001) by acting as a $GABA_A$ antagonist (Keefer et al., 2001; Kao et al.,1999; Higgins and Gardier, 1990). TMPP also caused a significant increase in the value of the first ($p < 0.05$), third ($p < 0.05$), fifth ($p < 0.01$), and seventh ($p < 0.05$) Laguerre coefficient (inset). Our results confirm earlier reports by showing that TMPP increased CA1 excitability by augmenting the first-order kernel, the second-order kernel, and the Laguerre coefficients in a similar manner to picrotoxin (Figure 12.8A,B).

TEA was used in a concentration of 4 mM because 12.5 mM and 25 mM caused seizure activity in preliminary experiments. TEA ($n = 6$) at 4 mM increased the value of k_1 by 21% (sd ±11%), decreased the peak facilitation value of k_2 by 60%, increased the early inhibitory phase (5 msec to 15 msec) and the late inhibitory phase (200 to 1200 msec) (Figure 12.8C). TEA also caused statistically significant change in the value of the first ($p < 0.01$), fourth ($p < 0.01$), sixth ($p < 0.05$), and ninth ($p < 0.05$) Laguerre coefficient (inset). The increase in the value of k_1 is consistent with previous reports where it was found that TEA caused an increase

A

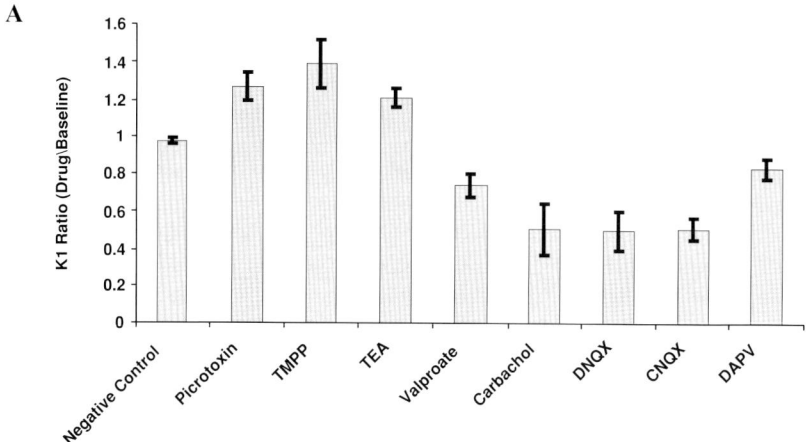

B

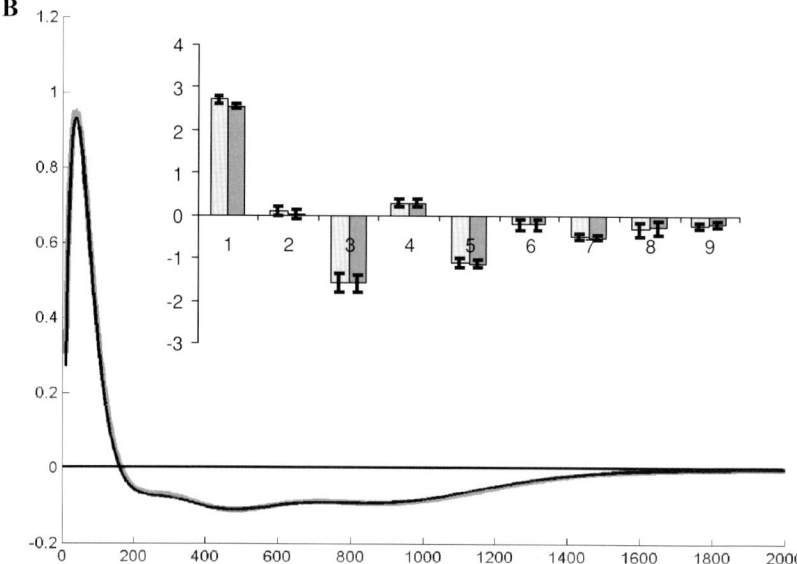

FIGURE 12.7. A: Effect of various chemical compounds on the k_1 ratio. B: Negative control study. Second-order kernel before (gray) and after addition of a negative control solution of aCSF (black). As expected there was no change in the STP properties under negative control testing conditions. Inset: Effect of negative control on the Laguerre coefficients (baseline: white; negative control: gray; bars: ±1 sd; Gholmieh et al., 2003).

in the CA1 field responses (Southan et al., 1997; Song et al., 2001). In addition the second-order kernel showed an increase in the early inhibitory phase and a decrease in peak facilitation value of k_2 (Figure 12.8C).

Carbachol ($n = 6$) at 1 mM caused the STP properties of CA1 to shift from facilitation to depression. In addition to causing the k_1 value to drop by 49%

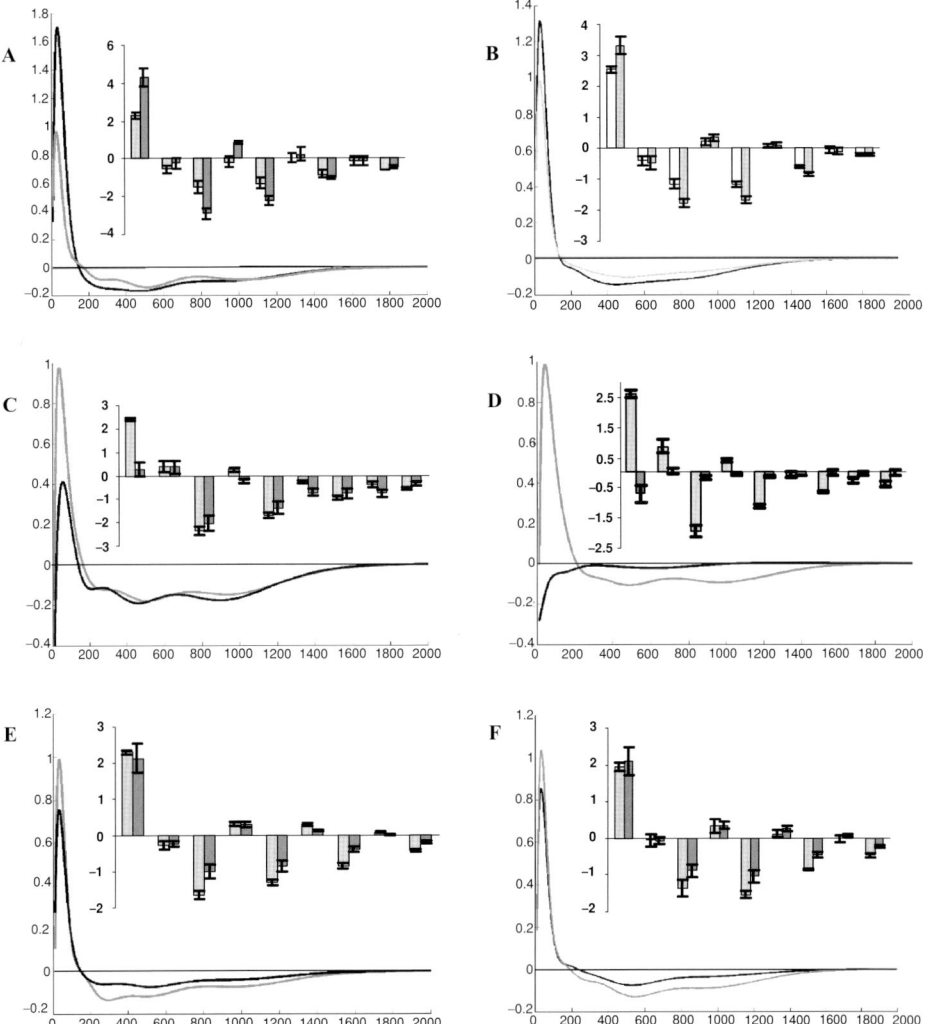

FIGURE 12.8. Effect of various chemical compounds on the second-order kernels and the corresponding Laguerre coefficients. Baseline second-order kernel and the corresponding Laguerre coefficients (inset) shown in gray. Chemical compound effect on the second-order kernel and the Laguerre coefficients shown in black and dark gray, respectively. A: Picrotoxin; B: TMPP; C: Tetraethylammonium; D: Carbachol; E: DNQX; F: CNQX (Gholmieh et al., 2003).

(sd $\pm 36\%$), carbachol caused the second-order kernel to become a negative exponential decaying function with peak inhibition value of -30% (Figure 12.8D). The effect is reflected in a statistically significant manner in the Laguerre coefficients (inset). In particular, the value of the first and fourth coefficient became negative, and the value of the second, third, fifth, and seventh Laguerre coefficient was markedly decreased ($p < 0.01$). The depressive effect on k_1 is consistent with previous reports of dose-dependent field potential suppression (Hesen et al., 1998; Yajeya et al., 2000).

AMPA receptors are thought to mediate the bulk component of field potentials at the CA1 glutamatergic synapses. DNQX and CNQX are well known for being AMPA receptor antagonists (Andreasen et al., 1989). DNQX at 3 µM ($n = 6$) caused a decrease in the value of the first-order kernel by 49% (sd ± 24%) and in k_2 peak facilitation value by 25% (Figure 12.8E). The effect of DNQX was further characterized through the changes in the Laguerre coefficients. Variations were observed in the value of the third ($p < 0.01$), fifth ($p < 0.05$), seventh ($p < 0.01$), and ninth ($p < 0.01$) Laguerre coefficient (inset). No significant change was seen in the value of the first coefficient. CNQX ($n = 5$) caused a decrease in the value of the first order kernel by 50% (sd ± 12%) and in k_2 peak facilitation value by 15% (Figure 12.8F). As with DNQX, no significant change was seen in the value of the first coefficient. There was also a decrease in the value of the third, fifth ($p < 0.05$), seventh ($p < 0.01$), and ninth ($p < 0.01$) coefficient.

Both compounds caused some changes in the Laguerre coefficients with no effect on the first coefficient and a depressive effect on the third, fifth, and seventh coefficients. The peak facilitation for DNQX and CNQX decreased by 23% and 14%, respectively without showing a statistically significant difference ($p < 0.05$). We also observed dose-dependent DNQX suppression of the first- and second-order kernels. The IC50 of DNQX on k_1 was found to be around 3 µM (consistent with a previous report; Andreasen et al., 1989) and the IC50 for the second-order kernel peak facilitation value was around 5 µM.

DAP5 is an NMDA receptor antagonist. DAP5 ($n = 7$) decreased k_1 value by 17% (sd ± 0.15) and k_2 peak facilitation value by 10% (Figure 12.9A). However, DAP5 did not cause any statistically significant changes in most of the Laguerre coefficients (inset) except for the second coefficient ($p < 0.05$). The modest effect of DAP5 on the STP descriptor is due to the relatively high concentration of magnesium in the aCSF and the low mean frequency of stimulation. Our results are consistent with a previous study (Muller et al., 1990) which showed that the NMDA component of the conditioning EPSP at 1 mM of Mg was around 10% and that the peak facilitation of the test EPSP response decreased by 10% upon the addition of NMDA channel blocker.

Valproate is anti-seizure medication that is known to block Ca^{++} and Na^+ channels. Valproate ($n = 6$) decreased k_1 value by 26% (sd ± 0.16), and increased k_2 peak facilitation value by 20% (Figure 12.9B). The effect on STP changed the value of only the first Laguerre coefficient ($p < 0.05$). A previous paired pulse study showed a decrease in the conditioning response (Franceschetti et al., 1986) consistent with the depressive effect of valproate on k_1. The effect of valproate on k_1 and k_2 partially mimicked those obtained with low calcium concentration (see Section 12.3) and are consistent with the calcium channel blocking effect of valproate.

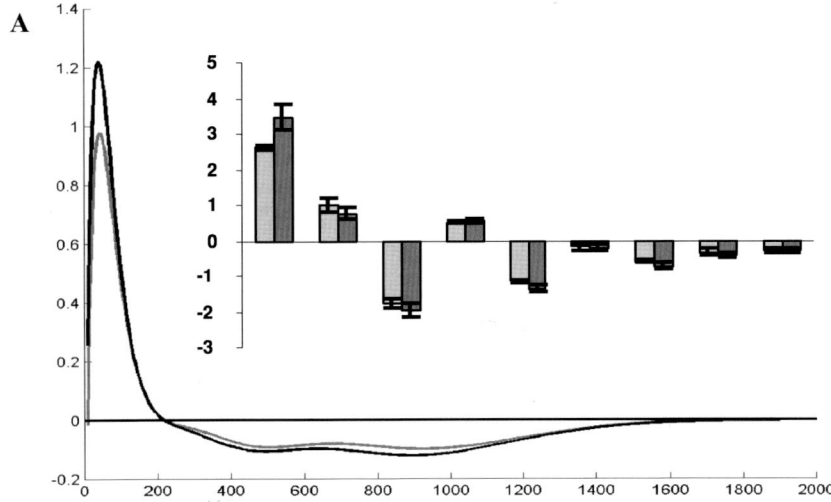

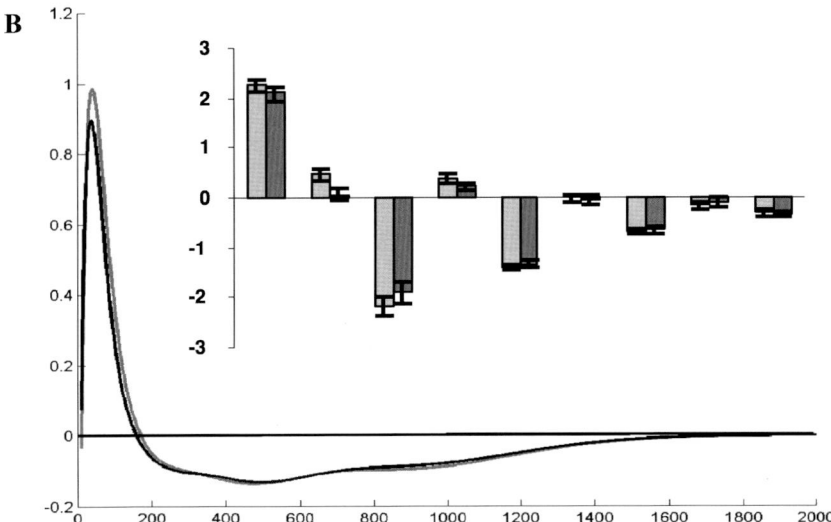

FIGURE 12.9. Effect of DAP5 and valproate on the second-order kernels and the corresponding Laguerre coefficients. Baseline second-order kernel and the corresponding Laguerre coefficients (inset) shown in gray. Chemical compound effect on the second-order kernel and the Laguerre coefficients shown in black and dark gray, respectively. A: DAP5; B: valproate (Gholmieh et al., 2003).

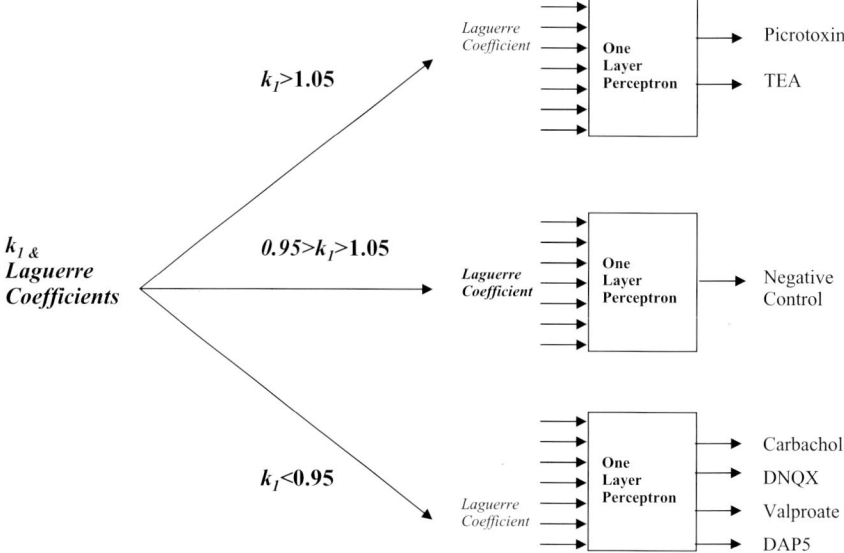

FIGURE 12.10. The classification algorithm. The first-order kernel ratio along with the Laguerre coefficients formed the input to the classifier. The compound was first classified through its ability to increase or decrease k_1 ratio. In the second stage, the Laguerre coefficients formed the input to a single layer of preceptrons that categorized each compound into its specific class (Gholmieh et al., 2003).

12.5.2   Classification of Neurotoxins Using an Artificial Neural Network (ANN)

The neurotoxins were classified using an artificial neural network (ANN). The classifier was trained using DNQX, DAP5, carbachol, picrotoxin, TEA, valproate, and the negative control. It was tested using CNQX and TMPP. The classification algorithm consisted of two steps (Figure 12.10) where the first-order kernel and the Laguerre coefficients formed the input to the ANN comprised of a single layer of perceptrons. The first step consisted of classifying the compounds using the k_1 ratio. TEA and picrotoxin caused an increase in k_1 and were classified in the first preliminary class. The second preliminary class consisted of only the negative control set where no change in k_1 ratio was observed. Valproate, DNQX, DAP5, and carbachol formed the third preliminary class through their depressive effect on k_1. The second step of the classification algorithm consisted of feeding the Laguerre coefficient into the ANNs.

Overall the ANN was able to classify 81% percent of the 43 training experiments. The results were as follows: picrotoxin (4/5), TEA (5/6), negative control (6/7), valproate (5/6), carbachol (5/6), DAP5 (5/7), and DNQX (4/6). The classifier classified TMPP and CNQX in the same class as picrotoxin and DNQX, respectively. This result was expected, inasmuch as TMPPP and picrotoxin belong to the same pharmacological class, and CNQX and DNQX also belong to the same pharmacological class. Likewise, there was a strong similarity between the corresponding second-order kernels.

12.6   Long-Term Monitoring: Slice Culture

The developed biosensor does not have long-term survivability because the half-life of acute hippocampal slices is around six hours. In addition, the biosensor is not usable outside the experimental room secondary to the cumbersome associated hardware (e.g., computer, amplifiers, and electrodes). These obstacles can be overcome using hippocampal slice culture over a multi-electrode array in a similar manner to the cell-culture-based biosensor (Pancrazio et al., 2003) where the hardware was modified in order to create a portable tissue-based biosensor.

Planar micro-electrode arrays offer two key advantages for long-term monitoring of the hippocampal slice culture: (1) simultaneous recording and stimulation, and (2) closed sterile housing that minimizes the risk for infection. In addition to the commercially available MMEP4 (8×8), we have developed a high-density micro-electrode array. This $4 \times 12 + 2 \times 8$ array conforms to the location of the input and output and offers several advantages over the 8×8 array (see Chapter 12).

While the conformal array technology was being developed, we carried out exploratory experiments in order to demonstrate that the acute slices and cultured slices share the same nonlinear dynamics. Essentially, a modified version of the roller technique originally described by Gahwiler (1981) was used to culture hippocampal slices on the MMEP setups. Hippocampal slices (400 μm thick) were prepared from seven-day-old rats with a McIllwain tissue slicer. Slices were glued over multi-electrode arrays using a combination of thrombin (10 μl, 50 units/ml, Sigma) and plasma (10 μl, Sigma). The slices were left for 5 min in order to form a rigid clot. The MEA array along with the slice was then positioned in a large covered 7 cm diameter Petri dish. Fourteen milliliter of culture medium was added. One liter of culture medium contained Basal Medium Eagle ((Sigma B-9638), 4.6 g), Earle Balanced Salts ((Sigma E-6132), 2.17 g), and 33% Horse Serum (Gemini-Bio-Products #100-105). The medium was supplemented by adding: NaCl, 15 mM; ascorbic acid, 0.4 mM; glucose, 36 mM; HEPES, 20 mM; $CaCl_2$, 150 μM; $MgSO_4$, 1.2 mM; glutamine, 2mM; insulin (1 mg/100 ml); and 3 ml penicillin/ streptomycin (Gibco BRL #15240-062). The arrays were then left to rotate at 12 revolutions per h with a 15° inclination angle to allow air/CO_2 and medium to alternatively cover the slice. The medium was changed every two to three days. Using this technique, we have been able to maintain cultures for over a month with good morphology.

Electrophysiological testing was carried using paired-pulse and random train stimulation. Figure 12.11 shows a two-week-old cultured slice that exhibited paired-pulse facilitation at 30 msec intervals. This observation is in accordance with the literature that showed that CA1 in vitro mature while retaining in vivo characteristics (Bahr, 1995). Figure 12.12 compares the second-order kernels from

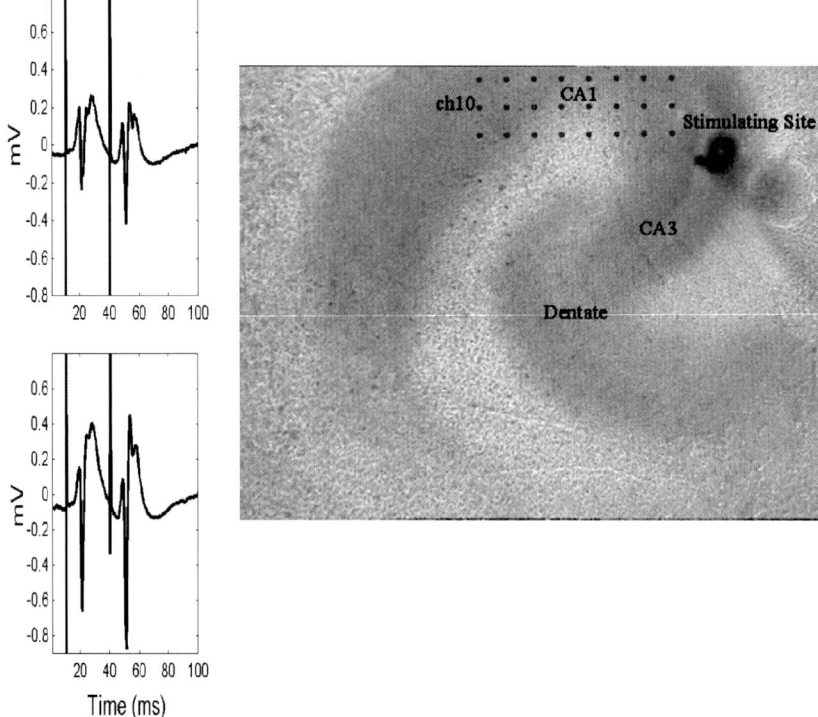

FIGURE 12.11. Paired-pulse facilitation of population spikes in CA1 of a two-week-old cultured slice. Paired pulse facilitation was observed in the CA1 of cultured slices, confirming earlier works that CA1 in vitro mature while retaining in vivo characteristics. Paired pulses were delivered at 30 msec intervals. Horizontal time is in msec and the vertical scale is in mV (Gholmieh et al., 2001).

acute (panel A) and cultured slices (panel B). The second-order kernels from both preparations have the same characteristics as described in Figure 12.3; that is, they have an initial phase (0 to 50 msec), a peak of facilitation (around 20 to 50 msec), and a late inhibitory phase (200 to 2000 msec).

The results of these pilot studies coupled with the discriminatory powers of the nonlinear dynamics (Volterra–Poisson kernels) established a solid ground for the future development of an organotypic hippocampal culture tissue-based biosensor for long-term neurotoxins detection.

12.7   Conclusion

In summary, the new modeling approach used for the biosensor development led to a computationally optimized, comprehensive, and compact representation of the nonlinear dynamics associated with STP. It also introduced a mathematical

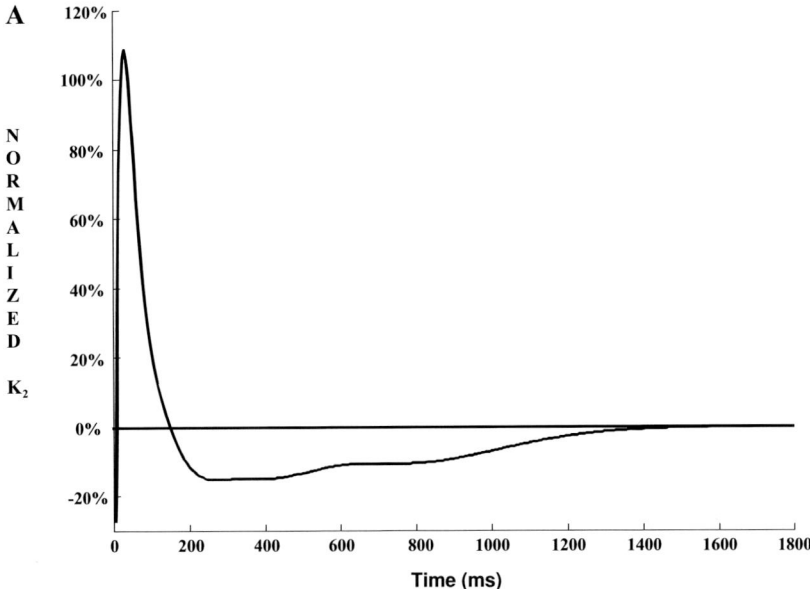

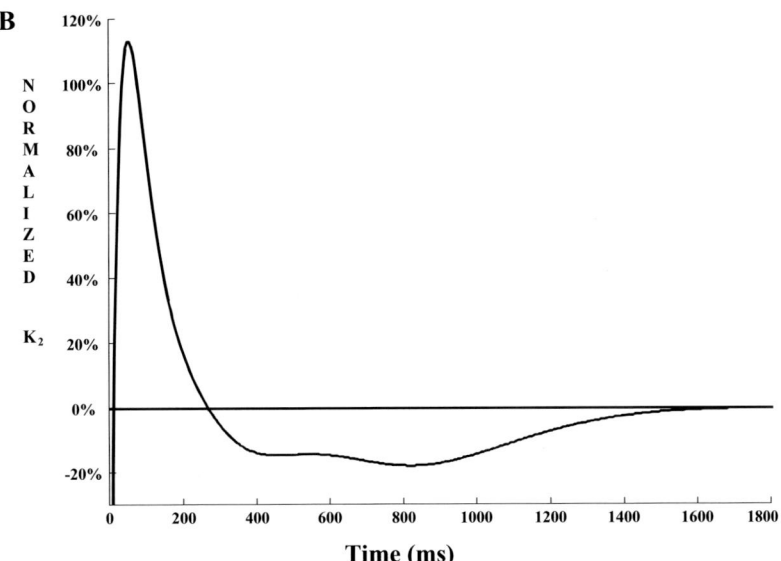

FIGURE 12.12. Comparison of estimated second-order kernels from acute (panel A) and cultured slices (panel B). The second-order kernels from both preparations have the same characteristics as described in Figure 12.4; that is, they have an initial phase (0 to 50 msec), a peak of facilitation (around 20 to 50 msec), and a late inhibitory phase (200 to 2000 msec; Gholmieh et al., 2001).

model that helped with evaluating the accuracy of the estimated STP descriptors (Volterra–Poisson kernels) and provided predictive capabilities.

Compared to existing methods in the literature, the developed method provided a better representation of STP with considerable improvement in experimental data collection time and in computational efficiency associated with the estimation of the STP descriptors.

The biosensor was tested using picrotoxin, trimethylopropane phosphate, tetraethylammonium, valproate, carbachol, DAP5, CNQX, and DNQX. Each chemical compound gave a specific profile corresponding to its pharmacological class. The second-order kernel was decomposed into nine Laguerre functions whose coefficients along with the first-order kernel were used as features for classification purposes. The first-order kernel and the Laguerre coefficients formed the input to an artificial neural network comprised of a single layer of perceptrons. The ANN was able to classify each tested compound into its respective class. More interestingly, the trained neural network correctly classified compounds outside the training sets, for example, TMPP and CNQX, into their respective pharmacological classes.

Finally, we have taken the extra step to demonstrate the long-term monitoring potential of the developed technology by culturing organotypic hippocampal slices and showing that they have similar nonlinear dynamics to acute hippocampal slices.

Acknowledgments. This research was supported by grants from the Office of Naval Research (ONR), the Defense Advanced Research Projects Agency (DARPA), and the Biomedical Simulation Resource (BMSR).

Appendix A

A.1   Matrix Rearrangement

This appendix demonstrates how to calculate the first- and second-order kernels using numerical examples. The amplitude of the output (evoked population spike amplitude) can be expressed as function of the k_1, k_2, and the time interval between the stimulation impulses. For example, the 50th electrical impulse will generate a population spike amplitude with an amplitude of y_{50} that can be estimated using a second-order model according to Equation (12.1) where $y_{50} = k_1 + k_2(n_{50} - n_{49}) + k_2(n_{50} - n_{48}) + k_2(n_{50} - n_{47})$.

For simplicity, let's assume k_2 is formed of only 3 Laguerre basis, then the above equation can be expanded using (Equation (12.3)): that is,

$$k_2(n_i - n_j) = \sum_{0}^{2} c_l L_l(n_i - n_j)$$

into:

$$
\begin{aligned}
y_{50} = {} & k_{1+} + c_0 L_0(n_{50} - n_{49}) + c_1 L_1(n_{50} - n_{49}) + c_2 L_2(n_{50} - n_{49}) \\
& + c_0 L_0(n_{50} - n_{48}) + c_1 L_1(n_{50} - n_{48}) + c_2 L_2(n_{50} - n_{48}) \\
& + c_0 L_0(n_{50} - n_{47}) + c_1 L_1(n_{50} - n_{47}) + c_2 L_2(n_{50} - n_{47}) \\
= {} & k_{1+} + c_0[L_0(n_{50} - n_{49}) + L_0(n_{50} - n_{48}) + L_0(n_{50} - n_{47})] \\
& + c_1[L_1(n_{50} - n_{49}) + L_1(n_{50} - n_{48}) + L_1(n_{50} - n_{47})] \\
& + c_2[L_2(n_{50} - n_{49}) + L_2(n_{50} - n_{48}) + L_2(n_{50} - n_{47})].
\end{aligned}
$$

Each random train includes 400 impulses, generating 400 equations with four unknowns (k_1 and the three Laguerre coefficients) that can be solved using a least squares approach:

$$
\begin{pmatrix} \cdot y_{50} \\ y_{51} \\ y_{52} \\ y_{53} \end{pmatrix} =
\begin{pmatrix}
1 & \sum\limits_{n_j < n_i < n_{50}} L_0(n_i - n_j) & \sum\limits_{n_j < n_i < n_{50}} L_1(n_i - n_j) & \sum\limits_{n_j < n_i < n_{50}} L_2(n_i - n_j) \\
1 & \sum\limits_{n_j < n_i < n_{51}} L_0(n_i - n_j) & \sum\limits_{n_j < n_i < n_{51}} L_1(n_i - n_j) & \sum\limits_{n_j < n_i < n_{51}} L_2(n_i - n_j) \\
1 & \sum\limits_{n_j < n_i < n_{52}} L_0(n_i - n_j) & \sum\limits_{n_j < n_i < n_{52}} L_1(n_i - n_j) & \sum\limits_{n_j < n_i < n_{52}} L_2(n_i - n_j) \\
1 & \sum\limits_{n_j < n_i < n_{53}} L_0(n_i - n_j) & \sum\limits_{n_j < n_i < n_{53}} L_1(n_i - n_j) & \sum\limits_{n_j < n_i < n_{53}} L_2(n_i - n_j)
\end{pmatrix}
* \begin{pmatrix} k1 \\ c0 \\ c1 \\ c2 \end{pmatrix}
$$

$Y = L * C \rightarrow C = \text{pseudo-inverse}(L) \, ^* Y$

References

Abbot, L.F., Varela, J.A., Sen, K., and Nelson, S.B. (1997). Synaptic depression and cortical gain control. *Science* 275: 220–224.

Alberston, T.E., Walby, W.F., Stark, L.G., and Joy, R.M. (1996). The effect of propofol on the CA1 pyramidal cell excitability and GABA-A mediated inhibition in the rat hippocampal slice. *Life Sci.* 58(26): 2397–2407.

Alger, B.E. and Teyler, T.J. (1976). Long-term and short-term plasticity in the CA1, CA3, and dentate regions of the rat hippocampal slice. *Brain Res.* 110(3): 463–480.

Ameri, A., Wilhelm, A., and Simmet, T. (1999). Effects of the endogenous cannbinoid anadamide on neuronal activity in rat hippocampal slices. *Brit. J. Pharmacol.* 126(8): 1831–1839.

Andersen, P., Bliss, T.V.P., and Skrede, K.K. (1971). Lamellar organization of hippocampal excitatory pathways. *Exp. Brain Res.* 13: 208–211.

Andreasen, M., Lambert, J.D., and Jensen, M.S. (1989). Effects of new non-N-methyl-D-aspartate antagonists on synaptic transmission in the in vitro rat hippocampus. *J. Physiol.* 414: 317–336.

Augustine, G.J., and Charlton, M.P. (1986). Calcium dependence of presynaptic calcium current and post-synaptic response at the squid giant synapse. *J. Physiology.* 381: 619–640.

Augustine, G.J., Charlton, M.P., and Smith, S.J. (1985). Calcium entry and transmitter release at voltage-clamped nerve terminals of squid. *J. Physiology.* 367: 163–181.

Bahr, B.A. (1995). Long-term hippocampal slices: A model system for investigating synaptic mechanisms and pathologic processes. *J. Neurosci. Res.* 42(3): 294–305, Oct 15.

Baudry, M., Arst, D., Oliver, M., and Lynch, G. (1981). Development of glutamate binding sites and their regulation by calcium in rat hippocampus. *Brain Res.* 227(1): 37–48.

Berger, T.W., Eriksson, J.L., Ciarolla, D.A., and Scalabassi, R.J. (1988a). Nonlinear systems analysis of the hippocampal perforant path-dentate projection. II Effects of random train stimulation. *J. Neurophsiol.* 60: 1077–1094.

Berger, T.W., Eriksson, J.L., Ciarolla, D.A., and Scalabassi, R.J. (1988b). Nonlinear systems analysis of the hippocampal perforant path-dentate projection. III Comparison of random train and paired impulse stimulation. *J. Neurophsiol.* 60: 1095–1109.

Berger, T.W., Harty, T.P., Barrionuevo, G., and Scalabassi, R.J. (1989). Modeling of neuronal networks through experimental decomposition In: Maramarelis, V.Z., ed., *Advanced Methods of Physiological System Modeling*, Vol. II. Plenum Press, New York, pp. 113–128.

Berger, T.W., Robinson, G.B., Port, R.L., and Scalabassi, R.J. (1987). Nonlinear systems analysis for the functional properties of the hippocampal formation. In: Maramarelis, V.Z., ed., *Advanced Methods of Physiological System Modeling*, Vol. I., Biomedical Simulation Resource, Los Angeles, 73–103.

Bliss, T.V.P. and Lomo, T. (1973). Long-lasting potentiation of synaptic transmission in the dentate area of the anaesthetized rabbit following stimulation of the perforant path. *J. Neurophysiol.* 232: 331–356.

Buonomano, D.V. (1999). Distinct functional types of associative long-term potentiation in the neocortical and hippocampal pyramidal neurons. *J. Neurosci.* 19(16): 6748–6754.

Buonomano, D.V. (2000). Decoding temporal information: A model based on short-term synaptic plasticity. *J. Neurosci.* 20(3): 1129–1141.

Castellucci, V.F. and Kandel, E.R. (1974). A quantal analysis of the synaptic depression underlying habituation of the gill-withdrawal reflex in Aplysia. *Proc. Natl. Acad. Sci. U. S. A.* 71(12): 5004–5008.

Castro-Alamancos, M.A., and Connors, B.W. (1997). Distinct forms of short-term plasticity at excitatory synapses of hippocampus and neocortex. *Proc. Natl. Acad. Sci. U. S. A.* 94: 4161–4166.

Colmers, W.F., Lukowiak, K., and Pittman, Q.J. (1985). Neuropeptide Y reduces orthidromically evoked population spike in rat hippocampal CA1 by a possibly presynaptic mechanism. *Brain Res.* 346(2): 404–408.

Courellis, S.H., Marmarelis, V.Z., and Berger, T.W. (2000). Modeling event-driven nonlinear dynamics in neuronal systems with multiple inputs, *Annual Conference Biomedical Engineering Society*, Seattle, WA.

Creager, R., Dunwiddie, T., and Lynch, G. (1980). Paired-pulse and frequency facilitation in the CA1 region of the in vitro rat hippocampus. *J. Neurophysiol.* 299: 409–424.

Davies, C.H., Davies, S.N., and Collingridge, G.L. (1990). Paired pulse depression of monosynaptic GABA-mediated inhibitory postsynaptic responses in rat hippocampus. *J. Neurophysiol.* 424: 513–31.

Dobrunz, L.E., Huang, E.P., and Stevens, C.F. (1997). Very short-term plasticity in hippocampal synapses. *Proc. Natl. Acad. Sci. U. S. A.* 94: 14843–14847.

Egert, U., Schlosshauer, B., Fennrich, S., Nisch, W., Fejtl, M., Knott, T., Muller, T., and Hammerle, H. (1998). A novel organotypic long-term culture of the rat hippocampus on substrate-integrated multielectrode arrays. *Brain Res. Protocols* 2: 229–242.

Fortune, E.S. and Rose, G.R. (2000). Short-term synaptic plasticity contributes to the temporal filtering of electrosensory information. *J. Neurosci.* 20(18): 7122–7130.

Fountain, S.B. and Teyler, T.J. (1995). Brain slice techniques in neurotoxicology. In Chang, L.W. and Slikker, W., eds., *Neurotoxicology: Approaches and Methods.* Academic Press, New York, pp. 517–535.

Franceschetti, S., Hamon, B., and Heinemann, U. (1986). The action of valproate on spontaneous epileptiform activity in the absence of synaptic transmission and on evoked changes in $[Ca2+]o$ and $[K+]o$ in the hippocampal slice. *Brain Res.* 386(1–2): 1–11.

Fueta, Y., Ohno, K., and Mita, T. (1998). Large frequency potentiation induced by 2 Hz stimulation in the hippocampus of epileptic El mice. *Brain Res.* 792(1): 79–88.

Gahwiler, B.H. (1981). Organotypic monolayer cultures of nervous tissue. *J. Neurosci. Meth.* 4, 329–342.

Galarreta, M. and Hestrin, S. (1998). Frequency-dependent synaptic depression and the balance of excitation and inhibition in the neocortex. *Natl. Am.* 1(7): 587594.

Gholmieh, G., Courellis, S., Dimoka, A., Marmarelis, V.Z., and Berger, T.W. (2004). A real time EPSP and spike amplitude extraction method. *J. Neurosci. Meth.* 136(2): 111–121.

Gholmieh, G., Courellis, S., Fakheri, S., Cheung, E., Marmarelis, V.Z., Baudry, M., and Berger, T.W. (2003). Detection and classification of neurotoxins using a novel short-term plasticity quantification method. *Biosens. Bioelectron.* 18(12): 1467–1478.

Gholmieh, G., Courellis, S., Marmarelis, V.Z., and Berger, T.W. (2002). An efficient method for studying short-term plasticity with random impulse train stimuli. *J. Neurosci. Meth.* 121(2): 111–127.

Gholmieh, G., Courellis, S., Marmarelis, V.Z., and Berger, T.W. (2005). Detecting CA1 short-term plasticity variations associated with changes in stimulus intensity and extracellular medium composition. *J. Neurocomput.*

Gholmieh, G., Soussou, W., Courellis, S., Marmarelis, V.Z., Berger, T.W, and Baudry, M. (2001). A biosensor for detecting changes in cognitive processing based on non-linear system analysis. *Biosens. Bioelectron.* 16(7–8): 491–50.

Gross, G.W., Rhoadas, B.K., Reust, D.L., and Schwalm F.U. (1993). Stimulation of monolayer networks in culture through thin film indium-tin oxide recording electrodes. *J. Neurosci. Meth.* 50 131–143.

Harris, K.M. and Teyler, T.J. (1983). Age differences in a circadian influence on hippocampal LTP. *Brain Res.* 261(1): 69–73.

Harris, E.W., Ganong, A.H., and Cotman, C.W. (1984). Long-term potentiation in the hippocampus involves activation of N-methyl-D-aspartate receptors. *Brain Res.* 323 (1): 132–137.

Hesen, W., Karten, Y.J., van de Witte, S.V., and Joels, M. (1998). Serotonin and carbachol induced suppression of synaptic excitability in rat CA1 hippocampal area: Effects of corticosteroid receptor activation. *J. Neuroendocrinol.* 10(1): 9–19.

Higgins, G.M. and Gardier, R.W. (1990). Gamma-Aminobutyric acid antagonism produced by an organophosphate-containing combustion product. *Toxicol. Appl. Pharmacol.* 105(1): 103–112.

Huang, Y.Y. and Malenka, R.C. (1993). Examination of TEA-induced synaptic enhancement in area CA1 of the hippocampus: The role of voltage-dependent Ca2+ channels in the induction of LTP. *J. Neurosci.* 13(2): 568–576.

Igelmund, P. and Heinemann, U. (1995). Synaptic transmission and paired pulse behaviour of CA1 pyramidal cells in hippocampal slices from a hibernator at low temperature: Importance of ionic environment. *Brain Res.* 689(1): 9–20.

Kao, W.Y., Liu, Q.Y., Ma, W., Ritchie, G.D., Lin, J., Nordholm, A.F., Rossi, J., 3rd, Barker, J.L., Stenger, D.A., and Pancrazio, J.J. (1999). Inhibition of spontaneous GABAergic transmission by trimethylolpropane phosphate. *Neurotoxicology* 20(5): 843–849.

Katz, B. and Miledi, R. (1968). The role of calcium in neuromuscular facilitation. *J. Physiol.* 195(2): 481–92.

Keefer, E.W., Gramowski, A., Stenger, D.A., Pancrazio, J.J., and Gross, G.W. (2001). Characterization of acute neurotoxic effects of trimethylolpropane phosphate via neuronal network biosensors. *Biosens. Bioelectron.* 16(7–8): 513–525.

Konig, P., Engel, A.K., and Singer, W. (1996). Integrator or coincidence detector? The role of the cortical neuron revisited. *Trends Neurosci.* 19: 130–137.

Krausz, H.I. and Friesen, W.O. (1975). Identification of nonlinear systems using random impulse train inputs. *Biol. Cybern.* 19: 217–230.

Krugers, H.J., Mulder, M., Korf, J., Havekes, L., DeKloet, E.R., and Joels, M. (1997). Altered synaptic plasticity in hippocampal CA1 area of apolipoprotein E deficient mice. *Neuroreport* 8(11): 2505–2510.

Landfield, P.W., Pitler, T.A., and Applegate, M.D. (1986). The effects of high Mg2+−to−Ca2+ ratios on frequency potentiation in hippocampal slices of young and aged rats. *J. Neurophysiol.* 56(3): 797–811.

Lee, Y.W. and Schetzen, M. (1965). Measurement of the kernels of a non-linear system by cross-correlation. *Int. J. Control* 2: 237–254.

Leung, L.S. and Fu, X.W. (1994). Factors affecting paired-pulse facilitation in the hippocampal CA1 neurons in vitro. *Brain Res.* 650: 75–84.

Lin, J., Cassell, J., Ritchie, G.D., Rossi, J., 3rd, and Nordholm, A.F. (1998). Repeated exposure to trimethylolpropane phosphate induces central nervous system sensitization and facilitates electrical kindling. *Physiol. Behav.* 65(1): 51–58.

Lin, J., Ritchie, G.D., Stenger, D.A., Nordholm, A.F., Pancrazio, J.J., and Rossi, J., 3rd, Ritchie (2001). Trimethylolpropane phosphate induces epileptiform discharges in the CA1 region of the rat hippocampus. *Toxi. Appl. Pharmacol.* 171(2): 126–134.

Marmarelis, V.Z. (1993). Identification of nonlinear biological systems using Laguerre expansions of kernels. *Ann. Biomed. Eng.* 21: 573–589.

Meyer, J.H., Lee, S., Wittenberg, G.F., Randall, R.D., and Gruol, D.L. (1999). Neurosteroid regulation of inhibitory synaptic transmission in the rat hippocampus. *Neuroscience* 90(4): 1177–1183.

Muller, D. and Lynch, G. (1990). Synaptic modulation of N-Methyl-D-aspartate receptor mediated responses in hippocampus. *Synapse* 5: 94–103.

Nelson, T.E., Ur, C.L., and Gruol, D.L. (1999). Chronic intermittent ethanol exposure alters CA1 synaptic transmission in rat hippocampal slices. *Neuroscience* 94(2): 431–442.

Novak, J.L. and Wheeler, B.C. (1988). Multisite hippocampal slice recording and stimulation using a 32 element microelectrode array. *J. Neurosci. Meth.* 23: 149–159.

Oka, H., Shimono, K., Ogawa R., Sugihara, H., and Taketani, M. (1999). A new planar multielectrode array for extracellular recording: Application to hippocampal acute slice. *J. Neurosci. Meth.* 93: 61–67.

Pananceau, M., Chen, H., and Gustafsson, B. (1998). Short-term facilitation evoked during brief afferent tetani is not altered by long-term potentiation in the guinea-pig hippocampal CA1 region. *J. Physiol.* 508.2: 503–514.

Pancrazio, J.J., Gray, S.A., Shubin, Y.S., Kulagina, N., Cuttino, D.S., Shaffer, K.M., Eisemann, K., Curran, A., Zim, B., Gross, G.W., and O'Shaughnessy, T.J. (2003). A portable microelectrode array recording system incorporating cultured neuronal networks for neurotoxin detection. *Biosens. Bioelectron.* (18) 11: 1339–1347.

Papatheodoropoulos, C. and Kostopoulos, G. (2000). Dorsal-ventral differentiation of short-term synaptic plasticity in rat CA1 hippocampal region. *Neurosci. Lett.* 286(1): 57–60.

Perrett, S., Dudek, S., Eagleman, D., Montague, P., and Friedlander, M. (2001). LTD induction in adult visual cortex: Role of stimulus timing and inhibition. *J. Neurosci.* 21(7): 2308–2319.

Rausche, G., Igelmund, P., and Heinemann, U. (1990). Effects of changes in extracellular potassium, magnesium and calcium concentration on synaptic transmission in area CA1 and the dentate gyrus of rat hippocampal slices. *Euro. J. Physiol.* 415(5): 588–593.

Rossi, J., 3rd, Ritchie, G.D., McInturf, S., and Nordholm, A.F. (2001). Reduction of motor seizures in rats induced by the ethyl bicyclophosphate trimethylolpropane phosphate (TMPP). *Prog. Neuro-Psychopharmacol. Biol. Psychiatry* 25(11.6): 1323–1340.

Sagratella, S., Proietti, M.L., Frank, C., and de Carolis, A.S. (1991). Effects of some calcium antagonists and of calcium concentration changes on CA1 paired pulse inhibition in rat hippocampal slices. *Gen. Pharmacol.* 22(2): 227–230.

Scalabassi, R.J., Eriksson, J.L, Port, R., Robinson, G., and Berger, T.W (1988). Nonlinear systems analysis of the hippocampal perforant path-dentate projection. i. Theoretical and interpretational considerations. *J. Neurophysiol.* 60: 1066–1076.

Song, D., Xiaping, X., Wang, Z., and Berger, T.W. (2001). Differential effect of TEA on long-term synaptic modification in hippocampal CA1 and dentate gyrus in vitro. *Neurobiol. Learn. Memory* 76: 357–387.

Southan, A.P. and Owen, D.G. (1997). The contrasting effects of dendrotoxins and other potassium channel blockers in the CA1 and dentate gyrus regions of rat hippocampal slices. *Brit. J. Pharmacol.* 122(2): 335–343.

Stanford, I.M, Wheal, H.V., and Chad, J.E. (1995). Bicuculline enhances the late GabaB receptor-mediated paired pulse inhibition observed in the rat hippocampal slices. *Euro. J. Pharmacol.* 277(2–3): 229–234.

Stringer, J.L., and Lothman, E.W. (1998). In vitro effects of extracellular calcium concentrations on hippocampal pyramidal cell responses. *Exp. Neurol.* 101(1): 132–146.

Turner, RW. and Miller J.J. (1982). Effects of extracellular calcium on low frequency induced potentiation and habituation in the in vitro hippocampal slice preparation. *Canad. J. Physiol. Pharmacol.* 60(3): 266–267.

Wheal, H.V., Lancaster, B., and Bliss, T.V. (1983). Long-term potentiation in Schaffer collateral and commissural systems of the hippocampus: In vitro study in rats pretreated with kainic acid. *Brain Res.* 272(2): 247–253.

Wheeler, B.C. and Novak, J.L. (1986). Current source density estimation using microelectrode array data from the hippocampal slice preparation. *IEEE Trans. Biomed. Eng.* 33(12): 1204–1212.

Xie, X., Liaw, J.S., Baudry, M., and Berger, T.W. (1997). Novel expression mechanism for synaptic potentiation: Alignment of presynaptic release site and postsynaptic receptor. *Proc. Natl. Acad. Sci. U. S. A.* 94(13): 6983–6988.

Yajeya, J., De La Fuente, A., Criado, J.M., Bajo, V., Sanchez-Riolobos, A., and Heredia, M. (2000). Muscarinic agonist carbachol depresses excitatory synaptic transmission in the rat basolateral amygdala in vitro. *Synapse* 38(2): 151–160.

Yamamoto, C., Matsumoto, K., and Takagi, M. (1980). Potentiation of excitatory postsynaptic potentials during and after repetitive stimulation in thin hippocampal sections. *Exp. Brain Res.* 38(4): 469–477.

Zucker, R.S. (1989). Short-term synaptic plasticity. *Ann. Rev. Neurosci.* 12: 13–31.

13

The Retinasensor: An In Vitro Tool to Study Drug Effects on Retinal Signaling

ELKE GUENTHER, THORALF HERRMANN, AND ALFRED STETT

13.1   Introduction

Recent advances in growing electrically active cells on substrate-integrated micro-electrode arrays (MEAs), either as cell cultures or in a tissue slice, have led to test systems in which cellular activity can be recorded acutely or up to several months. As the MEA technology can be applied to any electrogenic tissue (i.e., central and peripheral neurons, heart cells, and muscle cells), MEA biosensors are ideal in vitro systems to monitor both acute and chronic effects of drugs and toxins and to perform functional studies under physiological or induced pathophysiological conditions that mimic in vivo damages. By recording the electrical response at various locations on a tissue, a spatial map of drug effects at different sites may be generated, providing important clues about a drug's specificity.

In the following we describe a preparation of the vertebrate retina on microelectrode arrays to record local electroretinograms in vitro (microERG) under defined experimental conditions. We show that the so-called retinasensor is a suitable in vitro tool to easily and effectively assess effects of pharmacological compounds and putative therapeutics on retinal function. The major advantage of the preparation is that the retina together with the retinal pigment epithelium can be isolated as a whole by merely cutting the optic nerve without the necessity of slicing the tissue and thus damaging the cellular layers as in other in vitro preparations of the central nervous system.

13.2   The Retinasensor

13.2.1   The Electroretinogram

The retina is a peripheral, easily accessible part of the central nervous system that lines the back of the eye. All vertebrate retinas are composed of three layers of nerve cell bodies and two layers of synapses. The outer nuclear layer contains cell bodies of the photoreceptors, the rods and cones; the inner nuclear layer contains

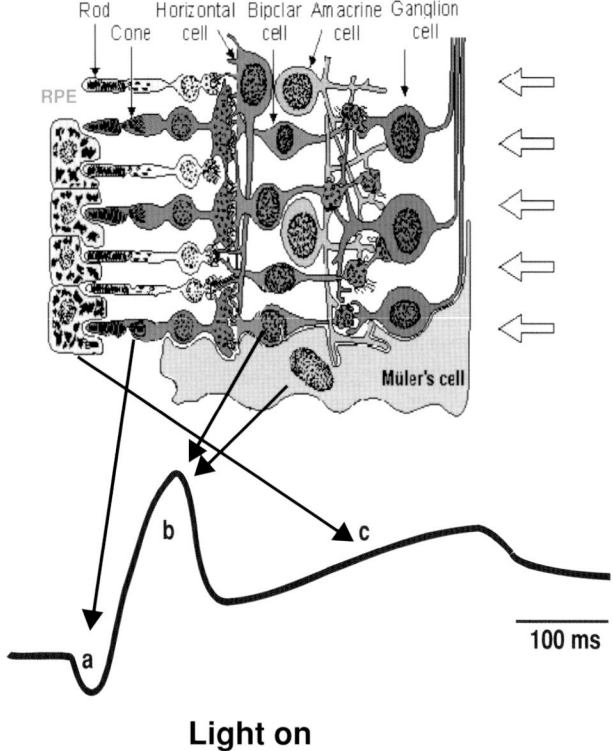

FIGURE 13.1. Top: Schematic representation of a retinal cross-section showing the assembly of all retinal neurons together with the retinal Müller glia cells and the retinal pigment epithelium. Bottom: Typical example of an electroretinogram (ERG) showing the characteristic waveform composed of the a-wave, the b-wave, and the c-wave induced by stimulating the retina with light. The different waves can be attributed to different retinal cell types or layers as indicated by the arrows.

cell bodies of the bipolar, horizontal, and amacrine cells; and the ganglion cell layer contains cell bodies of ganglion cells and displaced amacrine cells. Dividing these nerve cell layers are two neuropils where synaptic contacts occur (Polyak, 1941; Rodieck, 1973; Kolb, 1991; see Figure 13.1).

   Incidence of light results in a complex signaling within these retinal neurones that is reflected in the electroretinogram (ERG). The ERG is a mass electrical response of the eye that is made up of several components. Since the early twentieth century, numerous studies have been performed on the origin and characteristics of the ERG. Einthoven and Jolly (1908) separated the light-induced ERG response into three waves, called the a-, b-, and c-waves. An additional wave recorded at the termination of the light stimulus was called the d-wave. Granit (1933) demonstrated that the a-, b-, and c-waves are the summed responses of three underlying basic components, named PI to PIII. The component analysis of Granit that won him the

Nobel Prize for Physiology or Medicine in 1954 has been modified slightly over the years but remains the basis for our understanding of the ERG.

ERGs have been recorded in different animal species (Steinberg et al., 1970; Kline et al., 1985; Newman, 1985; Hood and Birch, 1995; Lei and Perlman, 1999; Kapousta-Bruneau, 2000; Jamison et al. 2001) and clearly differ in amplitude and pattern. Some of this variability is due to species differences, particularly the relative densities of rods and cones, whereas technical factors such as duration and intensity of photostimulation and method of recording also affect the waveform. Nevertheless, ERG responses recorded from vertebrate species are characterized by the basic features of a negative a-wave followed by a positive b-wave and c-wave.

Retinal function can be affected by acute injuries, intoxication, or retinal diseases, resulting in visual impairment or blindness. Under these conditions, the shape and amplitude of an ERG is altered, making it of clinical value as a diagnostic tool. The ERG is a test used worldwide in ophthalmology to assess the status of the retina in human patients (Marmor and Zrenner, 1998) and in laboratory animals used as models of retinal disease (Kueng-Hitz, 2000; Bolz et al., 2001).

The clinical value of the ERG as a diagnostic tool comes from the fact that the different ERG waves can be attributed to distinct retinal layers or even cell types (see Figure 13.1). It is known that the a-wave is generated by the photoreceptors (Tomita, 1950; Brown and Wiesel, 1961; Sillman et al., 1969) and the c-wave originates in the retinal pigment epithelium (Noell, 1954; Steinberg et al., 1970; Marmor and Hock, 1982). The b-wave is formed in retinal cells that are postsynaptic to the photoreceptors but its exact source is still under dispute. However, it is agreed that large parts are generated by ON-bipolar cells (Gurevich and Slaughter, 1993; Sieving et al., 1994), and there is also a participation of Mueller glia cells (Miller and Dowling, 1970). With a proper analysis of the ERG waves under physiological and pathophysiological conditions it is thus possible to dissect out the functional integrity of different retinal structures and to understand information-processing mechanisms and/or the sites of retinal disorders and dysfunction. In the following, a new experimental system is described that for the first time allows multi-site recording of ERGs in vitro.

13.2.2   Practical Approach

13.2.2.1   Tissue Preparation

One- to three-day-old chickens were light-adapted and killed by decapitation. Light adaptation ensures that the pigment epithelium (RPE) is tightly attached to the retina due to the protrusion of the photoreceptor outer segments into the RPE. The eyes were dissected immediately after enucleation by hemisection of the upper third of the eye-cup with a sharp razor blade. The lens and vitreous were carefully removed. Square segments of 4 mm^2 were cut out from the eye-cup and the neural retina and the pigment epithelium was gently removed from the underlying tissue. One sample was placed ganglion cell side down (Figure 13.2) onto a MEA chip

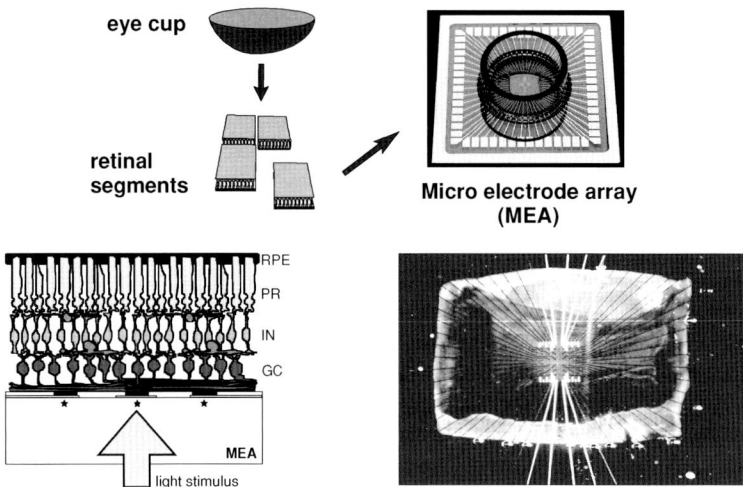

FIGURE 13.2. Preparation of samples of the chicken retina on a microelectrode array. The retina is placed ganglion cells down onto the MEA chip (lower left). The retina together with the retinal pigment epithelium (RPE) seen through the translucent MEA chip. The brownish color comes from the pigmentation of the RPE (lower right).

consisting of 60 surface-integrated planar electrodes (TiN, diameter 30 μm, rectangular arrangement, interelectrode spacing 200 μm). A recording chamber was glued onto the glass chip to allow superfusion of the tissue. The sample was superfused at a rate of 1 ml/min (syringe pump, flow 400 μl/m, TSE, Bad Homburg, Germany) with bubbled (95% O_2, 5% CO_2) standard Ringer solution (in mM: 120 NaCl, 5 KCl, 3 $CaCl_2$, 1 $MgSO_4$, 25 $NaHCO_3$, 25 glucose, 1 NaH_2PO_4; osmolarity 340 mOsm, pH 7.5). The MEA chip was mounted to a MEA stage with integrated heating (MCS, Reutlingen, Germany) and placed onto the table of an inverted microscope (Zeiss Axiovert 10, Zeiss AG, Oberkochen, Germany). Experiments were performed at a temperature of 35°C. The remaining samples were stored in bubbled ringer solution on a heating plate and kept in darkness for further use.

13.2.2.2 Setup

After at least 30 min of dark adaptation, an experiment was started. Full-field stimulation with light impulses of defined length and frequency was performed with the halogen lamp of the microscope (100 W) and an electromagnetic shutter. The light was projected through the objectives of the inverted microscope homogeneously onto the retinal sample. Light intensity and wavelength were controlled by neutral density and color filters (optical filters, Schott AG, Mainz, Germany) manually inserted into the optical pathway. Full-field light intensity on the retina level without a filter was 100 mW/cm^2 for white light, 20 mW/cm^2 for blue light, and 40 mW/cm^2 for red light. A schematic of the optical path and the recording setup is shown in Figure 13.3.

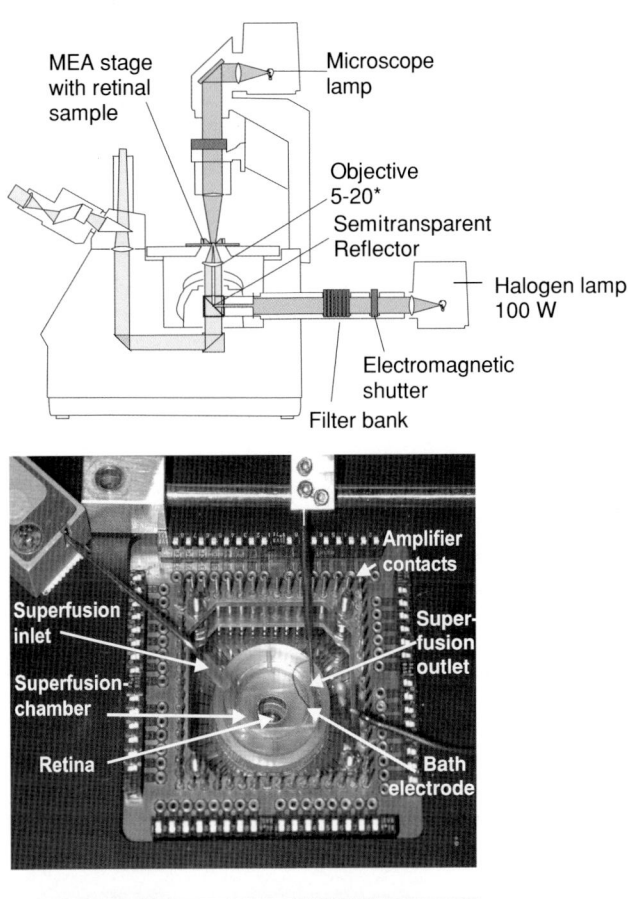

FIGURE 13.3. Schematic representation of the optical path through an inverted microscope (top). The MEA chip with the perfusion chamber containing the retina is placed on the microscope stage (middle). View on the recording setup (bottom).

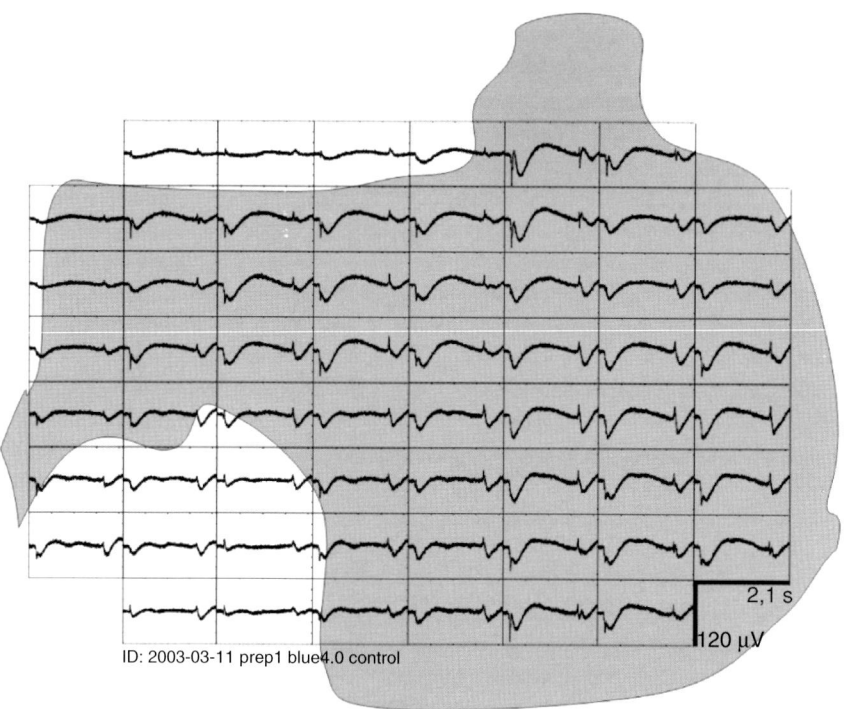

FIGURE 13.4. Micro-ERGs induced by light stimulation and recorded on 60 electrode sites in parallel. Shading outlines the retina/RPE complex whereas, in the white area, both tissues fell apart and only the retina was present. Typical micro-ERGs with good signal-to-noise ratios were only recorded in the central part within the retina/RPE complex.

13.2.2.3   Micro-ERG Recording

Micro-electrode recordings of retinal activity were obtained from up to 60 retinal sites with a MEA60 system (Multi Channel Systems MCS GmbH, Reutlingen, Germany) with a bandwidth of 0.5 to 2.8 kHz and a gain of 2300. For analysis, the responses to five consecutive stimuli were averaged at each site. Figure 13.4 shows a typical recording of local micro-ERGs at 60 electrode sites in parallel as displayed on the computer screen with the McRack software (MCS, Reutlingen, Germany). The whole electrode field (1.4×1.4 mm^2) was covered by the retinal sample, but only in the grey shaded area was the retinal pigment epithelium closely attached to the retina. The micro-ERGs are most pronounced in the central part of the retina/RPE preparation whereas the signals deteriorate at its marginal zones or in the area with retina only (white electrode area in Figure 13.4).

In Figure 13.5, light-induced signals on a single electrode are displayed. With adequate filter setting, either micro-ERGs (0.5 to 100 Hz) or ganglion cell spikes (200 Hz to 2.8 kHz) can be recorded. The micro-ERG in vitro clearly shows a light-on and light-off response and the light-on response is composed of the three main

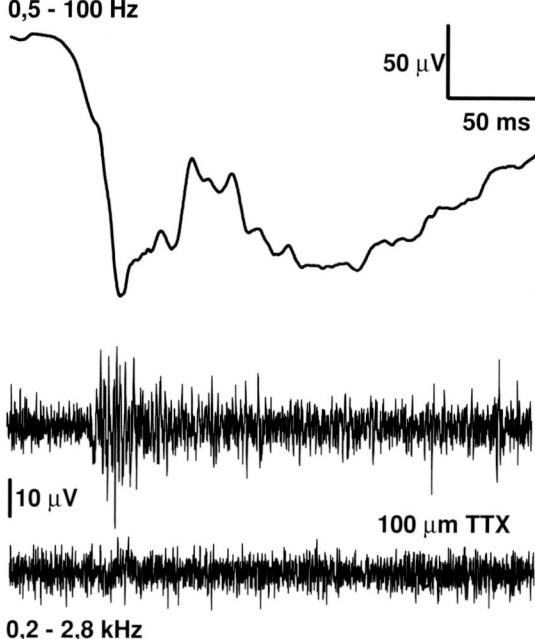

0,5 - 100 Hz

50 µV

50 ms

10 µV

100 µm TTX

0,2 - 2,8 kHz

FIGURE 13.5. ON responses of the chicken micro-ERG recorded at one electrode. Top: slow wave component (average of five sweeps); bottom: spike recording of ganglion cell activity, filter settings as indicated. The transient ON-response of ganglion cells was blocked by the sodium channel antagonist TTX.

components, the a-, b-, and c-waves (Figure 13.5 top) as in a typical ERG recording in vivo. Compared to ERG studies on the chicken performed in vivo, the amplitude of the b-wave is smaller in our in vitro preparation (Wioland et al., 1990; Schwahn et al., 2000). This may be attributed to differences in the light stimulation and the method of recording (Perlman, 2003). However, we have observed the relation of Mg^{2+} and Ca^{2+} in the bath solution to be a most crucial factor for the size of the b-wave. Largest b-waves in vitro were obtained with 3 mM Ca^{2+} and 1 mM Mg^{2+}. Light-induced micro-ERG recordings were performed up to two hours with one retinal sample.

In contrast with ERG recordings in vivo, the retinasensor preparation additionally allows the recording of retinal ganglion cell (RGC) activity (Figure 13.5, bottom). Because the retina is placed with the ganglion cell site down onto a MEA, the electrodes are in direct contact with either the RGC bodies or axons, and ON- and OFF-RGC spikes were recorded using a highpass filter. As in patch-clamp recordings of RGCs in a retinal slice, RGC spike activity was blocked with the sodium channel blocker TTX applied to the superfusion solution (Schmid and Guenther, 1998), additionally demonstrating the vitality of the preparation.

A clear dependence on the intensity of the light stimulus was observed for the micro-ERGs. Amplitudes of the micro-ERG components increased and kinetics

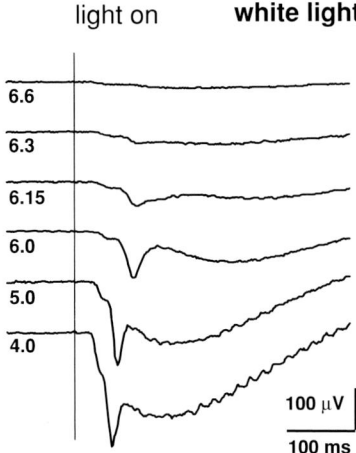

FIGURE 13.6. Light-dependence of micro-ERGs. Numbers on the right indicate the attenuation of light by neutral density filters in log units.

became faster with increasing intensity of a white light projected onto the retina (Figure 13.6). These light-dependent alterations are comparable to what has been reported for human and animal ERG recordings in vivo. However, overall light sensitivity depends on the species, the way the light is projected onto the retina, and, in vivo, how much of the light reaches the retina through the optic apparatus of the eye. Because light intensity in vivo is measured at the level of the cornea and not of the retina as in vitro, in vivo and in vitro data cannot directly be compared in terms of absolute sensitivities.

Lowering the temperature of the retina preparation from 35°C to 26°C had profound effects on the a-wave and b-wave amplitude whereas the c-wave amplitude was almost not affected (Figure 13.7). Again, this is in accordance with observations made in vivo in both chicken and mammals (Ookawa, 1970; Mizota and Adachi-Usami, 2002).

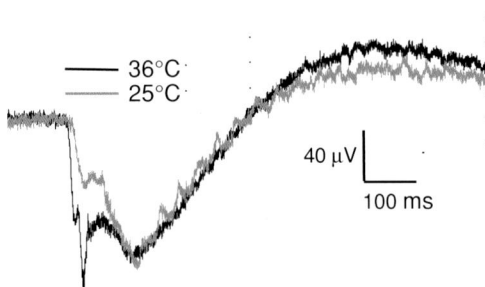

FIGURE 13.7. Temperature sensitivity of micro-ERG components.

13.2.3   Pharmacological Characterization of Micro-ERGs

In order to evaluate if drug-induced alterations of the ERG waveform can be detected with the retinasensor, 2-Amino-4-phosphonobutanoic acid (AP-4) was applied to the bath medium. AP-4 is a selective agonist at metabotrobic glutamate receptors of type 6 (mGluR6). In the retina, mGluR6 receptors are exclusively expressed on ON-bipolar cells. In darkness, when photoreceptors release glutamate, activation of mGluR6 produces a hyperpolarization of the postsynaptic ON-bipolar cells by closing a cation channel in the cell membrane. In contrast, with light exposure, when glutamate is no longer released, mGluR6 receptors are inactive, the cation channels open, and ON-bipolar cells are depolarized (Slaughter and Miller, 1981; Nawy and Jahr, 1990). If mGluR6 is activated with AP-4, ON-bipolar cells are hyperpolarized even in the presence of light. Thus, after a light stimulus, ON-bipolar cells should no longer contribute to the b-wave of the micro-ERG, and b-wave amplitude should be smaller with AP-4 than without. This is exactly what we observed when applying AP-4 to our in vitro preparation (Figure 13.8), indicating that the retinasensor is a suitable tool to analyze drug effects on retinal function. As in other retinal in vitro recordings, the AP-4 effect on the micro-ERG was reversible after washing the retina with normal ringer solution.

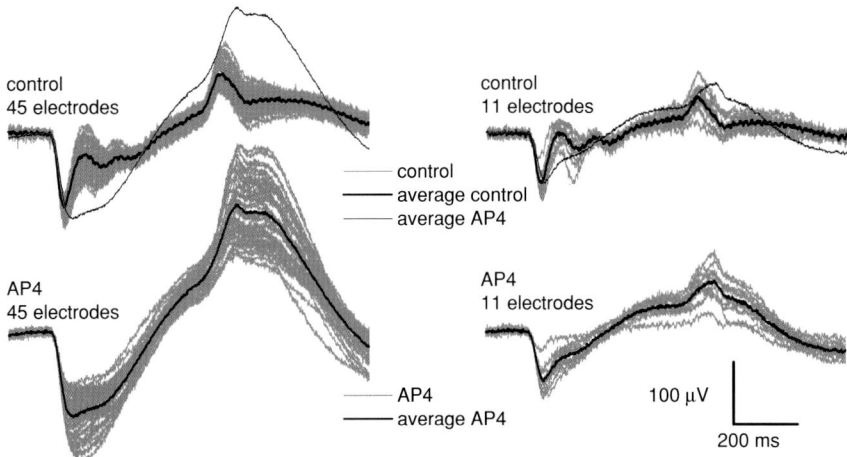

FIGURE 13.8. Effect of the mGluR6 agonist AP-4 on micro-ERG. Top row: control recordings obtained before application of 100 μM AP4; bottom row: recordings under exposure of the sample to AP4. In the given example, two principle waveforms of the recorded micro-ERG could be separated (left, right). The grey traces are individual recordings from a number of electrodes; each trace is the average of 30 recordings. The bold lines mark the average of the recordings from the given number of electrodes (black: average control, blue: average AP4). Stimulated with white light, 100 mW/cm². For comparison reasons, the averaged AP4 response is overlaid on the control recordings. The effect is reversible after washing (data not shown).

13.3   Summary and Conclusion

An in vitro preparation of the chicken retina was established on multi-electrode arrays. The so-called retinasensor allows the multi-site recording of micro-ERGs for hours and has been shown to be sensitive enough to assess drug effects on retinal function. The advantage of the system is that the retina can be isolated together with the retinal pigment epithelium as a whole by merely cutting the optic nerve. In contrast with other brain preparations, slicing of the tissue can be avoided and thus no damaged cells are present at the electrode–tissue interface that would interfere with the stimulation of healthy cell layers. A further advantage of the retinasensor is the recording of both micro-ERGs and ganglion cell activity. Because retinal ganglion cells do not contribute much to the ERG, additional information on the status of ganglion cells can be provided with the retinasensor as compared to ERG recordings in vivo. Taken together, the retinasensor offers a new assay system to reliably assess retinal function and dysfunction in vitro.

References

Bolz, H., von Brederlow, B., Ramirez, A., Bryda, E.C., Kutsche, K., Nothwang, H.G., Seeliger, M., C.-Salcedó Cabrera, M., Caballeró Vila, M., Pelaez Molina, O., Gal, A., and Kubisch, C. (2001). Mutations in cadherin-23 (otocadherin), a novel member of the cadherin gene family, cause Usher syn-drome type 1D. *Nat. Gene.* 27: 108–112.

Brown, K.T., and Wiesel, T.N. (1961a). Analysis of the intraretinal electroretinogram in the intact cat eye. *J. Physiol. (London)* 158: 229–256.

Brown, K.T., and Wiesel, T.N. (1961b). Localization of origins of electroretinogram components by intraretinal recording in the intact cat eye. *J. Physiol. (London)* 158: 257–280.

Einthoven, W., and Jolly, W.A. (1908). The form and magnitude of the electrical response of the eye to stimulation by light at various intensities. *Quaterly J. Exp. Physiol.* 1: 373–416.

Granit, R. (1933). The components of the retinal action potential in mammals and their relation to the discharge in the optic nerve. *J. Physiol. (London)* 77: 207–239.

Gurevich, L., Slaughter, M. M. (1993). Comparison of the waveforms of the ON bipolar neuron and the b-wave of the electroretinogram. *Vis. Res.* 33: 2431–2435.

Hood, D.C. and Birch, D.G. (1995). Phototransduction in human cones measured using the a-wave of the ERG. *Vis. Res.* 35: 2801–2810.

Jamison, J.A., Bush, R.A., Lei, B., and Sieving, P. (2001). Characterization of the rod photoresponse isolated from the dark-adapted primate ERG. *Vis. Neurosci.* 18: 445–455.

Kapousta-Bruneau, N.V. (2000). Opposite effects of $GABA_A$ and $GABA_C$ receptor antagonists on the b-wave of ERG recorded from the isolated rat retina. *Vis. Res.* 40: 1653–1665.

Kline, R.P., Ripps, H., and Dowling, J.E. (1985). Light-induced potassium fluxes in the skate retina. *Neuroscience* 14: 225–235.

Kolb, H. (1991). The neural organization of the human retina. In: Heckenlively, J.R. and Arden, G.B., eds., *Principles and Practices of Clinical Electrophysiology of Vision.* Mosby Year Book, St. Louis, pp. 25–52.

Kueng-Hitz, N., Grimm, C., Lansel, N., Hafezi, F., He, L., Fox, D.A., Reme, C.E., Niemeyer, G., and Wenzel, A. (2001). The retina of c-fos-/-mice: electrophysiologic, morphologic and biochemical aspects. *Invest. Ophthalmol. Vis. Sci.* 41(3): 909–916.

Lei, B. and Perlman, I. (1999). The contribution of voltage- and time-dependent poassium conductances to the electroretinogram in rabbits. *Vis. Neurosci.* 16: 743–754.

Marmor, M.F. and Zrenner, E. (1998). Standard for clinical electroretinography. *Doc. Ophthalmol.* 97: 143–156.

Miller, R.F., and Dowling, J.E. (1970). Intracellular responses of the Müller (glial) cells of the mudpuppy retina: their relation to the b-wave of the electroretinogram. *J. Neurophysiol.* 33: 323–341.

Mizota, A. and Adachi-Usami, E. (2002). Effect of body temperature on electroretinogram of mice. *Invest. Ophthalmol. Vis. Sci.* 43(12): 3754–3757.

Nawy, S. and Jahr, C.E. (1990). Suppression by glutamate of cGMP-activated conductance in retinal bipolar cells. *Nature* 346: 269–271.

Newman, E.A. (1985). Current source-density analysis of the b-wave of frog retina. *J. Neurophysiol.* 43: 1355–1366.

Noell, W.K. (1954). The origin of the electroretinogram. *Am. J. Ophthalmol.* 30: 78–90.

Ookawa, T. (1971). Effects of acute hypothermia on the chick ERG. *Experientia* 27(4): 405–407.

Perlman, I. (2003). The electroretinogram. http://webvision.med.utah.edu, Part XI.

Polyak, S.L. (1941). *The Retina.* University of Chicago Press, Chicago.

Rodieck, R.W. (1973). *The Vertebrate Retina: Principles of Structure and Function.* W.H. Freeman, San Francisco.

Schmid, S., and Guenther, E. (1998). Alterations in channel density and kinetic properties of the sodium current in retinal ganglion cells of the rat during in vivo differentiation. *Neuroscience* 85(1):249–258.

Schwahn, H.N., Kaymak, H., and Schaeffel, F. (2000). Effects of atropine on refractive development, dopamine release, and slow retinal potentials in the chick. *Vis. Neurosci.* 17(2): 165–176.

Sieving, P.A., Murayama, K., and Naarendorp, F. (1994). Push-pull model of the primate photopic electroretinogram: A role for hyperpolarizing neurons in shaping the b-wave. *Vis. Neurosci.* 11: 519–532.

Sillman, A.J., Ito, H., and Tomita, T. (1969a). Studies on the mass receptor potential of the isolated frog retina. I. General properties of the response. *Vis. Res.* 9: 1435–1442.

Sillman, A.J., Ito, H., and Tomita, T. (1969b). Studies on the mass receptor potential of the isolated frog retina. II. On the basis of the ionic mechanism. *Vis. Res.* 9: 1443–1451.

Slaughter, M.M. and Miller, R.F. (1981). 2-amino-4-phosphobutyric acid: A new pharmacological tool for retina research. *Science* 211: 182–184.

Steinberg, R.H., Schmidt, R., and Brown, K.T. (1970). Intracellular responses to light from cat pigment epithelium: Origin of the electroretinogram c-wave. *Nature* 227: 728–730.

Tomita, T. (1950). Studies on the intraretinal action potential. I. Relation between the localization of micropipette in the retina and the shape of the intraretinal action potential. *Jap. J. Physiol.* 1: 110–117.

Wioland, N. and Rudolf, G. (1991). Light and dark induced variations of the c-wave voltage of the chicken eye after treatment with sodium aspartate. *Vis. Res.* 31(4): 643–648.

14
Chronic Alcohol Effects on Hippocampal Neuronal Networks

LARRY P. GONZALEZ, KEN D. MARSHALL, PRASHANTHA D. HOLLA, AND ANAND MOHAN

14.1   Introduction

In vivo studies in our laboratory have shown that the function of hippocampal neurons is altered by both acute and chronic ethanol treatment, as is the sensitivity of these neurons to cholinergic agonists. Recently, we have begun to use substrate-embedded multi-electrode arrays (Med64, Panasonic) to investigate hippocampal neuronal network activity. Recordings of neuronal activity obtained through the use of multi-electrode arrays (MEAs) provide information about regional variations in treatment effects and allow examination of neuronal circuit activity. Because drug actions in the CNS may depend upon alterations in the function of complex neuronal networks, MEAs may allow evaluation of effects on emergent properties that may only be observed by monitoring network activity (Faingold 2004).

The long-term goal of the following research is to develop an understanding of the neuronal mechanisms that mediate the actions of ethanol in the central nervous system and the effects of withdrawal from a period of chronic ethanol treatment, with the aim of developing treatments for those suffering from this addiction. In particular, these studies examine the effects of ethanol on an identified population of neurons within the hippocampus using Med64 arrays for the evaluation of neuronal function. The rationale for such an investigation is based upon previous research that has suggested that ethanol alters the function and morphology of hippocampal cholinergic systems and upon studies that have implicated the hippocampus in many of the behavioral and physiological responses to acute and chronic ethanol.

Many of the effects of ethanol on behavior are similar to those resulting from experimental manipulations of hippocampal function (Devenport et al., 1981; Hughes, 1982; Arendt et al., 1989; Devenport and Hale, 1989), including impairments in operant responding, attentional deficits, and deficits in spatial memory. Chronic ethanol treatment can result in prominent morphological changes in the hippocampus of rats and humans, and these changes are correlated with a variety of deficits in cognitive functioning (Tarter, 1975; Walker et al., 1981; Arendt et al., 1983). Heavy use of alcohol is associated with the development of physical dependence, and withdrawal from chronic alcohol exposure results in a syndrome of physiological and psychological symptoms (Victor, 1953; Isbell et al., 1955;

Majchrowicz 1975). The symptoms of ethanol withdrawal can include motor convulsions and, in fact, alcohol abuse is the major cause of status epilepticus in patients over 20 years old (Kovanen et al., 1984). The hippocampus is important in the mediation of epileptiform activity (Buzsaki et al., 1989; Wheal, 1989), which may be relevant to the occurrence of ethanol withdrawal seizures, and many of the central actions of ethanol may be mediated by either direct or indirect effects in the hippocampus.

Durand and Carlen (1984) and Abraham et al. (1981) report long-lasting impairment in inhibitory neurotransmission in hippocampus after chronic ethanol exposure. Whittington and Little (1990) have reported increased paired-pulse potentiation in the hippocampus, decreased paired-pulse inhibition, and also decreased thresholds to induce orthodromic or antidromic population spikes during withdrawal from chronic ethanol. Hippocampal neurons are among the most sensitive to the effects of acute and chronic alcohol exposure.

14.2   Ethanol and Hippocampal Cholinergic Neurons

Although the mechanisms responsible for many of the acute and chronic effects of alcohol on the central nervous system remain largely unknown (Goldstein 1979), ethanol treatment is accompanied by specific alterations in the activity of several neurotransmitter systems (Kalant, 1975; Lahti, 1975; Hunt and Majchrowicz, 1979). Correlational studies of ethanol and central cholinergic activity suggest the involvement of cholinergic mechanisms in the mediation of ethanol-induced effects (Erickson and Graham, 1973; Morgan and Phillis, 1975; Parker et al., 1978; Sinclair and Lo, 1978). Cholinergic neurons, and more specifically those cholinergic neurons terminating in the hippocampus and their cholinoceptive targets, are particularly sensitive to the neurochemical, electrophysiological, and pathological effects of ethanol.

Because of the possible importance of the hippocampus and cholinoceptive hippocampal neurons in mediating the effects of ethanol, we have investigated the functioning of hippocampal cholinoceptive neurons following acute and chronic ethanol administration.

14.3   Ethanol and In Vivo Electrophysiology

In previous studies, we have reported a significant reduction in spontaneous hippocampal single-unit activity following the acute systemic administration of ethanol and alterations in responses of these neurons to locally applied acetylcholine (Gonzalez and Sun, 1992). Similarly, hippocampal responses to electrical stimulation of the medial septum, which increases hippocampal release of acetylcholine, are reduced by acute systemic ethanol administration (Gonzalez and Sun, 1992). In addition, research in our laboratory has demonstrated that hippocampal neurons are more sensitive to manipulation of cholinergic activity after chronic ethanol exposure and withdrawal. Following withdrawal from chronic ethanol treatment, the dose–response curve of hippocampal single-unit responses to locally applied acetylcholine is significantly shifted to the left and the threshold for responses to septal stimulation is significantly lower (Gonzalez and Sun, 1992).

A functional consequence of this withdrawal-induced increase in sensitivity is seen in our finding that intrahippocampal infusion of physostigmine, a cholinesterase inhibitor that enhances the action of acetylcholine, elicited more electrophysiological seizure activity in chronic ethanol-treated animals during ethanol withdrawal than in ethanol-naive control animals (Gonzalez, 1985). We have also reported in vivo studies indicating hippocampal effects of ethanol withdrawal evident in EEG spike and sharp wave activity (Veatch and Gonzalez, 1996), in hippocampal amino acid levels (Gonzalez and Gable, III 1990), and in the development of kindled seizures with hippocampal electrical stimulation (Veatch and Gonzalez, 1997, 1999, 2000). We interpret these latter findings to indicate that ethanol exposure and withdrawal results in a long-lasting disruption of the complex neuronal interactions that are necessary in the mediation of kindling and the propagation of coordinated neuronal activity either within the hippocampus or between the hippocampus and its afferent projections.

These studies all suggest that hippocampal neurons show a unique sensitivity to ethanol and indicate that significant alterations in hippocampal function follow chronic ethanol treatment and withdrawal. In the remainder of this report we describe our recent studies using the in vitro hippocampal slice preparation with the Med64 MEA.

14.4   In Vitro Hippocampus and Multi-Electrode Arrays

Our laboratory has begun to investigate regional differences in hippocampal effects of ethanol through the use of electrophysiological recordings using MEAs. The hippocampus has a number of advantages for studies of this type including the well-known morphology of hippocampal neurons, the maintenance of the primary internal circuitry within horizontal tissue slices, as well as the sensitivity of hippocampal neurons to chronic ethanol treatment (Little, 1999; Weiner, 2002).

We have performed a series of experiments with hippocampal tissue slices obtained from ethanol-naive animals and from animals after acute withdrawal from chronic ethanol treatment to investigate the effects of chronic ethanol treatment and withdrawal on cholinergic function in the hippocampus. In these studies we have examined evoked field potentials in the Schaffer collateral and mossy fiber pathways, and we have also investigated effects on carbachol-induced rhythmic activity.

14.5   Materials and Methods

14.5.1   Animals

The subjects in all of the following studies were adult, Sprague–Dawley rats, 60 to 90 days old, and weighing 200 to 224 g at the beginning of each experiment.

Animals were housed two per cage in an AAALAC-approved animal colony under a standard 12 h light–dark cycle (6:00 AM, on/6:00 PM, off) and had free access to food and water for the duration of the experiment. Animals were maintained for at least seven days in the same conditions of environment, diet, and daily handling before any experimental treatment.

14.5.2   Chronic Ethanol Administration

Animals received chronic exposure to ethanol in ethanol-vapor inhalation chambers. These are 24 in. plexiglas cubes subdivided vertically at 12 in. to provide two cages (24 in. × 24 in. × 12 in.). The air flow to each cage was regulated independently and provided individual housing for four animals. Food and water were available ad lib. Fresh air was flushed through the chamber continuously at the rate of two liters per min. to provide for the respiratory needs of the animals. To this fresh air flow was added ethanol vapor (0 to 30 mg/ml) which was obtained by pumping air through a one-liter aspirator bottle containing 1000 ml of 95% ethanol, at flow rates of 0 to 700 ml/min. The ethanol flow rate was adjusted every 12 h to maintain behavioral levels of intoxication between Ataxia-2 and Ataxia-3 as described by Majchrowicz (1975). The concentration of ethanol in the chambers was adjusted based upon this behavioral index of intoxication and upon periodic measurements of blood ethanol levels. Chamber ethanol concentration was adjusted to maintain blood ethanol levels between 250 to 350 mg/dL at the time of withdrawal. Using this procedure, blood ethanol levels increased gradually, approaching this range after several days of exposure.

The total length of ethanol exposure was 14 days. Control animals received similar handling and were housed in similar chambers for the same length of time as ethanol-exposed animals, but received no exposure to ethanol. Blood samples were also obtained periodically from chronic ethanol-naive controls so that they receive treatment similar to the chronic ethanol animals. Although this procedure for ethanol exposure does not allow for the control of nutritional deficits that might result from ethanol exposure, with very gradual increases in the ethanol vapor concentration (and subsequent increases in blood ethanol levels), ethanol animals are observed to maintain their weight within 10 to 15% of control animals for the entire exposure period. Previous studies from this laboratory (Gonzalez et al., 1989) have indicated that animals exposed to ethanol vapor with adjustments in the ethanol flow rate to obtain a gradual increase in blood ethanol levels as described above will exhibit a variety of withdrawal signs upon removal from the ethanol chamber. These signs include muscle tremor, teeth chattering, and increased audiogenic seizure sensitivity.

14.5.3   Blood Ethanol Determinations

Blood ethanol levels were determined periodically during chronic ethanol exposure and following ethanol withdrawal through the use of an Analox AM-1 Blood Alcohol Analyzer. Using this instrument, 20 µl samples of blood obtained from the dorsal tail vein are spun in a centrifuge, and 5 to10 µl of the blood plasma from each sample are then injected into the analyzer. The instrument is calibrated to known ethanol standards prior to each group of samples, and blood ethanol levels are read directly from the instrument. Blood samples are obtained from a small cut initially made at the tip of the tail, with subsequent samples obtained by gently re-opening the wound by handling the tail. Using this procedure, repeated samples can be obtained over the course of ethanol exposure without undue stress to the animal. Ethanol-naive control animals received similar treatment, so that groups did not differ with respect to this experience.

14.5.4   Ethanol Withdrawal

After chronic ethanol exposure as described previously, animals were removed from the inhalation chamber and placed in a plexiglas observation chamber, and animals were observed periodically for evidence of an ethanol withdrawal syndrome as determined by the presence or absence of the following behavioral signs (Majchrowicz, 1975; Gonzalez et al., 1989): muscle tremor, abnormal body posture, abnormal gait, pelvic and abdominal elevation, myoclonic jerks, tonic limb extension, and convulsive motor seizures. In addition, blood samples were obtained at the time of withdrawal to allow for a determination of blood ethanol concentrations as described previously. Control animals that received similar handling, but with no exposure to ethanol, were also examined. Special care was taken to avoid any unnecessary handling of the animals throughout the experimental period and to avoid exposing the animals to unnecessary afferent stimulation.

14.5.5   General Electrophysiological Recording Procedures

14.5.5.1   Hippocampal Slice Preparation

Six to eight hours after removal from the chamber, animals were anesthetized with halothane (3%) and killed by decapitation for tissue collection. Immediately after decapitation the brain was quickly ($\sim$1 min) and carefully removed to a beaker of ice-cold artificial cerebrospinal fluid (ACSF). The ACSF was prepared each morning and was composed of (in mM) 126 NaCl, 2.5 KCl, 1.24 NaH$_2$PO$_4$, 1.3 MgSO$_4$, 2.4 CaCl$_2$, 11 glucose, 26 NaHCO$_3$. The brain was blocked (in ice-cold ACSF) by removing the cerebellum and portions of the cerebral cortex and was then positioned on a vibrating microtome (Leica, Nussloch, Germany; VT-1000S). Several 350 μm transverse slices were cut from the medial part of the brain where the lamellar structure of the hippocampus was clearly visible. The slices were allowed to recover for at least one hour in a warmed (34°C), oxygenated bath of ACSF before recordings were made.

14.5.5.2   Electrophysiological Recording

After recovery in warmed ACSF, the slices were transferred to a multi-electrode recording dish (Panasonic; MED64 probe). Briefly, the MED64 probe has 64 planar recording electrodes, each 50×50 μm, arranged in an 8×8 array. The electrodes were embedded in the center of a transparent glass dish with an interelectrode distance of 300 μm. A tissue slice was placed on the floor of the dish, and extracellular recordings were obtained from the electrode array underneath the slice. After placement on the electrode probe, the slice was photographed and the probe was placed on a connector linking the MED64 probe to a multi-channel recording system. The slice, probe, and connector were then transferred into an incubator where the ambient temperature was maintained at 32 to 34°C.

Oxygenated ACSF was delivered to the tissue at a rate of 2.0 ml/min by a peristaltic pump and warmed to 34°C by an inline heater. For the remainder of the experiment, the ACSF level was maintained so that the slice could interface with a heated, humidified 95% O_2/5% CO_2 gas mixture that was blown in above the slice. During the course of recordings, the composition of the perfusion fluid was altered depending upon the specific experiment. Spontaneous activity and evoked field potentials were recorded from the 64-electrode array and the signals were filtered (1.0 to 10.0 kHz) and digitized for computer analysis at a sampling rate of 20 kHz.

14.5.5.3   Electrical Stimulation

Evoked field potentials were elicited in some experiments by application of biphasic, constant-current pulses of 0.2 ms duration through selected electrodes within the 64-electrode array. One electrode was selected in the CA_1 stratum radiatum area for stimulation of the Schaffer collateral pathway (CA_3 to CA_1 synapse) and a second stimulating electrode was selected in the perforant path near the dentate granule cell body layer (mossy fiber to CA_3 synapse). Figures 14.1 and 14.2 illustrate typical responses observed across the tissue slice with stimulation at each site. Input/output (I/O) curves were obtained by stimulating at each site with stimulus intensities between 10 μA and 60 μA. Subsequent stimulations were conducted at an intensity that initially elicited responses equal to 50% of the maximum response observed. This permitted observation of treatment effects that either increased or decreased response amplitude.

14.6   Ethanol and Presynaptic Inhibition

Cholinergic receptors mediate a variety of neuronal effects in the hippocampus, including presynaptic inhibition (Hounsgaard, 1978; Valentino and Dingledine, 1981; de Sevilla and Buno, 2003). Our in vivo studies have suggested that hippocampal cholinergic mechanisms are altered and hippocampal neuronal excitability is increased following withdrawal from chronic ethanol exposure (Gonzalez and Sun, 1992). Thus, chronic ethanol exposure and withdrawal might result in alterations in cholinergically mediated presynaptic inhibition that could contribute to hyperexcitability (Rothberg and Hunter, 1991; Rothberg et al., 1993). To investigate the possible involvement of cholinergic presynaptic inhibition

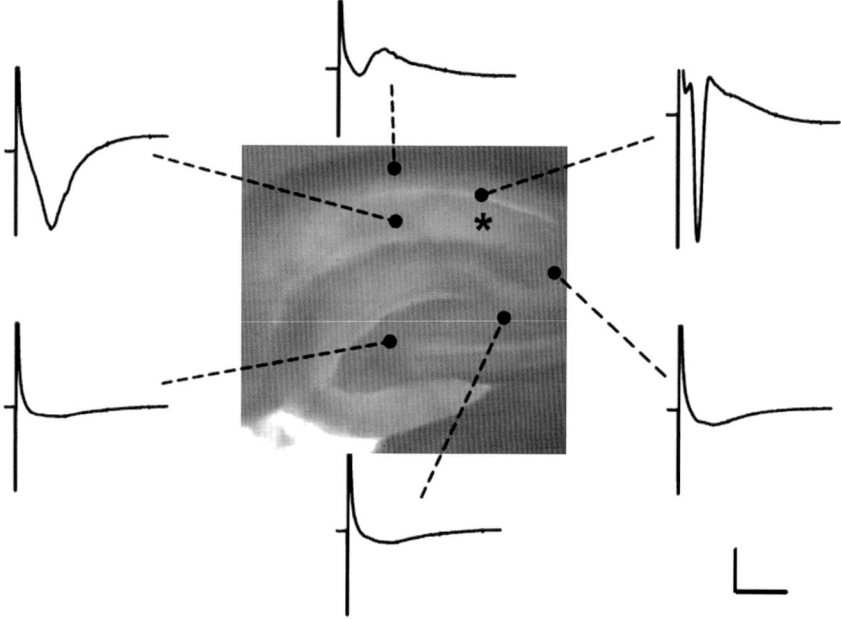

FIGURE 14.1. Sample MEA recordings from various areas across a slice of hippocampal tissue following single-pulse electrical stimulation of the Schaffer collateral pathway. The site of stimulation is indicated by an asterisk in the figure. Calibration bars indicate 500 µV and 10 msec.

in ethanol withdrawal effects, our laboratory examined the effects of perfusion with the muscarinic, cholinergic agonist carbachol. Field responses were elicited through electrical stimulation at selected sites within the hippocampus of ethanol-naive animals and of animals withdrawn after 14 days of chronic ethanol treatment. Carbachol is reported to enhance presynaptic inhibition within the hippocampus through its activation of presynaptic, muscarinic cholinergic receptors (Qian and Saggau, 1997; de Sevilla and Buno, 2003).

As a measure of the presynaptic nature of changes in hippocampal function following ethanol exposure and withdrawal, we examined the effects of carbachol perfusion on paired-pulse facilitation (PPF) in hippocampal slices. PPF was produced by presenting pairs of stimuli with interstimulus intervals between 10 and 200 msec. Responses to the second stimulus in the pair are facilitated relative to responses to the first stimulus and this effect is believed to reflect an increase in the probability of neurotransmitter release (Katz and Miledi, 1968; Miller, 1998; Zucker and Regehr, 2002). The PPF ratio (Response$_2$ amplitude/Response$_1$ amplitude) is sensitive to a variety of presynaptic manipulations and is commonly used as an indicator of a presynaptic locus of action.

In this experiment, hippocampal slices were prepared as described above and input/output curves and responses to paired-pulse stimulation were obtained during

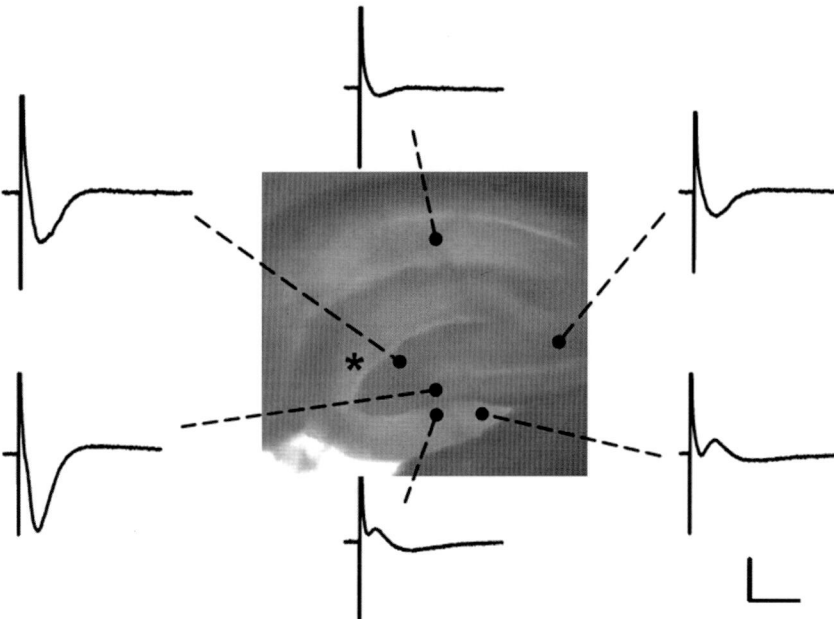

FIGURE 14.2. Distribution of field potentials following stimulation of the perforant path/mossy fiber pathways. Recordings are from the same slice illustrated in Figure 14.1. As in the previous figure, the site of electrical stimulation is indicated by an asterisk in the figure. Calibration bars indicate 225 μV and 10 msec.

perfusion with the standard ACSF mixture and also with the addition of various concentrations of carbachol to the perfusion fluid. Paired-pulse stimulation to the Schaffer collateral pathway produced a facilitation of the second response of the pair, as illustrated in Figure 14.3.

Carbachol diminished the response amplitude to Schaffer collateral stimulation but increased paired-pulse facilitation. The effects of 10 μM carbachol on the I/O curve in response to Schaffer collateral and mossy fiber stimulation are shown in Figure 14.4. The figure presents group means for data from a single electrode selected within the region of maximum response following stimulation of each of the two stimulation sites in each tissue slice. Slices from the two treatment groups (ethanol-naive and ethanol-withdrawn) did not differ in their initial responses to stimulation of either the Schaffer collateral or mossy fiber pathways.

Carbachol reduced response amplitudes following stimulation at each stimulation site. Comparison of responses observed in slices from naive and from ethanol-withdrawn animals indicated a significant reduction in the effect of carbachol on response amplitude following stimulation of the Schaffer collateral pathway in the slices from animals withdrawn from chronic ethanol treatment (Figure 14.4A). Carbachol effects on responses to mossy fiber stimulation, however, did not differ between groups (Figure 14.4B).

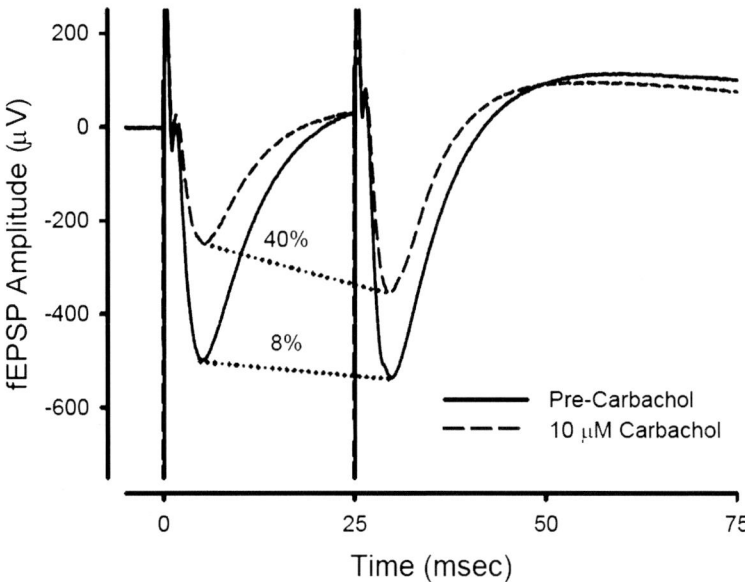

FIGURE 14.3. Evoked field EPSPs recorded from an electrode positioned within the hippocampal CA_1 field following stimulation of the Schaffer collateral pathway. Paired-pulse stimulation with an interstimulus interval of 25 msec resulted in the potentiation of the response to the second stimulus of the pair. The addition of carbachol to the perfusion fluid reduced the initial response, but further enhanced the second response.

Carbachol (10 µM), which reduced response amplitude to a single stimulus (Figure 14.4A), increased the facilitation observed in response to a second stimulus delivered to the Schaffer collateral pathway with interstimulus intervals of 50 msec or 200 msec (Figure 14.5A).

Because carbachol reduced the response to the first stimulus of each stimulus pair, this effect could also contribute to the percentage change seen after a second stimulus. Therefore, in some slices we increased the stimulus intensity after carbachol perfusion so that the initial response matched the response amplitude observed prior to carbachol perfusion. This manipulation did not alter the increase in PPF observed with carbachol.

Chronic ethanol treatment and withdrawal resulted in a significant reduction in the amount of carbachol-induced facilitation observed in response to paired-pulse stimulation of the Schaffer collateral pathway (Figure 14.5A). Paired-pulse stimulation of the perforant path/mossy-fiber pathway at these interstimulus intervals did not result in alterations in response amplitude and the paired-pulse response ratio did not change with carbachol perfusion (10 µM). Ethanol-naive and ethanol-withdrawn hippocampal slices did not differ in these latter effects; that is, neither group exhibited paired-pulse effects following mossy-fiber stimulation and this did not differ after carbachol perfusion.

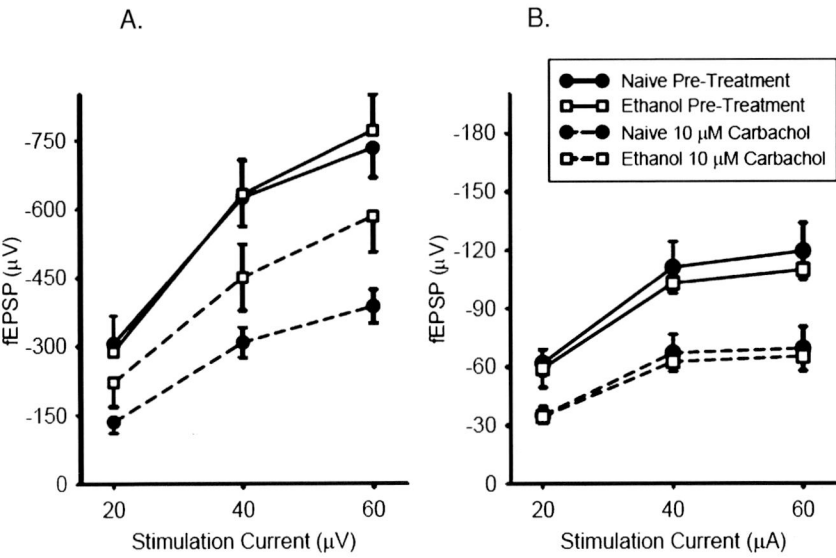

FIGURE 14.4. Input/output curves of responses from selected electrodes following (A) Schaffer collateral stimulation or (B) perforant path/mossy fiber pathway stimulation. Field EPSP amplitude increased with increasing stimulus intensity. The addition of carbachol (10 μM) to the perfusion fluid resulted in a decrease in response amplitude following stimulation at either site. Input/output curves of hippocampal slices obtained from animals following chronic ethanol treatment and withdrawal did not differ from those of ethanol-naive controls prior to carbachol treatment. The effect of carbachol following ethanol exposure and withdrawal, however, was significantly reduced in responses to stimulation of the Schaffer collateral pathway. No group differences were observed either before or after carbachol perfusion in the response to perforant path/mossy fiber pathway stimulation.

The effects observed with Schaffer collateral stimulation were also evident in alterations in the propagation of evoked responses across the slice, as illustrated in the contour plots shown in Figure 14.6. This figure presents the maximum response amplitude observed at each point across the slice following stimulation to the first and second stimulus of a pair and the normalized difference between the two, where difference is normalized relative to the amplitude of the first, control response. MATLAB (Mathworks) was used to calculate the interpolated response surface between the recordings from the 64 electrodes.

Figure 14.6 provides representations of the response to Schaffer collateral stimulation during perfusion with normal ACSF and then during perfusion with 5, 10, and 25 μM carbachol. The dose-related reductions in response amplitude and response propagation following carbachol are evident in the representations of the first response. At the same time, carbachol produced a dose-related increase in PPF as is evident in the normalized difference plots.

To quantify these effects, we determined the area of the tissue slice that responded to stimulation with a response amplitude of at least twice the

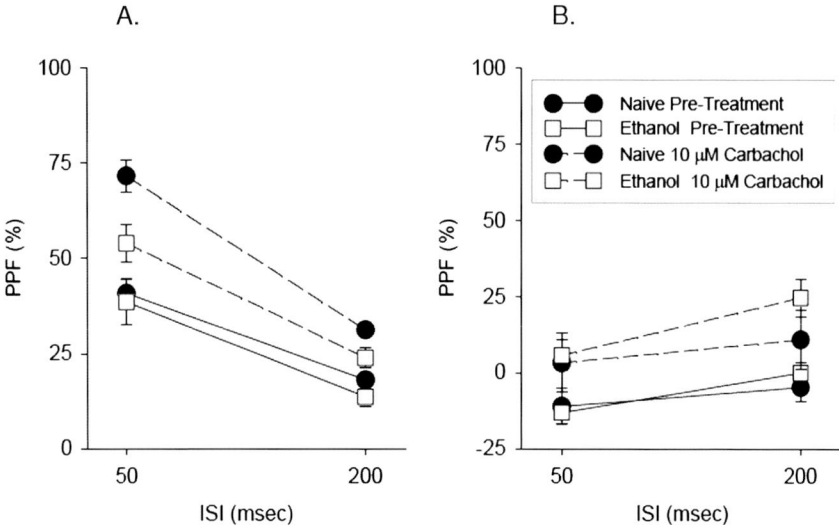

FIGURE 14.5. Paired-pulse facilitation before and after carbachol perfusion following (A) Schaffer collateral stimulation or (B) perforant path/mossy fiber pathway stimulation. PPF was primarily observed following Schaffer collateral stimulation at an interstimulus interval of 50 msec, with no facilitation seen following perforant path/mossy fiber pathway stimulation. Carbachol enhanced PPF, but this effect was reduced in slices from ethanol-exposed and withdrawn animals.

background noise level (typically <10 μV) in our recordings. This measurement provides an indication of the area of spread or amount of response propagation that was elicited by a particular stimulus. Paired-pulse effects were thus quantified in terms of the percentage of change in the amount of response propagation. Paired-pulse stimulation (ISI = 50 msec) to the Schaffer collateral pathway resulted in a relative increase in this measure of response propagation, and perfusion with carbachol resulted in a dose-related enhancement of this effect in hippocampal slices from ethanol-naive animals (Figure 14.7). Paired-pulse stimulation of the perforant path/mossy fiber projection at the same interstimulus interval did not result in a change in the area of response propagation (data not shown).

Slices from ethanol-withdrawn subjects did not differ from those of ethanol-naive animals in the area of tissue responding to stimulation prior to carbachol treatment, but showed a reduced effect of carbachol perfusion (Figure 14.7). These effects are similar to those observed in the measurements of response amplitude with paired-pulse stimulation of the Schaffer collateral pathway (Figure 14.5).

We also examined differences in response latency throughout the hippocampal slice following stimulation of the Schaffer collateral pathway. For this analysis, the locations of electrodes across the Med64 array were categorized as to their

Carbachol & PPF in Hippocampal CA₁

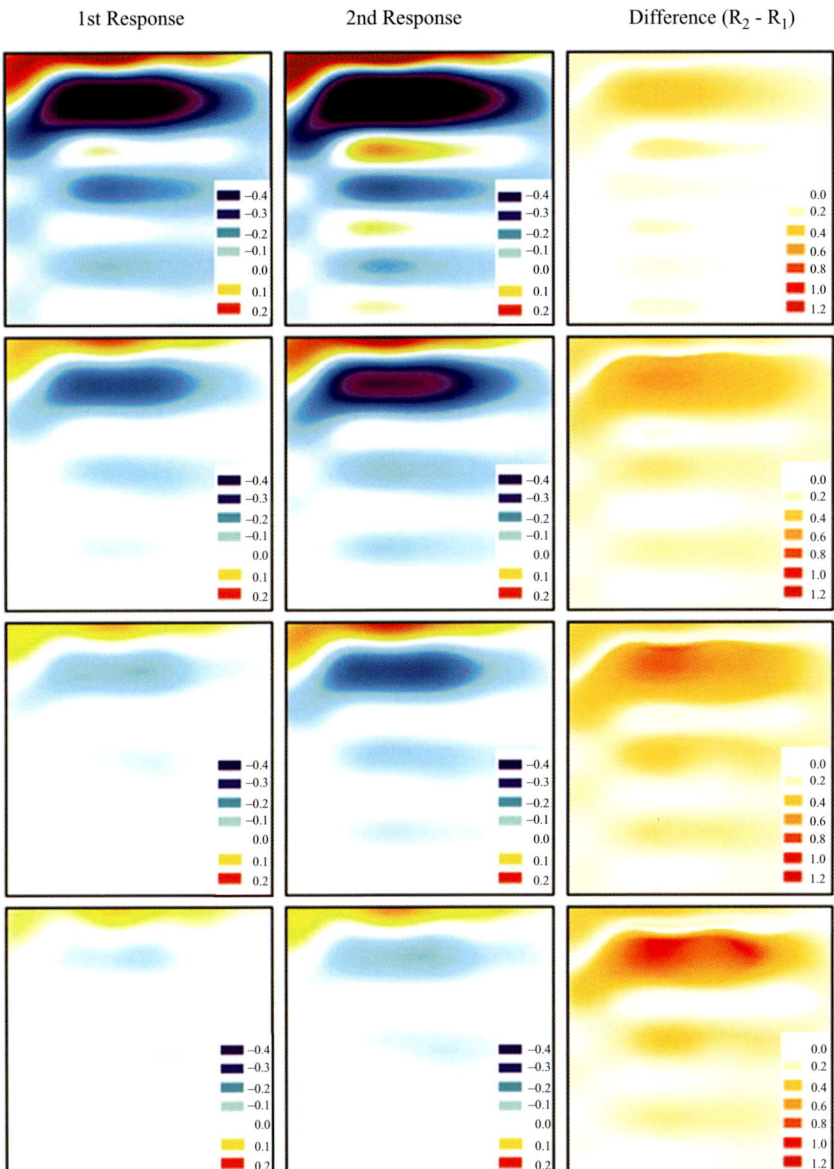

FIGURE 14.6. Contour map of maximum response amplitude (fEPSP) following Schaffer collateral stimulation across the 64-electrode array at various concentrations of carbachol. Responses to the first and second stimulus presented at an interstimulus interval of 50 msec, as well as the normalized difference between these, are represented in columns (left to right). Carbachol concentrations of 0, 5, 10, and 25 μM are presented in rows (top to bottom).

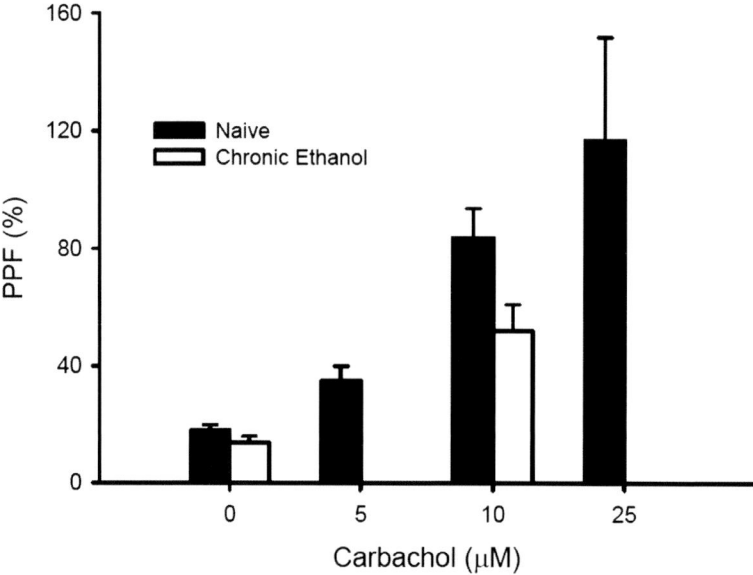

FIGURE 14.7. Percentage of change in the area of propagation following paired-pulse stimulation of the Schaffer collateral pathway with varying concentrations of carbachol. Hippocampal slices from animals withdrawn following chronic ethanol exposure and from ethanol-naive animals did not differ prior to carbachol treatment, but ethanol exposure and withdrawal significantly reduced the enhanced response propagation observed after carbachol perfusion.

position within the hippocampus for each tissue slice studied. Table 14.1 lists the subareas that were included. With Schaffer collateral stimulation, subregions within the CA_1 field showed the shortest response latency, followed by those within the CA_2, CA_3, and then dentate gyrus. Figure 14.8 shows a comparison of response propagation latency between hippocampal slices from ethanol-exposed and ethanol-naive animals.

In this figure, electrode location is presented on the x-axis in the order of short to long response latency as observed in control slices and mean response latency for electrodes located within these regions is presented for the two groups. Ethanol exposure and withdrawal did not significantly alter response propagation latencies.

14.7   Carbachol and Cholinergic Receptor Blockade

To determine whether carbachol's effect on evoked potential amplitude and paired-pulse facilitation was due to muscarinic mechanisms, we examined the effects of carbachol in the presence of the muscarinic antagonist atropine. Atropine (10 μM)

TABLE 14.1. Subregions within the
Hippocampus used in the identification of
electrode location.

Cerebral cortex	CC
Entorhinal cortex	ER
Fimbria	FI
Granular cells	GC
Molecular layer (inner)	IM
Molecular layer (outer)	OM
Polymorphic layer	PM
Pyramidal cells	PC
Stratum lacunosum Moleculare	SM
Stratum lucidum	SL
Stratum oriens	SO
Stratum radiatum	SR

when administered alone had no effect on evoked responses or on PPF (data not shown). However, this concentration of atropine completely reversed carbachol's effects on evoked potentials and paired-pulse facilitation when added to carbachol-containing perfusate (Figure 14.9). Thus, both the reduction in initial response amplitude during carbachol perfusion and the enhancement of PPF were prevented by the presence of atropine in the perfusion fluid.

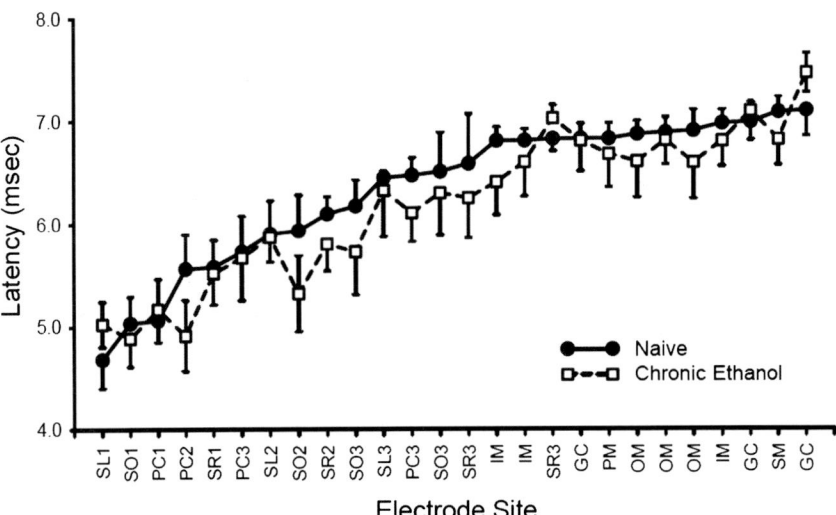

FIGURE 14.8. Latency to maximum response (fEPSP) at electrode sites across the Med64 MEA following Schaffer collateral stimulation. Electrode sites are labeled on the x-axis as in Table 14.1, and numbers appended to the site designation indicate electrodes in CA fields 1, 2, or 3. Electrode sites are sorted from left to right along the x-axis in order of ascending response latency based on the ethanol-naive group mean.

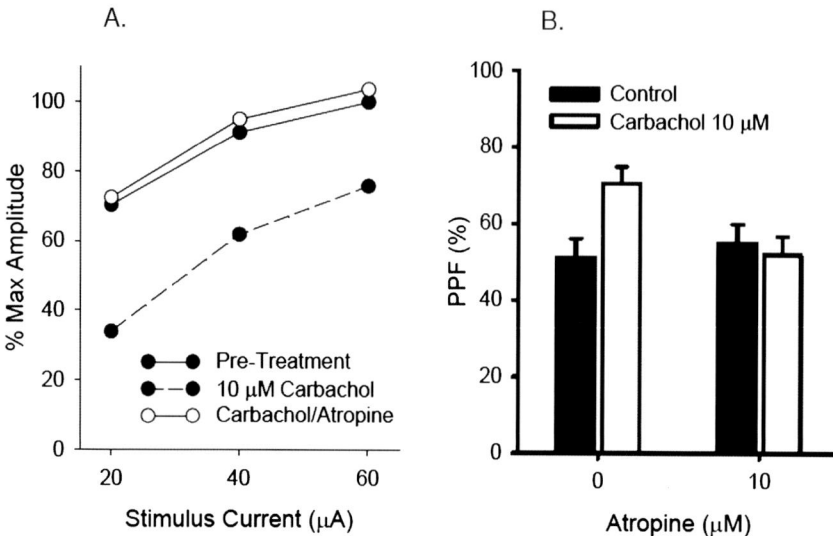

FIGURE 14.9. Effect of atropine (10 μM) and carbachol (10 μM) on (A) input/output curves and (B) paired-pulse facilitation after Schaffer collateral stimulation at an interstimulus interval of 50 msec. Addition of atropine to perfusion fluid containing carbachol reversed the effects of carbachol on both measures.

14.8   Chronic Ethanol and Extracellular Calcium

Although the specific mechanisms that mediate PPF are still under investigation, the most popular hypothesis suggests the involvement of residual calcium as a mediator of PPF (Miller, 1998; Zucker and Regehr, 2002). This hypothesis proposes that, following an increase in intracellular calcium through the activation of voltage-gated calcium channels by a neuronal impulse, the response to a second stimulus is enhanced because of the presence of residual calcium that increases the probability of neurotransmitter release. Because calcium entry is concentration-dependent, we investigated the involvement of calcium channel activity in the effects of ethanol withdrawal on PPF by manipulating the extracellular calcium concentration of the perfusion fluid. Lowering the concentration of calcium in the perfusion fluid applied to the hippocampus in tissue slices from ethanol-naive control animals produced effects similar to those of carbachol, that is, a reduction in response amplitude, but an enhancement of PPF (Figure 14.10).

These effects varied with the concentration of calcium in the perfusion fluid. Manipulation of the calcium concentration in the perfusion fluid applied to tissue slices from ethanol-exposed and withdrawn animals produced similar effects and these did not differ from the effects seen in ethanol-naive controls (Figure 14.11). These results suggest that differences in responses to carbachol in the hippocampus following ethanol withdrawal are not the result of alterations in extracellular calcium entry. Intracellular sources of calcium, however, might also contribute

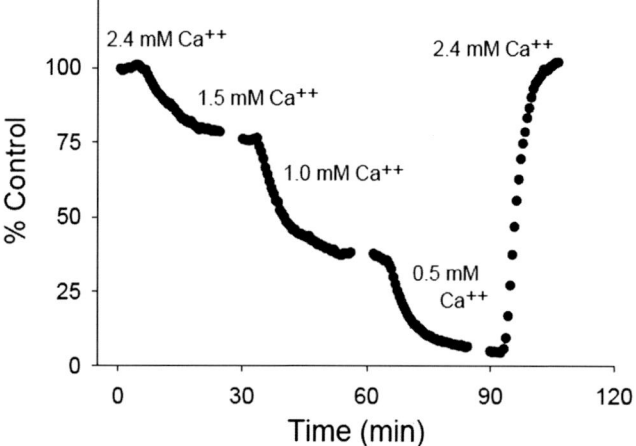

FIGURE 14.10. Effect of altering calcium concentration of perfusion fluid on fEPSP amplitude. Data are shown as a percentage of response amplitude during perfusion with standard ACSF (2.4 mM Ca^{++}).

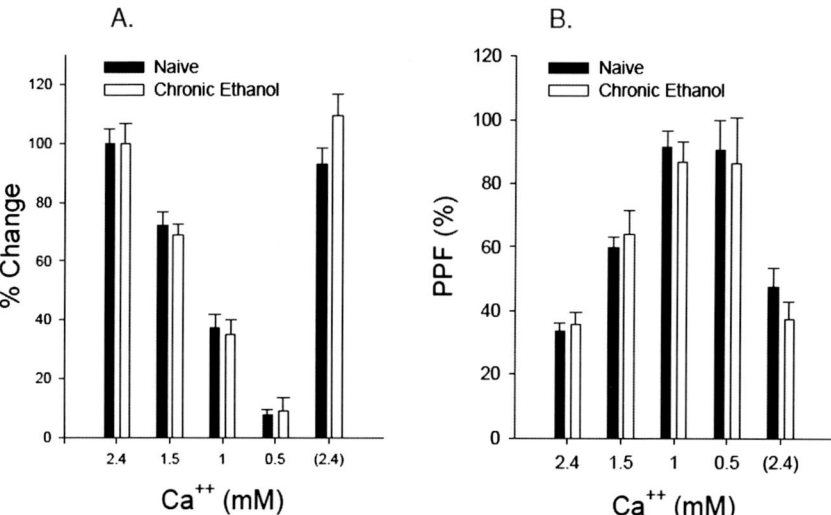

FIGURE 14.11. Calcium concentration effects and chronic ethanol treatment and withdrawal. Hippocampal slices from ethanol-naive animals were compared to slices from animals withdrawn from chronic ethanol treatment. The figure illustrates the effect of lowering the calcium concentration and a subsequent return to the baseline concentration (2.4 mM). Data are shown for effects on (A) percentage of change in fEPSP amplitude relative to baseline and (B) paired-pulse facilitation at an interstimulus interval of 50 msec.

to neurotransmitter release and mechanisms involving these sources of calcium might be altered following chronic ethanol exposure and withdrawal.

14.9   Gender Effects: Chronic Ethanol and Estradiol

Ethanol-naive female animals differ from males in several areas of hippocampal morphology (Juraska et al., 1989; Gould et al., 1990), in acetylcholine levels (Hortnagl et al., 1993), and in the activity of enzymes that regulate acetylcholine synthesis and metabolism (Loy and Sheldon, 1987). Furthermore, acetylcholine levels, hippocampal morphology, and function have been observed to vary with the estrus cycle of female animals and with circulating levels of sex hormones (Teyler et al., 1980; Lapchak et al., 1990; Woolley and McEwen, 1992; Hortnagl et al., 1993). Although most studies of the effects of chronic ethanol exposure and withdrawal have been conducted in male animals, it has been reported that females exhibit larger changes than do males in muscarinic cholinergic receptors in several brain areas, including the hippocampus, after chronic ethanol exposure (Witt et al., 1986). Gender differences have also been observed in bicuculline-induced seizure sensitivity following withdrawal from chronic ethanol exposure (Devaud et al., 1995). In addition, a variety of studies indicates significant gender differences in hippocampal function and development as well as in cholinergic neurotransmission, such that effects of ethanol may also be sensitive to gender and to levels of circulating hormones.

The study reported here used 60 day-old female Sprague–Dawley rats. Animals were ovariectomized and received implants of slow-release pellets containing either estradiol (17-beta-estradiol, 0.1 mg/pellet) or placebo 7 days prior to any treatment. To verify the hormonal status of female animals, vaginal epithelial cells were obtained through vaginal lavage and these were examined by light microscopy to determine the stage of estrus. Following ovariectomy, females receiving estradiol implants were maintained in a state of estrus and those receiving placebo implants were maintained in a state of diestrus. Animals then received 14 days exposure to ethanol through vapor inhalation. Blood ethanol concentrations during the last 7 days of ethanol exposure averaged 250 to 350 mg/dl. Additional groups received similar handling but no ethanol exposure.

Hippocampal tissue slices were obtained as described above, six hours after withdrawal from chronic treatment (ethanol or naive). Extracellular field potentials were recorded in response to brief electrical stimulation to several sites within the hippocampal network (mossy fibers and Schaffer collaterals), before and after perfusion with carbachol (10 μM). As we observed in male subjects, paired-pulse stimulation resulted in a facilitation of responses following Schaffer collateral pathway stimulation (Figure 14.12A).

Treatment groups did not differ significantly in these effects prior to carbachol treatment. Slices obtained from animals following chronic ethanol treatment and withdrawal showed a significant reduction in the carbachol enhancement of PPF, but only in those females that received estradiol replacement. Ovariectomy without

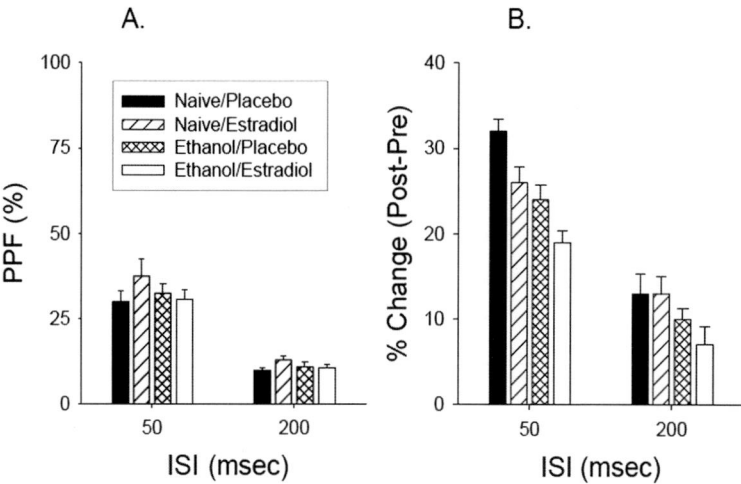

FIGURE 14.12. Chronic estradiol and chronic ethanol effects on paired-pulse facilitation with Schaffer collateral stimulation both (A) before and (B) after addition of carbachol (10 μM) to the perfusion fluid.

hormone replacement prevented the effect of ethanol exposure and withdrawal (Figure 14.12B) to reduce carbachol effects on PPF.

14.10   Ethanol Effects on Carbachol-Induced Rhythmic Oscillations

An additional test of chronic ethanol effects on hippocampal function examined group differences in carbachol-induced rhythmic activity in the tissue slice. Carbachol perfusion has sometimes been observed to induce rhythmic oscillations in the activity of the hippocampus (Fisahn et al., 1998; Shimono et al., 2000). This effect requires an intact hippocampal circuit such that oscillations can be prevented by acute disruption of the CA_3 to CA_1 pathways. An example of rhythmic oscillation induced by 20 μM carbachol added to the ACSF perfusion fluid is presented in Figure 14.13. This figure illustrates a carbachol-induced oscillation at approximately 30 Hz. Chronic ethanol treatment and withdrawal resulted in a significant reduction in the frequency of oscillation observed during perfusion with carbachol (Figure 14.14), suggesting that ethanol withdrawal is accompanied by alterations in hippocampal network properties and intercellular interactions.

14.11   Summary and Conclusions

Multi-electrode arrays provide a unique view of in vitro hippocampal function and permit evaluation of treatment effects in novel ways that may contribute to our

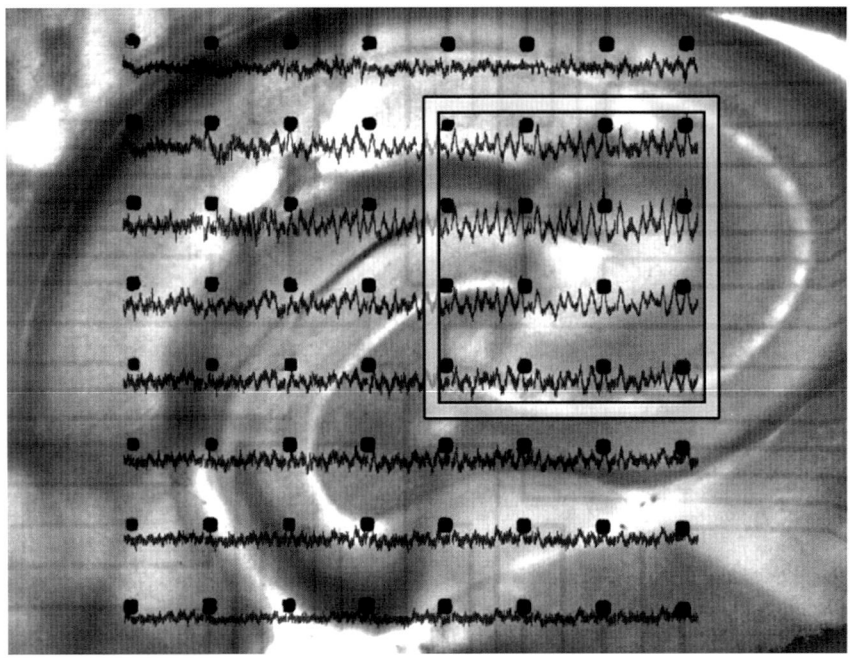

FIGURE 14.13. Sample recordings illustrating carbachol-induced rhythmic oscillation. Segments of electrical activity obtained at each electrode site are shown superimposed over the tissue slice. The box in the figure surrounds areas showing rhythmic oscillation.

A.

B.

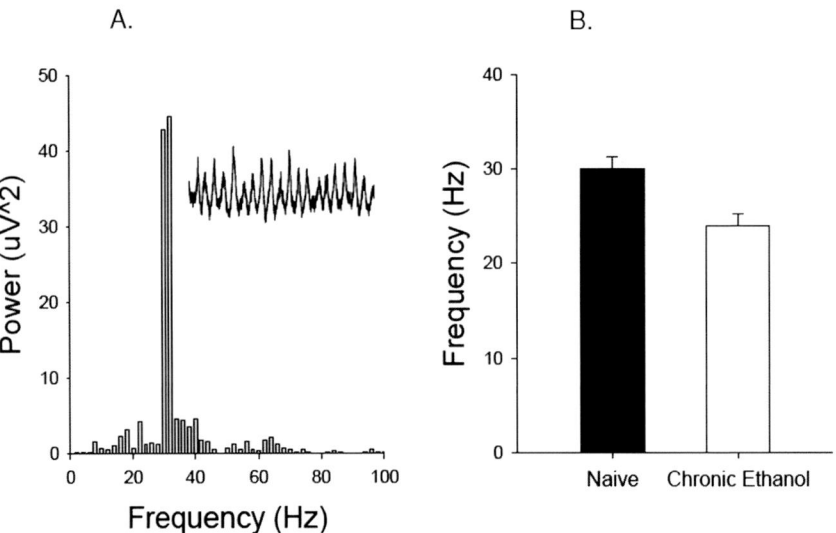

FIGURE 14.14. Quantitative analysis of carbachol-induced rhythmic oscillation and effects of chronic ethanol exposure and withdrawal. Fourier analysis of a segment of activity (A) indicates a dominant frequency in the illustrated sample of approximately 30 Hz. Hippocampal slices from animals withdrawn from chronic ethanol treatment (B) showed a lower frequency of oscillation compared to slices from ethanol-naive animals.

understanding of the central nervous system and the adaptations that occur with chronic alterations in the neuronal environment. Our studies comparing slices from ethanol-naive and ethanol-exposed animals support the hypothesis of altered hippocampal cholinergic function following withdrawal from chronic ethanol treatment. The use of the Med64 electrode array allows the examination of activity throughout the hippocampal trisynaptic circuit so that interactions between regions of the hippocampus can be investigated. The studies that we have described above suggest that a selective reduction occurs in presynaptic inhibition in the CA_1 field following acute withdrawal from chronic ethanol treatment. This is consistent with our earlier reports of electrographic hyperexcitability and altered sensitivity to cholinergic manipulations (Gonzalez, 1985; Gonzalez and Sun, 1992). The advantage of the multi-electrode array is the ability to examine regional and circuit pathway differences within a single tissue slice and to perform measurements of response propagation that are difficult in other in vitro recording preparations.

Acknowledgments. These studies were supported in part by NIAAA grants AA09959 and AA12283 to LPG.

References

Abraham, W.C., Hunter, B.E., Zornetzer, S.F., and Walker, D.W. (1981). Augmentation of short-term plasticity in CA1 of rat hippocampus after chronic ethanol treatment. *Brain Res.* 221: 271–287.

Arendt, T., Allen, Y., Marchbanks, R.M., Schugens, M.M., Sinden, J., Lantos, P.L., and Gray, J.A. (1989). Cholinergic system and memory in the rat: Effects of chronic ethanol, embryonic basal forebrain brain transplants and excitotoxic lesions of cholinergic basal forebrain projection system. *Neuroscience* 33: 435–462.

Arendt, T., Bigl, V., Arendt, A., and Tennstedt, A. (1983). Loss of neurons in the nucleus basalis of meynert in Alzheimer's disease, paralysis agitans, and Korsakoff's disease. *Acta Neuropathol.* 61: 101–108.

Buzsaki, G., Ponomareff, G.L., Bayardo, F., Ruiz, R., and Gage, F.H. (1989). Neuronal activity in the subcortically denervated hippocampus, A chronic model for epilepsy. Neuroscience 28(3): 527–538.

de Sevilla, F.D. and Buno, W. (2003). Presynaptic inhibition of Schaffer collateral synapses by stimulation of hippocampal cholinergic afferent fibres. *Eur. J. Neurosci.* 17(3): 555–558.

Devaud, L.L., Purdy, R.H., and Morrow, A.L. (1995). The neurosteroid, 3-hydroxy-5 pregnan-20-one, protects against bicuculline-induced seizures during ethanol withdrawal in rats. *Alcohol. Clin. Exper. Res.* 19: 350–355.

Devenport, L.D. and Hale, R.L. (1989). Contributions of hippocampus and neocortex to the expression of ethanol effects. *Psychopharmacology* 99: 337–344.

Devenport, L.D., Devenport, J.A., and Holloway, F.A. (1981). Necessity of the hippocampus for alcohol's indirect but not direct behavioral action. *Behav. Neural Biol.* 33: 476–87.

Durand, D. and Carlen, P.L. (1984). Decreased neuronal inhibition in vitro after long-term administration of ethanol. *Science* 224: 1359–1361.

Erickson, C.K. and Graham, D.T. (1973). Alteration of cortical and reticular acetylcholine release by ethanol in vivo. *J. Pharmacol. Exp. Ther.* 185: 583–593.

Faingold, C.L. (2004). Emergent properties of CNS neuronal networks as targets for pharmacology: Application to anticonvulsant drug action. *Prog. Neurobiol.* 72(1): 55–85.

Fisahn, A., Pike, F.G., Buhl, E.H., and Paulsen, O. (1998). Cholinergic induction of network oscillations at 40 Hz in the hippocampus in vitro. *Nature* 394(6689): 186–189.

Goldstein, A. (1979). Recent advances in basic research relevant to drug abuse. In: Dupont, R.I., Goldstein, A., O'Donnel, J., eds., *Handbook on Drug Abuse*. National Institute on Drug Abuse, Washington, DC, pp. 439–446.

Gonzalez, L.P. (1985). Changes in physostigmine-induced hippocampal seizures during ethanol withdrawal. *Brain Res.* 335: 384–388.

Gonzalez, L.P. and Gable, J.J., III. (1990). Ethanol increases free amino acid levels in hippocampus. *Pharmacol. (Life Sci. Adv.)* 9: 749–755.

Gonzalez, L.P. and Sun, R. (1992). Electrophysiological and neurochemical effects of ethanol on hippocampal CA3 neurons. In: Watson, R., ed., *Alcohol and Neurobiology: Brain Development and Hormone Regulation*. CRC Press, New York, pp. 201–220.

Gonzalez, L.P., Czachura, J.F., and Brewer, K.W. (1989). Spontaneous versus elicited seizures following ethanol withdrawal: Differential time course. *Alcohol* 6: 481–487.

Gould, E., Westlind-Danielsson, A., Frankfurt, M., and McEwen, B.S. (1990). Sex differences and thyroid hormone sensitivity of hippocampal pyramidal cells. *J. Neurosci.* 10(3): 996–1003.

Hortnagl, H., Hansen, L., Kindel, G., Schneide, B., Tamer, A.E., and Hanin, I. (1993). Sex differences and estrous cycle—Variations in the af64a- induced cholinergic deficit in the rat hippocampus. *Brain Res. Bull.* 31: 129–134.

Hounsgaard, J. (1978). Presynaptic inhibitory action of acetylcholine in area CA1 of the hippocampus. *Exp. Neurol.* 62(3): 787–797.

Hughes, R.N. (1982). A review of atropinic drug effects on exploratory choice behavior in laboratory rodents. *Behav. Neural Biol.* 34: 5–41.

Hunt, W.A. and Majchrowicz, E. (1979). Alterations in neurotransmitter function after acute and chronic treatment. In: Majchrowicz, E. and Noble, E.P., eds., *Biochemistry and Pharmacology of Ethanol*. Plenum Press, New York, pp. 167–185.

Isbell, H., Fraser, H.F., Wilker, A., Belleville, R.E., and Eisenman, A.J. (1955). An experimental study of the etiology of "rum fits" and delirium tremens. *J. Stud. Alcohol* 16: 1–33.

Juraska, J.M., Fitch, J.M., and Washburne, D.L. (1989). The dendritic morphology of pyramidal neurons in the rat hippocampal CA3 area. II. Effects of gender and the environment. *Brain Res.* 479: 115–119.

Kalant, H. (1975). Direct effects of ethanol on the nervous system. *Fed. Proc.* 34: 1930–1941.

Katz, B. and Miledi, R. (1968). The role of calcium in neuromuscular facilitation. *J. Physiol.* 195(2): 481–492.

Kovanen, J., Pilke, A., and Partinen, M. (1984). Status epilepticus and alcohol withdrawal. *Acta Neurol. Sci.* 69: 97–98.

Lahti, R.A. (1975). Alcohol, aldehydes and biogenic amines. In: Majchrowicz, E., ed. *Biochemical Pharmacology of Ethanol*. Plenum Press, New York, pp. 239–253.

Lapchak, P.A., Araujo, D.M., Quirion, R., and Beaudet, A. (1990). Chronic estradiol treatment alters central cholinergic function in the female rat: Effect on choline acetyltransferase activity, acetylcholine content, and nicotinic autoreceptor function. *Brain Res.* 525: 249–255.

Little, H.J. (1999). The contribution of electrophysiology to knowledge of the acute and chronic effects of ethanol. *Pharmacol. Therapeut.* 84(3): 333–353.

Loy, R. and Sheldon, R.A. (1987). Sexually dimorphic development of cholinergic enzymes in the rat septohippocampal system. *Dev. Brain Res.* 34: 156–160.

Majchrowicz, E. (1975). Induction of physical dependence upon ethanol and the associated behavioral changes in rats. *Psychopharmacologia* 43: 245–254.

Miller, R.J. (1998). Presynaptic receptors. *Ann. Rev. Pharmacol. Toxicol.* 38: 201–227.

Morgan, E.P. and Phillis, J.W. (1975). Effects of ethanol on acetylcholine release from brain of unanesthetized cats. *Gen. Pharmacol.* 6: 281–284.

Parker, T.H., Roberts, R.K., Henderson, G.I., Hoyumpa, A.M., Schmidt, D.E., and Schenker, S. (1978). The effects of ethanol on cerebral regional acetylcholine concentration and utilization. *Proc. Soc. Exper. Biol. Med.* 159: 270–275.

Qian, J. and Saggau, P. (1997). Presynaptic inhibition of synaptic transmission in the rat hippocampus by activation of muscarinic receptors: Involvement of presynaptic calcium influx. *Br. J. Pharmacol.* 122(3): 511–519.

Rothberg, B.S. and Hunter, B.E. (1991). Chronic ethanol treatment differentially affects muscarinic receptor responses in rat hippocampus. *Neurosci. Lett.* 132(2): 243–246.

Rothberg, B.S., Yasuda, R.P., Satkus, S.A., Wolfe, B.B., and Hunter, B.E. (1993). Effects of chronic ethanol on cholinergic actions in rat hippocampus: Electrophysiological studies and quantification of m1-m5 muscarinic receptor subtypes. *Brain Res.* 631(2): 227–234.

Shimono, K., Brucher, F., Granger, R., Lynch, G., and Taketani, M. (2000). Origins and distribution of cholinergically induced beta rhythms in hippocampal slices. *J. Neurosci.* 20(22): 8462–8473.

Sinclair, J.G. and Lo, G.F. (1978). Acute tolerance to ethanol on the release of acetylcholine from the cat cerebral cortex. *Canad. J. Physiol. Pharmacol.* 56: 668–670.

Tarter, R.E. (1975). Brain damage associated with chronic alcoholism. *Dis. Nerv. Syst.* 36: 185–187.

Teyler, T.J., Vardaris, R.M., Lewis, D., and Rawitch, A.B. (1980). Gonadal steroids: Effects on excitability of hippocampal pyramidal cells. *Science* 209: 1017–1018.

Valentino, R.J. and Dingledine, R. (1981). Presynaptic inhibitory effect of acetylcholine in the hippocampus. *J. Neurosci.* 1(7): 784–792.

Veatch, L.M. and Gonzalez, L.P. (1996). Repeated ethanol withdrawal produces site-dependent increases in EEG spiking. *Alcohol. Clin. Exper. Res.* 20: 262–267.

Veatch, L.M. and Gonzalez, L.P. (1997). Chronic ethanol retards kindling of hippocampal area CA3. *NeuroReport* 8(8): 1903–1906.

Veatch, L.M. and Gonzalez, L.P. (1999). Repeated ethanol withdrawal delays progression from focal to generalized seizures in hippocampal kindling. *Alcohol. Clin. Exper. Res.* 23(7): 1145–1150.

Veatch, L.M. and Gonzalez, L.P. (2000). Nifedipine alleviates alterations in hippocampal kindling following repeated ethanol withdrawal. *Alcohol. Clin. Exper. Res.* 24(4): 484–491.

Victor, M. (1953). The effect of alcohol on the nervous system. *Res. Publ. Assoc. Res. Nerv. Ment. Dis.* 32: 526–573.

Walker, D.W., Hunter, B.E., and Abraham, W.C. (1981). Neuroanatomical and functional deficits subsequent to chronic ethanol administration in animals. *Alcohol. Clin. Exp. Res.* 5: 267–282.

Weiner, J.L. (2002). Electrophysiological assessment of synaptic transmission in brain slices. In: Liu, Y. and Lovenger, D.M., eds., *Methods in Alcohol-Related Neuroscience Research*. CRC Press, New York, pp. 191–218.

Wheal, H.V. (1989). Function of synapses in the CA1 region of the hippocampus: their contribution to the generation or control of epileptiform activity. *Comp. Biochem. Physiol.* 93A: 211–220.

Whittington, M.A. and Little, H.J. (1990). Patterns of changes in field potentials in the isolated hippocampal slice on withdrawal from chronic ethanol treatment of mice in vivo. *Brain Res.* 523: 237–244.

Witt, E.D., Naitione, C.R., and Hanin, I. (1986). Sex differences in muscarinic receptor binding after chronic ethanol administration in the rat. *Psychopharmacology* 90: 537–542.

Woolley, C.S. and McEwen, B.S. (1992). Estradiol mediates fluctuation in hippocampal synapse density during the estrous cycle in the adult rat. *J. Neurosci.* 12(7): 2549–2554.

Zucker, R.S. and Regehr, W.G. (2002). Short-term synaptic plasticity. *Annu. Rev. Physiol.* 64: 355–405.

15
Applications of Multi-Electrode Array System in Drug Discovery Using Acute and Cultured Hippocampal Slices

MICHEL BAUDRY, MAKOTO TAKETANI, AND MICHAEL KRAUSE

15.1 Introduction

There are currently only limited technologies for discovering, classifying, and testing compounds that could significantly affect cognitive performance in humans. This chapter describes ways in which multi-electrode arrays can successfully be used in this context. The proposed approach emerged from collaborations between research groups at the University of California, Irvine (UCI), the University of Southern California (USC), and Tensor Biosciences. The academic groups have, for a number of years, been developing protocols and analytical software for activating and analyzing electrophysiological responses generated by complex networks in mammalian CNS. This approach represents an effort to record the nearly second-long events that are possibly the substrate of simple cognitive actions; it can also be seen as an attempt to create an experimental platform for practical applications of neural network research. Tensor Biosciences is a startup company that was founded by researchers at UCI and USC and is pursuing a multi-year contract with Matsushita Electric Industrial Co., Ltd. (Panasonic) to develop software for Panasonic's multi-electrode array system (MED64 System) and also build drug discovery platforms by using the MED64 System. Panasonic and its subsidiary, Alpha MED Sciences, have been gradually evolving turnkey hardware and software for stimulating and recording from 64 electrodes placed beneath a brain slice.

Biologically or chemically induced changes in network behavior are ultimately a reflection of effects on synaptic and extrasynaptic activity in brain networks. The latter are not easily predicted by the agent's physiological actions at the molecular and cellular levels. Rather, networks are premier examples of complex systems in which small changes in initial conditions can have large and unexpected consequences. It seems reasonable to assume that actions of psychoactive agents at the network level are substantially larger in magnitude than their actions on individual synapses. This point has been experimentally confirmed for ampakines, a class of compounds that positively modulate AMPA-type glutamate receptors. For instance, ampakine-induced changes on the hippocampal trisynaptic loop were severalfold greater in magnitude than changes in monosynaptic responses within this circuit (Sirvio et al., 1996). Moreover, compound concentrations necessary to enhance polysynaptic potentials proved to be about four times smaller than those required to increase monosynaptic field EPSPs (Sirvio et al., 1996).

In addition, ascending modulatory systems such as the cholinergic or the monoaminergic systems widely influence cortical network activity. In in vitro preparations, actions of modulatory systems on cortical networks can be mimicked by pharmacological compounds acting pre- or postsynaptically in the target areas of these neuromodulatory systems. We address this issue in this chapter, thereby emphasizing multi-electrode arrays as an important tool to identify and evaluate novel neuropharmacological compounds.

15.2   Brain-on-a-Chip^{TM}

The Brain-on-a-Chip^{TM} technology tests effects of pharmacological compounds on neuronal networks by using Panasonic's MED64 System to monitor network activity in acute or cultured brain slices. The selected preparations display striking similarities to certain network activities as seen in vivo. Thus, studying compounds using multi-electrode array approaches can be used to predict certain in vivo effects of compounds that could not be studied using alternative in vitro techniques. The generally higher throughput of in vitro approaches as compared to in vivo techniques makes multi-electrode array systems a suitable approach for screening larger numbers of compounds. Multi-electrode array technologies provide at least four key advantages over classical technologies.

1. *Use of cultured slices for long-term recordings:* Traditionally, slice electrophysiology experiments are performed in the short period of time during which recording can take place (typically up to 4 to 6 h after slice preparation), which contrasts sharply with the long chronic recordings obtained under in vivo conditions.
2. *Use of co-culture of slices from different brain regions incorporating modulatory systems to various networks:* Another limitation of slice technology is that, in most cases, slice networks have lost a number of inputs, and, in particular, inputs originating from diffuse neuromodulatory systems, such as noradrenergic, serotonergic, and cholinergic neurons, which profoundly modify the in vivo operation of networks. By using co-cultures of slices from various brain regions it is possible to incorporate modulatory influences to many networks.
3. *Use of acute slices exhibiting spontaneous or pharmacologically induced EEG-like rhythms:* In vitro preparations have traditionally been used to create conditions for studying a single synaptic or cellular response, preferably in a single neuron type in isolation. In vivo experiments on the other hand show that many compounds or behavioral situations clearly affect EEG activity. We have developed protocols to recreate EEG-like activity in brain slices on the multi-electrode array similar to those found in vivo. Such protocols therefore

represent a useful extension of in vitro electrophysiology, and may yield more realistic and predictive information regarding the effects of compounds on various network parameters (see also Chapter 18 by Colgin for further illustration of this point).

4. *Use of spatial information in in vitro recordings:* Finally, in vivo recordings have recently evolved toward recording larger and larger numbers of units in an attempt to understand the behavior of large ensembles of neurons. The technology described in this volume has also evolved toward recording a large ensemble of neurons through the use of complex arrays of electrodes (typically 8×8 arrays of geometrically positioned electrodes).

These points are illustrated below where we describe the use of cultured slices for first testing neuroprotective compounds against excitotoxicity, and then for studying long-term plasticity phenomena such as long-term potentiation.

15.3   Slice Culture on a Chip: Neuroprotective Effects of NMDA-R Antagonists

In order to address the question of how neurotoxic agents affect synaptic responses in hippocampal pathways, stable long-term monitoring of synaptic transmission is required. We cultured hippocampal slices directly on the surface of the MED probe, which allowed us to conduct long-term extracellular recordings in CA1 and other hippocampal subfields. Directly culturing hippocampal slices on polyethylenimine-coated probes resulted in a tight adhesion of the slices on the probes, and provided for stable maintenance of the stimulating and recording sites. Typically, one pair of electrodes (one stimulating and one recording electrode) was selected in CA1 to stimulate Schaffer collateral afferents and to extracellularly record excitatory postsynaptic field potentials (fEPSPs) with the best possible quality.

Culturing slices directly on the MED probes bears some significant advantages. For instance, in traditional electrophysiological approaches using glass electrodes it was impossible to track a specific excitatory pathway over many days in cultured slices because of the difficulty of positioning the electrodes in exactly the same manner day by day (although some recent work by Dean Buonomano's laboratory has used some clever approaches to address this issue). A cultured slice on the MED probe elegantly circumvents this problem, as the position of the electrodes relative to the fiber pathways does not change over days. Therefore, the MED probe is exquisitely suited for studying slice cultures.

In agreement with prior studies (Stoppini et al., 1991), slices prepared from postnatal day 10 to 12 rats provided optimal results when tested electrophysiologically after culturing on the MED probe in static conditions with interface levels of culture medium. An example of a slice cultured on the MED probe for one week in vitro as well as positions of the electrodes is depicted in Figure 15.1A. Under our experimental conditions, no obvious morphological changes were noticed during

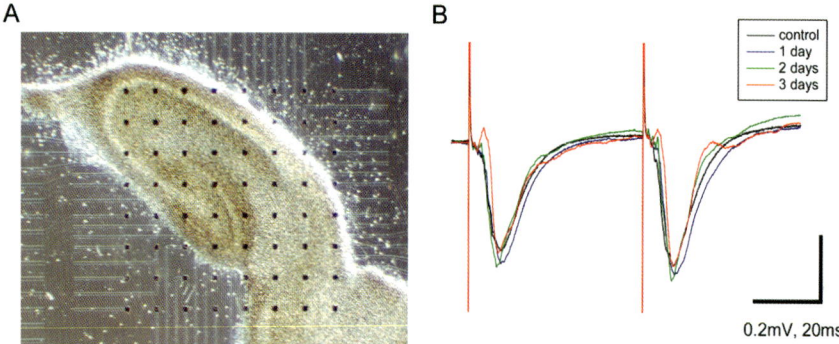

FIGURE 15.1. Experimental design used to perform chronic recording in cultured hippocampal slices: (A) hippocampal slices from 11-day-old rats were cultured and maintained on the MED-64 probes for extended periods of time; (B) paired-pulse stimulations were delivered to various electrodes and extracellular field potentials were recorded. (Reprinted with authorization from Shimono et al., 2001.)

at least two weeks in culture on MED probes. Slight to moderate migration of the cells out of the slice was observed, normally starting after 14 to 20 days. Some flattening of the slices also occurred after two weeks in vitro, which did not appear to interfere with fEPSP recordings. During the first 7 days of in vitro culturing, we observed some increase in the amplitude of fEPSPs; however after 7 to 10 days, the responses stabilized and remained stable over subsequent recording periods (Figure 15.1B). These results were consistent with the findings of Muller et al. As a result, all experiments reported here were done after at least 10 days of culturing slices on the MED probes.

Numerous studies have used selective glutamate receptor agonists for producing pathological conditions in the brain in order to study mechanisms of neuronal death and neurotoxicity. In our case, NMDA and AMPA were chosen as excitotoxic agents to compare electrophysiological parameters with traditional parameters of neurodegeneration, and to gain additional information regarding the mechanisms contributing to the loss of synaptic function in excitotoxic hippocampal injury. Long-term incubation of cultured hippocampal slices in the presence of either NMDA or AMPA resulted in concentration-dependent decreases in synaptic responses. Representative I/O relationships and fEPSP recordings before and at various times after incubation in the presence of NMDA (10 μM) or AMPA (1 μM) are shown in Figure 15.2. After 3 h of incubation in the presence of 10 μM NMDA, the maximal amplitude of synaptic responses was reduced to 18 ± 4% ($n = 3$) of control values and this decrease did not change significantly during subsequent incubation periods (Figure 15.2A). When 1 μM AMPA was applied for 3 h, the decrease in synaptic responses was smaller, and the responses continued to decrease at subsequent time points, and stabilized after one day (Figure 15.2B).

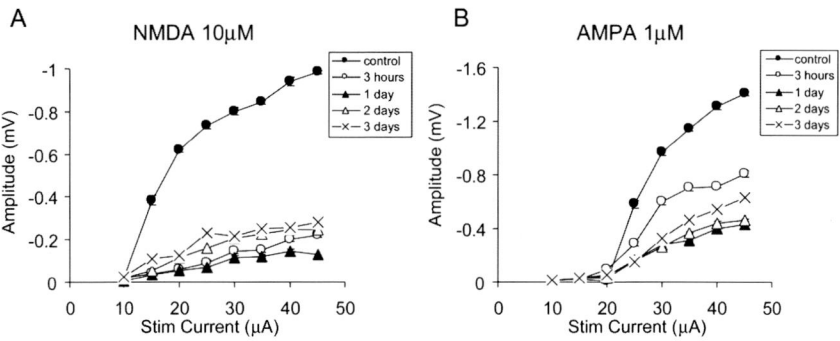

FIGURE 15.2. Chronic effects of NMDA and AMPA on synaptic responses in CA1. After collecting baseline responses, slices were treated with NMDA (10 μM; A) or AMPA (1 μM; B) for the indicated periods of time, and I/O relationships were obtained. The amplitudes of synaptic sponses were plotted as a function of the intensities of stimulation.

The mechanisms underlying the reduction in amplitude of synaptic responses produced by incubation with these two excitotoxins were further studied by applying various concentrations of NMDA and AMPA for two days. As in the control experiments, the plateau values of the respective I/O curves were used for the analysis (Figure 15.3). Fitting the concentration–response curves with the Hill function generated EC50 values of 5.2 ± 0.5 μM for NMDA and 0.81 ± 0.1 μM for AMPA, with Hill coefficients of 1.7 ± 0.2 and 2.6 ± 0.8, respectively. For comparison, we plotted in the same figure concentration–response curves obtained with the PI uptake method (Figure 15.3, open circles and dashed lines). EC50 values of PI uptake method were 21.6 ± 3.4 μM for NMDA and 3.87 ± 0.1 μM for AMPA and Hill coefficients of 3.7 ± 1.0 and 2.0 ± 0.1, respectively. The concentration–response curves obtained with electrophysiology were shifted toward the left and were also less steep (at least for NMDA) than those observed with the PI method, suggesting that these two methods reveal different cellular mechanisms activated by NMDA and AMPA.

To further address this question, we investigated the time-course for recovery of synaptic responses after agonist removal. At the end of incubation in the presence of 10 μM NMDA for 40 min, 3 h, 1 day, and 3 days, the amplitudes of synaptic responses were 48 ± 12, 22 ± 7, 13 ± 3, and $11 \pm 3\%$ of control values, respectively ($n = 3$). However, following 1 h of NMDA washout under the same conditions, synaptic responses were 109 ± 4, 66 ± 5, 52 ± 2, and $18 \pm 3\%$ of the initial amplitude, respectively, indicating that synaptic responses in fact gradually decrease and become irreversibly diminished only after 3 days of continuous treatment with NMDA. Similar results were obtained for AMPA: a 40 min incubation in the presence of 3 μM AMPA led to a large decrease in the amplitude of synaptic responses ($12 \pm 5\%$ of control values) and the responses recovered to up to $93 \pm 2\%$ of control after AMPA was removed for 1 h ($n = 3$); 3 days of incubation in the

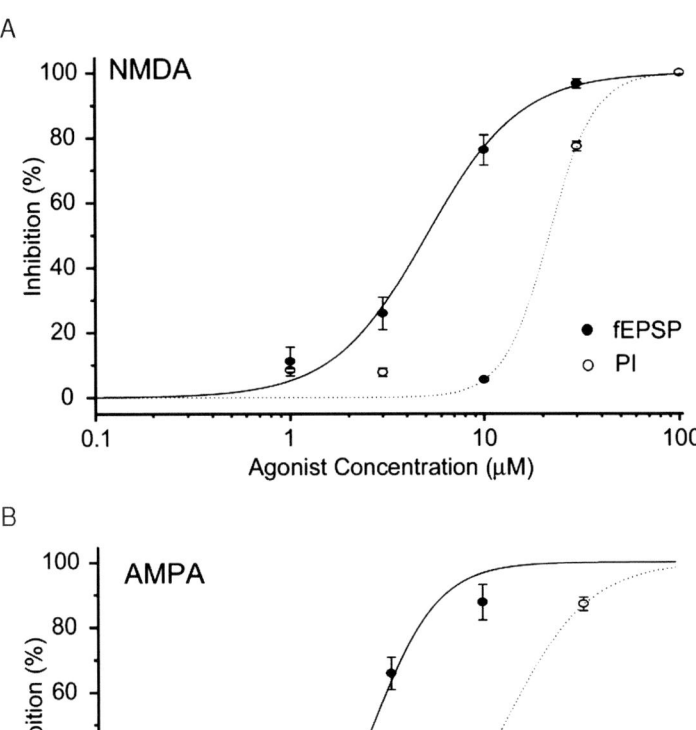

FIGURE 15.3. Chronic effects of various concentrations of NMDA and AMPA on synaptic responses and neuronal damage in CA1. Synaptic responses to maximum stimulation intensities were recorded 48 h after chronic treatment of cultured hippocampal slices with NMDA (A) or AMPA (B). Results were calculated as percentage of inhibition of baseline values and represent means 9 S.E.M. of three experiments (solid circles). Neuronal damage (open circles) was assayed with the PI uptake method. Results were calculated as percentage of maximal damage and represent means ± S.E.M of two experiments. Solid and dashed lines represent the best-fit curves for synaptic responses and PI uptake.

presence of 1 µM AMPA decreased synaptic response amplitudes to 46 ± 4% ($n = 3$) and no recovery was observed even after 24 h of AMPA removal.

To validate the ability of our system to discriminate drugs with different neuroprotective properties, the potency of two distinct noncompetitive NMDA receptor antagonists was determined. MK-801 is a known high-affinity antagonist, whereas

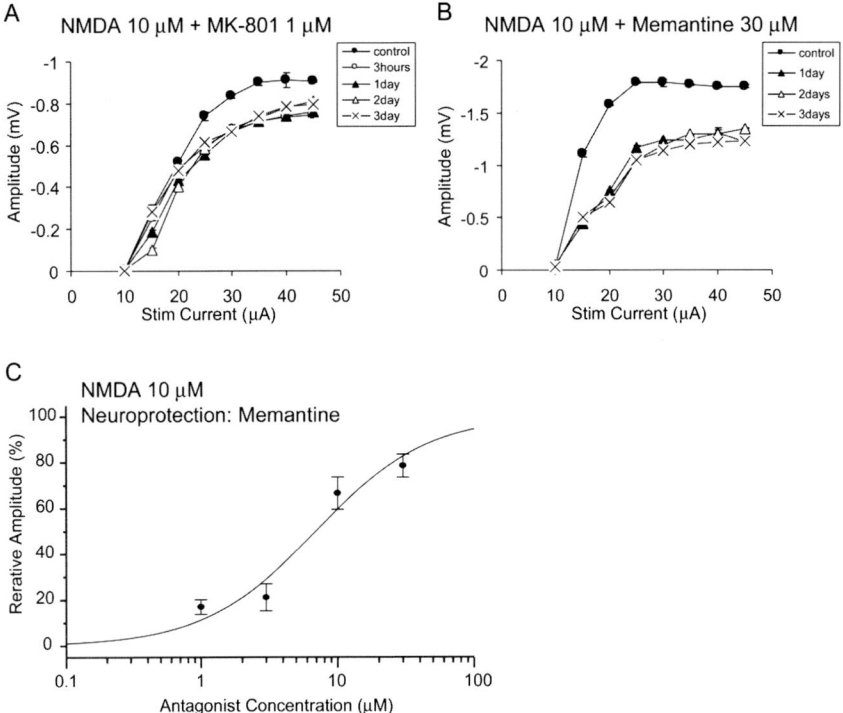

FIGURE 15.4. Effects of MK801 and memantine on NMDA-induced decrease in synaptic responses in CA1. After collecting baseline responses, slices were treated with 10 μM NMDA and 1 μM MK801 (A) or 10 μM NMDA and 30 μM memantine (B) for the indicated periods of time. Input/output relationships were obtained and the amplitudes of the synaptic responses were plotted as a function of the intensities of stimulation (typical cases are shown). Synaptic responses elicited by stimulation at maximum intensity were recorded 72 h after chronic treatment of cultured hippocampal slices with 10 μM NMDA in the presence of various concentrations of memantine (C). Results were calculated as percentage of responses recorded under control conditions (i.e., in the absence of NMDA) and represent means ± S.E.M. of three experiments.

memantine is an uncompetitive low-affinity antagonist and is currently used in clinical treatments for Alzheimer's disease (Parsons et al., 1999). Based on the concentration–response analysis performed above (Figure 15.3A), 10 μM NMDA was chosen as the concentration of agonist for the antagonist tests. Inasmuch as we were interested in evaluating the neuroprotective effects of these antagonists in chronic application, we did not study their effects before the first 3 h time point.

After performing control measurements, slices were incubated in culture media containing 10 μM NMDA and 1 μM MK-801 for various periods of time. Under these conditions, synaptic responses were only slightly depressed from 3 h up to 3 days of incubation (Figure 15.4A), indicating that 1 μM MK-801 almost completely protected synaptic transmission against NMDA-mediated neurotoxicity, a result in good agreement with previous studies using different markers of synaptic damage (Peterson et al., 1989; Pringle et al., 2000; Kristensen et al., 2001). Relative amplitude of synaptic responses measured at stimulation intensities corresponding to the plateau of I/O curves after 3 days of incubation was $91 \pm 6\%$ of control ($n = 3$).

When memantine was used as an antagonist, the pattern of protection was relatively similar, as synaptic responses were significantly protected for up to 3 days of incubation in the presence of 10 μM NMDA and 30 μM memantine. However, the degree of protection was much smaller than that provided by MK-801, as only $78 \pm 5\%$ ($n = 3$) of the initial amplitude was preserved under these conditions (Figure 15.4B). The concentration dependency of memantine protection was studied using 10 μM NMDA and various concentrations of memantine. The concentration–response curve was fitted with the Hill equation and provided values for IC50 of 6.9 ± 1.6 μM, with a Hill coefficient of 1.1 ± 3 (Figure 15.4C).

Despite its relatively high affinity, MK-801 is not used clinically to protect CNS from neurodegenerative and other kinds of disorders. One of the reasons for its inadequateness for medical treatment is related to the almost irreversible nature of its binding to the NMDA receptors (Parsons et al., 1999), and some of its neurotoxic properties. In our experiments, incubation of the slices in the presence of 1 μM MK-801 alone resulted in decreased synaptic responses after 1 day. Interestingly, synaptic responses were not affected after 3 h of incubation. However, after 1, 2, and 3 days of incubation, the amplitudes of responses were decreased to 83 ± 6, 79 ± 11, and $76 \pm 10\%$ ($n = 3$) of control, respectively. These decreases in amplitude were not significantly different from the decrease in amplitudes produced by 1 μM MK-801 in the presence of 10 μM NMDA. In contrast to MK-801, long-term incubation in the presence of memantine even at 30 μM did not result in a significant decrease in synaptic transmission.

Results obtained with chronic application of NMDA receptor agonist or antagonists are summarized in Figure 15.5. Control responses were measured in 23 slices and were compared with the results obtained after incubation during 3 days in the presence of NMDA, NMDA plus antagonists, or antagonists alone. Our results indicate that the large (70 to 80%) decrease in synaptic responses produced by incubation in the presence of 10 μM NMDA could be almost completely prevented ($91 \pm 6\%$, $n = 3$) by co-incubation with 1 μM MK-801 (not significantly different from control values). Protection was substantially weaker for memantine even at higher concentrations, as only $78 \pm 5\%$ ($n = 3$) of synaptic responses remained after treatment with NMDA and 30 μM memantine ($p < .05$ as compared to control values). However, it is important to note that incubation with memantine (30 μM) but not MK-801 (1 μM) in the absence of NMDA resulted in responses that were statistically identical to those obtained under control conditions (control vs. memantine: $p = .267$, control vs. MK-801: $p = .010$).

These data clearly demonstrate that chronic multi-site recordings in an in vitro preparation represent a new methodology to assess the properties and mechanisms of neuronal damage. This technology should be of interest to a wide range of

3 Day Drug Incubation

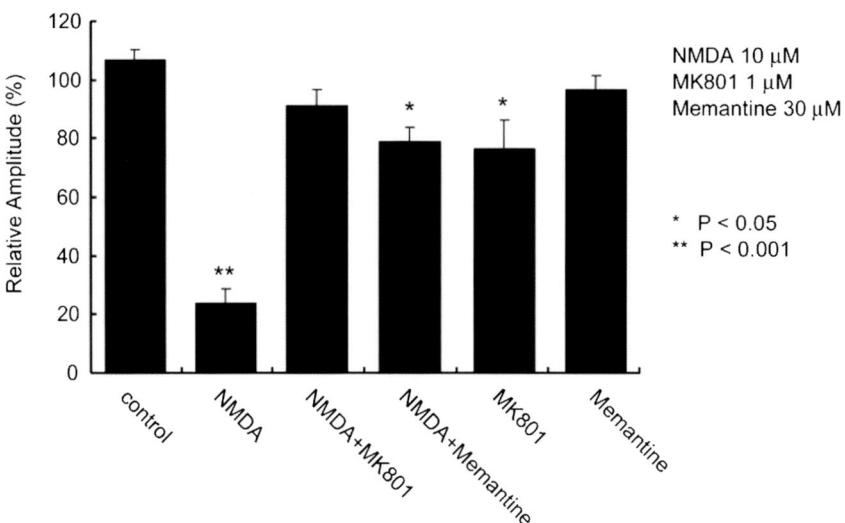

FIGURE 15.5. Comparison of the effects of various chronic treatments on synaptic responses in CA1. Synaptic responses to stimulation at maximal intensity were recorded in cultured hippocampal slices three days after treatment with various drugs or combinations of drugs. Results represent amplitudes of synaptic responses; they are expressed as percentage of the values recorded under control conditions and are means $\pm$ S.E.M. (control: $n = 23$; other cases: $n = 3$, $*p < 0.05$, $**p < 0.001$).

neuroscientists and should provide a new and powerful method to study the chronic effects of drugs or other experimental manipulations in an in vitro preparation.

15.4   Chronic LTP Assay for Memory Enhancers

15.4.1   Long-Term Recording of Long-Term Potentiation in Cultured Hippocampal Slices

As discussed above, direct culturing of hippocampal slices on MED probes resulted in a tight adhesion of the slices on the probes, thus providing for stable maintenance of the stimulating and recording sites and thereby long-term extracellular recording in hippocampal field CA1 (Shimono et al., 2000). LTP was induced by delivering trains of high frequency to one of two inputs converging on one set of pyramidal cells in CA1, with the other input being used as control (Figure 15.6). Field EPSPs evoked by stimulation on either test or control pathway were recorded before and 1 h, 1 day, and 2 days after delivering the tetanus stimulations to the test pathway. Although the fEPSPs evoked in the control pathway remained stable

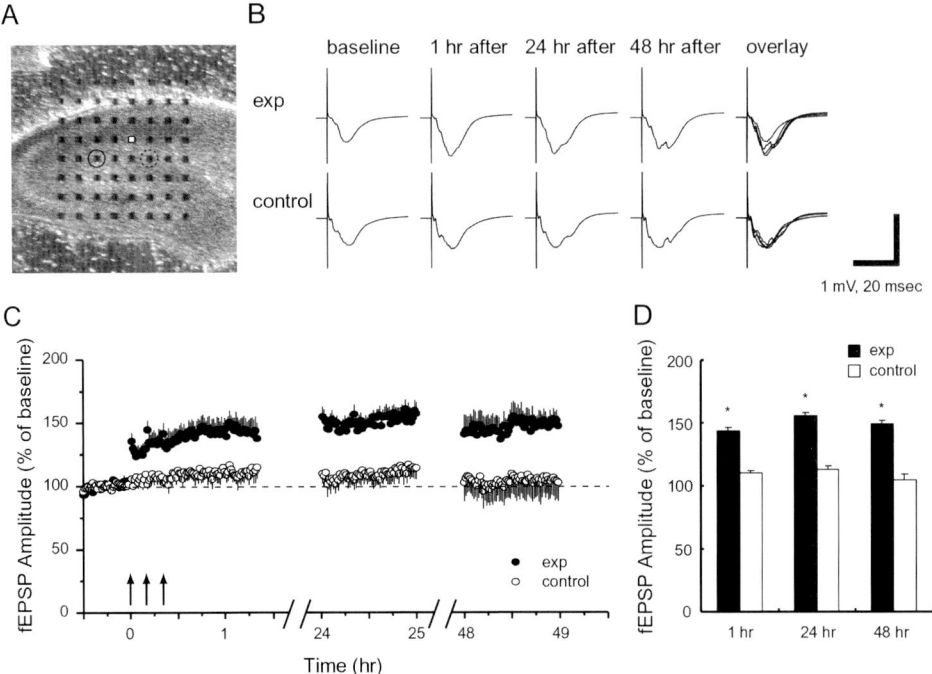

FIGURE 15.6. Long-lasting recording of long-term potentiation in cultured hippocampal slices. (A) Micrograph of a hippocampal slice cultured on an MED probe. Interelectrode distance is 150 μm. The recording electrode is indicated by a white square, and stimulation electrodes are indicated by a solid circle (tetanized pathway (exp)) and a dotted circle (control pathway). (B) fEPSPs evoked by Schaffer fiber were recorded in field CA1 before and after tetanus stimulation. fEPSPs evoked by stimulation of the tetanized pathway (exp) are shown on the top row and fEPSPs evoked by stimulation of the control pathway (control) are shown on the bottom row (each response is the average of 10 sweeps; typical examples are shown). (C) Summarized graph of long-lasting LTP recording. The maximum amplitude of fEPSP was determined and calculated as percentage of averaged baseline values (means ± S.E.M., $n = 8$). Arrows indicate time of high-frequency stimulation. (D) LTP amplitude was averaged over a 10 min period at 1 h, 24 h, and 48 h after high-frequency stimulation, and was expressed as percentage of baseline values.

during recording, potentiation of the fEPSPs evoked in the test pathway started immediately after tetanus and, in a subset of slices, lasted for more than 2 days without any significant decay over this time period (Figure 15.6B). Potentiation lasting more than 1 h occurred in 27 out of 38 slices tested, and the number of slices in which LTP lasted more than 1 day was 15 out of 27 slices. Data obtained from long-term recordings of LTP that lasted more than 2 days ($n = 8$) are summarized in Figure 15.6C. Field EPSP amplitudes in test and control pathways represented 143 ± 3% and 110 ± 2% of their respective pretetanus levels after 1 h, 156 ± 3% and 113 ± 2% after 1 day, and 149 ± 3% and 105 ± 4% after 2 days, respectively.

Differences between responses recorded in the test and control pathways at 1 h, 1 day, and 2 days after tetanus were highly significant ($p < .0001$, $n = 8$).

15.4.2   Effects of APV on LTP Induction in Cultured Hippocampal Slices

To determine whether long-lasting LTP induced in slices cultured on MED probes was similar to LTP typically induced in acute or cultured hippocampal slices, tetanus stimulation was applied in the presence of the NMDA receptor antagonist, D-APV. When APV (50 µM) was applied 30 min before and during high-frequency stimulation, no LTP was observed (relative amplitudes of test and control pathways were $114 \pm 1\%$ and $112 \pm 1\%$ 1 h after tetanus and $103 \pm 2\%$ and $102 \pm 1\%$ 3 h after APV washout; $n = 11$). On the other hand, when high-frequency stimulation was delivered 3 h after APV washout, LTP was again elicited and to the same degree as in naive slices ($147 \pm 3\%$ and $116 \pm 3\%$ in test and control pathways, respectively; $n = 5$; data not shown). Such potentiation also lasted more than one day.

To further assess the effects of APV on synaptic plasticity, synaptic responses were determined one day following high-frequency stimulation in the presence of APV, and APV washout. Synaptic responses remained unmodified as compared to their baseline values whether or not the slices had been treated with APV, and whether or not they had been tetanized in the presence of APV. This result clearly indicates that, under our conditions, high-frequency stimulation did not produce an NMDA receptor-independent form of long-term potentiation. To eliminate the possibility that the lack of modification of synaptic responses was due to some deleterious effects of tetanization in the presence of APV, we tested the effects of high-frequency stimulation one day after APV washout. LTP was elicited in six out of eight slices and the magnitude of LTP was similar to that produced in naive slices (average relative amplitudes expressed as percentage of initial baselines were $153 \pm 3\%$ and $116 \pm 2\%$ after 1 h in the test and control pathways, respectively; means $\pm$ S.E.M. of six experiments).

Trains of high-frequency stimulation in cultured hippocampal slices elicit a long-lasting form of LTP that, in a significant number of slices, can be recorded for several days. This form of LTP was completely blocked by APV, an antagonist of NMDA receptors. This blockade was completely reversible as LTP could be induced 3 hr or 24 hr after APV washout. Thus, this form of LTP exhibits similar features as typical LTP elicited in CA1 of acute or cultured hippocampal slices. In this regard, our data regarding the probability and the amplitude of LTP in hippocampal slices cultured for 10 days are in good agreement with previous results.

Our results also indicate that LTP in field CA1 of a subset of cultured hippocampal slices can occur without showing any significant decay over more than two days, a result reminiscent of what has been reported for in vivo LTP in field CA1 (Staubli and Lynch, 1987). This finding indicates that this experimental model might be ideally suited to study mechanisms underlying LTP maintenance and consolidation. Furthermore, as cultured slices have been successfully used with various mutant mice, our results open the way for a wide range of studies related to long-term recordings of synaptic modifications resulting from a broad spectrum of manipulations.

15.5   Co-Culture on a Chip: Effects of Neuromodulators

Many psychoactive agents act by modulating one or more of the diffuse ascending systems: that is, serotonin, acetylcholine, dopamine, and norepinephrine. These compounds facilitate or retard transmission but generally do not act as agonists or antagonists and often have only subtle effects on transmitter binding. Because of this, detection of modulatory agents is often best accomplished with an assay that incorporates synaptic transmission. This presents a severe problem in the case of the diffuse systems because the pertinent axons are both very sparse and disconnected from their cell bodies. It is thus difficult to establish that a stimulating electrode is in contact with the targeted fibers or that an evoked response reflects monosynaptic transmission involving the intended projections. One solution to these problems is to co-culture the cell bodies that give rise to the diffuse system of interest with an appropriate anatomical target of the system. However, co-cultures have not been widely used with the cultured slice technique recently introduced by Stoppini et al. (1991). The method we developed uses slices prepared from rat brains in the second postnatal week, a time point at which the major anatomical systems have been laid down. Moreover, the slices gradually take on a much more adultlike state than is the case with traditional organotypic cultures.

Figure 15.7 consists of micrographs of sections through two cultured hippocampal slices, one (left side) co-cultured with the median raphe nucleus of the brainstem and the other co-cultured with the medial septal nucleus. The raphe–hippocampal sections have been processed for immunocytochemistry using antibodies against serotonin and the septal–hippocampal sections were processed for acetylcholinesterase histochemistry. Serotonergic and cholinergic cell bodies can be seen in the boxes marked "B" in the survey micrographs in the top panels of the figure. Labeled fibers arising from these neurons are evident in the higher-power images in the middle panels. The bottom panels show the serotonergic and cholinergic fibers forming dense plexuses above and below the cell body layers within hippocampus, the pattern of innervation found in situ. The dense puncta in these figures correspond in size, appearance, and distribution to axon terminals. In all, cultured slices prepared relatively late in development provide excellent targets for the diffuse projections.

Success in the co-culture experiment opened the way to tests of how the diffuse ascending projections affect network-level operations in the hippocampus and retrohippocampal cortex. Figure 15.8A shows a septal/hippocampal culture sitting atop a 64-electrode array and Figure 15.8B provides typical recordings of spontaneous physiological activity. Note the triplets of large biphasic waves that

Raphe-Hippocampal Culture
5-HT

Septo-Hippocampal Culture
AChE

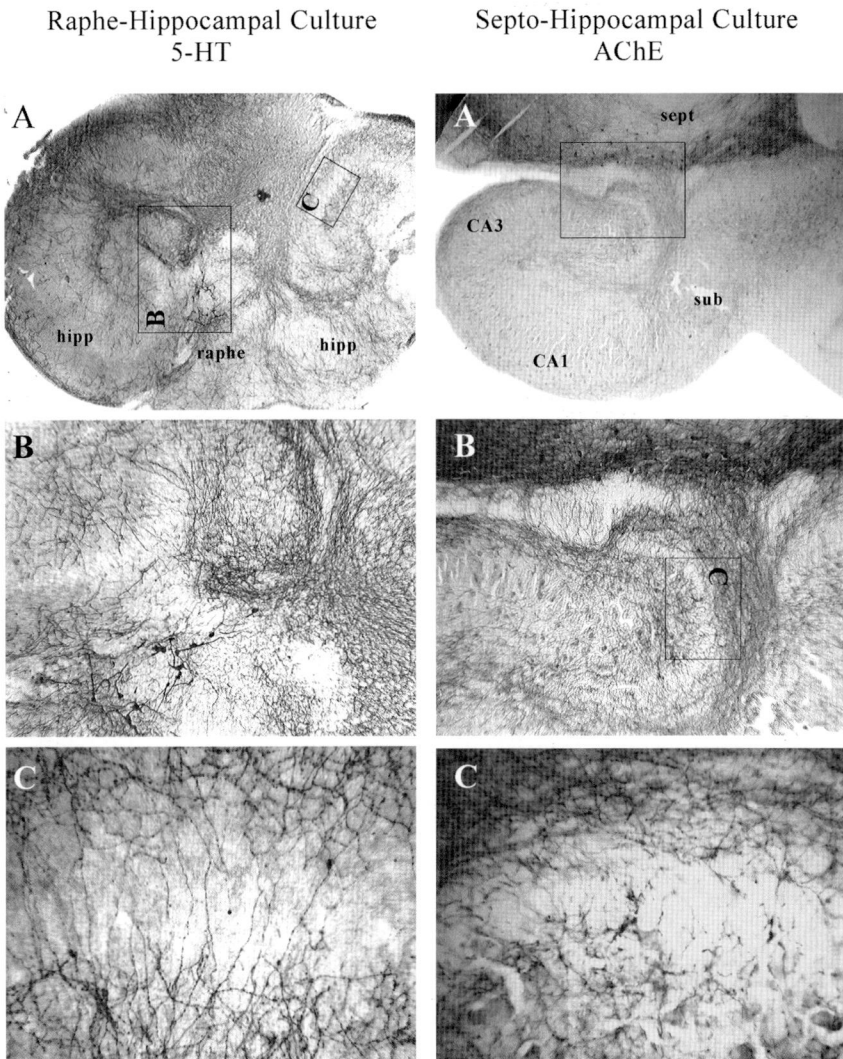

FIGURE 15.7. Co-cultures of slices. Micrographs of sections through two cultured hippocampal slices, one (left side) co-cultured with a slice from the median raphe nucleus of the brainstem and the other co-cultured with a slice from the medial septal nucleus. The raphe–hippocampal sections have been processed for immunocytochemistry using antibodies against serotonin and the septal–hippocampal sections were processed for acetylcholinesterase histochemistry. Serotonergic and cholinergic cell bodies can be seen in the boxes marked "B" in the survey micrographs in the top panels of the figure. Labeled fibers arising from these neurons are evident in the higher-power images in the middle panels. The bottom panels show the serotonergic and cholinergic fibers forming dense plexuses above and below the cell body layers within the hippocampus, the pattern of innervation found in situ. The dense puncta in these figures correspond in size, appearance, and distribution to axon terminals.

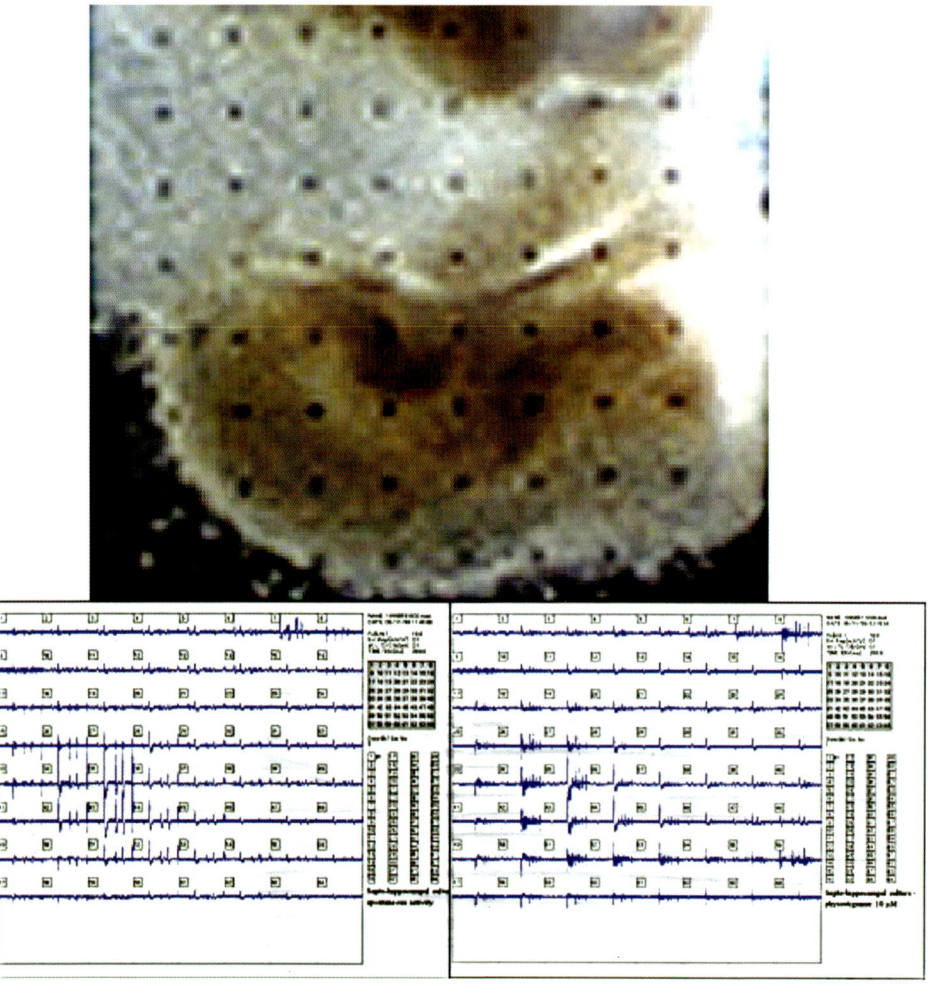

FIGURE 15.8. Recordings in co-cultures: (A) micrograph of a septal/hippocampal culture sitting atop a 64-electrode array. (B) Typical recordings of spontaneous physiological activity. Note the triplets of large biphasic waves that are particularly prominent at electrodes #36 and #44. Based on their shape and distribution, these potentials appear to be synchronous postsynaptic responses, that is, the result of a sizeable number of CA3 pyramidal neurons firing at about the same time. (C) Results obtained with physostigmine, a psychoactive drug that modulates cholinergic transmission by blocking the catalytic enzyme acetylcholinesterase. As is evident, the drug triggered rhythmic EEG activity at sites throughout the hippocampus, an effect similar to that it produces in vivo.

are particularly prominent at electrodes #36 and #44. Based on their shape and distribution, these potentials appear to be synchronous postsynaptic responses, that is, the result of a sizeable number of CA3 pyramidal neurons firing at about the same time. Figure 15.8C shows the first results obtained with physostigmine, a psychoactive drug that modulates cholinergic transmission by blocking the catalytic enzyme acetylcholinesterase. As is evident, the drug triggered rhythmic EEG activity at sites throughout the hippocampus, an effect similar to that it produces in vivo. Analyses now in progress indicate that multiple frequencies are promoted by physostigmine and that the response to the drug is regionally differentiated. But as they stand, the above results demonstrate the feasibility of using cultured slices to detect compounds that modulate ascending diffuse projections. Rhythmic activity of a type not seen in conventional slices has also been recorded in raphe–hippocampal preparations (data not shown) and experiments with modulatory drugs will begin shortly.

15.5.1 Studying the Modulation of Cholinergic Rhythmic Activity in Acute Ventral Hippocampal Slices

In this section we present results obtained with acute hippocampal slices recorded with the MED64 System. It has recently been shown that rhythmic activity in the gamma frequency band (20 to 40 Hz) could be induced and studied in the CA3 area of ventral hippocampal slices (Fisahn et al., 1998; Fellous and Sejnowski, 2000; for review see McBain and Fisahn, 2001). This was an interesting discovery as rhythmic activity in the slice showed some resemblance to that recorded in vivo, where it is observed during memory formation (Fell et al., 2001), sleep–wake states (Llinas and Ribary, 1993; Maloney et al., 1997), and visual activity (for review see Singer, 1999). To induce rhythmic activity in ventral hippocampal slices, we used the cholinergic agonist carbachol, allowing us to study rhythmic activity in isolation in the brain slice preparation.

Depending on the interelectrode space on the MED64 probe (most of our recordings were done using a probe with 150 μm interelectrode space), rhythmic activity can be monitored in many subregions of the hippocampal slice. It is our (Shimono et al., 2000) as well as other investigators' (Fisahn et al., 1998; Traub et al., 2000) experience that rhythmic activity shows the largest spectral power in the CA3 region. We therefore concentrated our investigation on this region. Because our recordings also contain data from other regions (e.g. the CA1 region), a comparison of compound effects on different regions would be possible. To record from acute slices, we adapted the MED64 System to a perfusion system, allowing the addition of compounds together with artificial cerebrospinal fluid. Slices in that system were kept in a warm humidified atmosphere of 95% O_2: 5% CO_2. In the presence of carbachol, rhythmic activity was persistent for hours, and 5 μM of the agonist in the bath was found to be a moderate concentration, where spectral power or frequency had not reached their maximal values. We therefore used this concentration in most of our recordings to study modulation of rhythmic activity by a variety of pharmacological compounds. Spectral power and frequency ranged approximately from 5 to 600 μV^2 and 18 to 40 Hz, respectively.

We were mostly interested in the aminergic modulation of cholinergic rhythmic activity in acute hippocampal slices, and therefore focus in this section on the serotonergic modulation of these rhythms. The hippocampus receives a moderate to high serotonergic innervation, originating from the dorsal and medial raphe nuclei (Moore and Halaris, 1975; Azmitia and Segal, 1978; Oleskevich and Descarries, 1990; McQuade and Sharp, 1997). Activating the serotonergic system in the raphe nucleus in vivo desynchronizes the hippocampal encephalogram (Assaf and Miller, 1978; Nitz and McNaughton, 1999). We therefore asked the question whether and how 5-HT modulates cholinergic-induced rhythmic activity in ventral hippocampal slices (Krause and Jia, 2005).

15.5.2 Involvement of 5-HT$_{1A}$ and 5-HT$_2$ Receptors in the Modulation of Cholinergic Hippocampal Rhythmic Activity

Experiments in our laboratory showed (Krause and Jia, 2005) that 5-HT decreased the spectral power, but not the frequency, of carbachol-induced rhythmic activity in a reversible manner (Figure 15.9). This modulation was concentration-dependent

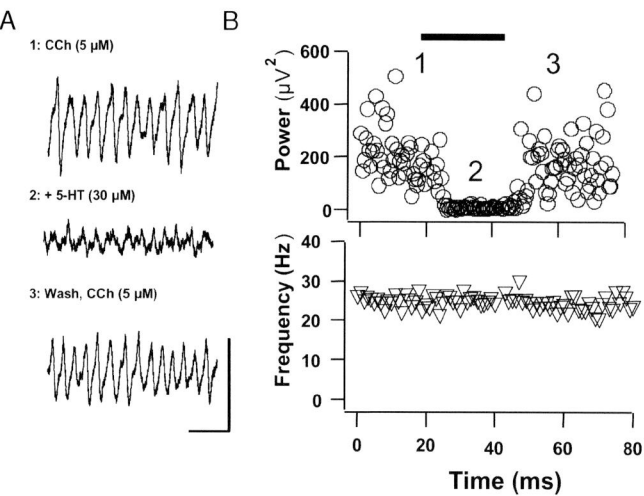

FIGURE 15.9. Carbachol-induced rhythmic activity in ventral hippocampal slices and its suppression by 5-HT: (A) carbachol (CCh, 5 μM) in the bath leads to sustained rhythmic activity (1). Adding 5-HT (30 μM) suppresses spectral power of this rhythmic activity (2). This effect is reversible (3). (B) Diagrams showing power and frequency of CCh-induced rhythmic activity before, during, and after 5-HT application (indicated by the black horizontal bar). The numbers indicate time points were sweeps were taken in A. Calibration: 50 μV, 100 msec. (Edited from Krause and Jia, 2005, with permission from Elsevier.)

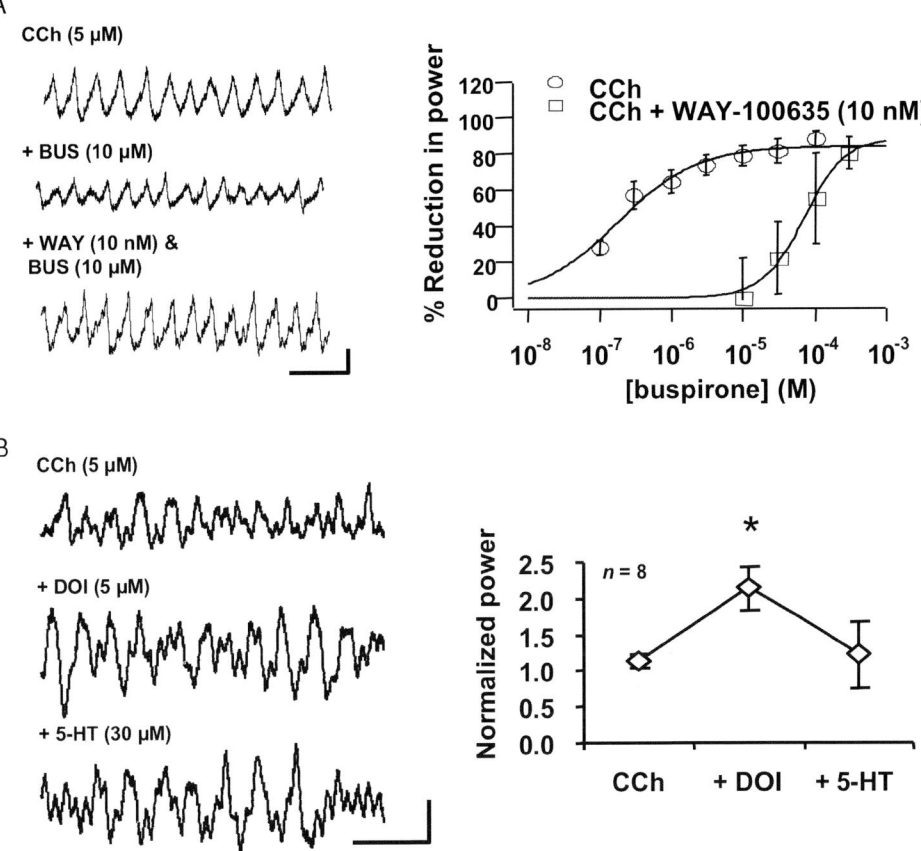

FIGURE 15.10. CCh-induced activity is differentially modulated by 5-HT$_{1A}$ and 5-HT$_2$ receptors. (A) The 5-HT$_{1A}$ agonist buspirone (BUS) suppresses CCh-induced rhythmic activity in a concentration-dependent manner. The suppression is reversed by the 5-HT$_{1A}$-selective antagonist WAY-100635 (WAY, 10 nM). (B) The 5-HT$_2$ agonist DOI (DOI, 5 μM) increases CCh-induced activity. 5-HT (30 μM) fails to decrease rhythmic activity in the presence of DOI. Calibration: 50 μV, 100 msec. (Edited from Krause and Jia, 2005, with permission from Elsevier.)

with an estimated IC$_{50}$ for 5-HT of 4 μM (data not shown). We next used receptor-specific agonists and antagonists to determine the receptor subtype(s) underlying the suppression of carbachol-induced rhythmic activity. Interestingly, we found a differential modulation of carbachol-induced activity by 5-HT$_{1A}$ and 5-HT$_2$ receptors (Figure 15.10). Whereas the 5-HT$_{1A}$ agonists buspirone (estimated IC$_{50}$: 0.25 μM) and 8-OH-DPAT (not shown) mimicked the effect of 5-HT on rhythmic activity, the 5-HT$_2$ receptor agonist DOI increased spectral power of the rhythms. In addition, 5-HT no longer decreased spectral power in the slice in the presence of DOI. The 5-HT$_{1A}$ antagonist WAY-100635 or the 5-HT$_2$ antagonist ritanserin alone had no effect on carbachol-induced rhythmic activity.

On the basis of the long-lasting effect of 5-HT showing no desensitization we excluded the involvement of 5-HT$_3$ receptors. However, we cannot exclude the involvement of other 5-HT receptors. From these results we concluded that 5-HT shows differential effects on carbachol-induced rhythmic activity in ventral hippocampal slices with 5-HT$_{1A}$ activation exerting a predominant effect to elicit a decreased spectral power of the rhythm.

15.5.3   Testing the Ability of MM223, a Novel 5-HT$_{1A}$ Agonist to Modulate Carbachol-Induced Rhythmic Activity

We next used our system to test and compare a novel selective 5-HT$_{1A}$ agonist, MM223 (Mokrosz et al., 1999; Bojarski et al., 2002), in its ability to modulate carbachol-induced rhythmic activity. Many 5-HT$_{1A}$ ligands such as buspirone, gepirone, or WAY-100635 are 1-arylpiperazines. In contrast, MM223 is a 1,2,3,4 tetrahydroisoquinoline amide derivative, representing a novel class of 5-HT$_{1A}$ ligands (Figure 15.11A). Therefore, we wanted to test MM223's ability to affect carbachol-induced rhythmic activity. This compound exhibited a Ki of 0.95 nM at the 5-HT$_{1A}$ receptor, and of 452 nM at the 5-HT$_2$ receptor, as assessed in radioligand binding studies (Bojarski et al., 2002). In contrast, buspirone shows a Ki of ~20 nM at the 5-HT$_{1A}$ receptor (Sharif et al., 2004). Using the lip retraction test (in rat), MM223 was found to be a partial agonist at postsynaptic 5-HT$_{1A}$ receptors (Mokrosz et al., 1999). MM223 mimicked the effect of buspirone and 5-HT in suppressing carbachol-induced rhythmic activity, and its action was antagonized by WAY-100635, thereby confirming its agonist behavior at a presumed 5-HT$_{1A}$-mediated physiological action (Figure 15.11B). Using a Scatchard analysis on the data shown in Figure 15.11C, we found Ki values of 3.44 µM for buspirone, and 0.73 µM for MM223, indicating that MM223 was more potent in decreasing carbachol-induced rhythmic activity than buspirone (Figure 15.11D). Our results using the MED64 System therefore confirmed the results from receptor binding studies mentioned above.

Taken together, these data show that carbachol-induced rhythmic activity in ventral hippocampal slices can be used to evaluate pharmacological features of novel compounds.

15.6   Conclusions

The Brain-on-a-Chip^{TM} technology bridges the gap between biochemical and single-cell testing and behavior by determining the effects of compounds on living slices of brain containing intact networks of neurons—a largely unexplored area essential to understanding drug effects on complex human behavior. The Brain-on-a-Chip^{TM} technology is an information-rich and physiologically relevant

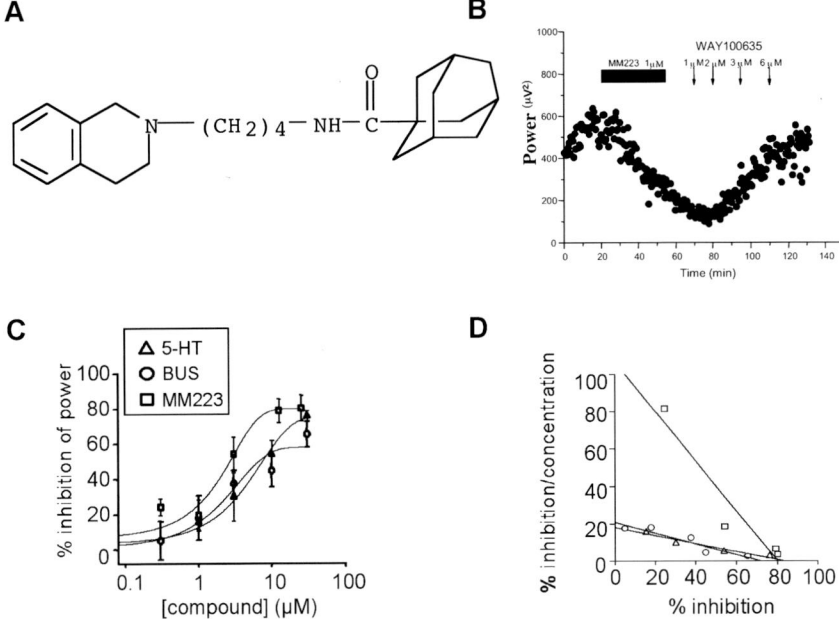

FIGURE 15.11. The novel 5-HT$_{1A}$-selective agonist MM223 suppresses CCh-induced rhythmic activity more potently than buspirone or 5-HT: (A) structure of MM223. (B) Suppression of CCh-induced activity by MM223 (1 μM) is reversed by WAY-100635. (C) Concentration-dependence of CCh-induced inhibition of spectral power by 5-HT, buspirone (BUS), and MM223. MM223 suppresses CCh-induced rhythmic activity more potently than BUS or 5-HT. (D) Scatchard plot of the data shown in (C) indicates linear relationships between the percentage of inhibition per concentration of agonist and the percentage of inhibition itself as a measure for the Ki value. See legend in (C) for symbols representing all three compounds.

in vitro brain assay system that addresses all four major drawbacks of conventional electrophysiology:

1. *Acute and chronic measurements:* We have shown that the MED64 can generate more temporal information about the activity of neuronal circuits than any conventional methods. We have developed methods to culture and maintain living brain tissue slices directly on the multi-electrode chips for up to several weeks. This enables us to study the chronic or long-term effects of drugs on brain activity. Whereas most conventional neurotoxicity tests can only measure the extent of cell death in response to potentially toxic drugs, we now have a functional assay capable of detecting a much wider range of drug side effects.

2. *Access to local and modulatory activity:* We can directly study the effects of drug candidates on all major neurotransmitter pathways present in the brain. We extended our chronic culture technology to co-cultures of various types of brain tissues to study the important modulatory projections from the serotonergic,

dopaminergic, cholinergic, and adrenergic neurons in the brainstem. For example, when the tissue from raphe nuclei containing serotonergic neurons was cultured next to hippocampal tissue for several days, the serotonergic neurons extended projections into the hippocampus, where they established functional serotonergic synapses. This allowed us to study the effects of drugs, such as selective serotonin reuptake inhibitors (SSRIs), on endogenously released modulatory transmitters, that is, 5-HT, in the hippocampus. This truly revolutionary approach enables us to study the effects of virtually all drugs for most major psychiatric disorders in a physiologically relevant in vitro system.

3. *Real signal:* Our data indicate that we can use a more realistic activity profile of living brain circuits than other, less complex methods. Our in vitro tissue slices exhibit endogenous dynamic oscillations similar to those found in living brains. In fact, we were able to chemically induce and manipulate these oscillations to produce the three major types of EEG waves: beta, gamma, and theta. Thus, the readout of our in vitro assay is physiologically relevant to the situation in a living brain. In addition, the dynamic oscillations are 10 to 100 times more sensitive to drugs than monosynaptic responses typically recorded in conventional slice electrophysiology. These features, for the first time, make it possible to study the effects of behavioral-relevant dosages of drugs in vitro.

4. *Network:* We generate much more spatial information about the activity of neuronal circuits than conventional methods. We use a 2-D array of 64 micro-electrodes embedded in a special chip (MED64 probe) to simultaneously study the entire oscillating neuronal network containing a number of different cell types, synapses, and localized receptor types, all in a single experiment. In fact, we routinely generate movies of neuronal activity within neuronal circuits on a millisecond time scale resolution to directly observe the effects of drugs on the various neurotransmitter pathways within the central nervous system.

We believe that all these features will be continuously improving and that this type of approach will become more and more widely used to bridge the gap between in vitro and in vivo approaches.

Acknowledgments. The authors wish to thank Yusheng Jia, Michael Lee, Frank Tsuji, Lam Ho, and Ken Shimono for their help with some of the experiments, and A. Bojarski for the generous gift of MM223.

References

Assaf, S.Y. and Miller, J.J. (1978) The role of a raphe serotonin system in the control of septal unit activity and hippocampal desynchronization. *Neuroscience* 3: 539–550.

Azmitia, E.C. and Segal, M. (1978). An autoradiographic analysis of the differential ascending projections of the dorsal and median raphe nuclei in the rat. *J. Comp. Neurol.* 179: 641–667.

Bojarski, A.J., Mokrosz, M.J., Minol, S.C., Koziol, A., Wesolowska, A., Tatarczynska, E., Klodzinska, A., and Chojnacka-Wojcik, E. (2002). The influence of substitution at aromatic part of 1,2,3,4-tetrahydroisoquinoline on in vitro and in vivo 5-HT(1A)/5-HT(2A) receptor activities of its 1-adamantoyloaminoalkyl derivatives. *Bioorg. Med. Chem.* 10: 87–95.

Fell, J., Klaver, P., Lehnertz, K., Grunwald, T., Schaller, C., Elger, C.E., and Fernandez, G. (2001). Human memory formation is accompanied by rhinal-hippocampal coupling and decoupling. *Nat. Neurosci.* 4: 1259–1264.

Fellous, J.M. and Sejnowski, T.J. (2000). Cholinergic induction of oscillations in the hippocampal slice in the slow (0.5–2 Hz), theta (5–12 Hz), and gamma (35–70 Hz) bands. *Hippocampus* 10: 187–197.

Fisahn, A., Pike, F.G., Buhl, E.H., and Paulsen, O. (1998). Cholinergic induction of network oscillations at 40 Hz in the hippocampus in vitro. *Nature* 394: 186–189.

Fix, A.S., Horn, J.W., Wightman, K.A., Johnson, C.A., Long, G.G., Storts, R.W., Farber, N., Wozniak, D.F., and Olney, J.W. (1993). Neuronal vacuolization and necrosis induced by the noncompetitive N-methyl-D-aspartate (NMDA) antagonist MK(+)801 (dizocilpine maleate): A light and electron microscopic evaluation of the rat retrosplenial cortex. *Exp. Neurol.* 123: 204–215.

Krause, M. and Jia, Y. (2005). Serotonergic modulation of carbachol-induced rhythmic activity in hippocampal slices. *Neuropharmacology.* 2005 Mar; 48(3): 381–390. Epub 2005 Jan 25.

Kristensen, B.W., Noraberg, J., and Zimmer, J. (2001). Comparison of exitotoxic profiles of ATPA, AMPA, KA and NMDA in organotypic hippocampal slice cultures. *Brain Res.* 917: 21–44.

Llinas, R. and Ribary, U. (1993). Coherent 40-Hz oscillation characterizes dream state in humans. *Proc. Natl. Acad. Sci. U. S. A.* 90: 2078–2081.

Maloney, K.J., Cape, E.G., Gotman, J., Jones, B.E. (1997). High-frequency gamma electroencephalogram activity in association with sleep-wake states and spontaneous behaviors in the rat. *Neuroscience* 76: 541–555.

McBain, C.J. and Fisahn, A. (2001). Interneurons unbound. *Nat. Rev. Neurosci.* 2: 11–23.

McQuade, R. and Sharp, T. (1997). Functional mapping of dorsal and median raphe 5-hydroxytryptamine pathways in forebrain of the rat using microdialysis. *J. Neurochem.* 69: 791–796.

Mokrosz, M.J., Bojarski, A.J., Duszynska, B., Tatarczynska, E., Klodzinska, A., Deren-Wesolek, A., Charakchieva-Minol, S., and Chojnacka-Wojcik, E. (1999). 1,2,3,4-tetrahydroisoquinoline derivatives: A new class of 5-HT1A receptor ligands. *Bioorg. Med. Chem.* 7: 287–295.

Moore, R.Y. and Halaris, A.E. (1975). Hippocampal innervation by serotonin neurons of the midbrain raphe in the rat. *J. Comp. Neurol.* 164: 171–183.

Nitz, D.A. and McNaughton, B.L. (1999). Hippocampal EEG and unit activity responses to modulation of serotonergic median raphe neurons in the freely behaving rat. *Learn. Mem.* 6: 153–167.

Oleskevich, S. and Descarries, L. (1990). Quantified distribution of the serotonin innervation in adult rat hippocampus. *Neuroscience* 34: 19–33.

Parsons, C.G., Danysz, W., and Quack, G. (1999). Memantine is a clinically well tolerated N-methyl-D-aspartate (NMDA) receptor antagonist—A review of preclinical data. *Neuropharmacology* 38: 735–767.

Peterson, C., Neal, J.H., and Cotman, C.W. (1989). Development of N-methyl-D-aspartate excitotoxity in cultured hippocampal neurons. *Dev. Brain Res.* 48: 187–195.

Pringle, A.K., Self, J., Eshak, M., and Iannotti, F. (2000). Reducing conditions significantly attenuate the neuroprotective efficacy of competitive, but not other NMDA receptor antagonists in vitro. *Eur. J. Neurosci.* 12: 3833–3842.

Sharif, N.A., Drace, C.D., Williams, G.W., and Crider, J.Y. (2004). Cloned human 5-HT1A receptor pharmacology determined using agonist binding and measurement of cAMP accumulation. *J. Pharm. Pharmacol.* 56: 1267–1274.

Shimono, K., Brucher, F., Granger, R., Lynch, G., and Taketani, M. (2000). Origins and distribution of cholinergically induced beta rhythms in hippocampal slices. *J. Neurosci.* 20: 8462–8473.

Singer, W. (1999). Neuronal synchrony: A versatile code for the definition of relations? *Neuron* 24: 49–65, 111–125.

Sirvio, J., Larson, J., Quach, C.N., Rogers, G.A,, and Lynch, G. (1996). Effects of pharmacologically facilitating glutamatergic transmission in the trisynaptic intrahippocampal circuit. *Neuroscience.* 74: 1025–1035.

Staubli, U. and Lynch, G. (1987). Stable hippocampal long-term potentiation elicited by 'theta' pattern stimulation. *Brain Res.* 435: 227–234.

Stoppini, L., Buchs, P.A., and Muller, D. (1991). A simple method for organotypic cultures of nervous tissue. *J. Neurosci. Meth.* 37: 173–182.

Traub, R.D., Bibbig, A., Fisahn, A., LeBeau, F.E., Whittington, M.A., and Buhl, E.H. (2000). A model of gamma-frequency network oscillations induced in the rat CA3 region by carbachol in vitro. *Eur. J. Neurosci.* 12: 4093–4106.

16
Rhythm Generation in Spinal Cultures: Is It the Neuron or the Network?

Jürg Streit, Anne Tscherter, and Pascal Darbon

16.1 Rhythm Generation in Neural Networks In Vitro: Worth Studying?

16.1.1 Rhythm Generation in Intact Preparations

16.1.1.1 Rhythms as an Important Feature of CNS Function

Neural networks of many regions of the CNS are able to generate synchronized rhythmic activity. In humans, rhythmic cortical activity has been recorded for years with electroencephalography (EEG). The various frequency bands that are observed in these recordings are associated with different states of consciousness (Steriade, 2001). In the hippocampus, rhythmic activity has been related to long-term potentiation and memory functions (Vertes and Kocsis, 1997). The release of neuropeptides in the hypothalamus is controlled through rhythmically active neural networks (Kwiecien and Hammond, 1998). Finally, repetitive muscle contractions that occur during respiration, locomotion, or scratching are controlled by rhythmically active neural networks in brainstem and spinal cord (Grillner et al., 1998; Rekling and Feldman, 1998).

All these examples show that rhythmic activity in neural networks underlies many of the specific CNS functions and also suggest that the capability for rhythm generation must be a fundamental property of neural networks. In principle there are two ways a network can generate rhythms: rhythm may be produced by a well-defined circuit, usually composed of excitatory and inhibitory cells, or rhythm generation depends mainly on the cellular properties of certain neurons in the network and the circuit structure is of lesser importance. A mechanism of the second type is thought to underlie population bursting: a network is activated through the positive feedback of recurrent excitation and silenced by one or several accommodation mechanisms. It is the aim of this chapter to present some results concerning the mechanisms involved in rhythm generation, which occurs in networks of cultured spinal neurons. Such networks are at least partially grown in vitro: in the case of dissociated cell cultures, they are entirely regrown from randomly seeded cells. In the case of organotypic slice cultures, networks develop out of premature networks in the fetal spinal cord. Both types of networks therefore have different levels of circuit structure. Comparing the types of rhythmic activity generated in these networks will reveal some emergent properties of "random" neural networks for rhythm generation.

What are random neural networks? We define them as networks of dissociated neurons, in contrast to networks in slice cultures, acute slices, and intact spinal cord, which have increasing levels of complexity of circuit structure grown in vivo. To evaluate the findings obtained in culture in terms of their relevance for rhythm generation in vivo, we first briefly review some of the most important findings about rhythm generation and fictive locomotion in intact spinal cord.

16.1.1.2   Fictive Locomotion in the Cat Spinal Cord

The neuronal system generating the stereotypic movements characteristic of locomotion is composed of three parts: first, of a supraspinal part that is responsible for initiating locomotion and for maintaining a certain degree of drive, second, of the spinal networks that generate motor patterns, and third, of sensory feedback that adapts the motor pattern to external events. More than 100 years ago, Sherrington (1898) observed that cats and other mammals can perform locomotor movements of the legs even after a complete transection of the spinal cord. He proposed that this activity might be produced in mammals by a chain of reflexes, requiring afferent inputs for its maintenance. However, in 1911 Brown observed locomotor movements in cats after spinal section even when the dorsal roots were cut bilaterally (Brown, 1911). With these experiments he demonstrated that neuronal networks in the spinal cord deprived of sensory inputs and supraspinal influences can generate a co-ordinated rhythmic motor output. Such rhythmic alternating activity in the motoneuron pools of flexor and extensor muscles and also on opposite sides of the isolated spinal cord is called fictive locomotion. It is now clear that the autonomous spinal networks providing this activity—later called central pattern generators (CPGs)—are found in all vertebrates, probably including humans (Dietz et al., 1998).

The CPGs have been activated experimentally in three different ways: first, by stimulation of sensory afferents, second, by supraspinal stimulation, and third, by pharmacological activation. Sensory stimuli can trigger CPGs in high decerebrated cats, because locomotion can simply be initiated by moving the treadmill belt on which the cat is standing (suspended with a harness). In addition, tonic stimulation of the dorsal roots can evoke locomotion (for review see Barbeau et al., 1999). Spinal CPGs can also be activated by descending reticulospinal pathways. In high decerebrate cats, locomotion can be initiated by electrical stimulation of the mesencephalic locomotor region (MLR; Shik and Orlovsky, 1976). The speed of locomotion as well as the preferred gaits (walking, trotting, or galloping) can be adjusted by modifying the strength of the stimulation or the speed of the treadmill. Lundberg and Jankowska (Jankowska et al., 1967) were the first to show that rhythms can also be evoked pharmacologically by the application of L-dopa. This dopamine precursor can activate fictive locomotion in the paralyzed spinal cat. Many other neurotransmitters have been shown to either activate or modulate CPG rhythms in a state-dependent way (for review see Rossignol et al., 2001).

From the early experiments in the cat, the half-center model was proposed to describe CPG function. In this model, rhythm generation and alternation are explained by the reciprocal inhibition of two half-centers (for the left and the right side or for flexors and extensors) through crossing inhibitory axons. It was later shown that rhythm generation but not alternation persisted in the presence of blockers of inhibitory synaptic transmission. These and other experiments led to the new hypothesis of coupled oscillators (Grillner and Zangger, 1979). According to this hypothesis, which is still favored, CPGs are composed of several oscillator networks, which are functionally independent in terms of rhythm generation. Pattern generation by such networks results from the appropriate phase coupling among the oscillator networks.

16.1.1.3   The Lamprey Model

One problem in the cat experiments is the difficulty of getting direct experimental access to the CPGs in the spinal cord. Therefore, other preparations such as the spinal cord of the turtle (Mortin and Stein, 1989), or embryonic preparations such as that of the tadpole (Dale, 1995), or the zebrafish embryo (Fetcho and O'Malley, 1995) were developed to obtain deeper insight into the cellular basis of CPGs. An ideal preparation is the lamprey, a primitive vertebrate with a flat spinal cord and brain stem, which can be maintained in vitro for several days. The lamprey swims by producing an undulating wave based on the alternating activation of motor units on the left and the right side of each segment along the body. Fictive swimming can be evoked in the isolated spinal cord of the lamprey by excitatory amino acids such as glutamate or N-methyl-D-aspartic acid (NMDA). The CPGs in the lamprey spinal cord have therefore been investigated and analyzed in great detail. Furthermore, the results from these studies have been used to design computer models of the lamprey CPG, the performance of which could be compared to experimental findings. Much of this work (experimental and computational) has been done in the group of S. Grillner and is reviewed in numerous papers (Grillner et al., 1991, 1998; Grillner, 2003). Here we mention just two of the new findings revealed in this model. First, it was shown that the CPG networks are composed of excitatory interneurons, which use glutamate as neurotransmitter and project to the ipsilateral side, and inhibitory interneurons, which use glycine and project to the contralateral side. Second, a cellular pacemaker mechanism was found, which is based on Ca^{2+}-influx through NMDA receptor channels as a depolarizing mechanism and the subsequent activation of Ca^{2+}-dependent K^+ currents as the hyperpolarizing mechanism. Other channels, such as voltage-dependent Ca^{2+} channels, contribute to this pacemaker mechanism. The resulting rhythm can be modulated by various agents including 5-hydroxytryptamine (5-HT).

16.1.1.4   Development of Pattern Generators: The Chick Spinal Cord

The chick spinal cord is the preparation mainly used to investigate the development of CPGs. The isolated spinal cord of the chick embryo displays spontaneous episodes of rhythmic activity. Such spontaneous activity is a characteristic feature of developing circuits in many parts of the CNS. It is remarkably similar in tissues as diverse as the hippocampus, the retina, and the spinal cord (O'Donovan and Rinzel, 1997). In the retina, this activity is known to be important for the formation and refinement of neuronal projections. In the spinal cord, however, little is known about its role in development.

Nevertheless, some effort has been made mainly by the group of Michael O'Donovan to reveal the mechanisms involved in such embryonic rhythm generation (O'Donovan, 1999). This group found that these rhythms can be mediated not only by glutamate receptors (of the $(+/-)$-α-amino-3-hydroxy-5-methylisoxazole-4-proprionic acid (AMPA) and NMDA types) but also by glycine and γ-aminobutyric acid (GABA) A receptors. These normally inhibitory receptors have excitatory effects in the embryo because the chloride equilibrium potential lies above threshold. This seems to be a general principle in development, which was first discovered in hippocampal networks (Ben-Ari, 2001). Later, the same phenomenon was demonstrated in the developing rat spinal cord, where rhythm generation first depends on cholinergic, then on $GABA_A$/glycine/NMDA and finally mainly on AMPA receptors (Milner and Landmesser, 1999). The spontaneous rhythms seen in the spinal cord of the chick embryo have been modeled by computer simulation based on the two parameters of hyperexcitability and activity-dependent depression (Tabak et al., 2001). Because these rhythms share many properties with the bursting induced in spinal cultures by disinhibition, we discuss this model more extensively later in this chapter.

16.1.1.5   CPGs in Rodent Spinal Cord (Rat and Mouse)

The isolated spinal cord of the neonatal rat, introduced in 1987 by Kudo and Yamada, has now become a standard preparation to study mammalian CPGs. The rat is quite immature at birth and a rapid maturation of motor behavior takes place during the first two postnatal weeks. Although rats younger than postnatal day 12 are unable to walk because of postural weakness (Westerga and Gramsbergen, 1990), their CPGs seem to function, because they can already swim a few hours after birth (Bekoff and Trainer, 1979). CPGs in neonatal rat spinal cord proved to share many of the properties previously described in cat and lamprey spinal cords. Rhythmic activity that alternates between the two sides of the spinal cord (and between ipsilateral flexors and extensors) could be activated by various neurotransmitters as well as by supraspinal or afferent stimulation (Atsuta et al., 1990; Cazalets et al., 1992; Kiehn and Butt, 2003; Kudo and Yamada, 1987; Magnuson and Trinder, 1997; Marchetti et al., 2001).

The most robust rhythms are induced by a combination of NMDA and 5-HT. The general scheme of lamprey CPGs with ipsilateral projecting excitatory glutamatergic and commissural inhibitory glycinergic interneurons seems to be maintained in the rat, although the situation is certainly more complicated (Beato and Nistri, 1999). New approaches for a detailed characterization of the mammalian loco-motor CPG have recently been introduced by combining genetic tools with elec-trophysiology and anatomy in the isolated mouse spinal cord (Kiehn and Butt, 2003). Using such methods, Kullander et al. (2003) showed that the regulation of commissural crossing of axons of interneurons in the spinal cord by Eph receptor and ligand molecules is crucial for the development of alternating patterns of ac-tivity. It is also clear that rhythm generation and alternation are different functions, inasmuch as alternation but not rhythmic activity is suppressed by midsagittal transections or by pharmacological blockade of glycinergic inhibition (Cowley and Schmidt, 1995; Kremer and Lev-Tov, 1997). Thus the model of the coupled oscillator network is also well suited for the rat spinal cord. As in the lamprey, NMDA induces membrane potential oscillations in interneurons and motoneurons of the rat spinal cord. Such oscillations, in combination with electrical coupling by gap junctions, can induce rhythmic activity (Tresch and Kiehn, 2000). Rhythms can also be induced by high K^+, zero Mg^{2+}, or disinhibition by a combination of the glycinergic blocker strychnine and the $GABA_A$ blocker bicuculline (Bracci et al., 1996, 1998). High K^+ induces alternating fictive locomotion patterns sim-ilar to those induced by NMDA and 5-HT, however, zero Mg^{2+} evokes rhythms with unstable phase shifts between left and right. Disinhibiton leads to episodes of synchronous rhythmic activity on both sides of the spinal cord that are similar to the spontaneous rhythms described in the embryonic chick (O'Donovan et al., 1998a).

16.1.2   Acute and Cultured Slices

16.1.2.1   Pattern Generators in Slices?

A widely used approach to combine in vivo grown neuronal networks with good experimental access to the levels of the network and individual cells is the use of acute slice preparations. However, although this is a widely used prepara-tion in other areas of the CNS such as brainstem, where pattern generators for respiration have been investigated (Rekling and Feldman, 1998), are few papers report on locomotor CPG functions in acute mammalian spinal slices (example is a recent paper by Demir et al., 2002). In contrast, acute slices of spinal cord are quite often used to study sensory neurons in the dorsal horns, indicating that although this preparation is technically quite demanding, there must be other rea-sons for its rare use. What then is the reason for this lack of motor studies in slices?

To answer this question, we must take a deeper look into the localization of the CPGs in rat and mouse spinal cords. Although there is no doubt that in lamprey and chick CPGs are distributed along the entire spinal cord, their localization in the rat spinal cord is still debated. CPGs for hindlimbs have been proposed to be restricted to the segments between T13 and L2, whereas the lower lumbar segments are passively driven by these CPGs (Cazalets et al., 1995). However, other groups found the capacity for rhythm generation in all lumbar and even in sacral segments (Cowley and Schmidt, 1997; Kjaerulff and Kiehn, 1996; Kremer and Lev-Tov, 1997; Nakayama et al., 1999), suggesting that locomotor rhythm generation is a more distributed spinal property. The basis for such discrepancies is probably the variations in the sensitivity of different segments to the agents used to induce rhythms (Kiehn and Kjaerulff, 1998). The CPG of the upper limbs seems to be located at C5 to T1, whereas thoracic segments T3 to T10 are driven by either of the two CPGs (Ballion et al., 2001).

The most important question with regard to the maintenance of CPGs in slices consists in the minimal size of a CPG. In the rat, this was found to be at least two segments (Ballion et al., 2001). Given that a 400 µm thick slice of spinal cord of neonatal rats contains about half a segment, it is not expected to contain a fully functional CPG. Nevertheless, alternating rhythmic activity was found in such transverse slices of neonatal rats (Demir et al., 2002). The reason for this discrepancy remains to be elucidated. In our hands, few preliminary trials with transverse slices on multi-electrode arrays (MEAs) showed asynchronous activity, but no rhythms. Rhythms could be induced by disinhibition in longitudinal slices of ventral horns (Tscherter, 2002).

16.1.2.2   Slice Cultures

The first attempt to maintain intact spinal slices (explants) in culture goes back to the 1960s. Crain and his group first reported on this new method (Crain and Peterson, 1967). They investigated and described the patterns of "bioelectrical" activity that spontaneously arise in this preparation. Slice cultures are usually prepared from fetal tissue, in our lab at embryonic age 14 (E14, one week before birth), and are kept in culture for up to four weeks. The development in culture therefore covers the last week of fetal development and the first weeks of postnatal development. Nevertheless, it is of course not clear to what extent development in vitro mimics in vivo development. This question cannot be answered in general but must be kept in mind for each phenomenon and parameter investigated. We therefore discuss it in the context of the specific findings presented in this chapter. As a reference we briefly present some key points of the in vivo development here.

Spontaneous activity of spinal motoneurons can be recorded in ventral roots as early as E13.5. These spontaneous bursts are synchronized and mediated between E13.5 and E15.5 by cholinergic and glycinergic synaptic transmission (Nishimaru et al., 1996; Ren and Greer, 2003). Later (E16.5 to E17.5) the spontaneous activity results from the combination of synaptic drive acting via non-NMDA glutamatergic, nicotinic acetylcholine, glycine, and $GABA_A$ receptors. Finally, at late stages (E18.5 to E 21.5) the glutamate system acting via non-NMDA receptors is the major drive for rhythm generation. The alternation between the left and right ventral roots is established between E16.5 and E18.5 (Kudo and Nishimaru, 1998). The commissural axons responsible for excitatory coupling and thus synchronization between both sides of the spinal cord at early stages and for the inhibitory coupling and thus alternation at late stages are GABAergic. Later, they are successively replaced by glycinergic projections (Kudo and Nishimaru, 1998). Thus the switch from excitatory to inhibitory effects of GABA (and glycine) seems to be crucial for the switch from synchronous activity to alternating activity.

Rhythmic activity can be evoked by bath application of 5-HT at E14.5 and by NMDA at E16.5 (Iizuka et al., 1998; Ozaki et al., 1996), at a stage when, interestingly, most of the descending projections are not yet functional. The 5-HT containing projections, for example, reach the lumbar cord at E15 to E16 (Schmidt and Jordan, 2000). The first postnatal week is characterized by changes in the electrical properties of the motoneurons (decrease in input resistance, increase in maximal firing rate), by a refinement of the reflex circuits, and by myelination (for review see Vinay et al., 2000).

In organotypic cultures of rat spinal slices, we have previously shown that a functional reflex arc between dorsal root ganglion cells and co-cultured skeletal muscle develops (Spenger et al., 1991; Streit et al., 1991). Myelination starts in the third week in culture. Disinhibition induces bursts of synchronized activity in the whole slice. The activity within the bursts usually oscillates at 4 to 5 Hz (Streit, 1993). Such oscillations usually start shortly after the onset of the bursts and slow down during the bursts. Similar patterns of bursting appear in the isolated spinal cord of the neonatal rat following disinhibition (Bracci et al., 1996; Cowley and Schmidt, 1995). Disinhibiton-induced bursting is driven through recurrent excitation via glutamate receptors, mainly of the non-NMDA type (Legrand et al., 2004). This finding shows that the developmental switch of the GABA and glycine system from excitatory to inhibitory effects did occur in the slice cultures.

We have proposed that the oscillations within the bursts are based on activity-dependent synaptic depression, which occurs in the cultures as well as in the isolated spinal cord (Pinco and Levtov, 1993; Streit et al., 1992). In a computational study we have shown that, indeed, depression leads to network oscillations in the observed frequency range (Senn et al., 1996). All these findings suggest that the slice cultures maintain important properties of rhythm generation of the in vivo spinal circuits. However, CPGs do not fully develop in slice cultures. NMDA and 5-HT are ineffective in evoking rhythms (Ballerini et al., 1999; Streit, 1996), and the rhythmic activity that is induced by high K^+/low Mg^{2+} is always synchronous in the left and the right side in spite of the inhibitory effects of the GABA/glycine system. Thus, fictive locomotion patterns cannot be evoked. Recent findings have shown that the alternation depends on the eph ligand/receptor system which prevents axons of excitatory interneurons from crossing to the other side of the spinal cord (Kullander et al., 2003). In light of this study it may well be that too many excitatory axons crossing the midline develop in the slice. Furthermore there seems to be a homeostatic regulation of circuits producing spontaneous activity: long-term blockade of spontaneous activity in the cultures leads to a suppression of GABAergic inhibitory synapses (Galante et al., 2000).

16.1.3   Random Networks

16.1.3.1   Networks of Dissociated Spinal Neurons

Neural networks can form entirely in vitro from dissociated and randomly seeded neurons in culture. Although molecular cues of individual cells as well as activity-dependent mechanisms may still structure such networks, we call them random networks because the network architecture is randomized at day 0 in culture. It was recognized early that such cultures develop patterns of spontaneous activity. In the 1970s P.G. Nelson's group described the receptors mediating such spontaneous activity in cultures of dissociated cells of the mouse spinal cord (Ransom et al., 1977). These authors found that GABAergic circuits develop earlier than glutamatergic circuits and that the patterns of activity observed at different in vitro ages corresponded to the ratio of GABAergic to glutamatergic transmission. Later on, some of these results were confirmed in rat cultures (O'Brien and Fischbach, 1986). At about the same time G. Gross and his group started to grow dissociated mouse neurons on MEAs and to extracellularly record activity simultaneously from many points in the network (Gross et al., 1982). Based on such data they investigated how several receptors and ion channels contribute to generate the activity patterns of the network. They found that, as in slice cultures and isolated spinal cords, rhythmic bursting could be induced either by high K^+/low Mg^{2+} or by disinhibition with bicuculline and strychnine. More recently they showed that such rhythms become highly regular when all except the NMDA receptors are blocked (Keefer et al., 2001).

Looking for the source of spontaneous activity in cultured networks Latham et al. (2000a,b) discovered that cultures of dissociated mouse spinal cord contain a percentage of intrinsically spiking neurons. In a theoretical study they showed how the number of intrinsically firing cells can determine whether spontaneous activity in the cultures is steady or bursting (Latham et al., 2000a). They could confirm their theoretical predictions by varying the number of intrinsic spikers by changing culturing conditions (Latham et al., 2000b).

16.1.3.2   Emergent Properties of Random Networks

The finding that at least some of the rhythms observed in networks in intact spinal cords can be reproduced by random cultures suggests that the network architecture is not a critical issue for such rhythm generation. This view is even strengthened by the fact that similar rhythms as described above are found in networks of neurons from other areas of the CNS such as the cortex or the hypothalamus (Muller and Swandulla, 1995; Robinson et al., 1993). This finding shows that this type of rhythm generation is not specific for networks of spinal neurons. Two possible hypotheses may explain such a general mechanism for rhythm generation. First, all these networks contain a class of robust pacemaker cells, which drive the network, or, second, rhythm generation may be an emergent property of neural networks, which does not require a specific network architecture. Emergent properties of networks are properties that are not immediately evident from the behavior of individual neurons (Faingold, 2004). In the remaining part of this chapter we present some evidence for the second hypothesis. Furthermore, we present some insights into the mechanisms involved in emergent properties from our studies of organotypic and dissociated cultures of mouse and rat spinal cord, combining MEA with whole cell recordings.

Recent studies are just starting to reveal the enormous complexity of circuits of interneurons in the spinal cord in terms of specific cell types (Kiehn and Kullander, 2004). In the context of these findings one may ask whether it makes sense to investigate rhythm generation in such artificial systems as cell cultures. We think that with this complexity in mind it is even more important to know which functions of the network emerge from the properties of their components. Such knowledge serves as a basis on which the more complex functions requiring specific network architecture can be understood.

16.2   Slice Cultures of Spinal Cord: Where Are the Pacemakers?

16.2.1   Collective Network Behavior Revealed by MEA Recordings

16.2.1.1   Slice Cultures on MEAs

To analyze ensemble activity of neuronal networks it is crucial to record from many points of the network at once. Several methods have been developed to enable such multi-site recording. They are either based on voltage-sensitive or calcium-sensitive dyes or extracellular electrodes. Dyes that are sensitive to voltage or calcium produce activity- dependent light signals, which can be detected by a camera or an array of photodiodes (Darbon et al., 2002a). They usually have good spatial but limited temporal resolution. Extracellular electrodes measure potential differences between the recording electrode and the ground. Such transients are produced by the current flow that is due to changes in membrane conductance of individual cells. They can be measured by needle electrodes, which are moved close to the cells with micromanipulators or, in a technically easier procedure, by electrodes, which are incorporated into the substrate of the cell culture. Such multi-electrode arrays have been used to record from cell cultures since the 1970s. Their spatial and temporal resolution depends on technical parameters such as the number of electrodes that can be packed into the array and the limitations of the analog-to-digital converter (A/D) card and the computer that acquires the data. Because the speed and the memory capacity of average lab computers has increased dramatically in the last ten years, handling large amounts of data and thus recording with high temporal and spatial resolution with MEAs became possible for many labs.

The MEAs used for our studies were developed in the Institute of Microsystems of the Ecole Polytechnique Fédérale de Lausanne (EPFL) and are now made

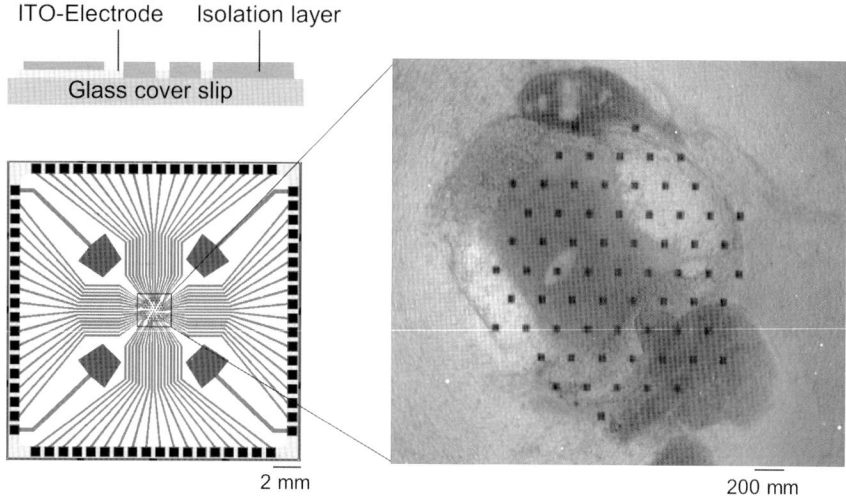

FIGURE 16.1. Spinal slice culture on MEA. Left side: cross-section and layout of the MEA chip. Note the four large ground electrodes, Right side: spinal slice culture on a hexagonal layout of black platinum electrodes after 14 days in vitro.

commercially available by Ayanda Biosystems, Lausanne, Switzerland. They are produced using standard photolithographic methods (for details see Heuschkel, 2001). They are composed of a glass substrate (700 μm thick, 21 × 21 mm), indium tin oxide (ITO) electrodes (100 nm thick, 40 × 40 μm) and leads, and a SU.8 polymer insulation layer (5 μm thick). In some of the arrays the electrodes are additionally covered by a layer of platinum. The recording site is composed of 68 electrodes arranged in several configurations (hexagon, rectangle, or four zones) with an interelectrode spacing of 200 μm. The electrodes have an impedance of 300 kohm (platinum electrodes) up to 1 Mohm (ITO electrodes) at 1 kHz in normal extracellular solution.

Slices of spinal cord from embryonic rats or mice at E14 were attached to the MEAs (see Figure 16.1) with coagulated chicken plasma and kept in plastic tubes that were placed in rotating drums in incubators for up to five weeks. The rotation caused an alternating exposure of the cultures to air (containing 5% CO_2 to maintain the pH at 7.4) and a nutrient medium. The medium was Dulbecco's MEM with glutamax, 10% fetal calf serum, and nerve growth factor. More details about the cultures are given in Tscherter et al. (2001).

16.2.1.2 Signals Recorded by MEAs

For the experiments, slice cultures with an age of 10 to 20 days in vitro (DIV) were used. The MEA with the culture was placed into a plexiglas chamber, mounted on an inverted microscope, and superfused with a bath solution of the following composition (mM): NaCl 145, KCl 4, $MgCl_2$ 1, $CaCl_2$ 2, HEPES 5, Na pyruvate 2, glucose 5 at pH 7.4. The bath solution was exchanged every 10 to 15 min during the experiments, which usually lasted for five to eight hours. Recordings were made at room temperature, in the absence of solution flow.

Each electrode was AC-coupled to an individual custom-made preamplifier and amplifier. The amplified signals were digitized at a rate of 6 kHz with 12 bit resolution and stored on hard disc for later offline analysis. The A/D card was controlled by a custom-made Labview® program. Three different signals were usually recorded by the electrodes (see Tscherter et al., 2001): fast, medium, and slow. The fast transients (see Figure 16.2), lasting less than 4 msec, correspond to single-action potentials in neuronal somata and axons (single-unit activity). They often appear in clusters (multi-unit activity), which probably originate from closely timed action potentials of several neurons seen by one electrode. The medium transients lasted for 100 to 500 msec and probably correspond to local field potentials that are caused by synaptically induced strong depolarization

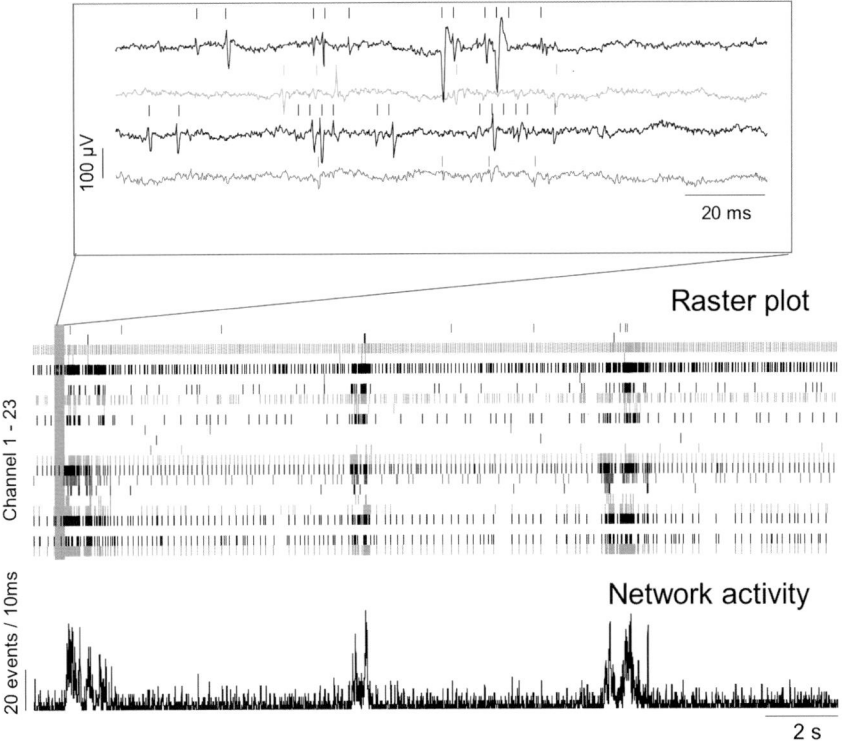

FIGURE 16.2. Spike detection: the upper graph shows original traces from four electrodes, from which the fast transients (spikes) were detected and indicated by time markers. Middle graph: the time markers of all electrodes (events) are displayed in a raster plot. Lower trace: all detected events in the network are counted in 10 msec bins and displayed as network activity plot.

in large groups of neurons. The slow signals last for several (5 to 20) seconds and are probably caused by changes in the composition or volume of the interstitial space.

Only the fast signals were considered for further analysis. The first step of such analysis was the detection of these signals. No attempt was made to sort the spikes seen at one electrode. Spike detection was based on computation of the standard deviation of the original traces. A threshold was set at three times the mean standard deviation, above which signals were detected. For single units, detection was easy and resulted in so-called events, which were directly related to the underlying spikes (see Figure 16.2). For multi-unit activity, detection of individual spikes was not possible due to the overlap of the signals. For such activity, an event rate of 333 Hz was defined. The selectivity of spike detection was assured in each experiment by using recordings obtained in the presence of the sodium channel blocker TTX as a zero reference. The detected events were plotted versus time for each electrode (raster plot, see Figure 16.2). Counting the number of events in bins of 10 msec resulted in plots of the network activity versus time (Figure 16.2). The methods of spike detection and presentation are described in more detail in Tscherter et al., (2001).

16.2.1.3   Spontaneous Activity in Slice Cultures

All cultures show a high amount of spontaneous activity. As shown in Figure 16.3, most of this activity spreads in the slice leading to simultaneous activity at many or even most of the electrodes. Such "waves" of network activation (see also Tscherter et al. 2001) last for about 100 msec and often appear repetitively at frequencies around 4 to 5 Hz (see Figure 16.3). During such waves, activity is most prominent in the ventral parts of the slice, around the central fissure. These areas are activated during virtually every wave, whereas the more dorsal parts of the slice are less active and are not reached in all waves. Between the waves, sporadic activity, which is restricted to one or a few electrodes (asynchronous background activity), appears. Activation of the slice during the waves is based on glutamatergic synaptic transmission through AMPA/kainate and NMDA receptors (Legrand et al., 2004), because they are completely blocked by a combination of blockers of these receptors (6-cyano-7-nitroquinoxaline-2-3-dione (CNQX) and (+/−)-2-amino-5-phosphonopentanoic acid (APV)). However, after blockade of excitatory synaptic transmission, asynchronous activity is still recorded at several electrodes.

In vivo, the output of the networks of interneurons in the ventral horns goes first to motoneurons and then to skeletal muscles. In slice cultures a similar output forms when skeletal muscle is co-cultured with the spinal slices. We have previously shown that such muscle fibers are indeed innervated by spinal neurons and that the patterns of muscle contractions follow the patterns of activity in spinal cord slices (Streit et al., 1991; Streit, 1996; Tscherter et al., 2001). Combining MEA recordings with an optical device to record muscle contractions, both the electrical and the mechanical activity of muscle fibers can be measured together with the activity in

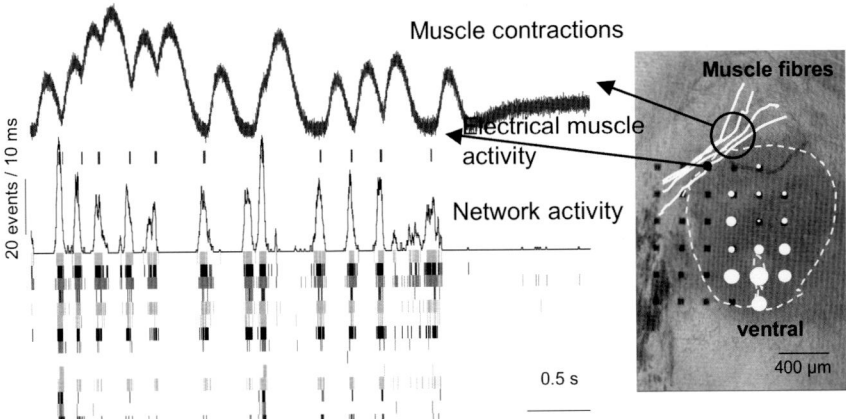

FIGURE 16.3. Spontaneous activity in spinal network and in co-cultured muscle. The raster and network activity plots (black) show synchronous network activity (population bursts). The distribution of activity in the slice is visualized by the points on the left. The size of the points is proportional to the activity at the electrodes. The population bursts in the spinal network induce contractions in co-cultured skeletal muscle fibers. The red event markers show the electrical activity and the red trace the contractions of the muscle (measured by an optical device).

the neuronal networks of the ventral horns. From such recordings it is clear that muscle contractions correlate with the population bursts in the spinal networks. (see Figure 16.3). Therefore correlated network activity seems to be necessary to activate motoneurons. In this way these in vitro networks even produce a simple form of behavior.

16.2.2 How Are Rhythms Induced?

16.2.2.1 Fast Oscillations

As mentioned before, the patterns of spontaneous activity are organized into short population bursts (waves) which often appear in short trains of four to five bursts following each other with intervals of around 200 msec. In some cultures they form persistent, highly regular oscillations of network activity at 4 to 5 Hz which last for several hours. In principle such oscillations can appear in slice cultures of the rat and the mouse and also in cultures of dissociated spinal neurons. Therefore, as with other rhythms described in this chapter, oscillations do not depend on a highly specific network architecture, which needs be preserved. Nevertheless the normal pattern of spontaneous activity seen in the cultures is quite irregular with only short episodes of oscillations (see Figure 16.3).

In slice cultures of rat spinal cord, similar oscillations as described above reliably appear under disinhibition. As pointed out in more detail in the next section, disinhibition means the pharmacological removal of synaptic inhibition from the network. Because the most prominent fast inhibitory neurotransmitters in the spinal cord are GABA and glycine, disinhibition is achieved by combining blockers of GABA$_A$ and glycine receptors, bicuculline, and strychnine. In slice cultures of rat spinal cord, disinhibition leads to a pattern of spontaneous activity that consists of long episodes (or bursts) of high network activity followed by silent intervals. During such bursts, activity is usually high at the beginning, then drops to a lower level of sustained activity for several hundreds of milliseconds and finally starts to oscillate for another several seconds (see Tscherter et al., 2001). The period of the oscillations becomes longer toward the end of the episodes. A very similar pattern of "intraburst" oscillations is seen in isolated spinal cords of neonatal rats after disinhibition (Bracci et al., 1996) and in chick embryo (O'Donovan et al., 1998b). However, they are usually not found in slice cultures of embryonic mouse and in cultures of dissociated spinal neurons after disinhibition, whereas, as mentioned before, they sometimes appear as innate rhythms (without pharmacology) in these preparations. The reasons for these discrepancies are not known.

We have previously proposed that the oscillations are based on repetitive network activation through recurrent excitation. The use-dependent fast depression of excitatory synaptic transmission acts as an accommodation mechanism. Such depression has been found in slice cultures of rats (Streit et al., 1992) as well as in isolated spinal cords of neonatal rats and chick embryo (Pinco and Levtov, 1993; Tabak and O'Donovan, 1998). A computer model with the main parameters of recurrent excitation and use-dependent synaptic depression reproduces oscillations at 4 to 5 Hz (Senn et al., 1996). It remains to be shown whether the differences in the generation of oscillations between preparations can be explained by differences in synaptic depression.

16.2.2.2   Disinhibition-Induced Slow Bursting

As mentioned before, disinhibition by bicuculline and strychnine reliably induces a slow bursting in slice cultures as well as in cultures of dissociated spinal neurons. This pattern is characterized by long-lasting 1 to 20 sec episodes of high activity in the whole network (bursts) followed by silent intervals with low and asynchronous activity. The activity during the bursts is either persistent (in dissociated cultures and mouse slice cultures) or at least toward the end of the bursts oscillating as described in the preceding section. Persistent activity decreases during bursts with a rapid decay at the beginning, a subsequent plateau phase with slow decay, and usually a second rapid decay at the end. This decay is a network effect inasmuch as it reflects the decaying number of spikes in the network, which causes a decreasing synaptic input to the individual cells. On the other hand, it is also caused by spike frequency adaptation in individual neurons as described later in this chapter.

During disinhibition-induced bursting, the resting membrane potential of the neurons is hyperpolarized compared to the innate spontaneous activity by more than 10 mV. This hyperpolarization is due to an increase in the activity of the electrogenic Na/K pump caused by the high level of activity (Darbon et al., 2003). Usually the intervals between bursts are too short for a visible recovery from such up-regulation of the pump. Therefore the membrane potential during the intervals is stable in many cells. Nevertheless, when the intervals are long (e.g., due to a partial block of excitatory transmission by CNQX), recovery from hyperpolarization becomes evident (Darbon et al., 2003).

Two more conclusions can be drawn from these disinhibition-induced patterns: First, in all cultures, GABA and glycine act as inhibitory neurotransmitters because they do not support but rather suppress rhythms. This shows that the developmental switch of GABA and glycine from excitatory to inhibitory system has occurred during in vitro development. Second, synaptic inhibition does not usually prevent population bursting, because the latter appears both in the presence and in the absence of functional synaptic inhibition. Nevertheless, synaptic inhibition partly contributes to the termination of population bursts because they are much shorter with functional synaptic inhibition than without.

16.2.2.3   Regular Fast Bursting

Disinhibition-induced bursts appear more frequently and more regularly when NMDA is added (see Figure 16.4 and Legrand et al., 2004). This typical pattern of regular fast bursting is additionally characterized by a high level of background activity during interburst intervals, a slowing of burst onsets, and a decrease in burst amplitudes. The same rhythms are induced when disinhibition is combined with elevated K^+ in the bath solution, when disinhibition is reduced using low concentrations of bicuculline and strychnine, or when elevated K^+ is combined with $0 \, Mg^{2+}$ in the bath solution (Streit et al., 2001). Transiently, such rhythms also appear when the Na/K pump is blocked by strophanthidin (Darbon et al., 2003). All these procedures lead to a depolarization of neurons in the network. We therefore propose that these regular fast rhythms are produced by recurrent excitation, as described before, but in the presence of a general depolarization of the network. In terms of frequency and shape, regular fast rhythms are similar to the fictive locomotion patterns observed in isolated spinal cord preparations. However, slice cultures lack the important feature of alternation between left and right, showing that only rhythm-generating networks (unit oscillators) but no complete pattern generators have developed in culture.

16.2.3   Where Are the Pacemakers?

16.2.3.1   Burst Sources

Some attempts have been made to localize the pattern generator networks in the transverse plane of intact spinal cords. The methods used include lesion studies, staining with dyes such as sulforhodamine, which are taken up by neurons in an activity-dependent way (Kjaerulff et al., 1994), and calcium imaging (Demir et al., 2002; McPherson et al., 1997). The outcome of these studies, although

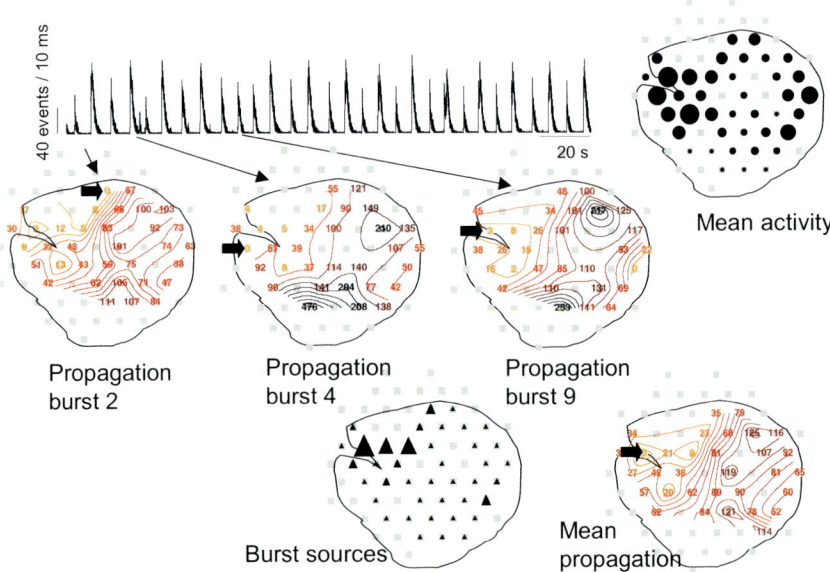

FIGURE 16.4. Burst sources and propagation: regular bursting induced by disinhibition and NMDA. The distribution of activity is shown by the black points on the right. Burst sources are determined from the raster plots at the onset of each burst. Propagation is analyzed from the distribution of the delays and the interpolated isochrones. The variability in source and propagation is shown for three bursts in the middle row. The distribution of burst sources (triangles, size is proportional to the bursts initiated at the electrode: smallest triangles show active electrodes with no sources), and the mean propagation over all bursts are shown in the lower row.

not entirely conclusive due to methodological limitations, points to a localization of the pattern generator networks around the central canal. Intracellular recordings from this area using sharp electrodes or patch-clamp have indeed shown that a high percentage of neurons are rhythmically modulated during fictive locomotion (Butt et al., 2002). However, these methods did not allow distinguishing between cells that drive rhythms and those that are driven by rhythms. Therefore the question of whether somewhere in the spinal cord there are one or several pacemakers for rhythm generation like the sinus node in the heart, or whether rhythm generation is an emergent property of the pattern-generating network, remained unsolved.

We have addressed this question using slice cultures of rat and mouse spinal cords on MEAs. Mapping the activity seen at the electrodes on the slices (see Figures 16.3 and 16.4, and Tscherter et al., 2001) reliably shows that activity is highest in the ventral parts of the slices around the central fissure. This finding strongly suggests that the rhythm-generating networks in spinal slice cultures in fact belong to the central pattern generators. We then looked for the origin of rhythmic activity by analyzing the propagation of the wavefronts of each burst (see Figure 16.4). We found that even when rhythms are highly regular (induced by disinhibition and NMDA) there is much variability in the origin and the propagation of the wavefronts from burst to burst. Nevertheless, there are a limited number (three to eight) of sites from which bursts originate (burst sources). These burst sources are usually grouped around the central fissure, although single "ectopic" sites are sometimes seen in the dorsal part of the slices, at the sites of the entrance of the dorsal roots.

In spite of the variability from burst to burst, there is also a tendency in the propagation of the wavefront in the slice: bursts start from a source at one side of the central fissure, propagate to the opposite ventral horn, and finally to the dorsal parts of the slice (see Figure 16.4 and Tscherter et al., 2001). In most slices, burst sources are found on both sides of the slice. Although bursts always reach both ventral horns of the slice, they sometimes fail to propagate to the dorsal parts. This general pathway of propagation is the same for innate spontaneous activity as well as for all three types of induced rhythms. The findings of several burst sources and the variability of propagation rule out the hypothesis of a single pacemaker as the source of rhythm generation. However, they are compatible with the existence of a network of pacemaker cells distributed around the central fissure.

16.2.3.2   Pacemaker Cells?

To identify intrinsic spiking cells it is necessary to block all synaptic transmission in the network. The activity that is left under such conditions can be attributed to intrinsic activity. It turns out to be sufficient to block glutamatergic receptors by a combination of CNQX and APV to suppress bursting in the network. Under such conditions asynchronous activity remains at about 30% of the electrodes (Figure 16.5). This activity is probably entirely due to intrinsic spiking, inasmuch as it is not further changed when synaptic release is totally blocked by a bath solution containing 0 Ca^{2+} and 3 Mg^{2+}. Looking at individual electrodes, the rate of such intrinsic activity varies between 0.1 and 10 Hz. At some of the electrodes, regular tonic activity is seen, whereas in others activity fluctuates. Rarely is clear bursting seen at one electrode, and bursting is restricted to one electrode and differs also in rate and regularity from bursting seen in the presence of synaptic transmission.

The interpretation of these data is compromised by the fact that more than one cell could contribute to the activity recorded by one electrode. Nevertheless we can conclude that up to 30% of neurons in the network are capable of intrinsic spiking, which, however, does not define rhythms observed in the presence of synaptic transmission. Therefore these neurons cannot be considered as true cellular pacemakers. To find out whether these cells may at least trigger bursts, we compared their spatial distribution to that of burst sources. We found a good correspondence between the distribution of intrinsic activity and burst sources (see Figure 16.5), suggesting that intrinsic spiking is indeed the source of population bursts.

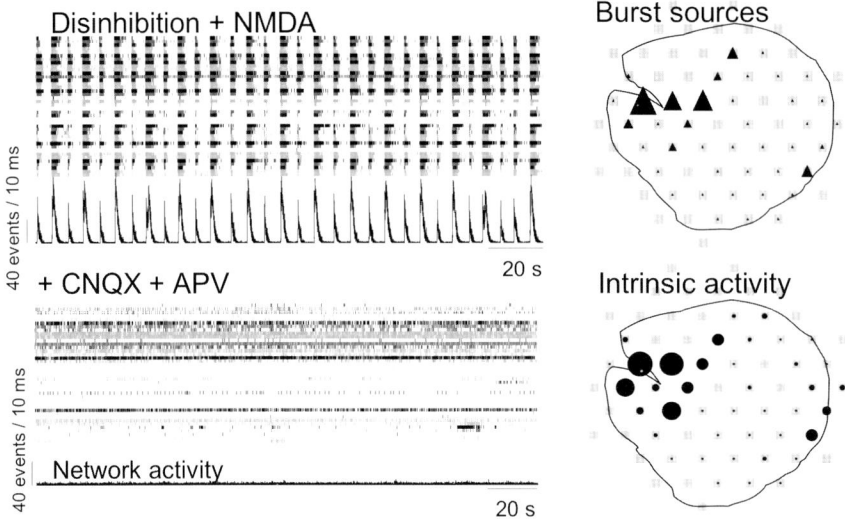

FIGURE 16.5. Burst sources and intrinsic activity: the upper row shows the distribution of burst sources during regular fast bursting (disinhibition + NMDA). The lower row shows the distribution of intrinsic activity after blockade of glutamatergic synaptic transmission by CNQX (10 μM) and APV (50 μM). Note the good correspondence between the two distributions.

16.3   Random Networks Grown In Vitro: What Can We Learn?

16.3.1   What Are Neurons Doing?

16.3.1.1   Rhythm Generation in Random Networks

The findings presented so far suggest that rhythm generation is based on repetitive recruitment of neurons in the network by intrinsic spiking cells through recurrent excitation. Such a mechanism does not depend on a specific network architecture. Therefore, the same rhythms as seen in organotypic slices should also appear in randomized networks, provided cellular properties are the same. Networks can be randomized by dissociation of neurons of the dissected slices on the day of preparation. When these dissociated cells are seeded out in cultures, they form a new network within several days. After three to four weeks in cultures, these networks reach a steady state in terms of patterns of spontaneous activity they produce. It has been known for some time that such networks are capable of rhythm generation. We therefore used them to test the hypothesis that random networks reproduce population bursting. Indeed, we found all three patterns of rhythmic activity in dissociated cultures as well that we have described before in slice cultures (compare Tscherter et al., 2001 and Streit et al., 2001).

Nevertheless, there are differences between the culture types in the protocols used to evoke rhythms and in the ease with which they appear. A major difference is that the fast oscillations during the bursts, which are reliably found during disinhibition in rat slice cultures, are rarely seen in dissociated cultures, where activity persists during the bursts (as shown before). The reason for this discrepancy is still unknown. It is, however, not related to structural differences between slice and dissociated cultures, because activity during the bursts in mouse slice cultures does not oscillate and thus resembles the rhythms found in rat dissociated cultures. Because the fast oscillations are probably based on synaptic depression, the observed differences between the culture systems may reflect differences in the frequency response of the synapses. This remains to be shown experimentally.

In addition to this difference, rhythms in dissociated and slice cultures are similar, suggesting a common mechanism of rhythm generation. As in slice cultures, highly regular fast rhythms appear with disinhibition and NMDA in dissociated cultures (Legrand et al., 2004). The bursts also originate from several sources and propagate in the network along variable paths, which, however, follow a general pattern from each source. Several sources can even share the same general pattern of network recruitment, suggesting that the intrinsic spiking neurons are strongly interconnected and thus form a "trigger network" (Yvon et al., 2005). The cells belonging to such trigger networks have short delays for recruitment. When they are distributed over the whole network, this leads to an uneven spatial distribution of delays. When they are concentrated at one site, the propagation of the wavefront from there through the network can be smooth, leading to an even distribution of latencies (see Streit et al., 2001, Yvon et al., 2005). The recruitment of the whole network requires on average 50 to 100 msec for a network covering a rectangle of 3×1 mm. Because the conduction velocity in axons of the cultured cells is around 0.3 m/sec, it becomes clear that conduction time is only a small percentage of the total recruitment time, which depends strongly on mean synaptic density and excitability of neurons. In about half of the dissociated cultures, two or occasionally even more trigger networks are present.

16.3.1.2   Intrinsic Spiking Neurons

Cultures of dissociated cells have the advantage over slice cultures that the networks form a monolayer and therefore individual cells are more directly accessible for intracellular single-cell recordings. We therefore used this preparation to combine network activity recordings by MEA with single-cell recordings by the whole-cell patch-clamp method (see Figure 16.6). We usually found a good correlation between network activity and postsynaptic potentials and action potentials in neurons. In the example shown in Figure 16.6, the recorded neuron responds to increased network activity with trains of spikes. This example illustrates the typical case in which the neuron is driven by the network. However, in some neurons, episodes of repetitive spiking appear that are unrelated to network activity. Such spiking is based on a slowly depolarizing membrane potential and persists when glutamatergic synaptic transmission is blocked by CNQX and APV (Figure 16.7).

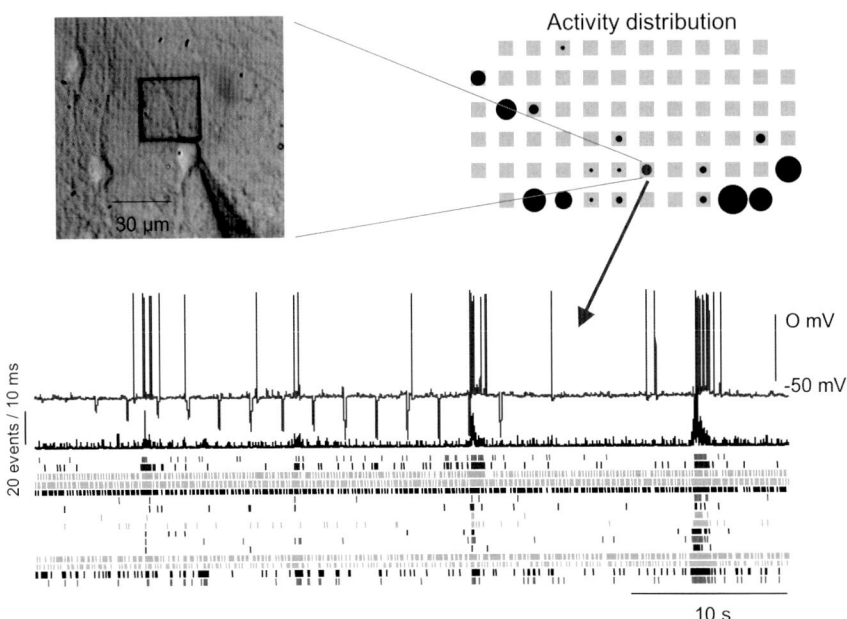

FIGURE 16.6. MEA recordings combined with whole cell recording from one cell. The picture on the left shows one MEA electrode (ITO only to keep it transparent) together with a patch-clamped neuron. Below, the raster and network activity plot of spontaneous activity are shown together with the single-cell recording (upper trace). Note the correlation of the firing of the cell with the ensemble activity of the network. The distribution of the activity and the location of the recorded neuron in the network are shown in the upper right graph.

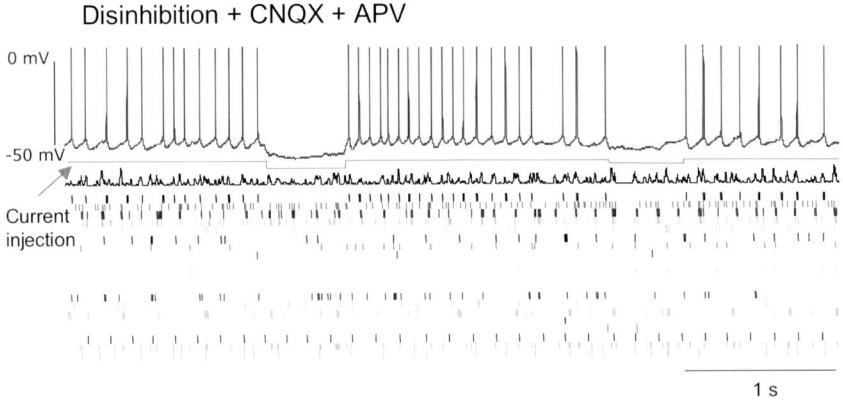

FIGURE 16.7. Intrinsic spiking persists after block of fast synaptic transmission. Raster and network activity plots are shown together with the intracellular recording from one intrinsic spiking neuron (upper trace) and the activity distribution (lower trace). All recordings were made after the blockade of fast synaptic transmission with strychnine, bicuculline, CNQX, and APV. Note that the intrinsic spiking in the neuron is stopped during hyperpolarizing pulses.

It can be switched off by a slight hyperpolarization of the membrane without leaving an oscillating membrane potential. On the other hand, spiking frequency can be increased by a depolarization of the membrane. Thus even small fluctuations of the membrane potential cause immediate changes in spike rate. This explains the irregular tonic rates of intrinsic activity recorded by the MEA electrodes (see Figure 16.7).

We are currently investigating the conductance underlying the slow depolarization that causes intrinsic spiking. Our findings suggest that the major component is a TTX-sensitive persistent Na^+-current (I_{NaP}), which is activated at a membrane potential between –60 and –20 mV. In addition there is a co-operative effect of the hyperpolarization-activated cation current (I_h), which is activated at potentials more negative than –60 mV. Both currents are present in only some of the spinal neurons, but I_{NaP} is found exclusively in intrinsic spiking cells (Darbon et al., 2004).

A critical question is whether the intrinsic spiking cells are indeed capable of driving the network, that is, recruiting enough cells to start a population burst. This is clearly not the case for most of these neurons. Even when they fire intrinsically, the network does not follow. This can be easily demonstrated by spikes that are artificially induced by injection of current pulses. In the majority of experiments in which this has been done, the network did not follow these spikes, not even repetitive spiking at high frequency. However, we occasionally found cells that were able to drive the network on their own, even with single spikes. From these findings we conclude that only a small percentage of intrinsic spiking cells are able to recruit the network. These cells are functionally connected to one or several highly excitable trigger networks, thus explaining the reliable repetitive network recruitment during fast regular bursting. In addition, one has to consider that fast regular bursting occurs when the network is depolarized (see above). Under such conditions, both the number of intrinsic spiking cells as well as the excitability of the trigger networks are increased. The receptors underlying the recruitment of the trigger networks are mainly AMPA/kainate, and, less importantly, NMDA receptors (Legrand et al., 2004). This seems to be different from networks of dissociated neurons of mouse spinal cord, in which regular oscillatory activity can be maintained entirely by NMDA receptors (Keefer et al., 2001).

16.3.2   What Shapes Rhythms?

16.3.2.1   Spike Frequency Adaptation

Rhythms are composed of a state of high network activity and a state of low network activity (intervals). In the previous sections, we have seen that intrinsic spiking drives the network through recurrent excitation from the low state to the high state. What brings it back then to the low state? In the original half-cycle model of pattern generation in the spinal cord, this is achieved by mutual synaptic inhibition. In innate spontaneous activity, population bursts are indeed much shorter (around 100 msec) than during disinhibition, suggesting that synaptic inhibition is indeed involved in burst termination. Nevertheless, we have shown that the most reliable and regular rhythms appear under disinhibition, indicating that mechanisms other than synaptic inhibition are involved. We have previously mentioned that use-dependent synaptic depression is involved in shaping fast oscillations. The recovery time constant of synaptic depression is around 200 msec, which agrees well with the frequency range of fast oscillations of around 5 Hz; however, it is too fast to shape the slower patterns.

Looking at the persistent network activity during the slow and regular fast bursts induced by disinhibition in dissociated cultures, activity decreases in three phases during the burst: a first rapid decrease is followed by a plateau with only a slight decrease and finally a second rapid fall terminates the burst (see Figure 16.8). In individual cells, network activity causes a depolarization with a concomitant increase in spike rate. Spike frequency is initially high and decreases during the burst. Often, spiking even ceases during the burst. In some cells (about 30%), synaptic depolarization is so strong during bursts that spikes immediately die out due to an inactivation of Na^+ channels (see Figure 16.9). These cells thus respond to network bursts mainly with a depolarized plateau potential (Darbon et al., 2002a,b). The decrease in spike frequency during bursts is certainly due to decrease in synaptic input current. On the other hand, when a stable input current

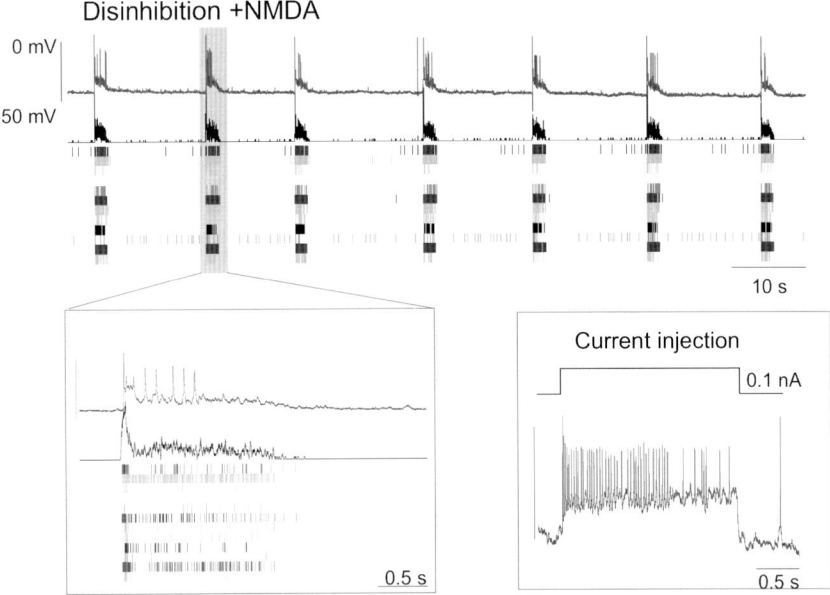

FIGURE 16.8. Spike frequency adaptation. Raster and network activity plots combined with intracellular recording from one neuron (upper trace) show the decrease in network activity and spike rate during the bursts. A similar decrease in spike rate occurs during constant pulses of depolarizing current injections of similar length and amplitude as the bursts.

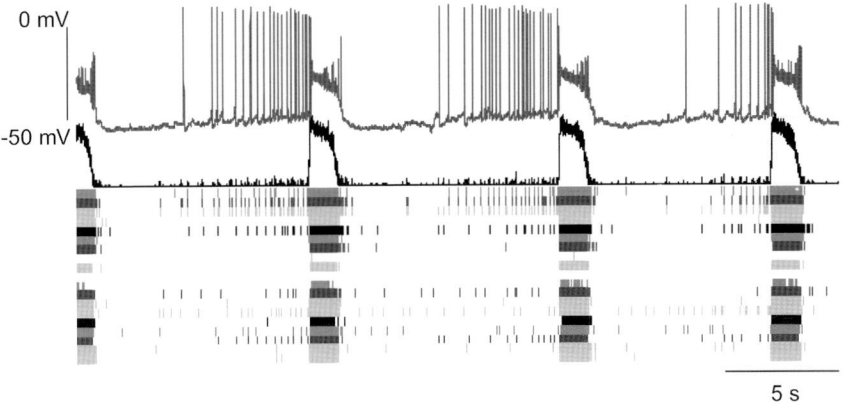

FIGURE 16.9. Intrinsic spiking is suppressed following bursts. Raster and network activity plots are shown together with the intracellular recording of an intrinsic spiking cell (upper trace) during disinhibition-induced bursting. Note the hyperpolarization combined with a suppression of spiking following the burst. In parallel, asynchronous background activity in the intervals is low following the burst and recovers during the interval. During the bursts spiking ceases in the recorded neuron due to the strong depolarization.

of the same length is injected into the cell, the frequency of the evoked spikes also decreases and can cease after seconds during the pulse, much as during the bursts. Furthermore, plateau potentials can also be induced by injection of large currents. From these observations, we conclude that such slow spike frequency adaptation is the primary mechanism leading to termination of these bursts. The early rapid decrease in network activity may additionally be caused by synaptic depression.

Spike frequency adaptation can be based on accumulation of Ca^{2+}-dependent K^+ currents or on a slow inactivation of Na^+ currents (I_{Na}) (Ellerkmann et al., 2001). We have previously shown that bursts cannot be terminated by an artificial increase in intracellular Ca^{2+} concentration (Darbon et al., 2002a). In addition, apamin and charybdotoxin, blockers of the Ca^{2+}-dependent K^+ currents, have only minor effects on bursting. Therefore we suggest that the slow spike frequency adaptation is due to a progressive inactivation of I_{Na}. In line with this hypothesis is the finding that spike frequency adaptation is accompanied by a decrease in spike amplitude and a slowing of spikes, both characteristic of a smaller I_{Na}. Furthermore, slow inactivation of I_{Na} seems to be enhanced by riluzole, because it enhances spike frequency adaptation (unpublished experiments). Thus we conclude that slow inactivation of I_{Na} is the major mechanism involved in the termination of long-lasting persistent bursts.

16.3.2.2   Auto-Regulation of Intrinsic Spiking

When the level of intrinsic activity at individual MEA electrodes is compared to the asynchronous activity in the intervals of slow bursting at the same electrodes (the first measured in the absence of synaptic transmission, the second during disinhibition), there is usually less asynchronous activity than intrinsic activity. This suggests that intrinsic spiking is suppressed following bursts. Indeed, at some electrodes, the background activity is very low following the burst and increases with time during the interval. In intrinsic spiking cells, spontaneous spiking is suppressed following bursts and slowly recovers during the interval (see Figure 16.9). The silent period following the burst often goes parallel with a transient hyperpolarization of the cell. As mentioned before, even those cells that show no such transient hyperpolarization following bursts are hyperpolarized during the intervals relative to the state before disinhibition (innate spontaneous activity). This suggests that bursts cause a hyperpolarization of the cells (by up to 10 mV), which slowly recovers during the intervals.

Only when the intervals are long enough (due to a low burst rate) does the transient nature of the hyperpolarization become evident; otherwise it appears as a persistent hyperpolarization. In such experiments the transient nature of the hyperpolarization is revealed when the intervals are prolonged by low concentrations of CNQX (Darbon et al. 2002b). The suppression of intrinsic spiking and the hyperpolarization can be induced in a neuron in the absence of synaptic transmission, when bursts are mimicked by repetitive long-lasting pulses of current injection (see Darbon et al., 2002b). During trains of such pulses, intrinsic spiking disappears and slowly recovers after the train. This shows that the hyperpolarization is not synaptically mediated but is an intrinsic property of the neuron itself. It can be regarded as a negative feedback mechanism that stabilizes cellular excitability, because intensely spiking neurons hyperpolarize and thus in turn decrease their spike rate. The mechanism involved in this auto-regulation is an up-regulation of the electrogenic Na/K pump by a Na^+ load of the cell, as occurs during frequent spiking (Darbon et al., 2003). This mechanism is not only present in intrinsic spiking neurons because we also found the hyperpolarization in intrinsically silent cells. In these cells we also observed a decrease in excitability following bursts (Darbon et al., 2002b).

16.3.3   Network Refractoriness

16.3.3.1   Pacing the Network with Electrical Stimuli

MEA electrodes cannot only be used for recording but also for stimulation at any site of the network. Single stimuli of 1 to 2 V and 0.1 msec duration applied at one electrode usually cause activity at several other electrodes. This activity includes spikes, which are due to direct electrical stimulation of neurons, and those that are evoked through synaptic transmission. Most of the directly stimulated spikes occur in the first ten milliseconds following stimulus. They appear at several electrodes, which are usually located close to the stimulating electrode, but which can be distributed over the whole network for strong stimuli (Darbon et al., 2002b). Such a wide distribution is probably due to the activation of several axons that cross the stimulation electrode.

In line with this hypothesis, such activity cannot be evoked from all electrodes, especially not from those that, when used for recording, show no activity. Electrodes are manually switched from recording to stimulation. The number of spikes in the first ten milliseconds following stimulus can be taken as a measure of cellular excitability, because this parameter does not involve synaptic transmission. Delays after stimulus of more than ten milliseconds usually point to activity that is mediated through excitatory synaptic transmission. In consequence, such activity is suppressed by CNQX and APV. Its intensity depends on the amount of spontaneous activity. During innate spontaneous activity and during fast regular bursting, when intrinsic activity is high, stimuli had almost no effect on the patterns or on the total amount of activity. This shows that under these conditions, intrinsic spiking is so intense that the additional spiking induced by external stimulation does not have much influence on the total activity in the network. Spontaneous activity is thus dominated by the various sources of refractoriness in the network (including synaptic inhibition) and not by the amount of activity (intrinsic plus stimulated) that is driving the network.

In contrast, when neurons are hyperpolarized during slow population bursting, intrinsic activity is reduced. Under such conditions, external stimulation can evoke bursts and pace the network if its frequency is properly adjusted. External stimuli do not evoke bursts when they are applied immediately after a burst, showing that bursts are followed by a network refractory period. Only after this refractory period can stimuli evoke bursts. Therefore, when stimulated at too high frequencies, every second or more stimuli fail to induce a burst (see Figure 16.10). The highest

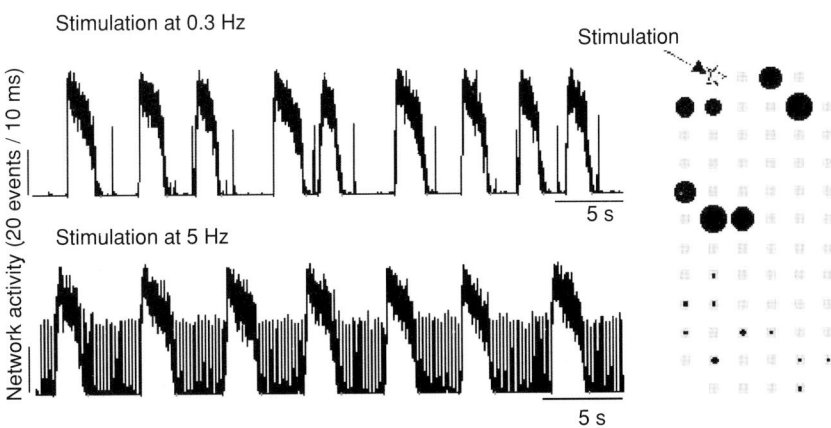

FIGURE 16.10. Network refractory period. Electrical stimuli (1 to 2 V, 0.5 msec) at one MEA electrode (star) evoke activity at several other electrodes. The size of the black points in the graph on the right side is proportional to the amount of activity, which is evoked in the first 10 msec following stimuli. The traces show the response of the network to two different frequencies of stimulation. Note that the network responds with burst rate, which is defined by the network refractory period and not by the frequency of stimulation. Stimuli, which are falling into the refractory period, fail to trigger a burst.

frequency of stimulation to which the network reliably responds with bursts (the critical frequency), is around 10/min (range 5 to 20/min; see Darbon et al., 2002b). In dissociated cultures, the frequency of spontaneous bursting (after disinhibition) is often close to the critical frequency, showing that it is determined by the refractory period and not by the amount of intrinsic spiking. In these cultures it is difficult to pace the rhythm, because at too high frequencies of stimulation, failures occur and at too low frequencies spontaneous bursts tend to disrupt bursting from stimulation. In slice cultures of the rat, however, bursting is usually slower than in dissociated cultures and the spontaneous rates are lower than the critical frequency. In these cultures pacing the rhythm is easily possible because the rate of intrinsic firing is obviously the critical parameter defining burst rate.

16.3.3.2   Network Refractory Period

Several factors may contribute to network refractoriness. We have seen before that the slow inactivation of I_{Na} and the up-regulation of the Na/K pump both contribute to a decrease in cellular excitability. In addition, a slowly recovering component of synaptic depression has been proposed to be involved in network refractoriness (Tabak et al., 2000). In spinal cultures, we know that intracellular Ca^{2+} concentration $[Ca^{2+}]_i$ increases during bursts and slowly recovers during the intervals (Darbon et al., 2002a). In parallel, the rate of spontaneous postsynaptic currents is transiently increased following bursts (Darbon et al., 2002b). This observation can be interpreted as an increased rate of spontaneous release of transmitter due to elevated $[Ca^{2+}]_i$. We have previously proposed that these experiments argue against a slow component of synaptic depression being present during the intervals. Nevertheless they do not exclude this possibility because evoked and spontaneous synaptic release need not necessarily behave the same way.

It is certain that the network refractory period is based on recovery processes. Therefore, it is relative in terms of network activation. This means that the network can be activated at different levels of network excitability by a sufficiently strong stimulus (be it intrinsic or stimulus-evoked spiking). This has been shown previously in slice cultures, in which bursting was paced at different frequencies by electrical stimulation. In these experiments, several parameters, which reflect network excitability (burst duration, time to peak, cellular excitability), correlate with the burst rate in the sense that the shorter the interval is, the lower is the network excitability (Darbon et al., 2002b). It was, in fact, recognized even earlier that during spontaneous population bursting, burst duration is positively correlated to the preceding but not to the following interval duration (Streit, 1993; Tabak and O'Donovan, 1998). Tabak et al. (2001) have shown in a theoretical study that such a correlation points to a stochastic process for burst initiation and a deterministic process for burst termination.

We have shown here that burst initiation is based on network recruitment by intrinsic spiking. This is a stochastic process if the number of intrinsic spiking neurons and their spike frequency is mainly dependent on stochastic fluctuations of the membrane potential in individual neurons, as is the case during slow population bursting. When the number of intrinsic spiking neurons and their spike frequency becomes higher and more reliable during depolarization (as is the case during fast regular bursting), burst initiation becomes a deterministic process, inasmuch as it is now mainly controlled by the periodic suppression and relaxation of excitability in neurons (Giugliano et al., 2004). Thus, we found a strong correlation between burst duration and the preceding interval in disinhibition-induced slow population bursting. This correlation was lost when in the same culture slow bursting was switched to fast regular bursting by NMDA (Legrand et al., 2004), showing that burst initiation is now a deterministic process.

16.4   Summary: Is it the Neuron or the Network?

The mechanism involved in rhythm generation as proposed here is similar to the group pacemaker model, which has been proposed to explain the generation of respiratory rhythms in the pre-Bötzinger Complex of the brain stem (Rekling and Feldman, 1998). In this model, rhythm generation is based on properties of neurons that are not pacemakers themselves, but that form a pacemaker network through mutual excitatory coupling. We have identified important properties of neurons as an intrinsic spiking mechanism based on I_{Nap} and I_h (present in about 30% of the neurons), spike frequency adaptation based on a slow inactivation of I_{Na}, the regulation of the activity of the Na/K pump, and use-dependent synaptic depression.

An important parameter that regulates rhythms is the membrane potential. If neurons are in general hyperpolarized (as during high activity of the Na/K pump induced by disinhibition), population bursting is slow with long bursts and low levels of intrinsic spiking in the intervals. If neurons are in general depolarized (as with NMDA or high $[K^+]_e$), population bursting is fast with short bursts and high levels of intrinsic spiking in the intervals. Also the regularity of the rhythms depends on membrane potential: the more depolarized the neurons are, the more intrinsic spiking there is, the more reliably bursts are initiated, and the more regular bursting becomes. However, two intrinsic accommodation mechanisms, spike frequency adaptation and synaptic depression, also define the rate and regularity of rhythms. Probably the stronger one of the two mechanisms dominates rhythms, the more regular they are. Synaptic depression, due to its shorter relaxation time constant, produces faster rhythms than spike frequency adaptation. The interaction of both mechanisms may increase the variability and thus decrease the regularity of rhythms; on the other hand it underlies the more complex rhythms such as bursting with intraburst oscillations.

Mutual excitation of neurons can be mediated by any receptors that have an excitatory function. In our cultures (as in the isolated spinal cord of the neonatal rat), AMPA/kainate and NMDA receptors seemed to be the major components. However, NMDA receptors alone, GABA/glycine or acetylcholine can also support recurrent excitation during defined periods of development.

Because the proposed mechanisms are based on fundamental properties of neurons they should apply to networks at various levels of organization as found not only in cultures but also acute in slices and in vivo.

References

Atsuta, Y.E., Garcia-Rill, E., and Skinner, R.D. (1990). Characteristics of electrically induced locomotion in rat in vitro brain stem-spinal cord preparation. *J. Neurophysiol.* 64: 727–735.

Ballerini, L., Galante, M., Grandolfo, M., and Nistri, A. (1999). Generation of rhythmic patterns of activity by ventral interneurones in rat organotypic spinal slice culture. *J. Physiol.* 517: 459–475.

Ballion, B., Morin, D., and Viala, D. (2001). Forelimb locomotor generators and quadrupedal locomotion in the neonatal rat. *Eur. J. Neurosci.* 14: 1727–1738.

Barbeau, H., McCrea, D.A., O'Donovan, M., Rossignol, S., Grill, W.M., and Lemay, M.A. (1999). Tapping into spinal circuits to restore motor function. *Brain Res. Brain Res. Rev.* 30: 27–51.

Beato, M., and Nistri, A. (1999). Interaction between disinhibited bursting and fictive locomotor patterns in the rat isolated spinal cord. *J. Neurophysiol.* 82: 2029–2038.

Bekoff, A. and Trainer, W. (1979). The development of interlimb co-ordination during swimming in postnatal rats. *J. Exp. Biol.* 83: 1–11.

Ben-Ari, Y. (2001). Developing networks play a similar melody. *Trends Neurosci.* 24: 353–360.

Bracci, E., Ballerini, L., and Nistri, A. (1996). Spontaneous rhythmic bursts induced by pharmacological block of inhibition in lumbar motoneurons of the neonatal rat spinal cord. *J. Neurophysiol.* 75: 640–647.

Bracci, E., Beato, M., and Nistri, A. (1998). Extracellular K+ induces locomotor-like patterns in the rat spinal cord in vitro: Comparison with NMDA or 5-HT induced activity. *J. Neurophysiol.* 79: 2643–2652.

Brown, T.G. (1911). The intrinsic factors in the act of progression in the mammal. *Proc. R. Soc. Lond. (Biol.)* 84: 308–319.

Butt, S.J., Harris-Warrick, R.M., and Kiehn, O. (2002). Firing properties of identified interneuron populations in the mammalian hindlimb central pattern generator. *J. Neurosci.* 22: 9961–9971.

Cazalets, J.R., Borde, M., and Clarac, F. (1995). Localization and organization of the central pattern generator for hindlimb locomotion in newborn rat. *J. Neurosci.* 15: 4943–4951.

Cazalets, J.R., Sqalli-Houssaini, Y., and Clarac, F. (1992). Activation of the central pattern generators for locomotion by serotonin and excitatory amino acids in neonatal rat. *J. Physiol.* 455: 187–204.

Cowley, K.C. and Schmidt, B.J. (1995). Effects of inhibitory amino acid antagonists on reciprocal inhibitory interactions during rhythmic motor activity in the in vitro neonatal rat spinal cord. *J. Neurophysiol.* 74: 1109–1117.

Cowley, K.C. and Schmidt, B.J. (1997). Regional distribution of the locomotor pattern-generating network in the neonatal rat spinal cord. *J. Neurophysiol.* 77: 247–259.

Crain, S.M. and Peterson, E.R. (1967). Onset and development of functional interneuronal connections in explants of rat spinal cord-ganglia during maturation in culture. *Brain Res.* 6: 750–762.

Dale, N. (1995). Experimentally derived model for the locomotor pattern generator in the xenopus embryo. *J. Physiol.* 489: 489–510.

Darbon, P., Pignier, C., Niggli, E., and Streit, J. (2002a). Involvement of calcium in rhythmic activity induced by disinhibition in cultured spinal cord networks. *J. Neurophysiol.* 88: 1461–1468.

Darbon, P., Scicluna, L., Tscherter, A., and Streit, J. (2002b). Mechanisms controlling bursting activity induced by disinhibition in spinal cord networks. *Eur. J. Neurosci.* 15: 671–683.

Darbon, P., Tscherter, A., Yvon, C., and Streit, J. (2003). Role of the electrogenic Na/K pump in disinhibition-induced bursting in cultured spinal networks. *J. Neurophysiol.* 90: 3119–3129.

Darbon, P., Yvon, C., Legrand, J.-C., and Streit, J. (2004). INaP underlies intrinsic spiking and rhythm generation in networks of cultured rat spinal cord neurons. *Europ. J. Neurosci.* 20: 976–988.

Demir, R., Gao, B.X., Jackson, M.B., and Ziskind-Conhaim, L. (2002). Interactions between multiple rhythm generators produce complex patterns of oscillation in the developing rat spinal cord. *J. Neurophysiol.* 87: 1094–1105.

Dietz, V., Wirz, M., Colombo, G., and Curt, A. (1998). Locomotor capacity and recovery of spinal cord function in paraplegic patients: A clinical and electrophysiological evaluation. *Electroencephalogr. Clin. Neurophysiol.* 109: 140–153.

Ellerkmann, R.K., Riazanski, V., Elger, C.E., Urban, B.W., and Beck, H. (2001). Slow recovery from inactivation regulates the availability of voltage-dependent Na(+) channels in hippocampal granule cells, hilar neurons and basket cells. *J. Physiol.* 532: 385–397.

Faingold, C.L. (2004). Emergent properties of CNS neuronal networks as targets for pharmacology: Application to anticonvulsant drug action. *Prog. Neurobiol.* 72: 55–85.

Fetcho, J.R. and O'Malley, D.M. (1995). Visualization of active neural circuitry in the spinal cord of intact zebrafish. *J. Neurophysiol.* 73: 399–406.

Galante, M., Nistri, A., and Ballerini, L. (2000). Opposite changes in synaptic activity of organotypic rat spinal cord cultures after chronic block of AMPA/kainate or glycine and GABAA receptors. *J. Physiol.* 3: 639–651.

Giugliano, M., Darbon, P., Arsiero, M., Luescher, H.R., and Streit, J. (2004). Single-neuron discharge properties and network activity in dissociated cultures of neocortex. *J. Neurophysiol.* (March 24, 2004) 10–1152/jn.00067.2004.

Grillner, S. (2003). The motor infrastructure: From ion channels to neuronal networks. *Nat. Rev. Neurosci.* 4: 573–586.

Grillner, S. and Zangger, P. (1979). On the central generation of locomotion in the low spinal cat. *Exp. Brain Res.* 34: 241–261.

Grillner, S., Parker, D., and el Manira, A. (1998). Vertebrate locomotion—A lamprey perspective. *Ann. N. Y. Acad. Sci.* 860: 1–18.

Grillner, S., Wallen, P., Brodin, L., and Lansner, A. (1991). Neuronal network generating locomotor behavior in lamprey: Circuitry, transmitters, membrane properties, and simulation. *Annu. Rev. Neurosci.* 14: 169–199.

Gross, G.W., Williams, A.N., and Lucas, J.H. (1982). Recording of spontaneous activity with photoetched microelectrode surfaces from mouse spinal neurons in culture. *J. Neurosci. Meth.* 5: 13–22.

Heuschkel, M. (2001). Fabrication of multi-electrode array devices for electrophysiological monitoring of in-vitro cell/tissue cultures. Thesis EPFL, Lausanne.

Iizuka, M., Nishimaru, H., and Kudo, N. (1998). Development of the spatial pattern of 5-HT-induced locomotor rhythm in the lumbar spinal cord of rat fetuses in vitro. *Neurosci. Res.* 31: 107–111.

Jankowska, E., Jukes, M.G., Lund, S., and Lundberg, A. (1967). The effect of DOPA on the spinal cord. 6. Half-centre organization of interneurones transmitting effects from the flexor reflex afferents. *Acta Physiol. Scand.* 70: 389–402.

Keefer, E.W., Gramowski, A., and Gross, G.W. (2001). NMDA receptor-dependent periodic oscillations in cultured spinal cord networks. *J. Neurophysiol.* 86: 3030–3042.

Kiehn, O. and Butt, S.J. (2003). Physiological, anatomical and genetic identification of CPG neurons in the developing mammalian spinal cord. *Prog. Neurobiol.* 70: 347–361.

Kiehn, O. and Kjaerulff, O. (1998). Distribution of central pattern generators for rhythmic motor outputs in the spinal cord of limbed vertebrates. *Ann. N. Y. Acad. Sci.* 860: 110–129.

Kiehn, O. and Kullander, K. (2004). Central pattern generators deciphered by molecular genetics. *Neuron* 41: 317–321.

Kjaerulff, O. and Kiehn, O. (1996). Distribution of networks generating and coordinating locomotor activity in the neonatal rat spinal cord in vitro: A lesion study. *J. Neurosci.* 16: 5777–5794.

Kjaerulff, O., Barajon, I., and Kiehn, O. (1994). Sulphorhodamine-labelled cells in the neonatal rat spinal cord following chemically induced locomotor activity in vitro. *J. Physiol.* 478: 265–273.

Kremer, E. and Lev-Tov, A. (1997). Localization of the spinal network associated with generation of hindlimb locomotion in the neonatal rat and organization of its transverse coupling system. *J. Neurophysiol.* 77: 1155–1170.

Kudo, N. and Nishimaru, H. (1998). Reorganization of locomotor activity during development in the prenatal rat. *Ann. N. Y. Acad. Sci.* 860: 306–317.

Kudo, N. and Yamada, T. (1987). N-methyl-D,L-aspartate-induced locomotor activity in a spinal cord-hindlimb muscles preparation of the newborn rat studied in vitro. *Neurosci. Lett.* 75: 43–48.

Kullander, K., Butt, S.J., Lebret, J.M., Lundfald, L., Restrepo, C.E., Rydstrom, A., Klein, R., and Kiehn, O. (2003). Role of EphA4 and EphrinB3 in local neuronal circuits that control walking. *Science* 299: 1889–1892.

Kwiecien, R. and Hammond, C. (1998). Differential management of Ca2+ oscillations by anterior pituitary cells: A comparative overview. *Neuroendocrinology* 68: 135–151.

Latham, P.E., Richmond, B.J., Nelson, P.G., and Nirenberg, S. (2000a). Intrinsic dynamics in neuronal networks. I. Theory. *J. Neurophysiol.* 83: 808–827.

Latham, P.E., Richmond, B.J., Nirenberg, S., and Nelson, P.G. (2000b). Intrinsic dynamics in neuronal networks. II. Experiment. *J. Neurophysiol.* 83: 828–835.

Legrand, J.C., Darbon, P., and Streit, J. (2004). Contributions of NMDA receptors to network recruitment and rhythm generation in spinal cord cultures. *Eur. J. Neurosci.* 19: 521–532.

Magnuson, D.S. and Trinder, T.C. (1997). Locomotor rhythm evoked by ventrolateral funiculus stimulation in the neonatal rat spinal cord in vitro. *J. Neurophysiol.* 77: 200–206.

Marchetti, C., Beato, M., and Nistri, A. (2001). Alternating rhythmic activity induced by dorsal root stimulation in the neonatal rat spinal cord in vitro. *J. Physiol.* 530: 105–112.

McPherson, D.R., McClellan, A.D., and O'Donovan, M.J. (1997). Optical imaging of neuronal activity in tissue labeled by retrograde transport of Calcium Green Dextran. *Brain Res. Brain Res. Protoc.* 1: 157–164.

Milner, L.D. and Landmesser, L.T. (1999). Cholinergic and GABAergic inputs drive patterned spontaneous motoneuron activity before target contact. *J. Neurosci.* 19: 3007–3022.

Mortin, L.I. and Stein, P.S.G. (1989). Spinal cord segments containing key elements of the central patterns generators for three forms of scratch reflex in the turtle. *J. Neurosci.* 9: 2285– 2296.

Muller, W. and Swandulla, D. (1995). Synaptic feedback excitation has hypothalamic neural networks generate quasirhythmic burst activity. *J. Neurophysiol.* 73: 855–861.

Nakayama, K., Nishimaru, H., Iizuka, M., Ozaki, S., and Kudo, N. (1999). Rostrocaudal progression in the development of periodic spontaneous activity in fetal rat spinal motor circuits in vitro. *J. Neurophysiol.* 81: 2592–2595.

Nishimaru, H., Iizuka, M., Ozaki, S., and Kudo, N. (1996). Spontaneous motoneuronal activity mediated by glycine and GABA in the spinal cord of rat fetuses in vitro. *J. Physiol.* 497: 131–143.

O'Brien, R.J. and Fischbach, G.D. (1986). Excitatory synaptic transmission between interneurons and motoneurons in chick spinal cord cell cultures. *J. Neurosci.* 6: 3284–3289.

O'Donovan, M.J. (1999). The origin of spontaneous activity in developing networks of the vertebrate nervous system. *Curr. Opin. Neurobiol.* 9: 94–104.

O'Donovan, M.J. and Rinzel, J. (1997). Synaptic depression: A dynamic regulator of synaptic communication with varied functional roles. *Trends Neurosci.* 20: 431–433.

O'Donovan, M.J., Chub, N., and Wenner, P. (1998a). Mechanisms of spontaneous activity in developing spinal networks. *J. Neurobiol.* 37: 131–145.

O'Donovan, M.J., Wenner, P., Chub, N., Tabak, J., and Rinzel, J. (1998b). Mechanisms of spontaneous activity in the developing spinal cord and their relevance to locomotion. *Ann. N. Y. Acad. Sci.* 860: 130–141.

Ozaki, S., Yamada, T., Iizuka, M., Nishimaru, H., and Kudo, N. (1996). Development of locomotor activity induced by NMDA receptor activation in the lumbar spinal cord of the rat fetus studied in vitro. *Brain Res. Dev. Brain Res.* 97: 118–125.

Pinco, M. and Levtov, A. (1993). Modulation of monosynaptic excitation in the neonatal rat spinal cord. *J. Neurophysiol.* 70: 1151–1158.

Ransom, B.R., Christian, C.N., Bullock, P.N., and Nelson, P.G. (1977). Mouse spinal cord in cell culture. II. Synaptic activity and circuit behavior. *J. Neurophysiol.* 40: 1151–1161.

Rekling, J.C. and Feldman, J.L. (1998). PreBotzinger complex and pacemaker neurons: Hypothesized site and kernel for respiratory rhythm generation. *Annu. Rev. Physiol.* 60: 385–405.

Ren, J. and Greer, J.J. (2003). Ontogeny of rhythmic motor patterns generated in the embryonic rat spinal cord. *J. Neurophysiol.* 89: 1187–1195.

Robinson, H.P., Torimitsu, K., Jimbo, Y., Kuroda, Y., and Kawana, A. (1993). Periodic bursting of cultured cortical neurons in low magnesium: cellular and network mechanisms. *Jpn. J. Physiol.* 43: S125–130.

Rossignol, S., Giroux, N., Chau, C., Marcoux, J., Brustein, E., and Reader, T.A. (2001). Pharmacological aids to locomotor training after spinal injury in the cat. *J. Physiol.* 533: 65–74.

Schmidt, B.J. and Jordan, L.M. (2000). The role of serotonin in reflex modulation and locomotor rhythm production in the mammalian spinal cord. *Brain Res. Bull.* 53: 689–710.

Senn, W., Wyler, K., Streit, J., Larkum, M., Lüscher, H.-R., Mey, H., Müller, L., Stainhauser, D., Vogt, K., and Wannier, T. (1996). Dynamics of a random neural network with synaptic depression. *Neur. Netw.* 9: 575–588.

Sherrington, C. (1898). Decerebrate rigidity and reflex coordination of movements. *J. Physiol.* 22: 319–332.

Shik, M.L. and Orlovsky, G.N. (1976). Neurophysiology of locomotor automatism. *Physiol. Rev.* 56: 465–501.

Spenger, C., Braschler, U.F., Streit, J., and Lüscher, H.-R. (1991). An organotypic spinal cord-dorsal root ganglion-skeletal muscle coculture of embryonic rat. I. The morphological correlates of the spinal reflex arc. *Eur. J. Neurosci.* 3: 1037–1053.

Steriade, M. (2001). Impact of network activities on neuronal properties in corticothalamic systems. *J. Neurophysiol.* 86: 1–39.

Streit, J. (1993). Regular oscillations of synaptic activity in spinal networks in vitro. *J Neurophysiol.* 70: 871–878.

Streit, J. (1996). Mechanisms of pattern generation in cocultures of embryonic spinal cord and skeletal muscle. *Int. J. Dev. Neurosci.* 14: 137–148.

Streit, J., Lüscher, C., and Lüscher, H.-R. (1992). Depression of postsynaptic potentials by high frequency stimulation in embryonic motoneurons grown in spinal cord slice cultures. *J. Neurophysiol.* 68: 1793–1803.

Streit, J., Spenger, C., and Lüscher, H.-R. (1991). An organotypic spinal cord-dorsal root ganglia-skeletal muscle coculture of embryonic rat. II. Functional evidence for the formation of spinal reflex arcs in vitro. *Eur. J. Neurosci.* 3: 1054–1068.

Streit, J., Tscherter, A., Heuschkel, M.O., and Renaud, P. (2001). The generation of rhythmic activity in dissociated cultures of rat spinal cord. *Europ. J. Neurosci.* 14: 191–202.

Tabak, J. and O'Donovan, M.J. (1998). Statistical analysis and intersegmental delays reveal possible roles of network depression in the generation of spontaneous activity in the chick embryo spinal cord. *Ann. N. Y. Acad. Sci.* 860: 428–431.

Tabak, J., Rinzel, J., and O'Donovan, M.J. (2001). The role of activity-dependent network depression in the expression and self-regulation of spontaneous activity in the developing spinal cord. *J. Neurosci.* 21: 8966–8978.

Tabak, J., Senn, W., O'Donovan, M.J., and Rinzel, J. (2000). Modeling of spontaneous activity in developing spinal cord using activity-dependent depression in an excitatory network. *J. Neurosci.* 20: 3041–3056.

Tresch, M.C. and Kiehn, O. (2000). Motor coordination without action potentials in the mammalian spinal cord. *Nat. Neurosci.* 3: 593–599.

Tscherter, A. (2002). Rhythmic activity in cultured spinal cord networks: A multielectrode array study. Thesis, University of Bern, Bern.

Tscherter, A., Heuschkel, M.O., Renaud, P., and Streit, J. (2001). Spatiotemporal characterization of rhythmic activity in rat spinal cord slice cultures. *Eur. J. Neurosci.* 14: 179–190.

Vertes, R.P. and Kocsis, B. (1997). Brainstem-diencephalo-septohippocampal systems controlling the theta rhythm of the hippocampus. *Neuroscience* 81: 893–926.

Vinay, L., Brocard, F., Pflieger, J.F., Simeoni-Alias, J., and Clarac, F. (2000). Perinatal development of lumbar motoneurons and their inputs in the rat. *Brain Res. Bull.* 53: 635–647.

Westerga, J. and Gramsbergen, A. (1990). The development of locomotion in the rat. *Brain Res. Dev. Brain Res.* 57: 163–174.

Yvon, C., Rubli, R. and Streit, J. (2005). Patterns of spontaneous activity in unstructured and minimally structured spinal networks in culture. *Exp. Brain Res.* 165: 139–151.

17
Monitoring the Clock Neuron's Tick: Circadian Rhythm Analysis Using a Multi-Electrode Array Dish

Sato Honma, Wataru Nakamura, Tetsuo Shirakawa, and Ken-ichi Honma

17.1   Introduction

Almost all physiological and biochemical functions in the body oscillate with a period of approximately 24 h. These rhythms are not direct reactions to either environmental temperature fluctuations or the 24 h light–dark cycle, but are driven by an endogenous oscillator called the circadian clock, or simply "biological clock." Bodily functions even oscillate under unchanging environmental conditions such as constant darkness, though with a period close to, but significantly different from, 24 h. In fact the etymology of the term "circa (about)-dian (a day) rhythm" is derived from these many observations. For most organisms environmental light resets the intrinsic period exactly to 24 h. In mammals the master circadian clock resides in the suprachiasmatic nucleus of the hypothalamus (SCN; Moore and Eichler, 1972; Stephan and Zucker, 1972; Figure 17.1a). SCN neurons generate circadian periodicity and receive photic signals to adjust the intrinsic period to the 24 h day–night cycle. Thus they coordinate and orchestrate peripheral clocks in most organs and tissues throughout the body, expressing circadian rhythms in physiology and behavior as manifested by the sleep and wake cycle.

In traditional slice physiology, cellular or network oscillations occur on the order of milliseconds; however, measuring circadian periodicity of SCN neurons requires analysis on the order of days or even weeks. The multi-electrode array dish (MED) probe enabled us to monitor the activity of single SCN neurons long enough for circadian rhythm analysis. In this chapter we review the functions of "clock neurons", the SCN neurons with an intrinsic circadian oscillator, as unveiled by the use of the MED probe. We also include the methods we developed for the long-term culture of neurons and the analysis of time-series data.

17.2   Master Circadian Clock in the SCN

The SCN comprises bilaterally paired nuclei, each containing only ~8000 neurons within a volume of ~0.15 mm^3 (van den Pol, 1991). It receives direct photic

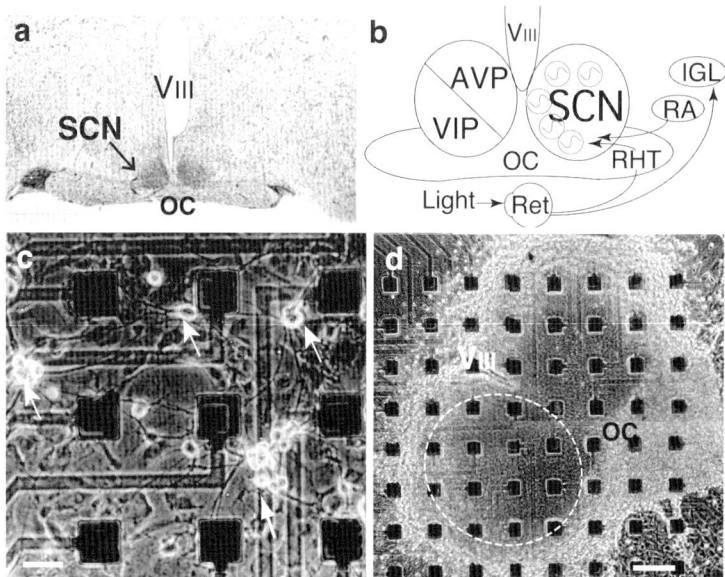

FIGURE 17.1. Photomicrograph and scheme of the SCN. (a) Coronal section of the rat hypothalamus stained with cresyl violet at the level of the SCN (arrow). OC: optic chiasm, V_{III}: third ventricle. (b) Schematic drawing of the rat SCN showing peptidergic neuron localization in the SCN and major afferent projections (arrows). ◎: clock neurons, AVP: arginine vasopressinergic neuron (predominated region), VIP: vasoactive intestinal polypeptidergic neurons (predominated region), IGL: intergeniculate leaflet of the thalamus, Ret: retina, RHT: retinohypothalamic tract, RA: raphe nucleus in the midbrain. (c) Dispersed cell culture of the rat SCN on the MED probe at day 14 in culture. In this dish, SCN cells were seeded at the low concentration of 1000 vial cells/mm^3. White arrows: SCN neurons. Black squares: electrodes (50 μm × 50 μm). Scale: 50 μm. (d) An organotypic slice culture of the rat SCN at day 11 in culture. Broken oval lines: border of the SCN. Scale: 150 μm. (Honma et al., unpublished data).

inputs from the retina through the retinohypothalamic tract (Moore, 1973). The major photoreceptor responsible for the resetting of the circadian clock was recently found to be melanopsin, an opsin-based photopigment in retinal ganglion cells (Berson et al., 2002). The SCN is histologically and functionally divided into two subdivisions, the dorsomedial (shell) and ventrolateral (core). Major external afferents, those from the retina, intergeniculate leaflet, and raphe nucleus, project into the ventrolateral SCN where vasoactive intestinal polypeptide (VIP)-containing neurons predominate. On the other hand, in the dorsomedial SCN where arginine vasopressin (AVP)-containing neurons predominate, afferents from outside the SCN are limited, but there is dense innervation from the ventrolateral SCN. Major efferents from the SCN project to the subparaventricular zone (SPZ), though dorsomedial and ventrolateral neurons project differentially (van den Pol, 1991; Moore, 1996). Interestingly, efferents project mostly to the immediate periphery of the SCN, whereupon circadian signals seem to be integrated and transmitted through secondary neurons.

In the SCN there are robust circadian rhythms in the synthesis and release of neuropeptides in vitro (Shinohara et al., 1995) as well as in vivo (Schwartz and Reppert, 1985). The multi-unit activity of the SCN continuously monitored from freely moving animals exhibits a robust circadian rhythm with a peak during the subjective day irrespective of the subjects' diurnality or nocturnality. If the SCN is surgically isolated as a hypothalamic island from the rest of the brain, the rhythm in the SCN continues, but the firing rhythm outside the hypothalamic island and also behavioral rhythms are both abolished. This indicates that the circadian pacemaker resides within the SCN itself (Inouye and Kawamura, 1979; Sato and Kawamura, 1984).

Until recently, it had been difficult to demonstrate whether the circadian rhythm was driven by single pacemaker neurons or emerged from neural networks within the SCN, because single-unit activity could not be monitored for sufficiently long durations for circadian rhythm analysis. This problem was finally solved by the use of the multi-electrode array system. Welsh et al. (1995) reported that activity rhythms of individual dispersed SCN neurons free-ran with a period close to 24 h. He also showed that tetrodotoxin (TTX), a sodium channel blocker, suppressed spontaneous firing of single SCN neurons but did not affect intracellular circadian oscillation, for the firing rhythm re-emerged after the washout of TTX at the phase expected by the rhythm before the TTX treatment. The data clearly demonstrated that the circadian rhythm in the SCN emerges not from a neuronal network but from individual pacemaker neurons. With the advent of the MED probe, the dream of assessing the mammalian biological clock via week-long or month-long analysis of neuronal SCN rhythms had finally become a reality.

17.3   Long-Term Monitoring of Clock Neurons' Tick: Two Types of SCN Cultures

To examine the characteristics of clock neurons, we cultured rat and mouse SCNs on the MED probes using two different culture systems: the dispersed cell culture for examining the functions of single SCN neurons and cell-to-cell interactions, and the organotypic slice culture for assessing regional specificity within the SCN and the role of SCN cell assemblages. The following sections describe these two systems.

17.3.1   Initial Preparation of the MED Probes

For initial preparation, careful cleaning and sterilization are necessary. Thereafter, the probe can be kept under clean conditions in a sterilized disposable Petri dish (100 mm diameter). On the day before the culture, the probe's dish surface is treated with brief repeated exposures to blue flame to increase its hydrophilicity.

For the dispersed cell culture, the dish surface is then precoated with 0.02% poly-L-ornithine overnight in 4°C to facilitate cell attachment (Honma et al., 1998a; Shirakawa et al., 2000). After rinsing with sterile distilled water (SDW), the probe is kept covered in culture medium supplemented with 2~5% serum for at least 2 h in a CO_2 incubator.

On the other hand, for the organotypic slice culture, after blue-flaming, the probe surface is coated with a collagen gel sheet (Type IC collagen, Nitta Gelatin, Japan), which is then firmed by incubating at 37°C for 2~3 h.

17.3.2   Dispersed SCN Cell Culture on the MED Probe

For the dispersed cell culture (Figure 17.1b), rat and mouse pups 3 to 5 days old are decapitated under hypothermic anaesthesia. Coronal hypothalamic slices of 600 μm and 350 μm thick for rats and mice, respectively, are obtained using a tissue chopper (McIlwain), and the SCN is dissected in ice-cold Preparation Buffer (Table 17.1a) under a dissecting microscope (Honma et al., 1998a; Shirakawa et al., 2000). Depending on experimental design and litter size, SCN blocks of roughly 3 to 30 pups are pooled together. The SCNs are incubated in 0.03% trypsin in Preparation Buffer at 37°C for 15 min. After rinsing the SCNs with Preparation Buffer containing 0.022% trypsin inhibitor and 0.01% DNase, SCN cells are dissociated by flushing with a fire-polished Pasteur pipette. The resulting cell suspension is filtrated with a #200 stainless mesh and centrifuged at 1500 rpm for 5 min. The dispersed SCN cells are resuspended with DMEM supplemented with 5% fetal bovine serum at a concentration of $\sim 10^6$ cells/ml and seeded in the central area of the probe at the relatively high density of 5000~7500 vial cells/mm^2. Figure 17.1b shows dispersed SCN cells seeded on a MED probe at a low density in order to show the neurons clearly.

Each dish is cultured with 1 ml of DMEM supplemented with 20× Supplement Solution (Table 17.1b) at 5% and fetal bovine serum at ~2%. Spontaneous

TABLE 17.1. Composition of preparation buffer and supplements for culture medium.

a. Preparation Buffer (pH 7.3/0°C)

NaCl	8.6 (g/liter ultrapure water)
KCl	0.3
HEPES	4.7
Na HCO$_3$	3.0
Kanamycin	0.02

b. 20× Supplement Solution (Mix with culture medium at 5% immediately before use)

Apotransferrin	2 mg/ml
Insulin (water soluble)	10 μg/ml
Putrescine hydrochloride	2 mM
Progesterone (water soluble)	20 nM
Sodium selenite	0.6 μM

discharges from dispersed SCN cell cultures are monitored from 4 days up to 5 months. By supplementing the culture medium with a low concentration of serum (0.5∼2.0%), cell mobility can be minimized over the duration of the culture. To reduce serum-induced glial overgrowth, the culture is additionally treated with antimitotics (1 μM each of cytosine arabinoside, uridine, and fluorodeoxyuridine) for 24 h near the beginning of the culture (∼3 to 7 days).

17.3.3   Organotypic SCN Slice Culture on the MED Probe

For the organotypic slice culture (Figure 17.1c), an SCN pair is dissected from a coronal hypothalamic slice 200∼250 μm thick for mice and 250∼350 μm for rats. The explants are transferred to the collagen-coated MED probe using a transfer pipette. After removing the excess medium from the probe, the SCN slice is incubated in a CO_2 incubator with 100% humidity for 1∼2 h until the slice attaches to the collagen gel. Then the slices are incubated with ∼250 μl DMEM/F12 medium supplemented with 5% fetal bovine serum for the first two days, and thereafter with no serum. Of critical importance, slices are kept wet but not submerged in medium. Within a few days, neurites extend from the periphery of the slice, and slices become thinner (Nakamura et al., 2001; Figure 17.1c). Spontaneous discharges of single SCN neurons are monitored from five days up to two months after the start of the culture.

17.3.4   Long-Term Monitoring of Spike Counts

For circadian rhythm analysis, spike numbers are monitored continuously throughout the duration of the experiment. To maintain the constant conditions this requires, half the medium is exchanged every day during recording. The number of spike discharges is counted by discriminating the spikes using a Time-Window Slicer (Nihon Kohden) and feeding them into a computer every 10 sec. The circadian periodicity ($p < 0.01$) and circadian period are computed using a chi-square periodogram (Sokolove and Bushell, 1978).

17.4   Circadian Periods of Clock Neurons in Dispersed Cell Cultures and Organotypic Slice Cultures of the SCN

SCN neurons exhibited robust circadian rhythms in their spontaneous discharges (Figure 17.2), and their periods were specific to each neuron (Welsh et al., 1995; Honma et al., 1998b). We recently compared the distributions of circadian periods in single SCN neurons between the two culture types (Honma et al., 2004). Of 308 neurons in 54 dishes examined in the dispersed cell culture, 220 neurons (71.4%) exhibited significant circadian rhythms in their firing. In the slice culture, out of 204 neurons from 24 slice cultures examined, 185 neurons (82.6%) exhibited significant circadian firing rhythms. The percentage of rhythmic neurons was significantly higher in the slice culture than in the dispersed cell culture.

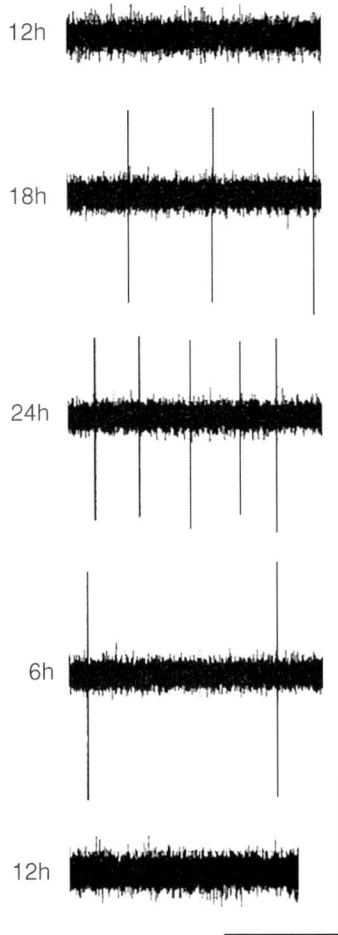

FIGURE 17.2. Spontaneous firing rhythms of a rat SCN neuron. Neuronal activity was recorded from a single SCN neuron every 6 h. Almost no firing occurs at 12:00 h, but regular firing is apparent at 24:00. Scale: 50 μV, 1 sec.

The circadian periods followed a Gaussian distribution in both cultures, and their mean circadian periods were not significantly different from each other. However, the distribution range was significantly narrower in the slice culture than in the dispersed cell culture. The circadian periods of spontaneous firing rhythms varied between 20.0 to 30.9 h with an average period of 24.1 ± 1.4 h (mean ± SD) (Figure 17.3a) in the dispersed cell culture, and between 22.4 to 26.7 h, with an average of 24.2 ± 0.7 h (Figure 17.3b) in the slice culture.

Within each dish of the dispersed cell cultures, periods of neuronal rhythms varied widely, whereas for the slice cultures, the rhythms monitored from different neurons were basically synchronized. Furthermore, in the dispersed cell cultures,

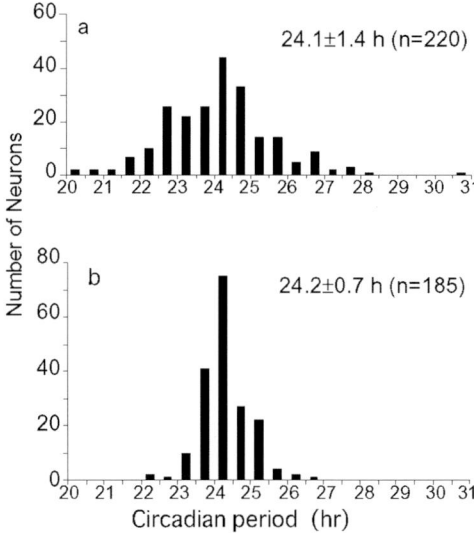

FIGURE 17.3. Diversity of circadian periods in individual SCN neurons depends on the culture method. Circadian period distribution in (a) the dispersed cell culture and (b) slice culture. The mean periods are not significantly different, but the distribution ranges differ significantly between the two cultures. (Adapted with permission from Honma et al., 1998b.)

but not in the slice cultures, there was a significant correlation between the period length and variation of circadian rhythm: the more the mean circadian period in a culture dish deviated from the overall mean, the larger the standard deviation of period in a dish became. These results suggest that in addition to the SCN cell assemblage with its close cell–cell apposition, the intrinsic circadian period plays a significant role in synchronizing the constitutional oscillators in the SCN. Furthermore, the entire SCN structure, including afferents from extra-SCN areas, seems to be necessary for the SCN neurons to express the rhythms with physiological periodicity, because the free-running periods of behavioral rhythms in rats (24.4 ± 0.2 h, $n = 147$) were distributed within a very narrow range of 24.0 and 24.8 h.

17.5   How Do Clock Neurons Talk to Each Other: Cross-Correlation Analysis to Assess Neuronal Interactions

Because the SCN is composed of many clock neurons with slightly but significantly different circadian periods, it is critical for these neurons to communicate with each other in order to synchronize and express a single circadian periodicity within the SCN. There are at least three modes of communication for SCN neurons to convey circadian information: chemical synapses, gap junctions (Shinohara et al., 2000), and diffusible factors (Silver et al., 1996). Using the dispersed SCN cell culture on the MED probe, we computed cross-correlation analyses of spike timings to determine if synaptic coupling existed, and, if so, whether it played an excitatory or inhibitory role.

In addition, the coupling through gap-junctions was also evaluated by computing cross-correlograms. In principle if diffusible factors could couple neuronal rhythms, neurons located nearby would possess the same period, although this would not be the case in the dispersed SCN cell culture. Our data (Honma et al., 1998b; Shirakawa et al., 2000) indicated that, in fact, neuronal rhythms could not be coupled by diffusible factors except those which were degraded or diluted immediately.

We analyzed the relationship between synchronization of circadian rhythms and synaptic communication in 310 neuron pairs in the dispersed cell culture of the rat SCN (Shirakawa et al., 2000). Seventy-eight neuron pairs (24%) displayed circadian firing rhythms in both neurons, and among them 35 (45% of neuron pairs with circadian rhythms) showed synchronized circadian rhythms. In those neuron pairs with synchronized circadian rhythms, we detected either excitatory (32) or inhibitory (3) communication without exception (Figures 17.4 and 17.5). The interspike interval (ISI) between the neuron pairs exhibiting excitatory communications was 1.28 ± 0.23 msec (mean $\pm$ SE), and 25 of the 32 pairs had an ISI shorter than 2 msec. Among the neuron pairs with inhibitory communication, the neuronal firing of the postsynaptic cell was suppressed for 100 to 140 msec

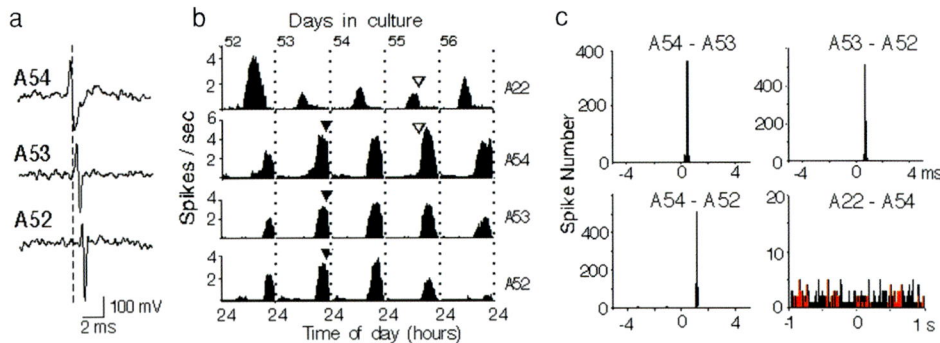

FIGURE 17.4. Excitatory synaptic communication between neuron pairs with synchronized circadian rhythm. (a) Spike waveforms of neurons A54, A53, and A52 on the same MED probe. The interpolar distances from A54 to A53 and to A52 are 150 and 300 μm, respectively. (b) Synchronization and desynchronization of circadian firing rhythms in the four neurons. The circadian rhythm is expressed as the mean firing rate (spikes/sec) in 15 min. Closed triangles indicate the time when cross-correlations were analyzed between A52, A53, and A54; and open triangles, between A22 and the other three neurons. (c) Cross-correlograms indicating excitatory synaptic communication between the neurons (A54, A53, A52) whose circadian rhythms were synchronized; and absence of functional communication between A22 and A54. Cross-correlation was not detected between A22 and A52 or A53 (data not shown). The abscissa is the timing of spikes with bins of 100 μs (except for A22 to A54) and of 10 ms (A22 to A54), and the ordinate is the cumulative number of spikes. (Adapted with permission from Shirakawa et al., 2000.)

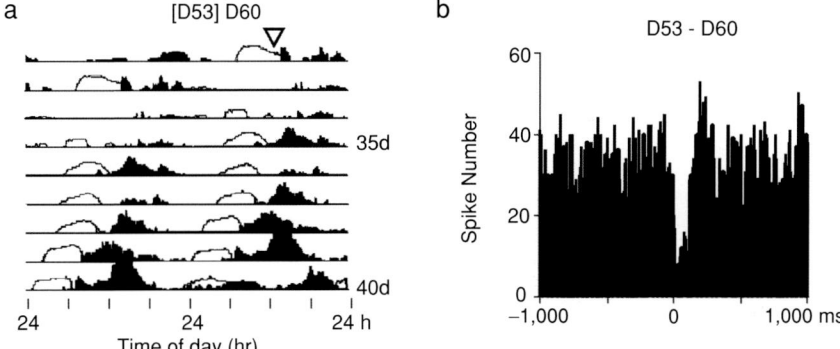

FIGURE 17.5. Inhibitory interaction between neurons showing synchronized circadian firing rhythm but with a phase-lag of about 6 h. (a) Double-plotted circadian firing rhythm of cell D53 (outlines) superimposed on that of cell D60. The circadian periods of both neurons are 23.6 hr. Cross-correlation was analyzed at the time marked by the open triangle. The numbers in the right margin indicate the number of days in cell culture. The full scale of firing rate: 8 (Hz, D53) and 6 (D60). (b) Cross-correlogram between cells D53 and D60 showing that D53 exerted a synaptic inhibition on D60. The two neurons were located on adjacent electrodes. (Adapted with permission from Shirakawa et al., 2000)

after the firing of the presynaptic cell (Figure 17.5). In 43 pairs out of 78 examined, circadian rhythms were not synchronized, and no significant correlation was detected in the timing of spikes in any of the desynchronized pairs (Figure 17.6).

These cross-correlation data strongly suggest that the circadian rhythms of SCN neurons are coupled by synaptic communication, at least in the dispersed cell

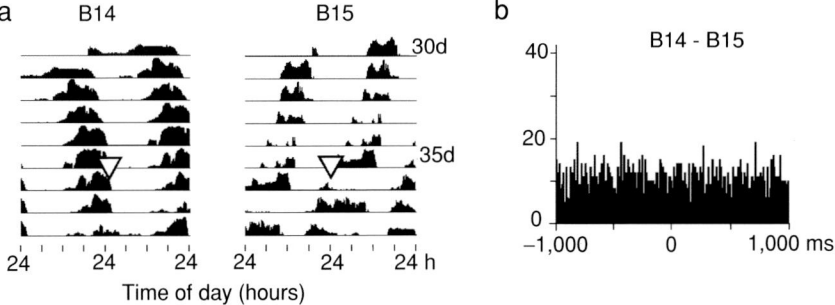

FIGURE 17.6. A lack of cross-correlation between neurons exhibiting circadian firing rhythms with different circadian periods. (a) Double-plotted circadian firing rhythms of two neurons located on adjacent electrodes. Circadian periods of cells B14 and B15 are 25.2 h and 22.8 h, respectively. Open triangles indicate the time when cross-correlation (b) was analyzed. The full scale of firing rate: 7 (B14) and 4 (B15). (b) Cross-correlogram between two neurons showing no significant correlation in spike timing. (Reproduced with permission from Shirakawa et al., 2000.)

culture. It remains unknown why only a few inhibitory communication neuronal pairs were found in spite of the fact that most SCN neurons contain gamma-aminobutyric acid (GABA), and GABA is reported to be responsible for inhibitory neurotransmission in the SCN (Moore and Speh, 1993; Strecker et al., 1997).

One possibility is that other inhibitory pairs were undetected if their constituent cell firings were not sufficiently concurrent (i.e., both occurring within the sampling/recording window used). Circadian firing rhythms so far recorded from inhibitory communication neuronal pairs were significantly out of phase as depicted in Figure 17.5a, suggesting difficulty in detecting inhibitory communication via cross-correlation analysis. Wagner et al. (1997, 2001) reported that GABA depolarized SCN neurons during the subjective day (i.e., the active phase) but hyperpolarized them during the subjective night (i.e., the inactive phase), and they attributed this to shifts in the chloride equilibrium potential. We thus examined the involvement of GABAergic communication in our SCN cell cultures. In dispersed SCN cell cultures, about 60% of the neurons were immunoreactive to antibodies against either GABA or glutamic acid decarboxylase (GAD). Bicuculline, the $GABA_A$ receptor antagonist, increased the firing rate in 36 neurons and decreased it in 7 out of 43 examined at 10 μM; and increased it in 40 neurons while decreasing it in 3 at 50 μM (Shirakawa et al., 2000). In some neurons, bicuculline dose-dependently decreased the firing rate during the active phase (Figure 17.7a), and increased it during the inactive phase (Figure 17.7b). This suggests that, at least for some SCN neurons, GABA can play either an excitatory or inhibitory role depending on the circadian phase.

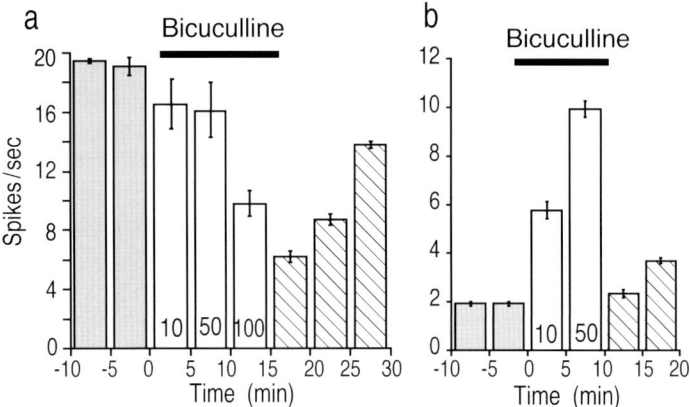

FIGURE 17.7. Effects of bicuculline on spontaneous firing activity of a single SCN neuron. A dose-dependent decrease in the firing rate during the active phase (a) or increase during the inactive phase (b) was observed in the same neuron measured 8 days apart. Numbers in the bars indicate concentrations of bicuculline (μM). Data are spike rates before (gray bars), during (open bars), and after (hatched bars) bicuculline application (mean $\pm$ SE). Bicuculline application begins at abscissa time 0. (Reproduced with permission from Shirakawa et al., 2000.)

The results of cross-correlation analysis suggest that excitatory communication contributes to the synchronization of circadian oscillators by delivering signals on the order of milliseconds, whereas inhibitory communication does so by modulating the excitability or coupling strength of the oscillating SCN neurons.

Recently Yamaguchi et al. (2003) studied cultured coronal SCN slices from transgenic mice expressing a promoter-driven luciferase reporter of the clock gene, *Period1* (*Per1*), and they established that individual SCN neurons exhibited synchronized, rhythmic *Per1* expression. When slices were subsequently dissected horizontally at the upper one-third, the dorsal third became desynchronized and the larger ventral section remained in synchrony. Their results suggest that it is not the close apposition of SCN neurons per se, but rather the intactness of larger regions—or even simply the total number of cell assemblages—that is necessary for synchronized, rhythmic gene expression. Further studies are needed to monitor simultaneously both neuronal and gene expression rhythms.

17.6   Sodium Channel-Dependent Cell Communication

Because the circadian rhythms of SCN neurons are thought to synchronize with functional synapses, we examined the effects of TTX on the coupling of the circadian firing rhythms. TTX suppressed the spontaneous firing of a single SCN neuron, but did not suppress its internal circadian oscillator (Figure 17.8a), which was consistent with the previous report of Welsh et al. (1995). We also examined the effects of TTX treatment, applied over seven consecutive days, on the synchronization of neuronal circadian firing (Honma et al., 2000). Of seven pairs whose circadian rhythms were initially synchronized, TTX induced desynchronization and loss of cross-correlation in five of them. In the remaining two pairs, rhythms maintained synchrony, but their ISIs were elongated.

In short, suppressing synaptic communication gradually desynchronized circadian firing rhythms of individual neurons. The effects of TTX were also examined in organotypic SCN slice cultures, where most neuronal rhythms were already synchronized. After the washing out of TTX, neuronal firing recovered in all slices examined; however, in most slices, circadian firing rhythms were desynchronized or had altered periods (Honma et al., 2002; Figure 17.8b). These results of TTX treatment suggest that the synaptic coupling of circadian rhythms is dynamic and may be activity-dependent.

17.7   Circadian Firing Rhythm in the SCN in Behaviorally Arrhythmic Clock Mutant Mice

Cell and tissue cultures on the MED probes provide excellent tools to examine a particular gene function. By culturing fetal or newborn brain tissues, neuronal activity can be monitored for long extents, even when obtained from mutant or

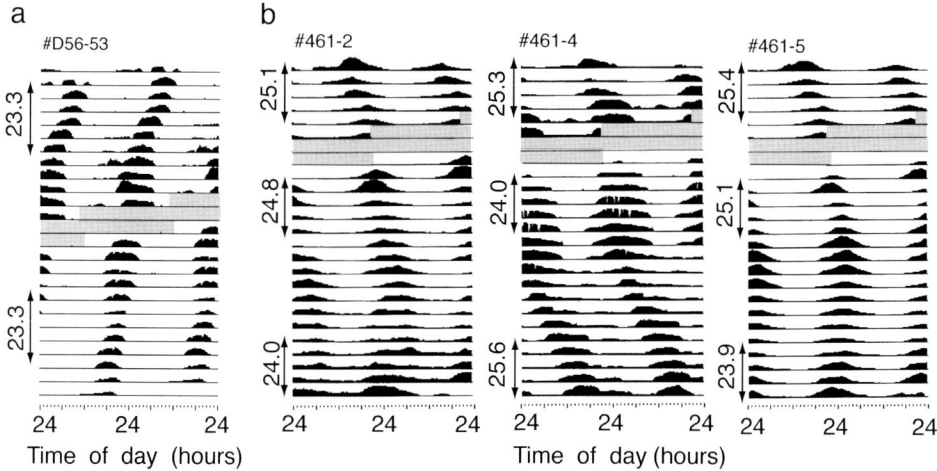

FIGURE 17.8. Effects of tetrodotoxin (TTX) on the circadian firing rhythms of a dispersed SCN neuron (a) and a slice culture (b). (a) Double-plotted actogram showing the circadian firing rhythm of a single SCN neuron in a dispersed cell culture before and after the application of 500 μM of TTX for 2 days (shadowed areas in the actogram). The number at the top of each actogram is the neuron identifier. Numbers along the left margin indicate the circadian periods over the days spanned by arrows. (b) Double-plotted circadian firing rhythms of 3 neurons from a single SCN slice. Before the TTX treatment (200 μM), 8 neurons on a probe expressed synchronized circadian firing rhythms which gradually desynchronized after treatment. (Honma et al., 2002, Society for Neuroscience Meeting.)

gene knockout animals with lethal defects. In addition, modifying gene expression via adenovirus vectors, antisense oligonucleotides, or siRNA is much easier with a culture system than with whole animals. By culturing tissue on the MED probes, we could examine gene functions while simultaneously monitoring neuronal activity.

We studied the role of the clock gene, *Clock*, by measuring spontaneous firing rhythms in the SCN of *Clock* mutant mice (Nakamura, et al., 2002). Under constant darkness (DD), behavioral rhythms of heterozygotes (*Clock/+*) free-ran with periods longer than wild-types, and those of homozygotes (*Clock/Clock*) became arrhythmic (Vitaterna, 1994; Figure 17.9d–f). *Clock*, which codes a basic helix–loop–helix transcription factor, is regarded as a crucial component of the molecular feedback loop generating the circadian rhythm intracellularly (Gekakis, 1998). In organotypic slice cultures of *Clock* mice SCNs, significant circadian rhythms were detected in 95% of neurons recorded from wild-types, 83% from *Clock/+* and 77% from *Clock/Clock*. The maximum firing rate differed from neuron to neuron but was remarkably constant for any given neuron, and the amplitude of firing rhythms did not vary among genotypes either (Figure 17.9g–i). The mean circadian period was 23.5 h in wild type, 24.8 h in *Clock/+*, and 27.2 h in *Clock/Clock*, respectively. The periods of wild-type and *Clock/+* corresponded

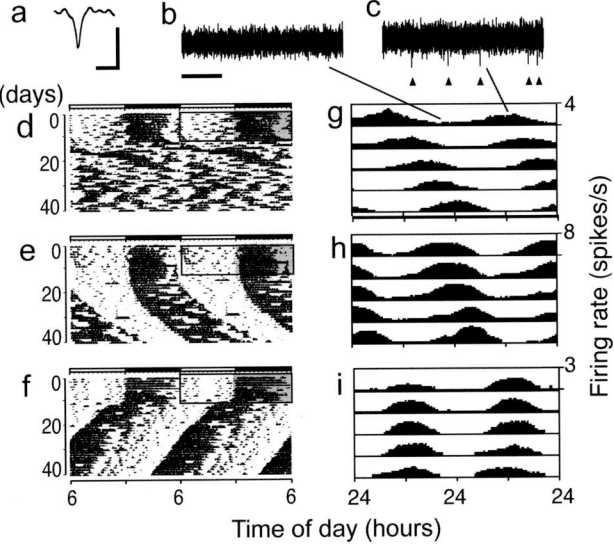

FIGURE 17.9. Significant circadian rhythms in neuronal firing contrasts with arrhythmic wheel-running rhythms in *Clock* mutant mice. (a) Averaged waveform of ten spikes from a *Clock/Clock* neuron. Scale: 100 µV, 2 ms. (b)–(c) Raw voltage trace of a *Clock/Clock* neuron recorded (b) during rest time, and (c) during activity time as indicated in actogram. Triangles indicate discriminated spikes. (d)–(f) Wheel-running activity rhythm of (d) *Clock/Clock*, (e)*Clock/+*, and (f) wild-type mice. White and black bars above each actogram, and open and shaded squares within each actogram, designate the light and dark phases of the 12 h light and 12 h dark cycle, respectively. After 10 days of recording under the light–dark cycle, mice were released into complete darkness. (g)–(i) Histograms of circadian firing rates of (g) *Clock/Clock*), (h) *Clock/+*, and (i) wild-type mice. Numbers in the top right corner of each panel indicate the scales of firing rates. (Reproduced with permission from Nakamura et al., 2002.)

well with that of behavioral rhythms in each genotype. On the other hand, SCN neurons of behaviorally arrhythmic *Clock/Clock* mice exhibited robust firing rhythms.

The findings indicate that the *Clock* mutation lengthens the circadian period but does not abolish the circadian oscillation itself. Surprisingly, in the dispersed cell culture, only 15% of neurons exhibited circadian firing rhythms in *Clock/Clock* SCNs, and their periods were lengthened. By contrast this was observed in 46% of wild-type SCNs. In short, the percentage of neurons with circadian rhythms was affected by both culture method and genotype. The results also suggest that the number of clock neurons is much less than had been previously expected. Indeed, the behavioral arrhythmicity observed in *Clock/Clock* mice might be attributable to these two findings involving clock neurons: their relative scarcity and their extremely long circadian periods. Such a combination could result in weakened intercellular communication and synchronization between SCN neurons.

17.8   Perspectives

By culturing SCN neurons on the MED probe, we have examined the properties of the circadian firing rhythms and the mechanisms with which they talk to one another. Autoregulatory transcription and translation feedback loops involving several clock genes and their products are proposed as a molecular machinery of the circadian rhythm generation (Reppert and Weaver, 2001). However, it remains basically unknown how the molecular oscillation is transmitted to cellular functions such as neuronal discharges (Honma and Honma, 2003). Recently Ikeda et al. (2003) simultaneously measured intracellular Ca^{++} levels and neuronal discharges by culturing SCN slices transfected with the Ca^{++}-binding fluoroprotein Cameleon on the MED probe and suggested that Ca^{++} is one of the key substances which links the molecular clock to cellular functions.

Most peripheral tissues express robust circadian rhythms in *Clock* gene expression in vitro, yet in vivo SCN lesioning results in the total loss of circadian rhythms in both behavior and physiology. The discrepancy suggests that the SCN serves to coordinate peripheral clocks via both neural and humoral mediators, but the mechanisms for this also remain mostly undetermined (Yoo et al., 2004).

How the SCN functions as the body's pacemaker still remains a mystery, but our goal of solving it is within reach, for we have now in hand excellent leading-edge tools at our disposal, including bioluminescent and fluorescent reporter genes, and, of course, the MED probe.

References

Berson, D.M., Dunn, F.A., and Takao, M. (2002). Phototransduction by retinal ganglion cells that set the circadian clock. *Science* 295: 1070–1073.

Gekakis, N., Staknis, D., Nguyen, H.B., Davis, F.C., Wilsbacher, L.D., King, D.P., Takahashi, J.S., and Weitz, C.J. (1998). Role of the CLOCK protein in the mammalian circadian mechanism. *Science* 280: 1564–1569.

Honma, S. and Honma, K. (2003). The biological clock: Ca^{2+} links the pendulum to the hands. *Trends Neurosci.* 26: 650–653.

Honma, S., Katsuno, Y., Tanahashi, Y., Abe, H., and Honma., K. (1998a). Circadian rhythms of arginine vasopressin and vasoactive intestinal polypeptide do not depend on cytoarchitecture of dispersed cell culture of rat suprachiasmatic nucleus. *Neuroscience* 86: 967–976.

Honma, S., Nakamura, W., Shirakawa, T. and Honma, K. (2002). Determination of circadian period by synaptic interactions in a multi-oscillatory system in the suprachiasmatic nucleus. *Abstracts of Society for Neurosciences 32nd Annual Meeting,* Prog. No.77.5.

Honma, S., Nakamura, W., Shirakawa, T., and Honma, K. (2004). Diversity in the circadian periods of single neurons of the rat suprachiasmatic nucleus depends on nuclear structure and intrinsic period. *Neurosci. Lett.* 358: 173–176.

Honma, S., Shirakawa, T., Katsuno, Y., Namihira, M., and Honma, K. (1998b). Circadian periods of single suprachiasmatic neurons in rats. *Neurosci. Lett.* 250: 157–160.

Honma, S., Shirakawa, T., Nakamura, W., and Honma, K. (2000). Synaptic communication of cellular oscillations in the rat suprachiasmatic neurons. *Neurosci. Lett.* 294: 113–116.

Ikeda, M., Sugiyama T., Wallace, C.S., Gompf , H.S., Yoshioka, T., Miyawaki, A., Allen, C.N. (2003). Circadian dynamics of cytosolic and nuclear Ca^{2+} in single suprachiasmatic nucleus neurons. *Neuron* 38: 253–263.

Inouye, S.T. and Kawamura, H. (1979). Persistence of circadian rhythmicity in a mammalian hypothalamic "island" containing the suprachiasmatic nucleus. *Proc. Natl. Acad. Sci. U.S.A.* 76: 5962–5966.

Michels, K.M., Morin, L.P., and Moore, R.Y. (1990). GABAA/benzodiazepine receptor localization in the circadian timing system. *Brain Res.* 531: 16–24.

Moore, R.Y. (1973). Retinohypothalamic projection in mammals: A comparative study. *Brain Res.* 49: 403–409.

Moore, R.Y. (1996). Entrainment pathways and the functional organization of the circadian system. *Prog. Brain Res.* 111: 103–119.

Moore, R.Y., and Eichler, V.B. (1972). Loss of a circadian adrenal corticosterone rhythm following suprachiasmatic lesions in the rat. *Brain Res.* 42: 201–206.

Moore, R.Y. and Speh, J.C. (1993). GABA is the principal neurotransmitter of the circadian system. *Neurosci. Lett.* 150: 112–116.

Nakamura, W., Honma, S., Shirakawa, T., and Honma, K. (2001). Regional pacemakers composed of multiple oscillator neurons in the rat suprachiasmatic nucleus. *Eur. J. Neurosci.* 14: 666–674.

Nakamura, W., Honma, S., Shirakawa, T., and Honma, K. (2002). Clock mutation lengthens the circadian period without damping rhythms in individual SCN neurons. *Nat. Neurosci.* 5: 399–400.

Reppert, S.M. and Weaver, D.R. (2001). Molecular analysis of mammalian circadian rhythms. *Annu. Rev. Physiol.* 63: 647–676.

Sato, T. and Kawamura, H. (1984). Circadian rhythms in multiple unit activity inside and outside the suprachiasmatic nucleus in the diurnal chipmunk (Eutamias sibiricus). *Neurosci. Res.* 1: 45–52.

Schwartz, W.J. and Reppert, S.M. (1985). Neural regulation of the circadian vasopressin rhythm in cerebrospinal fluid: A pre-eminent role for the suprachiasmatic nuclei. *J. Neurosci.* 5: 2771–2778.

Shinohara, K., Hiruma, H., Funabashi, T., and Kimura, F. (2000). GABAergic modulation of gap junction communication in slice cultures of the rat suprachiasmatic nucleus. *Neuroscience* 96: 591–596.

Shinohara, K., Honma, S., Katsuno, Y., Abe, H., and Honma, K. (1995). Two distinct oscillators in the rat suprachiasmatic nucleus in vitro. *Proc. Natl. Acad. Sci. U. S. A.* 92: 7396–7400.

Shirakawa, T., Honma, S., Katsuno, Y., Oguchi, H., and Honma, K. (2000). Synchronization of circadian firing rhythms in cultured rat suprachiasmatic neurons. *Eur. J. Neurosci.* 12: 2833–2838.

Silver, R., LeSauter, J., Tresco, P.A., and Lehman, M.N. (1996). A diffusible coupling signal from the transplanted suprachiasmatic nucleus controlling circadian locomotor rhythms. *Nature* 382: 810–813.

Sokolove, P.G. and Bushell, W.N. (1978). The chi square periodogram: Its utility for analysis of circadian rhythms. *J. Theor. Biol.* 72: 131–160.

Stephan, F.K. and Zucker, I. (1972). Circadian rhythms in drinking behavior and locomotor activity of rats are eliminated by hypothalamic lesions. *Proc. Natl. Acad. Sci. U. S. A.* 69: 1583–1586.

Strecker, G.J., Wuarin, J.P., and Dudek, F.E. (1997). $GABA_A$-mediated local synaptic pathways connect neurons in the rat suprachiasmatic nucleus. *J. Neurophysiol.* 78: 2217–2220.

van den Pol, A.N. (1991). The suprachiasmatic nucleus: Morphological and cytochemical substrates for cellular interaction. In: Klein, D.C., Moore, R.Y., and Reppert, S.M., eds., *Suprachiasmatic Nucleus, the Mind's Clock*. Oxford University Press, New York, pp.17–50.

Vitaterna, M.H., King, D.P., Chang, A.M., Kornhauser, J.M., Lowrey, P.L., McDonald, J.D., Dove, W.F., Pinto, L.H., Turek, F.W., and Takahashi, J.S. (1994). Mutagenesis and mapping of a mouse gene, *Clock*, essential for circadian behavior. *Science* 264: 719–725.

Wagner, S., Castel, M., Gainer, H., and Yarom, Y. (1997). GABA in the mammalian suprachiasmatic nucleus and its role in diurnal rhythmicity. *Nature* 387: 598–603.

Wagner, S., Sagiv. N., and Yarom, Y. (2001). GABA-induced current and circadian regulation of chloride in neurones of the rat suprachiasmatic nucleus. *J. Physiol.* 537: 853–869.

Welsh, D.K., Logothetis, D.E., Meister, M., and Reppert, S.M. (1995). Individual neurons dissociated from rat suprachiasmatic nucleus express independently phased circadian firing rhythms. *Neuron* 14: 697–706.

Yamaguchi, S., Isejima, H., Matsuo, T., Okura, R., Yagita, K., Kobayashi, M., and Okamura, H. (2003). Synchronization of cellular clocks in the suprachiasmatic nucleus. *Science* 302: 1408–1412.

Yoo, S.H., Yamazaki, S., Lowrey, P.L., Shimomura, K., Ko, C.H., Ethan, D., Buhr, E.D., Siepka, S.M., Hong, H.K., Oh, W.J., Yoo, O.J., Menaker, M., and Takahashi, J.S. (2004). Period2::Luciferase real-time reporting of circadian dynamics reveals persistent circadian oscillations in mouse peripheral tissues. *Proc. Natl. Acad. Sci. U. S. A.* 101: 5339–5346.

18
Investigation of Network Phenomena in Hippocampal Slices Using Multi-Electrode Recording Arrays

LAURA LEE COLGIN

18.1   Introduction

It is commonly thought that higher cognitive functions in the brain are performed by groups of neurons working together as a network. Presumably, information in brain networks is represented by spatiotemporal patterns of neural activity. A technology that is useful for recording time-varying neuronal signals across spatially distributed brain regions is multi-electrode recording from brain slices. The traditional method of electrophysiological recording from brain slices is single-electrode recordings. Although single-electrode recordings are highly useful for studying synaptic physiology, their utility in network analyses is limited. Yet recording from multiple electrodes introduces additional complexities that may or may not, depending on the experimental conditions, be justified by the amount of information that they provide. The majority of this chapter is dedicated to two specific examples of fields of study that benefited from the use of multi-electrode recording: hippocampal rhythms, and temporally sustained and spatially distributed evoked responses. The second part of the chapter briefly addresses difficulties associated with multi-electrode recording and suggests ways that they may be overcome.

18.2   Applications of Multi-Electrode Recording

18.2.1   EEG Rhythms in Brain Networks

Rhythmic oscillations can be measured as classical electroencephalographic (EEG) waves that reflect the synchronized activity of ensembles of cells and thus provide a measure of neural activity at the level of neural networks. Rhythms have long intrigued researchers because they seem to be important for understanding how the forebrain becomes coordinated during different types of behaviors. Several lines of evidence support the idea that rhythms play an important role in cognitive processing and that particular frequencies subserve different functions.

The hippocampus exhibits three main bandwidths of oscillations: theta (4 to 7 Hz), beta (13 to 30 Hz), and gamma (~40 Hz; Leung et al., 1982). Theta is the best understood and most widely studied of the three, and results across different levels of investigation link theta to memory processing in the hippocampus. In a study of inhibitory avoidance learning, the magnitude of theta rhythm during the posttraining period was positively correlated with degree of retention (Landfield et al., 1972). Subsequent work showed that elimination of theta by lesioning or inactivating the lower brain region (medial septum/diagonal band) that generates the rhythm caused profound memory deficits in rats (Winson et al., 1978; Mizumori et al., 1990; M'Harzi and Jarrard, 1992). Afferent stimulation delivered in a manner that mimics theta is ideally suited for inducing long-term potentiation, a form of synaptic plasticity that is widely regarded as the substrate of many forms of memory (Larson et al., 1986; Larson and Lynch, 1986; Stäubli and Lynch, 1987). Stimulation on the positive phase of theta rhythm generates LTP in hippocampal slices and intact preparations (Pavlides et al., 1988; Bramham and Srebro, 1989; Huerta and Lisman, 1993; Hyman et al., 2003). Recently, magnetoencephalographic (MEG) arrays have recorded hippocampal theta oscillations in humans performing a well-known memory task (Tesche and Karhu, 2000; Jensen and Tesche, 2002).

The gamma rhythm has been the focus of much experimental work, yet its function remains controversial. Walter Freeman first proposed that gamma rhythms play an important role in sensory processing by allowing coherence to develop between spatially discrete areas (Freeman, 1975). Similarly, Wolf Singer and colleagues suggested that gamma rhythms transiently synchronize cells with disparate receptive fields, enabling individual aspects of a stimulus to be assembled into a coherent whole (Gray and Singer, 1989). Gamma rhythms occur in the entorhinal cortex in vivo (Chrobak and Buzsaki, 1998) and in vitro (van der Linden et al., 1999; Shimono et al., 2000), and gamma rhythms in the hippocampal pyramidal cell fields can be entrained by the entorhinal cortex (Charpak et al., 1995).

Beta rhythms have received less experimental attention. Both metabotropic glutamate receptor activation (Boddeke et al., 1997) and carbachol infusion (Shimono et al., 2000) have been reported to induce beta-frequency rhythms in hippocampal slices. Roger Traub and colleagues have published several reports of beta rhythms in hippocampal slices following tetanic stimulation delivered at twice-threshold levels (see Traub et al., 1999 for an example). It is unclear how closely these in vitro forms of beta correspond to beta waves in live animals. One problem is that some of the same experimental manipulations that induce beta activity in slices (i.e., carbachol infusion, metabotropic glutamate receptor activation) produce gamma rhythms under slightly different experimental conditions (Whittington et al., 1995; Fisahn et al., 1998). Still, mechanisms of beta rhythm production in the intact hippocampus fit relatively well with the characteristics of the in vitro waves. According to the explanation proposed by Leung (1992), pyramidal cells are excited by a combination of afferent stimulation and tonic depolarizing input provided by lower brain regions such as the medial septum. The excited pyramidal cells

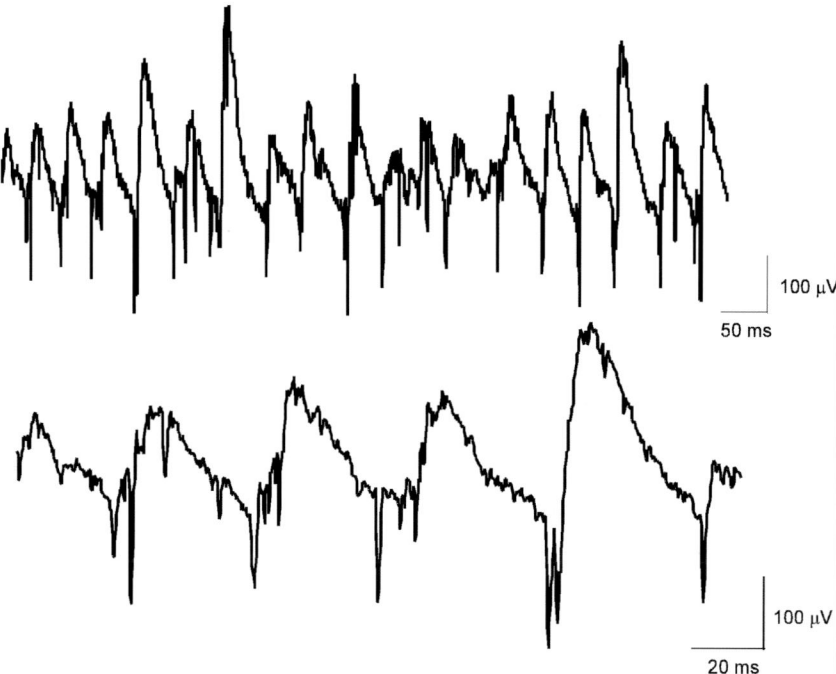

FIGURE 18.1. Carbachol-induced beta waves in hippocampal slices. Carbachol was infused at a concentration of 20 μM. Recordings were obtained from CA3 stratum pyramidale using conventional single-electrode recording methods (see Colgin et al., 2003 for a description of experimental methods).

then activate interneurons, which in turn feed back to inhibit the pyramidal cells after some delay. This model is consistent with the short excitatory phase/slow inhibitory phase cycle that is observed for beta rhythms in slices (Figure 18.1; see also Boddeke et al., 1997, Figure 2b and Traub et al., 1999, Figure 5B).

Regardless of how closely these rhythms correspond to their in vivo counterparts, beta oscillations remain an important dynamic of network function and the evidence described above supports the conclusion that the hippocampus retains its ability to produce these activity patterns in vitro. Beta rhythms are thought to be generated by local hippocampal circuitry (Leung, 1992), but they are also reported to facilitate long-range synchrony across cortical regions (Kopell et al., 2000). Little experimental work has been done to elucidate how beta rhythms are generated or how they synchronize disparate brain areas. The following section describes experiments that implemented multi-electrode recording to investigate the distribution, propagation, and mechanisms of production of beta rhythms in hippocampal slices.

Figure 18.2 depicts an example of cholinergically driven beta waves. A hippocampal slice placed upon an array of 64 electrodes is shown in Figure 18.2A. A

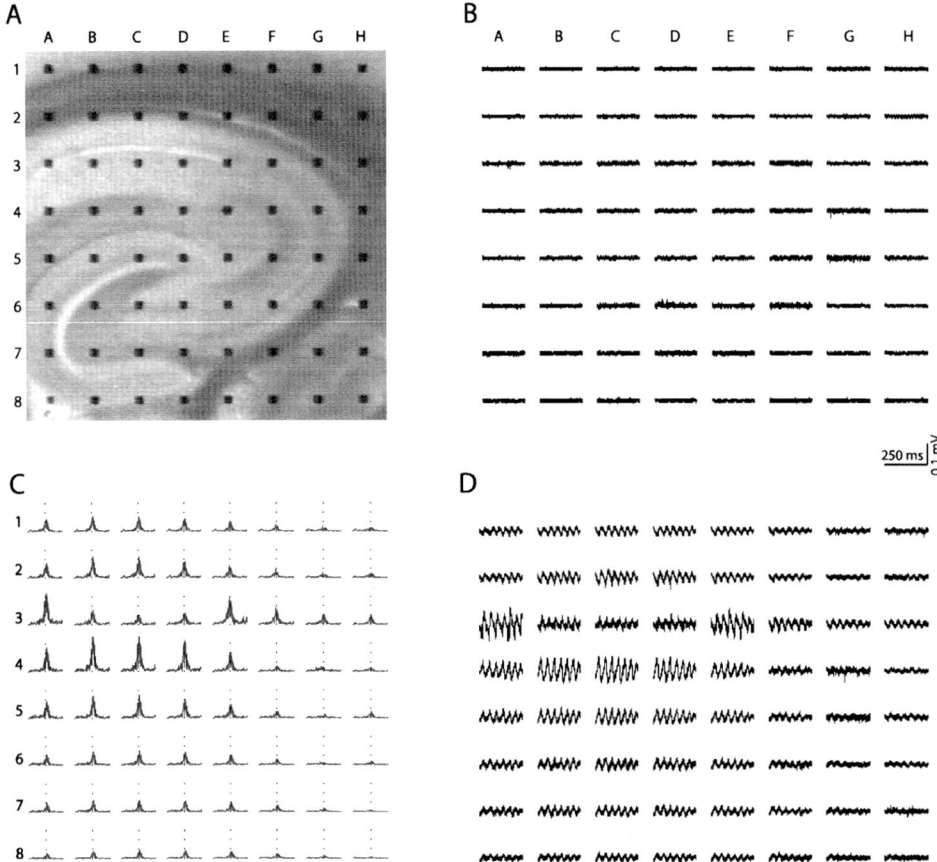

FIGURE 18.2. Carbachol-induced beta rhythms across 64 recording sites in a hippocampal slice. (A) Hippocampal slice placed upon an array of 64 electrodes (interelectrode spacing: 300 μm). (B) Spontaneous activity prior to carbachol infusion. Some minor single-unit activity was detectable, but no synchronized field activity was present. (C) Normalized power spectra computed for all 64 electrodes following infusion of 25 μM carbachol indicate that the dominant frequency was in the beta range and power was maximal in the apical dendritic field. Dotted line represents 26 Hz, the frequency at which power was maximal for this example. (D) Beta rhythms recorded after carbachol infusion. The waveform of the rhythms reversed across the cell body layer (e.g., electrode C2 vs. C4), indicating that they were locally generated. Amplitudes were large in the apical dendrites (e.g., electrode C4) and directly on the cell bodies (e.g., electrode E3). Calibration bars: 250 msec, 100 μV.

control recording of background activity from this slice reveals some spontaneous single-unit activity but no synchronized field potential oscillations or large amplitude activity of any kind (Figure 18.2B). Panel C shows the estimated power spectra across 64 electrodes for a two-second sample from the same slice following infusion of 25 μM carbachol. The dotted vertical line indicates the frequency at which power was maximal for this particular case (26 Hz). Panel D illustrates the field potential oscillations for the time sample corresponding to the power spectra in panel C. As shown in both panels C and D, power was greatest in the apical dendritic fields and directly on the pyramidal cell bodies of field CA1. As expected, the phase of the rhythms was reversed across the cell body layer; this is evident in the traces from electrodes C2 and C4 in panel D.

The origins and distributions of cholinergic beta rhythms in vitro were investigated in an earlier study using two-dimensional current source density analysis of recordings from 64-electrode arrays (Shimono et al., 2000). Figure 18.3 was reproduced from that report and shows estimated currents in the hippocampus during one cycle of a beta wave. Initially, a weak current source (depicted in yellow) emerged in the stratum radiatum of field CA3 accompanied by diffuse, low-magnitude sinks (depicted in blue) near the stratum pyramidale. Over the next 6 milliseconds, a strong current sink developed in the apical dendritic field while intense current sources were simultaneously observed in the pyramidal cell layer and basal dendritic field. This pattern of activity then spread to field CA1 and was sustained for the next 12 milliseconds.

Six milliseconds later, at approximately 25 milliseconds from the start of the cycle, a current source surfaced in the apical dendritic region of CA3 with associated current sinks near the cell bodies and stratum oriens. This apical source–basal sink dipole then spread across the extent of CA1 and persisted for the next 15 milliseconds. By 42 milliseconds after the start of the cycle, much of the coordinated activity had begun to dissipate. The brief apical sink-delayed apical source pattern described above was commonly observed across a large group of slices. The degree to which beta rhythms were more prominent in CA1 versus CA3 varied from slice to slice, however, supporting the idea that the waves are generated by local circuits.

Results from single-cell studies of carbachol's actions (Behrends and ten Bruggencate, 1993; Benson et al., 1988) and anatomical localization of muscarinic-type acetylcholine receptors in the hippocampus (Levey et al., 1995; Hajos et al., 1998) provide clues as to what the current sinks and sources shown in Figure 18.3 might represent. Carbachol blocks potassium currents in hippocampal pyramidal cells (Madison et al., 1987; Benson et al., 1988) thereby making these cells more excitable (i.e., more likely to spike). Carbachol additionally suppresses evoked inhibitory postsynaptic currents (IPSCs), an effect that probably contributes to the increased excitability of the pyramidal cell population, but at the same time enhances spontaneously occuring IPSCs (Pitler and Alger, 1992; Behrends and ten Bruggencate, 1993). These results suggest that cholinergic stimulation differentially affects two populations of inhibitory interneurons, an idea that is consistent with anatomical studies showing that M2 muscarinic receptors are located on the dendrites and soma of interneurons that innervate the apical dendrites of pyramidal cells and on axon terminals near the pyramidal cell layer (Levey et al., 1995; Hajos et al., 1998).

Taken together, the above results support the following model of carbachol-induced beta rhythm generation (introduced in Shimono et al., 2000). Pyramidal

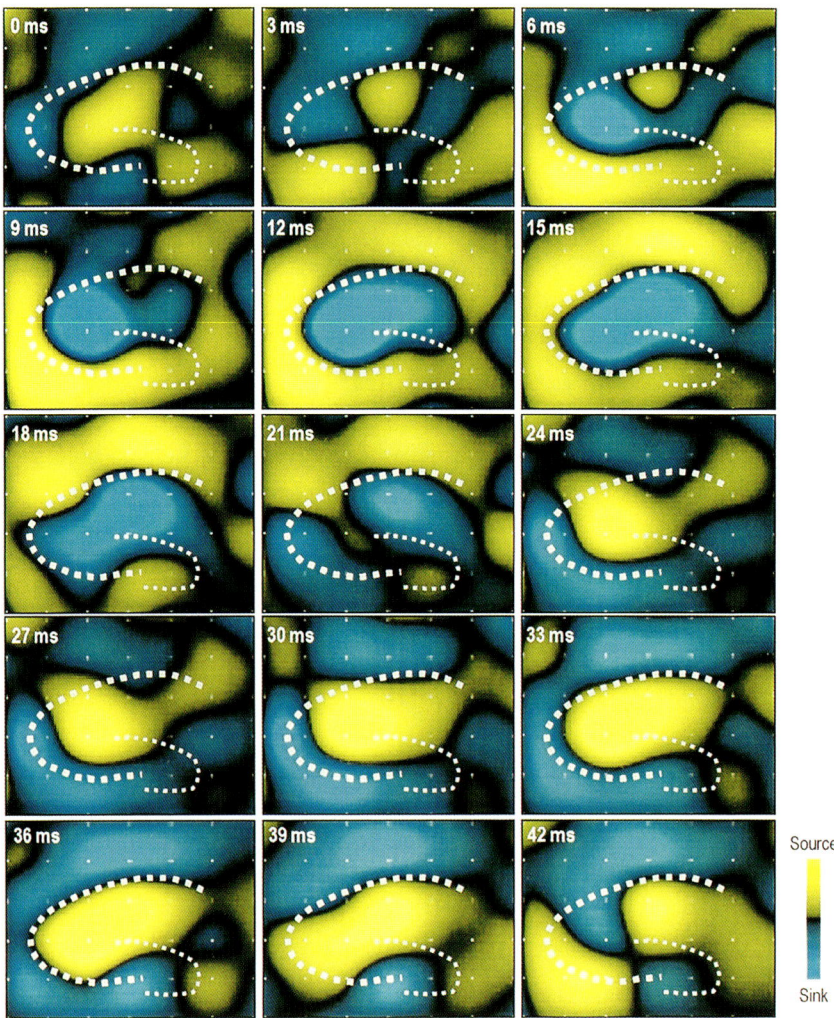

FIGURE 18.3. Evolution of current source density over time during one cycle of a beta wave. Each frame shows the instantaneous current source density computed at a particular time (indicated at the top left of each frame). Recordings were taken from 64 sites in the hippocampal network using a broad array of electrodes (interelectrode spacing: 450 μm). Blue indicates current sinks and yellow indicates sources. The positions of the pyramidal and granule cell body fields are represented by white, dashed lines. At $t = 6$ msec, an intense sink formed in the apical dendritic field of CA3, accompanied by a current source in the basal dendrites. The sink spread into the apical dendritic field of CA1 at $t = \sim 12$ msec. Activity dissipated, and then at $t = 27$ msec, an intense current source began to form in the apical dendritic field of CA3, accompanied by a current sink in the basal dendrites. The apical source coupled with its basal sink then spread to CA1 (~ 33 msec). The dipoles dissipated at 42 msec, and the cycle began anew. (Reproduced with permission from Shimono et al., 2000; © 2000 by the Society for Neuroscience.)

cells, tonically excited by blockade of potassium currents and suppression of inhibition at the soma, fire and spread their activity via the dense associational projections in field CA3 and collateral projections to CA1, resulting in the excitatory phase of the beta wave. Pyramidal cell firing activates a population of highly excitable interneurons that feed back to the apical dendrites of pyramidal cells, resulting in a strong inhibition in that region reflected as the inhibitory phase of the beta cycle. It is important to note that this model of beta rhythm production in slices fits well with mechanisms proposed for beta rhythm generation in freely moving animals (Leung, 1992).

The above description raises a critical experimental question regarding rhythms and their function in the hippocampal network: does the induction of rhythms shift the relative potencies of hippocampal afferents? As outlined above and described previously (Shimono et al., 2000), beta oscillations appear to be driven in part by inhibitory postsynaptic potentials (IPSPs) in the apical dendrites of pyramidal cells. This may result in a situation in which more distal afferents, such as the perforant path, are placed at a relative disadvantage to afferents terminating near the cell body and in the basal dendrites. This is likely not the operating mode when beta rhythms are absent. This suggests that the presence of rhythmic states serves to control the processing state of the hippocampal network. Empirical exploration of these topics using multi-electrode recording is described in the next section.

18.2.2   Beta Waves Can Affect the Size, Duration, Distribution, and Variability of Hippocampal Responses to Excitatory Inputs

The great majority of synapses in the hippocampus arise from associational and cortical afferents that use glutamate as a transmitter. As with other telencephalic areas, the hippocampus also receives significant projections from several subcortical structures that utilize an array of transmitters other than glutamate. The largest and best studied of subcortical projections to hippocampus is the cholinergic input from the medial septum/diagonal bands. Despite their relative sparseness, these afferents play an important role in the production of various rhythms in the hippocampus (Leung, 1992; Vertes and Kocsis, 1997; Leung, 1998). Less is known about how these inputs modulate hippocampal responses to activation of glutamatergic pathways and furthermore how responses are affected by ongoing rhythms. Cholinergic stimulation reduces the size of synaptic responses throughout most of the excitatory hippocampal pathways including the perforant path (Yamamoto and Kawai, 1967; Konopacki et al., 1987; Kahle and Cotman, 1989; Foster and Deadwyler, 1992; Colgin et al., 2003), CA3 associational system (Hasselmo et al., 1995), and Schaffer collaterals (Qian and Saggau, 1997; Hasselmo and Fehlau, 2001; Colgin et al., 2003), but the exact mechanisms responsible for the effects remain a source of controversy. Little to no work has been done to establish how cholinergic modulation of synaptic responses relates to cholinergically induced oscillations in hippocampus.

Several studies have shown that infusion of cholinergic agonists in hippocampal slices causes the near immediate appearance of rhythmic oscillations, but there is disagreement regarding the dominant frequency of the activity. Theta (4 to 7 Hz), beta (13 to 30 Hz), and gamma (~40 Hz) rhythms have each been reported to be triggered by application of carbachol (Konopacki et al., 1987; Huerta and Lisman, 1993; Williams and Kauer, 1997; Fisahn et al., 1998; Fellous and Sejnowski, 2000; Shimono et al., 2000; Colgin et al., 2003). In the Shimono et al. study using multi-electrode recording (described above), carbachol elicited regionally discrete beta activity in the majority of slices. Results of two-dimensional current source density analyses pointed to the conclusion that bursts of pyramidal cell discharges, spread of excitation through collateral projections, and activation of apically directed feedback interneurons generated the beta waves (Shimono et al., 2000; see above for an in-depth description).

The above arguments make several predictions with regard to how cholinergic activation would affect hippocampal responses to stimulation of glutamatergic pathways. Minimally, responses occurring against a background of fluctuating excitation and inhibition should be notably different than those recorded under control conditions. This assumes that the oscillatory circuits proposed to be responsible for the beta activity actually engage a significant proportion of the hippocampal neuronal population. In addition, responses evoked in the presence of cholinergic oscillations should be much more variable than conventional field excitatory postsynaptic potentials (EPSPs) and in some regular fashion be affected by the phase of the beta wave. The experiments detailed in this section were intended to test these particular predictions and as well to provide specific information on how cholinergic activity alters the size, shape, and duration of excitatory responses. The results accord with the hypothesis that cholinergic inputs impose timing requirements on hippocampal afferents. Experimental methods and procedures used in the experiments described below were similar to those detailed previously by Shimono et al. (2000).

The widely hypothesized representation of information in brain networks is a spatially distributed, self-sustained pattern of activity. Yet in conventional hippocampal slice experiments, widely used for studying the physiology of the hippocampal network, evoked responses do not reverberate across time throughout the circuit. A traditional hippocampal evoked excitatory postsynaptic potential is a well-characterized, stereotyped response lasting on the order of tens of milliseconds (Figure 18.4). It consists of a fast, negative-going, depolarizing waveform resulting from glutamatergic excitation, and an after-hyperpolarization due to GABAergic inhibition. It is difficult to imagine how networks could retain information using this response, when many task cues are encountered serially with delays of hundreds of milliseconds between stimuli. An explanation is needed to bridge the gap between the time course of synaptic responses and the longer time course necessary for network operations. Novel results, described below, support the idea that stimulation in the presence of beta rhythms can elicit responses that extend across a substantial portion of the hippocampal network and are sustained for hundreds of milliseconds.

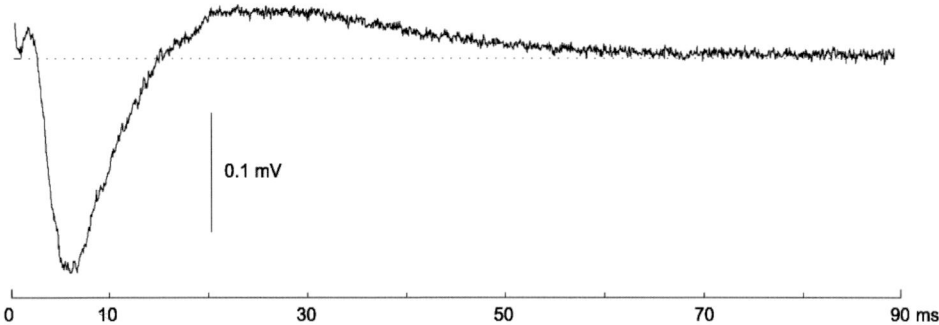

FIGURE 18.4. Example of a standard evoked potential in hippocampus. The initial negative-going waveform lasts approximately 10 msec, and is followed by a positive-going after-potential. This recording was taken from CA1 stratum radiatum in response to Schaffer collateral stimulation.

18.2.2.1 Laminar Profile and Regional Distribution of Sustained Network Response

Figure 18.5 compares an evoked response in the absence of rhythmic activity to a temporally sustained and spatially extended response evoked during cholinergically driven beta waves. As shown in Figure 18.5B, the dominant response in the control case was a rapidly developing and short-lived negative-going potential in the apical dendrites of field CA1 (e.g., electrode D2). This corresponds to the conventional field EPSP described in past physiological studies of the Schaffer-commissural (S-C) projections (see Figure 18.4). Responses elsewhere in the slice were much smaller than those recorded in field CA1. With the addition of carbachol, and subsequent generation of beta waves, two types of responses to single pulse stimulation emerged. In 16 of 53 recordings from eight slices, responses appeared similar to the control responses described above and are depicted in Figure 18.5C. However, in 37 of 53 traces from the same eight slices, carbachol produced two evident changes: single pulses triggered bursts of spikes in the stratum pyramidale of field CA3, and after-potentials became prominent features in both CA1 and CA3 (Figure 18.5D). An additional set of four slices was excluded from analyses because the delayed response was not observed in field CA1, most likely due to slight differences in slice preparation producing a weak CA3 → CA1 projection.

These effects can be seen more clearly in the single traces shown in Figures 18.6 and 18.7. The control response in field CA1, shown in gray, involved a typical biphasic response that reversed in polarity between the stratum radiatum (the terminus of the stimulated fibers; Figure 18.6, bottom) and the stratum oriens (Figure 18.6, top). The Schaffer-commissural response in field CA1 was greatly elaborated in the presence of carbachol (black traces, Figure 18.6). A sequence composed of a slow negative followed by a slow positive potential was added to the short latency field EPSP recorded in the apical dendrites (e.g., electrode E3 in Figure 18.5A).

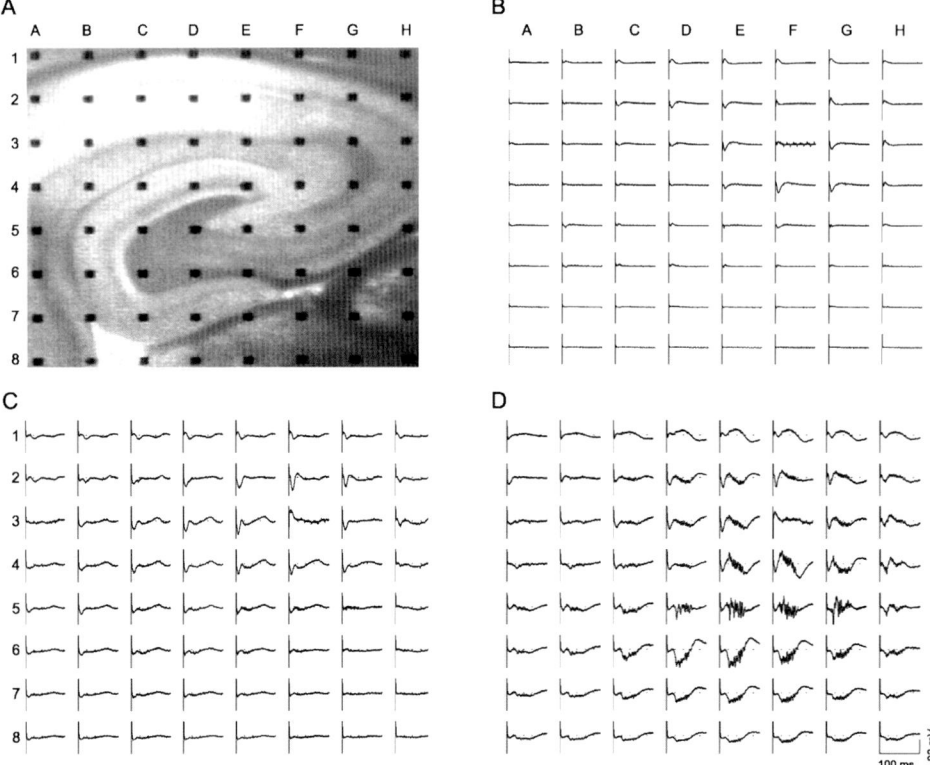

FIGURE 18.5. Evoked responses throughout the hippocampal network following stimulation of Schaffer collaterals in the presence and absence of carbachol-driven beta rhythms. Stimulation artifacts appear as a vertical line at the far left of each trace. (A) Hippocampal slice placed upon medium array of electrodes (interelectrode spacing: 300 μm). Electrode F3 was chosen for stimulation. (B) Evoked potentials across all 64 sites in the control condition. Note that responses did not propagate throughout the entire network. Activity was limited to the apical dendritic fields of CA3 (e.g., electrode F4) and CA1 (e.g., electrode E3). Phase reversals were prominent across the cell body layer in CA1 (e.g., electrodes E1 vs. E3). (C) Evoked potentials during carbachol-induced beta oscillations in an example in which a complex response was not generated. Responses and regional distribution were similar to the control condition. Background beta rhythm activity resumed almost immediately following stimulation. (D) Evoked response in the presence of cholinergic beta waves in an example in which a complex, reverberating response was generated. Initial fast negative-going potentials were observed in the apical dendritic fields of CA3 and CA1. However, instead of a prompt return to rhythmic activity, a sustained response was observed across the network. High-frequency cell spiking was recorded from CA3 pyramidal cells (e.g., electrode G5), presumably driven by an associated negative-going waveform in the basal dendritic field of CA3 (e.g., electrode E6). The apical dendritic field of CA3 exhibited a slow positive-going potential followed by a slow negative-going potential. The opposite apical-basal slow potential phase relationships were found in CA1. A delayed negative-going slow potential was recorded in the apical dendritic field of CA1 (e.g., electrode E3) with a corresponding delayed positive-going slow potential in the basal dendritic field (e.g., electrode E1). Note the increased spread of activation across the entire network during this complex response. Calibration bars: 100 msec, 200 μV.

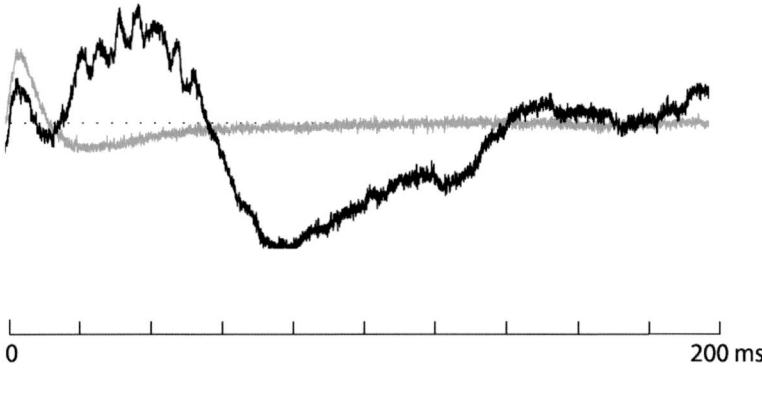

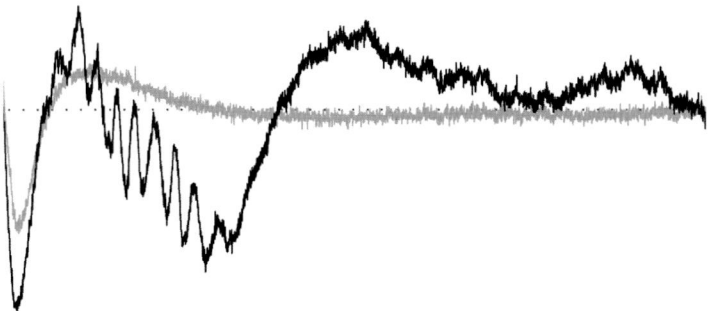

FIGURE 18.6. Evoked potentials in CA1 following stimulation of the Schaffer collateral pathway in the presence (black) and absence (gray) of carbachol-induced beta waves. The horizontal dotted line in the center of each trace denotes 0 μV. Stimulation in the absence of any rhythmic activity resulted in a stereotyped response (shown in gray), consisting of a fast negative-going potential followed by a positive-going after-potential in CA1 stratum radiatum (bottom) with a phase reversal of the response recorded in CA1 stratum oriens (top). The entire event was finished by 40 msec poststimulation. In the presence of cholinergically induced rhythms, a markedly different response was recorded (shown in black). In the apical dendritic field of CA1 (bottom), the primary fast response was followed by a sequence of slow potentials. The first was negative-going and was associated with high-frequency (~100 to 150 Hz) spiking. The second slow potential was positive-going and did not return to baseline until 140 msec after stimulation. The response in CA1 stratum radiatum was phase-reversed in stratum oriens, such that a positive-going slow potential was followed by a negative-going slow potential and a return to baseline at ~140 msec poststimulation.

The negative-going component, which began about 10 msec after the initial response and 25 msec after stimulation, was accompanied by evidence of high-frequency spiking. This was probably a reflection of activity in the pyramidal cell bodies in field CA3 because the spikes were largest in that region. Presumably, the delayed response in the apical dendrites of CA1 was due to prolonged spiking of CA3 pyramidal cells reactivating the Schaffer collateral/commissural fibers (discussed in greater detail below). All components of the elaborated response in the apical dendritic terminal field of the Schaffer collateral/commissural fibers reversed polarity in the stratum oriens (black trace, Figure 18.6, top).

Carbachol-induced rhythms also had very large effects on evoked potentials recorded in field CA3 (Figure 18.7). The control response in the apical dendrites of CA3a was a typical field EPSP (Figure 18.7, bottom left, gray trace) and the corresponding basal dendritic area, stratum oriens, showed the expected polarity reversal (Figure 18.7, top left, gray trace). In the presence of the cholinergic beta rhythms, stimulation of Schaffer collateral fibers produced a complex polysynaptic response in CA3 that consisted of a positive-going slow potential in stratum radiatum that began while the monosynaptic EPSP was still present followed by a negative-going slow potential (Figure 18.7, bottom left, black trace). Note that the apical dendritic sequence of the slow potentials in CA3 (positive–negative) was opposite that observed in CA1. A phase-reversed version of the apical dendritic response occurred in the basal dendrites of CA3 (i.e., negative-going slow potential followed by positive-going slow potential; Figure 18.7, top left, black trace). Bursts of spikes accompanied the first of the carbachol-dependent slow potentials in CA3 (Figure 18.7, right panels, black traces); these were pronounced in the cell body layer and became reduced in amplitude with distance from that layer. The bursts had a frequency of at least ~100 Hz and began in the earliest phase of the first of the postmonosynaptic EPSP slow waves. Because bursts of cell firing must be accompanied by nearby depolarizing currents, the negative-going waveform in the CA3 basal dendrites associated with the cell spiking was most likely a depolarizing waveform (supported by current source density analysis; see below) driving the CA3 cells to fire repetitively. Prolonged firing of CA3 pyramidal cells presumably caused a secondary activation of Schaffer collateral fibers, resulting in the delayed, slow, negative-going potential recorded in CA1 stratum radiatum. Note that there was no cell spiking in the stratum pyramidale of CA3 under control conditions (Figure 18.7, right panels, gray traces).

Although detectable responses were virtually absent in fields CA3b and CA3c in the control condition, responses reverberated throughout the entirety of CA3 in the presence of cholinergic beta oscillations. Cell spiking was observed across the entirety of CA3 stratum pyramidale during an elaborated carbachol response (Figure 18.7, right panels, black traces). Not only were network responses more spatially distributed than control responses, they extended across a much longer time period. This topic is addressed in greater detail below.

The records shown in Figures 18.6 and 18.7 suggest that the slow responses driven by Schaffer collateral stimulation during cholinergic activation build up in field CA3 and then propagate into field CA1. Tests of this were carried out using the bursts of cell spikes described earlier. Highpass filtering the responses to remove slow synaptic potentials revealed that the bursts were larger in field CA3 than in field CA1 (Figure 18.8). Cross-correlations showed that the spikes in CA1 and CA3 were well correlated when the former were delayed by 2 to 4 msec from the latter, a value that accords well with the known conduction velocity of

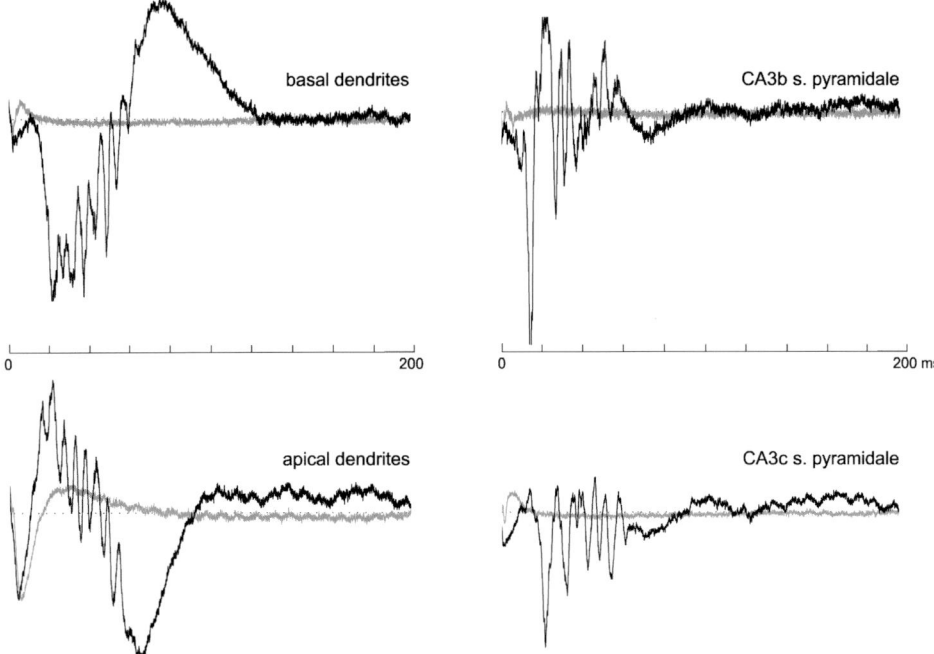

FIGURE 18.7. Evoked potentials in field CA3 in the presence (black) and absence (gray) of carbachol-driven beta rhythms following stimulation to the Schaffer collateral pathway. The dotted horizontal line in each trace indicates 0 μV. In the basal dendritic field of CA3 (top left), the control response (gray) was negligible, as was the fast component of the carbachol response. The initial phase of the slow potential of the carbachol response (shown in black) was a negative-going waveform with high-frequency spiking visible. The negative-going slow potential was followed by a positive-going potential, which returned to baseline at approximately 125 msec. In the apical dendrites (bottom left), a typical control response was recorded (shown in gray), consisting of a fast negative-going waveform followed by an after-hyperpolarizing potential. In the presence of carbachol (black trace), the apical dendritic response resembled the control response for a short time (<5 msec poststimulation) before veering off into a positive-going waveform (~10 msec poststimulation). Again, high-frequency spikes were visible during this phase. Note that the apical response was a phase-reversal of the basal response, such that the initial positive-going waveform was followed by a negative-going waveform. The source of the high-frequency spiking appeared to be the CA3 pyramidal cells (right traces, top and bottom), and spiking was observed across the entire extent of CA3 stratum pyramidale. Note that control responses (gray) in the cell body were insignificant. The negative-going waveform in the CA3 basal dendritic field was likely driving the high-frequency firing of CA3 pyramidal cells through the dense associational system of CA3.

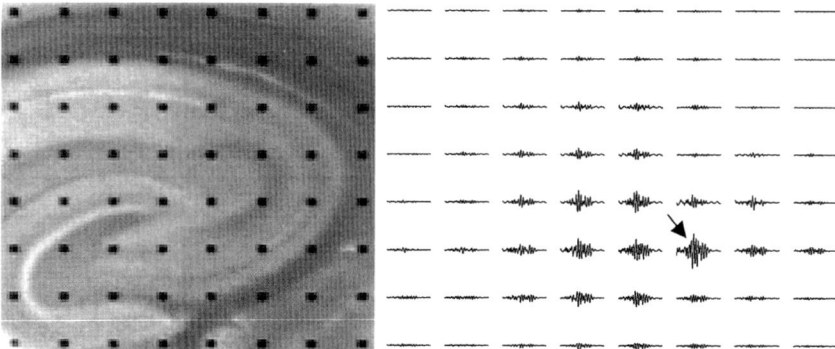

FIGURE 18.8. High-frequency bursting during complex response. A single stimulation pulse was delivered to the Schaffer collateral pathway following induction of beta rhythms by infusion of 25 µM carbachol. Responses recorded from the hippocampal slice depicted in the left panel were highpass filtered at 100 Hz to remove the slow potentials during the time segment from 10 to 80 msec after stimulation (right panel). It was evident that bursting was most prominent in field CA3, especially in stratum pyramidale (indicated by arrow). Calibration bars: 50 msec, 0.5 mV.

the Schaffer collateral projections from CA3 to CA1. These results confirm the idea that spiking bursts triggered by stimulation originate in field CA3 and then are propagated from there to CA1.

18.2.2.2   Time Course of Response

The minimum negative-going potential of the fast component occurred in the apical dendrites of field CA3 on average at 5.5 ± 0.7 msec after stimulation and in CA1 apical dendrites at 5.3 ± 1 msec (mean $\pm$ S.D., $n = 37$ stimulation trials from 8 slices). The minimum of the slow component of the CA3 basal dendritic negative-going potential occurred at 24 ± 10.5 msec after stimulation. In CA1, the minimum value of the slow, negative-going potential in the apical dendritic field occurred significantly later than its CA3 counterpart at 52.4 ± 19 msec poststimulation (paired t-test, two tails, $p < 0.0001$). The high degree of variance in the slow potentials was attributable to individual variations across slices. Also, the delayed potentials appeared to integrate multiple synaptic events, unlike the control monosynaptic EPSPs, thereby introducing additional degrees of variation.

The peak positive-going potential in CA3 apical dendrites corresponding to the slow negative-going potential in CA3 basal dendrites occurred at 17 ± 7.4 msec following stimulation. This peak was probably a source for the depolarizing currents in the basal dendrites but also likely reflected hyperpolarizing currents generated by feedback interneurons. Its maximum occurred at a significantly shorter time after the response than did the minimum negative-going slow potential in the basal dendritic field of CA3 (paired t-test, two tails, $p < 0.001$), indicating that the positive-going waveform was not merely a reversal of the basal dendritic waveform. In addition, the time course of its peak was consistent with IPSPs generated by feedback interneurons in response to the primary fast response in the CA3 apical dendrites.

18.2.2.3   Response Size

The minimum amplitude of the fast negative-going potential was -60 ± 40 μV in the CA3 apical dendritic field and -70 ± 50 μV in the CA1 apical dendritic field. Neither of these values was significantly correlated with the minimum amplitude of the slow negative-going potential in CA1 apical dendrites (-105 ± 43 μV) nor the slow negative-going potential in CA3 basal dendrites (-180 ± 88 μV). On average, the minimum amplitudes of the slow components were more negative than the fast component minimum amplitudes (paired t-test, two tails, $p < 0.01$) for both CA3 and CA1. The average maximum amplitude of the large positive-going waveform in CA3 apical dendrites was 170 ± 135 μV. None of the amplitude measures were found to be significantly correlated with the power of the oscillations.

18.2.2.4   Current Source Density Analysis

Current source density analysis is a technique that is used to estimate the locations of the synaptic currents underlying field potentials (Nicholson, 1973; Haberly and Shepherd, 1973; Nicholson and Freeman, 1975; Nicholson and Llinas, 1975). The method was employed for the results described below in its continuous, two-dimensional form. This variant of current source density analysis has been previously described in significant detail (Shimono et al., 2000), so the description that follows is brief. The two-dimensional array of electrodes allowed for simultaneous estimation of current flows in any direction within the plane of the slice. Data were lowpass filtered at 100 Hz and spatially smoothed by a 3 × 3-weighted average kernel (0 1/8 0, 1/8 1/2 1/8, and 0 1/8 0). The result was then convolved with a 3 × 3 Laplacian kernel to obtain a discrete approximation of the second spatial derivative. Lastly, currents were obtained for all electrodes by calculating the 8 × 8 current source density for each time step and then estimating the value at each electrode location using bilinear interpolation.

One limitation of this technique is that only large spatial patterns with radii ≥ one-half of the interelectrode distance can be accurately resolved. Another problem is that lowpass filtering removes high-frequency data, thereby limiting the amount of fine detail that can be observed. Nevertheless, the method is quite useful for investigating spatially distributed events that change relatively slowly across time, and it could be argued that beta rhythms fall into this category as they are thought to involve synchronized activity across a large number of neurons.

Figure 18.9 shows the results of two-dimensional current source density analysis for selected time points in a 30 msec time window following stimulation to the Schaffer collateral pathway. The estimated current sinks (blue) and sources (yellow) are shown in the top four panels for the sustained network response in the presence of beta rhythms induced by carbachol and in the bottom four panels for an evoked response in the absence of cholinergic activity. At 4 msec poststimulation,

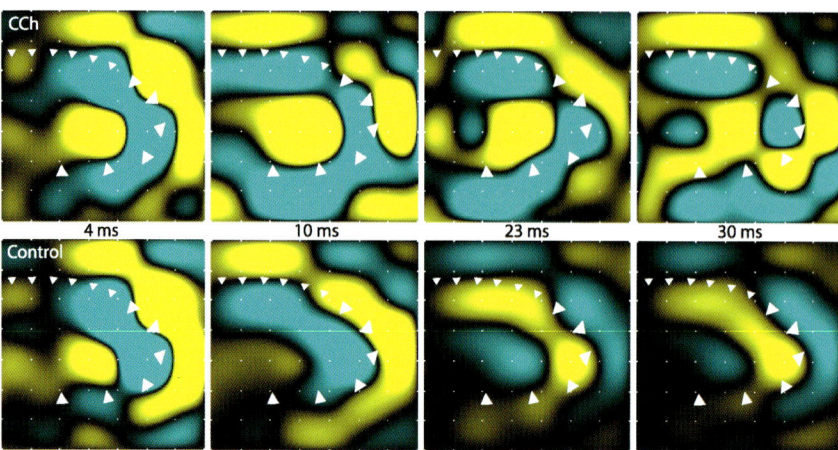

FIGURE 18.9. Two-dimensional current source density estimates for evoked responses in the presence (top) and absence (bottom) of carbachol-induced rhythms. Depolarizing current sinks are depicted in blue with hyperpolarizing current sources shown in yellow. The outline of the pyramidal cell bodies is depicted as white triangles, with larger triangles delineating field CA3 and smaller triangles for CA1. At 4 msec after a stimulation pulse was delivered to the Schaffer collateral pathway, differences between control and carbachol (CCh) responses were minimal. At 10 msec poststimulation during the CCh response, a large current source appeared in the apical dendrites of CA3 and a current sink appeared simultaneously in the basal dendrites, especially apparent in CA3c. In the control response, activity did not spread across the entirety of CA3 and the CA3 apical source/basal sink dipole was not present. At 23 msec, in the CCh case, the CA3 cell bodies and basal dendritic fields continued to be dominated by current sinks, and the CA3 apical source also remained. A prolonged depolarizing current sink was observed in the apical dendritic field of CA1 with a corresponding current source in CA1 stratum oriens. The control response, on the other hand, displayed weak activation at this time, with an after-hyperpolarization in the CA1 apical dendrites as the only distinguishing feature. This weak hyperpolarizing current source lingered in the CA1 apical dendrites of the control response at 30 msec, and stood in contrast to the strong depolarizing current sink observed in the same location at the same time point during the CCh response.

the responses were quite similar, although stimulation in the presence of cholinergic rhythms evoked a response that extended across slightly more of the network. In both cases, the major depolarizing current sinks occurred in the apical dendritic fields of CA1 and CA3. Corresponding current source dipoles were recorded across the cell body layers, but were not as prominent in the CA3 basal dendritic field for the carbachol response. Instead, it appeared that depolarizing currents were arising in the cell bodies of CA3 and beginning to produce a current sink. By 10 msec after stimulation, the entire network had fully mobilized in the carbachol case.

The two major differences between the response in the presence of carbachol-induced rhythms and the control response at this time were the appearance of a large current source in the CA3 apical dendrites and a well-formed current sink in the CA3 basal dendrites during the carbachol response. In the control condition, the dipoles in CA3 were the same as in CA1 (apical sinks/basal sources), whereas the dipole relationships reversed in CA3 when cholinergic beta rhythms were present (CA3 apical source/basal sink, CA1 apical sink/basal source). It is also evident that the spread of activity was far greater when carbachol-induced rhythms were present than that seen without carbachol, particularly in field CA3 where the activity reached to the terminus of the pyramidal cell layer (i.e., field CA3c) in the carbachol case. Certainly related to greater spread, the intensity of activity was markedly increased in the presence of beta rhythms. At 23 msec following stimulation, the excitatory components of the evoked response were absent in the control condition, replaced by hyperpolarizing current sources in the apical dendritic fields of CA1 and CA3a. In contrast, the cholinergic response exhibited sustained excitatory current sinks in the apical dendrites of field CA1. Current sinks were also observed in the basal dendrites and cell bodies of CA3 at this time. At 30 msec poststimulation, the sustained apical current sink and its corresponding basal source in CA1 remained robust in the carbachol response, and only weak apical current sources lingered in the control condition.

Current source density analyses yielded consistent patterns in 7 of the 8 slices at time points approximately 5, 10, 20, and 30 msec following stimulation. At later time points, a large degree of temporal variability in the estimated currents developed across slices, probably due to timing variations in the slow potentials and the emergence of prominent high-frequency components likely reflecting cell spiking in CA3. Thus, simultaneous current source densities computed after 30 msec poststimulation are not discussed here.

18.2.2.5   Response Variability

There was an additional degree of variability that has not yet been discussed. Stimulation to the Schaffer collateral system in the presence of cholinergic rhythms elicited a sustained network response in approximately 70% of the stimulation trials, but in the other stimulation trials a response similar to a control EPSP was evoked. Field potential oscillations reflect alternating waves of excitatory and inhibitory currents. It follows that the oscillation phase, either excitatory or inhibitory, would predict whether a cell fires. As discussed above, depolarizing currents in the basal dendrites appear to be generating the sustained network response. Closer inspection of evoked responses in the presence of cholinergic beta waves indicated that the phase of the oscillations on which the stimulation pulse was delivered in the CA3 stratum oriens may have been responsible for the observed response variability. Analyses of seven slices revealed that in 34 out of 47 stimulation trials in which the network response was evoked, 18 stimulation pulses clearly landed on a local minimum in the basal dendritic field of CA3. In 11 stimulation trials in which the network response occurred, stimulation arrived when potentials in the CA3 basal dendrites were close to zero. Only 2 out of 13 recordings in which a sustained response was not generated in the presence of carbachol showed the stimulation pulse arriving on a local minimum in the basal dendrites. In 5 recordings, it was difficult to determine where the stimulation pulse landed in the basal dendritic field because of degraded recording resolution likely due to volume conduction of activity. One slice was excluded from phase analysis because of this problem.

Figure 18.10 illustrates two responses to Schaffer collateral stimulation recorded from stratum radiatum of field CA1. The evoked response in the top panel was short lasting, and oscillatory activity resumed approximately 10 msec following stimulation. The stimulation pulse landed on a valley in the apical dendrites, which corresponded to a peak in the basal dendrites. In the bottom panel, the stimulation pulse landed on an apical peak in CA1, and a sustained network response was clearly evoked, disrupting rhythmic activity for over 200 msec. One measure of the magnitude of the prolongation of the response is the time it takes for the background beta oscillations to re-establish themselves. That is, physiological activity initiated within a group of neurons by stimulation of excitatory pathways should interfere with the production of rhythmic behavior by the same neurons. It is interesting to note that when stimulation arrives at a particular phase of ongoing rhythmic activity, one stimulation pulse to the Schaffer commissural projections is sufficient to block beta rhythms for at least 200 msec (Figure 18.10, bottom).

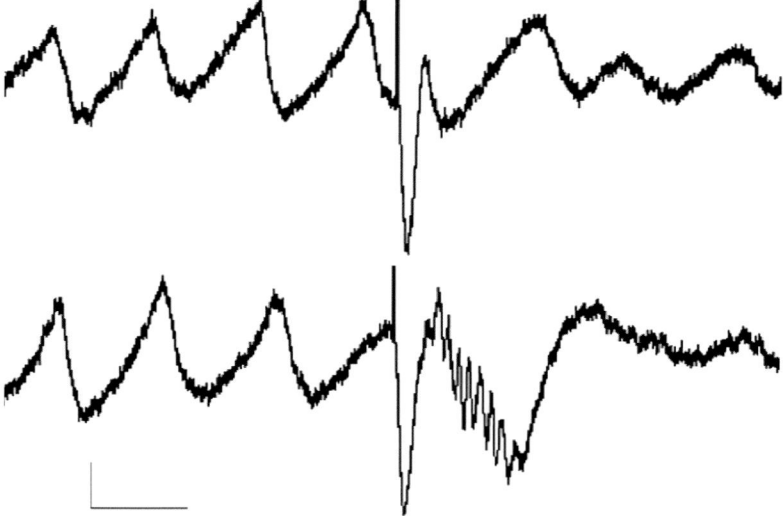

FIGURE 18.10. Within-slice variability of evoked responses in the presence of carbachol-induced oscillations. Recordings from CA1 stratum radiatum illustrate two examples of different responses to the same Schaffer collateral stimulation site (stimulation artifact seen as vertical line in center of trace). Note that in the top trace, stimulation arrived on an apical valley, and the response was not temporally sustained. In the bottom trace, stimulation was delivered on an apical peak, resulting in a temporally sustained response and subsequent disruption of rhythmic activity for ~200 msec. Calibration bars: 50 msec, 200 μV.

18.2.2.6 Cholinergic Activation Can Increase Complexity of Responses to Glutamatergic Inputs

The above results suggest that cholinergic input to the hippocampus can profoundly affect responses generated by glutamatergic pathways in a way that has not been reported previously and in addition provide information as to the nature of the interaction. Responses to stimulation of the Schaffer collateral projections, the principal associational system of hippocampus, became much more complex than the standard field EPSP following the introduction of cholinergic rhythms. Although there was variability within slices, the predominant change was the addition of two slow potentials to the conventional field EPSP. In field CA1, the first of these was negative in the apical dendrites whereas the second was positive; both reversed across the cell body layer, indicating that they were locally generated. High frequency ($\sim$100 Hz) spike bursts accompanied the first wave. Given this and its location, it is reasonable to assume that the negative slow wave in CA1 reflected depolarization resulting from secondary activation of the Schaffer collateral projections. Additional evidence on this point is considered below. If the first slow wave was due to EPSPs sufficient to spike the CA1 pyramidal cells, then the delayed positive wave is likely to be an inhibitory postsynaptic potential (IPSP) set in motion by the excitatory inputs.

Slow waves corresponding to those just described were also recorded in field CA3 in the presence of carbachol-driven oscillations, but these did not have the same laminar profile observed in field CA1. That is, the first wave of the slow potential was positive in the apical dendrites and negative in the basal dendrites. The second wave remained opposite in polarity from the first, and the two potentials were reversed across the CA3 cell body layer. Somatic spikes riding on the first wave were of greater magnitude than those in CA1. From these observations it can be concluded that Schaffer collateral stimulation pulses in the presence of cholinergic activation can, at least on some trials, trigger secondary EPSPs that are sufficiently potent to cause high-frequency discharges of the CA3 pyramidal neurons.

By far the most probable source of the excitatory input is the massive CA3 associational system that, together with its commissural counterpart, generates the great majority of synapses in the subfield (Swanson et al., 1978; Ishizuka et al., 1990). The associational system densely innervates both apical and basal dendrites of the CA3 pyramidal neurons (Hjorth-Simonsen, 1973). It appeared that in the presence of beta rhythms, the basal dendritic system became dominant in CA3 during the slow component of the response. Anatomical and physiological evidence provide clues as to how this may have taken place. Muscarinic acetylcholine receptors are found on the dendrites and soma of a class of interneurons that projects to the stratum radiatum and also on axon terminals of basket cell interneurons at the stratum pyramidale/oriens border (Levey et al., 1995; Hajos et al., 1998). Physiological results support the conclusion that cholinergic stimulation facilitates the former group of interneurons while suppressing the latter (Pitler and Alger, 1992; Behrends and ten Bruggencate, 1993). Due to the carbachol-induced increase in excitation of the apically projecting feedback interneurons, it is likely that the apical dendrites were hyperpolarized at the time the slow potential was initiated. This may have resulted in a situation where the basal dendritic associational system, released from basket cell inhibition by cholinergic stimulation, took control. This raises the possibility that subsequent inputs arriving in the basal dendritic field during this time may be treated preferentially over activation in the apical dendritic field. It is conceivable that sustained activation of this type, maintained primarily by reverberatory activity in the basal dendritic associational system of CA3, could serve to hold inputs for tens or hundreds of milliseconds.

As expected, the monosynaptic field EPSPs elicited by stimulation of Schaffer collateral fibers emerged at nearly the same time in CA3 and CA1 but cross-correlations showed that spike bursts associated with the temporally extended responses in CA1 were delayed by about 3 msec from those in CA3. This interval aligns with conduction and transmission delays for the Schaffer projections from CA3 to CA1. These observations combined with the location of the spike-associated slow wave in CA1 make it very likely that events in CA1 after the initial fast EPSP are driven by the delayed activation of the CA3 pyramidal neurons.

It will be noted that spike bursts emerged after, rather than in conjunction with, the monosynaptic CA3 response. Field EPSPs in CA3 were small and restricted in the "medio-lateral" axis, indicating that the fibers stimulated in CA1c did not form dense synaptic beds with a significant portion of the CA3 pyramidal cells. It appears then that a primary effect of cholinergic stimulation is to allow a relatively small starting population of neurons to set off spiking in cells throughout the CA3 region.

The dense associational system, to which each spiking cell would contribute, is well suited for generating reverberating activity but, as seen in the control cases, typically it does not. The spread of spikes and slow wave activity into those portions of CA3 adjacent to the dentate gyrus (CA3c) confirmed that the associational system had been activated by the Schaffer collateral/commissural projections in the presence of carbachol-induced beta waves. Physiological and anatomical studies together point to two muscarinic effects that are appropriate to produce such an outcome: a direct increase in the excitability of pyramidal neurons due to blockade of potassium currents (Nakajima et al., 1986; Madison et al., 1987; Benson et al., 1988), and suppression of release from feedback basket cells (see above).

It should be noted that the above arguments relate to the complex potential seen in the 10 to 100 msec interval following the monosynaptic field EPSP. The response often continued past this period as evidenced by a disruption of beta oscillations that on some trials lasted for 500 msec. Additional work is needed to identify the types of slow potentials associated with the extended response and to determine how they influence processing of glutamatergic inputs.

It is also important to point out that the effects described above have not been reported previously in a substantial body of literature concerning the effects of cholinergic agonists on responses elicited by Schaffer collateral stimulation. Instead, field excitatory postsynaptic responses to Schaffer collateral stimulation are widely reported to be suppressed by application of cholinergic agonists in hippocampal slices, an effect that is thought to be presynaptically mediated (i.e., due to reduced glutamate release; Valentino and Dingledine, 1981; Sheridan and Sutor, 1990; Qian and Saggau, 1997; Colgin et al., 2003). In contrast, as discussed above, the effects reported here are likely to result from a combination of the postsynaptic effects of carbachol (i.e., tonic depolarization of pyramidal cells (Nakajima et al., 1986; Madison et al., 1987; Benson et al., 1988) and decreased GABA release from basket cell terminals (Behrends and ten Bruggencate, 1993). Potential explanations as to why these effects of cholinergic stimulation would not be detected in the majority of field-recording studies in slices but could be observed in the current work is discussed below.

18.2.2.7   Cholinergic Stimulation May Impose Timing Requirements on Glutamatergic Inputs

As described in earlier studies, and shown here, cholinergic stimulation elicits rhythmic oscillations in hippocampal slices. Various experimental approaches have established that the rhythms are produced by EPSP–IPSP sequences. The sequences are generated by the two effects noted above (i.e., increased excitability of pyramidal cells and decreased GABA release from basket cell terminals) acting together with a third cholinergic action, namely, increased excitability of a population of feedback interneurons that innervate the apical trees of the pyramidal neurons. Tonic excitation of the pyramidal cells (effect 1 above) occurring in the absence of basket cell inhibition (effect 2) results in increased spiking that quickly amplifies itself, via the CA3 collateral system, to the point that it triggers apically directed inhibitory feedback (effect 3). Spiking is shut down by the last event but then resumes as the feedback IPSPs dissipate.

This description does not describe in other than general terms the manner and degree to which the oscillatory activity might affect the primary hippocampal pathways. The novel results described above address this point and thereby raise the issue of whether, as has often been proposed, cholinergically driven rhythms impose timing requirements on the arrival of excitatory inputs. Post hoc analyses suggest that this may indeed be the case. That is, the magnitude and duration of the response to Schaffer collateral stimulation varied greatly across trials following the infusion of carbachol and subsequent production of beta rhythms with much of the variability attributable to the timing of stimulation pulses with respect to the beta rhythm phase. Thus, synchronization associated with beta waves may be a strategy for creating reliable and predictable time windows within which afferents arriving at opportune moments can exploit the processing capabilities of the network. These results suggest that some degree of synchronization would be needed between the hippocampus and its excitatory inputs.

The superficial layers of entorhinal cortex are the primary hippocampal afferent and are reported to generate beta waves when infused with cholinergic agonists (Shimono et al., 2000). Interestingly, the deep layers of entorhinal cortex, an important target of hippocampus and subiculum, produce gamma rhythms under the same conditions (van der Linden et al., 1999). Iijima and colleagues (1996) reported that transfer of activity from the entorhinal cortex to the hippocampus was frequency-dependent, proposing that frequency-dependent transfer may be involved in selectively gating the entry of information into the hippocampus. Similarly, gamma activity has been reported to occur in hippocampus following carbachol infusion (Fisahn et al., 1998; Fellous and Sejnowski, 2000), and gamma coherence between hippocampus and entorhinal cortex is relatively high in vivo (Charpak et al., 1995; Chrobak and Buzsaki, 1998). In any event, hippocampus and retrohippocampal cortex appear to have sufficiently similar local circuits so that the cholinergic septal projections can provide the synchronization required by the beta activity.

18.2.2.8   Contributions of Beta Oscillations to Hippocampal Operations

Cortical rhythms are usually thought to synchronize afferents and thereby allow them to be more effective than would be the case if they arrived in a temporally scattered manner (e.g., Csicsvari et al., 1999). Other studies have shown that certain naturally occurring rhythms have deep relationships with synaptic plasticity (Larson et al., 1986; Staubli and Lynch, 1987; Pavlides et al., 1988; Huerta and Lisman, 1995; Lisman et al., 2001). The presently described results add to the list of potential functions of oscillations by showing that beta rhythms present windows of opportunity for afferents such that properly timed arrival of a modest input can result in a spatially and temporally extended response. With regard to the former, it is possible that the synchronization of large populations of cells that occurs during rhythms allows activity to spread more easily across a greater extent of a spatially distributed brain network.

As mentioned above, the results described here, obtained with a multi-electrode recording array, have not been reported in the numerous papers describing evoked responses in the presence of cholinergic stimulation. Cholinergic agents have only been reported to reduce monosynaptic evoked responses (Yamamoto and Kawai, 1967; Konopacki et al., 1987; Kahle and Cotman, 1989; Foster and Deadwyler, 1992; Qian and Saggau, 1997; Hasselmo and Fehlau, 2001; Colgin et al., 2003), but none of these studies involved stimulation in the presence of cholinergically driven rhythms. Thus, it may be the case that the cellular dynamics inherent in the rhythmic activity are in some way important to the production of temporally extended, polysynaptic responses of the type reported here. Alternatively, it is possible that some aspect of the multi-electrode array recording technique allowed these patterns to be seen. The worst-case scenario is that the slices used in the studies described above were not healthy and thus readily exhibited pathological responses for some reason potentially associated with the recording method.

However, it can be argued that an alternative explanation concerning the level of observation is more likely. Traditional slice-recording studies measure activity at the synaptic level whereas the phenomena described above are likely to reflect activity at a more macroscopic level involving ensembles of neurons acting together as a network. There are several reasons to suspect that this may be the case. For one, the microelectrodes in the 64-electrode recording device (Panasonic; MED64) used in the experiments described here are 50 μm $\times$ 50 μm in size, substantially larger than the pulled glass micro-pipettes widely used in slice-recording studies. Also, the slice is placed on top of the electrodes in a small, fluid-filled dish. In this situation, the recording electrodes are separated from the source of the currents by several materials: the coating that the electrodes are treated with for slice adhesion, a thin layer of artificial cerebrospinal fluid (ACSF), and a layer of dead cells on the surface of the slice. This is in contrast to recordings from fine-tipped ("sharp") glass electrodes in which the recording electrode is lowered into the tissue as close as possible to the place where the currents are being generated thereby allowing for recordings that are detailed and spatially specific. In the case of multi-electrode arrays, the area between the electrodes and the source is filled with resistive media (i.e., layers of dead cells, electrode coatings). Under these conditions, currents diffuse out in this space, and thus only low frequency spatial features of the population activity will be recovered by the recording electrodes.

Because of these characteristics (large electrode size, distance between current source and recording site), a substantial degree of spatial averaging is part of the recording process. It may be possible that some low-frequency population events that are picked up by the large metal electrodes are obscured by random activity of individual cells in sharp electrode recordings. This explanation is more believable in the case of cholinergic stimulation in which individual cells have increased spontaneous firing rates. In any event, these arguments raise an important point: applications that are concerned with broadly occurring population events are more conducive to the use of multi-electrode recording arrays than are investigations at more microscopic levels.

It is important to be aware of these and other issues when conducting experiments with multi-electrode recording arrays. The previous section offered two examples of applications in which the small local activity that was lost in the diffuse and spatially smoothed signals recorded with multi-electrode recording arrays was compensated by the additional information gained by simultaneous recording of population events across spatially distributed sites. The next section addresses additional complications to be aware of when using multi-electrode recording arrays and suggests ways that these problems can be overcome.

18.3 Technical Considerations Associated with Multi-Electrode Recording

18.3.1 Low Spatial Resolution

The origins of signals recorded from multi-electrode arrays can be complicated to determine. After leaving their original source, currents diffuse out through many layers of resistive media including dead cells, ACSF, and electrode coatings that essentially act as a lowpass filter. Much of the signal that remains will then be conducted with little to no degradation through the relatively large diameter, low impedance, metal recording electrode. Under these conditions, much of the locally occurring activity will be lost. On the other hand, broad synchronized population activity will be picked up, and in some cases will likely summate across time. Because of these factors, multi-electrode recording is best suited to applications involving investigation of large-scale ("network level") phenomena.

18.3.2   Volume Conduction

As noted above, there is a thin layer of conductive medium, namely, ACSF, between the source of the currents being measured and the recording electrode. Signals spread out through this medium away from their source and may be picked up by neighboring electrodes. This contributes to the aforementioned difficulty of local-izing the sources of the recorded activity. It is thus necessary to examine recordings post hoc and verify that clear dipoles (i.e., phase reversals of the recorded signals) are observed and that negligible activity is recorded from electrodes that are placed under areas where one would expect to see little to no activity (e.g., hippocampal fissure). Volume conduction problems can be greatly minimized by taking all pos-sible steps to ensure that the slice properly adheres to the electrodes (see Shimono et al., 2000 for details).

Current source density analysis can be used to alleviate both problems dis-cussed above. It eliminates the effect of volume conduction and to a large extent reveals local synaptic activations. However, as was noted above, only spatially distributed currents with radii ≥ one-half of the interelectrode distance can be confidently estimated. Thus, in order to study more locally occurring currents, ar-rays with densely packed electrodes should be used (see Shimono et al., 2002, for an example).

18.3.3   Analysis Issues

Data collected with multi-electrode recording arrays can often be cumbersome. Visual inspection is required as a preprocessing step to ensure that signals have not been degraded as a result of volume conduction and/or improper adhesion of the slice to the electrodes. Even when signals are of high quality, each experiment gen-erates a large amount of data: multiple signals (e.g., 64 in the examples above) for each time step. Inspection of complex spatiotemporal patterns present in the data is difficult because recording with the two-dimensional electrode grids generates three dimensions of information: the two spatial dimensions across time, a third di-mension. Thus, it is often useful to collapse the various time series across disparate spatial locations into a single visual pattern, which may be viewed and analyzed more easily. Two-dimensional current source density analysis can accomplish this; with this method, all 64 signals are transformed into one coherent image for each instantaneous time point sampled. However, it can still be difficult to compare these images across time and get a sense of the temporal order inherent in the signals.

Another option for simplification of multi-electrode data analysis is to employ 3-D visualization software, such as OpenDX, which enables transformation of the three dimensions of data into a single image. Although small details are lost with this method, meaningful patterns can emerge that are difficult to discern through other means. An example of 3-D visualization that was performed using OpenDX visualization software is shown in Figure 18.11.

In this example, each image represents 1500 msec-long, simultaneous recordings across 64 electrodes. The colors blue and yellow symbolize negative and positive voltages, respectively, with magnitudes of ±0.012 mV. Image (1) shows a response to Schaffer collateral stimulation in the absence of cholinergic rhythms. Little to no spontaneous activity is present, and the response to stimulation is brief and spatially restricted. Image (2) is a visualization of a recording obtained in the presence of cholinergically induced beta rhythms. Rhythmic activity is apparent before and after delivery of the stimulation pulse, and the evoked response is spatially elaborate and temporally sustained. The bottom image (3) illustrates an example of a case in which a drug that positively modulates AMPA-type glutamate receptors (i.e., an "ampakine") was infused in combination with the cholinergic agonist carbachol, resulting in enhanced rhythmic activity. In this case, the response appears even more complex and prolonged and additionally is accompanied by high-frequency components. Comparison across the three experimental conditions in this example demonstrates the usefulness of 3-D

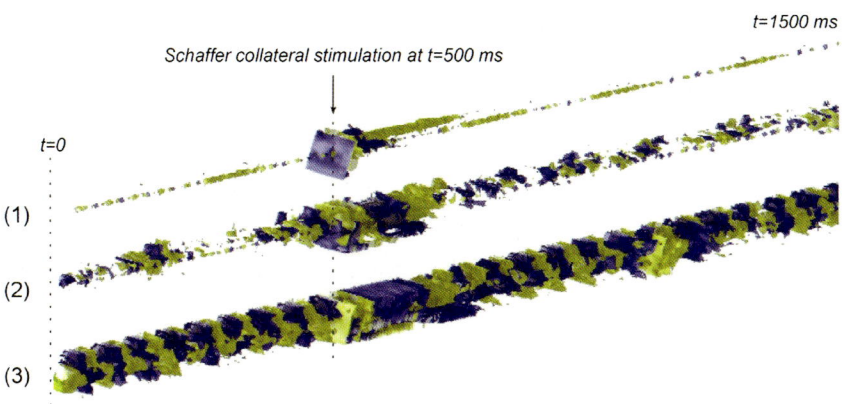

FIGURE 18.11. Overview of three complete time series (1500 msec in duration) visualized in three dimensions using OpenDX software. The three images represent: (1) control, (2) carbachol, and (3) carbachol + ampakine. Blue symbolizes −0.012 mV and yellow symbolizes +0.012 mV. The intermittent strand of positive voltage (yellow) seen in all three cases and especially apparent in (1) is an artifact due to the loss of recording capabilities at the site chosen for the stimulation electrode.

visualization methods. That is, through the use of 3-D visualization, one can tell in a glance that the three images are dramatically different.

18.4   Concluding Remarks

Tools that are capable of sampling electrophysiological signals across spatially distributed brain networks, such as multi-electrode recording arrays, will likely be used increasingly often as more and more researchers strive to bridge the gap between synaptic interactions and computations across neuronal ensembles. This chapter pointed out specific studies in which the increased dimensions of information provided by multi-electrode recording allowed spatiotemporal patterns to be observed that would have been difficult to discern using conventional recording methods. Although technical difficulties can complicate the use of multi-electrode recording, they can largely be overcome through sophisticated analysis techniques. As computational methods and the electronic devices themselves are advanced, further insights into the complex operations of brain networks are to be expected.

Acknowledgments. The writing of this chapter was funded by NIA Training Grant 5T32 AG00096-21. The author thanks Fernando A. Brucher for assistance with current source density analysis and helpful discussions, Gary Lynch for valuable comments on earlier versions of this work, Ken Shimono for help with multi-electrode recordings, and Cheryl A. Cotman and Linda C. Palmer for providing the 3-D visualizations depicted in Figure 18.11.

References

Behrends, J.C. and ten Bruggencate, G. (1993). Cholinergic modulation of synaptic inhibition in the guinea pig hippocampus in vitro: Excitation of GABAergic interneurons and inhibition of GABA-release. *J. Neurophysiol.* 69: 626–629.

Benson, D.M., Blitzer, R.D., and Landau, E.M. (1988). An analysis of the depolarization produced in guinea-pig hippocampus by cholinergic receptor stimulation. *J. Physiol.* 404: 479–496.

Boddeke, H., Best, R., and Boeijinga, P.H. (1997). Synchronous 20 Hz rhythmic activity in hippocampal networks induced by activation of metabotropic glutamate receptors in vitro. *Neuroscience* 76: 653–658.

Bramham, C.R. and Srebro, B. (1989). Synaptic plasticity in the hippocampus is modulated by behavioral state. *Brain Res.* 493: 74–86.

Charpak, S., Pare, D., and Llinas, R. (1995). The entorhinal cortex entrains fast CA1 hippocampal oscillations in the anaesthetized guinea-pig—Role of the monosynaptic component of the perforant path. *Euro. J. Neurosci.* 7: 1548–1557.

Chrobak, J.J. and Buzsaki, G. (1998). Gamma oscillations in the entorhinal cortex of the freely behaving rat. *J. Neurosci.* 18: 388–398.

Colgin, L.L., Kramar, E.A., Gall, C.M., and Lynch, G. (2003). Septal modulation of excitatory transmission in hippocampus. *J. Neurophysiol.* 90: 2358–2366.

Csicsvari, J., Hirase, H., Czurko, A., Mamiya, A., and Buzsaki, G. (1999). Oscillatory coupling of hippocampal pyramidal cells and interneurons in the behaving rat. *J. Neurosci.* 19: 274–287.

Fellous, J.M. and Sejnowski, T.J. (2000). Cholinergic induction of oscillations in the hippocampal slice in the slow (0.5–2 Hz), theta (5–12 Hz), and gamma (35–70 Hz) bands. *Hippocampus* 10: 187–197.

Fisahn, A., Pike, F.G., Buhl, E.H., and Paulsen, O. (1998). Cholinergic induction of network oscillations at 40 Hz in the hippocampus in vitro. *Nature* 394: 186–189.

Foster, T.C. and Deadwyler, S.A. (1992). Acetylcholine modulates averaged sensory evoked responses and perforant path evoked field potentials in the rat dentate gyrus. *Brain Res.* 587: 95–101.

Freeman, W. (1975). *Mass Action in the Nervous System*, Academic Press, New York.

Gray, C.M. and Singer, W. (1989). Stimulus-specific neuronal oscillations in orientation columns of cat visual cortex. *Proc. Natl. Acad. Sci. U. S. A.* 86: 1698–1702.

Haberly, L.B. and Shepherd, G.M. (1973). Current-density analysis of summed evoked potentials in opossum prepyriform cortex. *J. Neurophysiol.* 36: 789–802.

Hajos, N., Papp, E.C., Acsady, L., Levey, A.I., and Freund, T.F. (1998). Distinct interneuron types express m2 muscarinic receptor immunoreactivity on their dendrites or axon terminals in the hippocampus. *Neuroscience* 82: 355–376.

Hasselmo, M.E. and Fehlau, B.P. (2001). Differences in time course of ACh and GABA modulation of excitatory synaptic potentials in slices of rat hippocampus. *J. Neurophysiol.* 86: 1792–1802.

Hasselmo, M.E., Schnell, E., and Barkai, E. (1995). Dynamics of learning and recall at excitatory recurrent synapses and cholinergic modulation in rat hippocampal region Ca3. *J. Neurosci.* 15: 5249–5262.

Hjorth-Simonsen, A. (1973). Some intrinsic connections of the hippocampus in the rat: An experimental analysis. *J. Comp. Neurol.* 147: 145–161.

Huerta, P.T. and Lisman, J.E. (1993). Heightened synaptic plasticity of hippocampal CA1 neurons during a cholinergically induced rhythmic state. *Nature* 364: 723–725.

Huerta, P.T. and Lisman, J.E. (1995). Bidirectional synaptic plasticity induced by a single burst during cholinergic theta oscillation in CA1 in vitro. *Neuron* 15: 1053–1063.

Hyman, J.M., Wyble, B.P., Goyal, V., Rossi, C.A., and Hasselmo, M.E. (2003). Stimulation in hippocampal region CA1 in behaving rats yields long-term potentiation when delivered to the peak of theta and long-term depression when delivered to the trough. *J. Neurosci.* 23: 11725–11731.

Iijima, T., Witter, M.P., Ichikawa, M., Tominaga, T., Kajiwara, R., and Matsumoto, G. (1996). Entorhinal-hippocampal interactions revealed by real-time imaging. *Science* 272: 1176–1179.

Ishizuka, N., Weber, J., and Amaral, D.G. (1990). Organization of intrahippocampal projections originating from CA3 pyramidal cells in the rat. *J. Comp. Neurol.* 295: 580–623.

Jensen, O. and Tesche, C.D. (2002). Frontal theta activity in humans increases with memory load in a working memory task. *Eur. J. Neurosci.* 15: 1395–1399.

Kahle, J.S. and Cotman, C.W. (1989). Carbachol depresses synaptic responses in the medial but not the lateral perforant path. *Brain Res.* 482: 159–163.

Konopacki, J., Maciver, M.B., Bland, B.H., and Roth, S.H. (1987). Theta in hippocampal slices: Relation to synaptic responses of dentate neurons. *Brain Res. Bull.* 18: 25–27.

Kopell, N., Ermentrout, G.B., Whittington, M.A., and Traub, R.D. (2000). Gamma rhythms and beta rhythms have different synchronization properties. *Proc. Natl. Acad. Sci. U. S. A.* 97: 1867–1872.

Landfield, P.W., McGaugh, J.L., and Tusa, R.J. (1972). Theta rhythm: A temporal correlate of memory storage processes in the rat. *Science* 175: 87–89.

Larson, J. and Lynch, G. (1986). Induction of synaptic potentiation in hippocampus by patterned stimulation involves two events. *Science* 232: 985–988.

Larson, J., Wong, D., and Lynch, G. (1986). Patterned stimulation at the theta frequency is optimal for the induction of hippocampal long-term potentiation. *Brain Res.* 368: 347–350.

Leung, L.W., Lopes da Silva, F.H., and Wadman, W.J. (1982). Spectral characteristics of the hippocampal EEG in the freely moving rat. *Electroencephalog. Clin. Neurophysiol.* 54: 203–219.

Leung, L.S. (1992). Fast (beta) rhythms in the hippocampus: A review. *Hippocampus* 2: 93–98.

Leung, L.S. (1998). Generation of theta and gamma rhythms in the hippocampus. *Neurosci. Biobehav. Rev.* 22: 275–290.

Levey, A.I., Edmunds, S.M., Koliatsos, V., Wiley, R.G., and Heilman, C.J. (1995). Expression of m1–m4 muscarinic acetylcholine receptor proteins in rat hippocampus and regulation by cholinergic innervation. *J. Neurosci.* 15: 4077–4092.

Lisman, J., Jensen, O., and Kahana, M. (2001). Toward a physiologic explanation of behavioral data on human memory—The role of theta-gamma oscillations and NMDAR-dependent LTP. In: Holscher, C., ed., *Neuronal Mechanisms of Memory Formation: Concepts of Long-Term Potentiation and Beyond*, Cambridge University Press, Cambridge, pp.195–223.

Madison, D.V., Lancaster, B., and Nicoll, R.A. (1987). Voltage clamp analysis of cholinergic action in the hippocampus. *J. Neurosci.* 7: 733–741.

M'Harzi, M. and Jarrard, L.E. (1992). Effects of medial and lateral septal lesions on acquisition of a place and cue radial maze task. *Behav. Brain Res.* 49: 159–165.

Mizumori, S.J., Perez, G.M., Alvarado, M.C., Barnes, C.A., and McNaughton, B.L. (1990). Reversible inactivation of the medial septum differentially affects two forms of learning in rats. *Brain Res.* 528: 12–20.

Nakajima, Y., Nakajima, S., Leonard, R.J., and Yamaguchi, K. (1986). Acetylcholine raises excitability by inhibiting the fast transient potassium current in cultured hippocampal neurons. *Proc. Natl. Acad. Sci. U. S. A.* 83: 3022–3026.

Nicholson, C. (1973). Theoretical analysis of field potentials in anisotropic ensembles of neuronal elements. *IEEE Trans. Biomed. Eng.* 20: 278–288.

Nicholson, C. and Freeman, J.A. (1975). Theory of current source-density analysis and determination of conductivity tensor for anuran cerebellum. *J. Neurophysiol.* 38: 356–368.

Nicholson, C. and Llinas R. (1975). Real time current source-density analysis using multi-electrode array in cat cerebellum. *Brain Res.* 100: 418–424.

Pavlides, C., Greenstein, Y.J., Grudman, M., and Winson, J. (1988). Long-term potentiation in the dentate gyrus is induced preferentially on the positive phase of theta-rhythm. *Brain Res.* 439: 383–387.

Pitler, T.A. and Alger, B.E. (1992). Cholinergic excitation of GABAergic interneurons in the rat hippocampal slice. *J. Physiol.* 450: 127–142.

Qian, J. and Saggau P. (1997). Presynaptic inhibition of synaptic transmission in the rat hippocampus by activation of muscarinic receptors—Involvement of presynaptic calcium influx. *Brit. J. Pharmacol.* 122: 511–519.

Sheridan, R.D. and Sutor, B. (1990). Presynaptic M1 muscarinic cholinoceptors mediate inhibition of excitatory synaptic transmission in the hippocampus in vitro. *Neurosci. Lett.* 108: 273–278.

Shimono, K., Brucher, F., Granger, R., Lynch, G., and Taketani, M. (2000). Origins and distribution of cholinergically induced beta rhythms in hippocampal slices. *J. Neurosci.* 20: 8462–8473.

Shimono, K., Kubota, D., Brucher, F., Taketani, M., and Lynch, G. (2002). Asymmetrical distribution of the Schaffer projections within the apical dendrites of hippocampal field CA1. *Brain Res.* 950: 279–287.

Staubli, U. and Lynch, G. (1987). Stable hippocampal long-term potentiation elicited by 'theta' pattern stimulation. *Brain Res.* 435: 227–234.

Swanson, L.W., Wyss, J.M., and Cowan, W.M. (1978). An autoradiographic study of the organization of intrahippocampal association pathways in the rat. *J. Comp. Neurol.* 181: 681–715.

Tesche, C.D. and Karhu, J. (2000). Theta oscillations index human hippocampal activation during a working memory task. *Proc. Natl. Acad. Sci. U. S. A.* 97: 919–924.

Traub, R.D., Whittington, M.A., Buhl, E.H., Jefferys, J.G.R., and Faulkner, H.J. (1999). On the mechanism of the gamma -> beta frequency shift in neuronal oscillations induced in rat hippocampal slices by tetanic stimulation. *J. Neurosci.* 19: 1088–1105.

Valentino, R.J. and Dingledine, R. (1981). Presynaptic inhibitory effect of acetylcholine in the hippocampus. *J. Neurosci.* 1: 784–792.

van der Linden, S., Panzica, F., and de Curtis, M. (1999). Carbachol induces fast oscillations in the medial but not in the lateral entorhinal cortex of the isolated guinea pig brain. *J. Neurophysiol.* 82: 2441–2450.

Vertes, R.P. and Kocsis, B. (1997). Brainstem-diencephalo-septohippocampal systems controlling the theta rhythm of the hippocampus. *Neuroscience* 81: 893–926.

Whittington, M.A., Traub, R.D., and Jefferys, J.G.R. (1995). Synchronized oscillations in interneuron networks driven by metabotropic glutamate receptor activation. *Nature* 373: 612–615.

Williams, J.H. and Kauer, J.A. (1997). Properties of carbachol-induced oscillatory activity in rat hippocampus. *J. Neurophysiol.* 78: 2631–2640.

Winson, J. (1978). Loss of hippocampal theta rhythm results in spatial memory deficit in the rat. *Science* 201: 160–163.

Yamamoto, C. and Kawai, N. (1967). Presynaptic action of acetylcholine in thin sections from the guinea pig dentate gyrus in vitro. *Exp. Neurol.* 19: 176–187.

19

Exploring Fast Hippocampal Network Oscillations: Combining Multi-Electrode Recordings with Optical Imaging and Patch-Clamp Techniques

EDWARD O. MANN AND OLE PAULSEN

19.1   Introduction

Cortical processing depends on orchestrated activity across distributed neuronal assemblies, and network rhythms may provide a temporal structure relative to which individual neurons within these assemblies can be coordinated. Network oscillations in the gamma-frequency range ($\sim$30 to 100 Hz) have received particular attention because they are prominent in the awake brain, and have been implicated in cognitive processes, such as sensory binding (Singer, 1993), selective attention (Fries et al., 2001), and consciousness (Llinas et al., 1998). Gamma rhythms can be observed in the hippocampus, where they have been implicated in memory processing (Lisman and Idiart, 1995; Jensen and Lisman, 1996), and in vivo multi-electrode techniques have already uncovered some of the mechanisms underlying these hippocampal gamma oscillations in the behaving animal (Bragin et al., 1995; Csicsvari et al., 2003). Adapting such multi-electrode techniques to an in vitro model of hippocampal gamma oscillations would enable a rigorous pharmacological and physiological dissection of the cellular and synaptic mechanisms underlying these rhythms. Furthermore, as hippocampal gamma oscillations represent the coordinated activity of assemblies of neurons, such an in vitro model would provide a convenient screen for potential psychoactive drugs at the network level (Gill et al., 2002; Weiss et al., 2003).

Fast network oscillations can be induced in the hippocampal slice in vitro through a variety of paradigms, including patterned afferent stimulation (Whittington et al., 1995; Traub et al., 1996; Whittington et al., 1997), local application of drugs or solutions with altered ionic composition, for example, high potassium (LeBeau et al., 2002; Towers et al., 2002), bath application of kainate (Hajos et al., 2000; Hormuzdi et al., 2001), or drugs that activate metabotropic glutamate receptors (Whittington et al., 1995; Boddeke et al., 1997; Gillies et al., 2002) or muscarinic receptors (Fisahn et al., 1998; Shimono et al., 2000). The frequency of drug-induced oscillations is temperature-dependent. At room temperature, these oscillations are often in the beta-frequency range as defined in vivo ($\sim$10 to 30 Hz),

454

but commonly fall in the gamma-frequency band when recorded at or above 32°C (Ecker et al., 2001; Dickinson et al., 2003), and may therefore provide models for in vivo gamma oscillations.

Although planar multi-electrode arrays offer an attractive opportunity to explore the underlying mechanisms and physiological relevance of these network oscillations, the majority of previous studies have been performed in interface-style chambers. Shimono et al. (2000) succeeded in inducing fast network oscillations in hippocampal slices mounted on planar multi-electrode arrays, by lowering the fluid level to create semi-interface conditions and using an atmosphere above the chamber of humidified carbogen gas (95% O_2/5% CO_2). This development has facilitated the detailed study of spatiotemporal patterns of cellular and network events underlying fast hippocampal network oscillations in vitro (Shimono et al., 2000).

In this chapter, we discuss how planar multi-electrode arrays can be used to study fast hippocampal network oscillations induced by activation of kainate, muscarinic, and metabotropic glutamate receptors. This includes practical guidelines for recording and analysis, as well as a discussion of current source density analysis, and emphasizes throughout how to optimize the conditions for combining multi-electrode recordings with optical imaging using voltage-sensitive dyes and patch-clamp recordings from single neurons.

19.2   Practical Guidelines

19.2.1   Slice Preparation and Mounting

The methods for preparing hippocampal slices are discussed extensively elsewhere, but there are a few details of our methods that we find facilitate the successful recording of fast network oscillations.

• Following dissection, slices are stored for one hour in an interface chamber, between humidified carbogen gas and artificial cerebrospinal fluid (ACSF). Slices can be stored in a submerged chamber, but this tends to reduce adhesion to the multi-electrode probes and/or reduce the power of the recorded network oscillations.

• Good electrical contact between the slice and multi-electrode array is required for the detection of small oscillations in the field potential. This can be achieved by mechanically anchoring the slice. For our experiments, slices are adhered to the multi-electrode probes by complete removal of the surrounding ACSF, with a pipette and filter paper. ACSF is then reintroduced and the slices left to recover for a further hour in a chamber filled with humidified carbogen gas (Oka et al., 1999). This negates the requirement for a mechanical anchor over the slice during recording. For reasons that are detailed below, this technique has the advantage of increasing laminar flow across the slice during perfusion and facilitates optical imaging of the entire cortical preparation. A photo of a slice mounted on a microelectrode array is seen in Figure 19.1A.

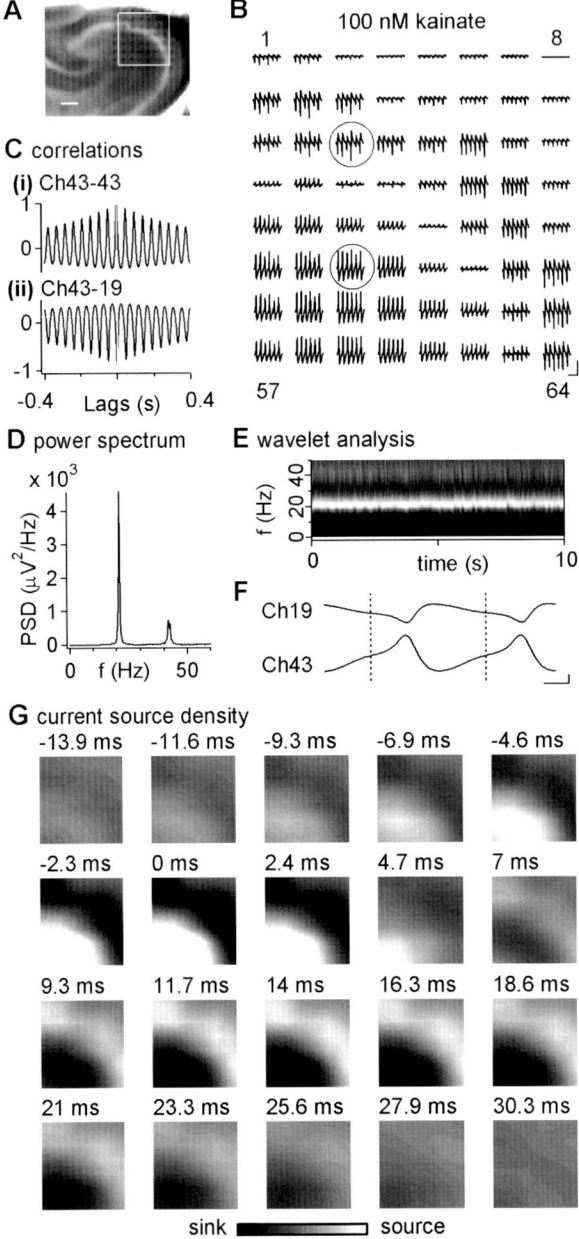

A

B 100 nM kainate

C correlations

(i) Ch43-43

(ii) Ch43-19

-0.4   Lags (s)   0.4

D power spectrum

E wavelet analysis

F Ch19

Ch43

G current source density
-13.9 ms   -11.6 ms   -9.3 ms   -6.9 ms   -4.6 ms

-2.3 ms   0 ms   2.4 ms   4.7 ms   7 ms

9.3 ms   11.7 ms   14 ms   16.3 ms   18.6 ms

21 ms   23.3 ms   25.6 ms   27.9 ms   30.3 ms

sink   source

FIGURE 19.1. Current source density analysis. (A) Planar multi-electrode arrays with an interpolar distance of 100 μm (Panasonic MED-P2105; Tensor Biosciences, Irvine, CA, USA) were used to record field potentials across all layers of the hippocampal CA3. The white box marks the area of 36 electrodes used in the construction of two-dimensional CSD profiles in (G). Scale bar, 200 μm. (B) Fast network oscillations induced by 100 nM kainate

19.2.2   Fast Network Oscillations in Submerged-Type Chambers

In order to combine multi-electrode recordings with visualized recordings from individual neurons under differential interference contrast (DIC) microscopy or optical imaging techniques, such as two-photon imaging, it is currently necessary to perform recordings in submerged conditions where immersion objectives can be used. This poses a problem for the study of fast hippocampal network oscillations that depend on interface-type conditions (Shimono et al., 2000). However, it has been found that a high laminar flow rate of oxygenated ACSF across the slice is sufficient to provide the conditions for drug-induced fast network oscillations in the hippocampus in vitro (Hajos et al., 2004; Mann et al., 2005b). To achieve such conditions for hippocampal slices mounted on planar micro-electrode arrays, it appears necessary to reduce the well volume of the probes by using inserts, which can be fabricated with a silicone elastomer such as Sylgard.

19.2.3   Noise Considerations

Combining planar multi-electrode recordings with other electrophysiological techniques increases the potential problems of noise pickup and ground loops, for which there are normally simple, but frustratingly time-consuming empirical solutions (see Axon Guide; http://www.axon.com/manuals/Axon_Guide.pdf). A major problem that can be encountered is large and irregular shifts in the bath potential, which may be due to air bubbles causing fluctuations in the leak currents along the

---

FIGURE 19.1. (*cont.*)

were recorded at 63 electrodes. Channel 8 was used to record a reference synchronization signal for simultaneous optical imaging (see Figure 19.2). Channels 19 and 43 are circled. Scale bars, 200 μV and 100 ms. (C) (i) Autocorrelation of the signal recorded on Channel 43, and (ii) cross-correlation between the signals recorded on Channels 43 and 19. (D) Power spectrum of the signal recorded on Channel 43 (PSD, power spectral density). (E) Wavelet analysis of a 10 min epoch of the signal on Channel 43. The magnitude of the wavelet transform (normalized Morlet wavelet; $\omega_0 = 6$; scales chosen to reflect unit frequencies (*f*) between 1 and 50 Hz; scale $= (\omega_0 + \sqrt{(2 + \omega_0^2)})/4\pi f$) was plotted as a function of time and frequency (corresponding to scale), with lighter shades representing increasing magnitude. (F) Peak-to-peak cycle averages of fast network oscillations recorded in the CA3 stratum pyramidale (Ch19) and stratum radiatum (Ch43). Dotted vertical lines represent the points over which two-dimensional profiles in (G) were calculated. Scale bars, 50 μV and 10 ms. (G) Two-dimensional CSD profiles of the current sinks and sources preceding and following the positive peak in the field potential recorded in the stratum radiatum (0.0 ms). CSD profiles were constructed from cycle averages passed through a $3 \times 3 \times 3$ median filter, and convolved with a 3×3 Laplacian kernel $(0 -1\ 0, -1\ 4 -1, 0 -1\ 0)$ to derive the second spatial derivative (Shimono et al., 2000). This analysis demonstrates that kainate-induced fast network oscillations are accompanied by alternating current sink/source pairs in the strata pyramidale and radiatum.

in-flow and out-flow tubes, and is accentuated by high perfusion rates. Such noise does not appear to be eliminated by grounding the in-flow and out-flow tubes, but disappears when the multi-electrode array is electrically isolated with two bubble traps close to the chamber.

19.3   Analysis of Network Oscillations Recorded Using Multi-Electrode Arrays

19.3.1   The Temporal Structure of Oscillations Recorded on Individual Channels

Oscillations can be analyzed in the time and frequency domains. There are several excellent textbooks on signal processing that give a rigorous treatment of this topic (Oppenheim and Willsky, 1997; Phillips and Parr, 1999; Proakis and Manolakis, 1996). Here, only a brief overview of practical analysis techniques is given.

Before any digital signal analysis is done it is paramount to ensure that the signals are appropriately filtered. Thus, the Nyquist theorem must be respected, the practical consequence being that the sampling frequency should be greater than twice the bandwidth limit of the signal, or conversely, that the signal should be lowpass-filtered at a frequency less than half the sampling frequency.

Inspection of raw traces usually gives an indication of any periodicity in the signal (Figure 19.1B), but the degree of periodicity might be difficult to judge from the raw trace alone if the signal consists of several components, or is corrupted with noise, which is often the case in biology. Autocorrelation is a useful technique to visualize periodicity in a signal during the epoch to be analyzed. Periodicity is seen in the autocorrelogram as repeated equidistant peaks (Figure 19.1Ci).

To convert a temporal signal into the frequency domain, a Fourier transform is computed. The fast Fourier transform (FFT) algorithm on computer is an efficient tool of digital signal processing. It is important to be aware, however, that different software packages might give subtle but important differences in the indexing and scaling of the output. The FFT of a real time-domain input results in a complex output with information about the magnitude and phase of the different frequency components. The Fourier transform of the autocorrelation function is the power spectrum, which is equivalent to the magnitude squared of the FFT of the signal itself (Figure 19.1D). Thus, in the power spectrum, information regarding the relative phase of different frequency components of the signal is lost. Moreover, neither the FFT nor the power spectrum preserves time information, and thus it is not possible to see when oscillations occurred. This limitation is particularly significant when analyzing nonstationary signals, which are common in biology.

To partially preserve such information, the short-time Fourier transform, or spectrogram, is often computed. This function maps a signal into a two-dimensional function of time and frequency using a sliding window and is displayed as the magnitude of the short-term Fourier transform versus time on the x-axis for different frequencies on the y-axis. The resolution is determined by the width of the time window. With several or varying frequency components, it may be difficult to choose an optimal time window, and in such circumstances, this analysis could, and maybe should, be replaced by wavelet analysis (Mallat, 1999; Torrence and Compo, 1998). Wavelet analysis is a windowing technique with variable-sized regions, which represents a signal as wavelet coefficients at different scales versus time. Low scales correspond to high frequencies and vice versa (Figure 19.1E). Wavelet techniques are a powerful tool for analyzing biological oscillations.

19.3.2   The Temporal Structure of Oscillations Recorded on Multiple Channels

Moving from single signals to signals from multiple electrodes, similar types of analyses can be made. First, by inspecting the traces, obvious phase relations, such as phase reversal, can be identified (e.g., traces 19 and 43 in Figure 19.1B). Second, cross-correlation techniques are used in the temporal domain to look at the temporal relations between signals at different electrodes (Figure 19.1Cii). Thus, one can analyze whether data are consistent with phase reversal across a boundary or propagation of oscillations from one area to another. Third, in the frequency domain, cross-spectral densities can be estimated, and the coherence between two signals computed. Unfortunately, coherence measures are often complicated by common noise recorded on two channels, making interpretations difficult or meaningless. Partial coherence analysis can be used to identify whether oscillations recorded from different electrodes have a common source, or whether separate sources of oscillatory activity exist (Kocsis et al., 1999). For nonstationary oscillations, the wavelet cross-spectrum is gaining increasing popularity as a tool to study transient phase locking of oscillations recorded from multiple channels (Lee, 2002).

Rather than merely measuring coherent signals between different recording electrodes, if the electrodes are sufficiently close, detailed spatiotemporal analysis of the oscillations can reveal insights into the underlying mechanisms. Current source density (CSD) analysis can be used to extract information about the local flow of current during oscillations, which may be linked mechanistically to the oscillatory activity.

19.3.3   Current Source Density Analysis

Current source density analysis provides information regarding the location of current sinks and sources in the extracellular space (for review, see Mitzdorf, 1985). If the anatomical and functional connectivity within the network is known, such analysis can help elucidate the mechanisms underlying hippocampal network oscillations (Leung, 1984; Buzsaki et al., 1986; Brankack et al., 1993; Bragin et al., 1995; Ylinen et al., 1995; Charpak et al., 1995; Shimono et al., 2000; Buzsaki, 2002; Csicsvari et al., 2003). The simplest method for the construction of CSD profiles from planar multi-electrode recordings is the calculation of the second spatial derivative of the field potential, in either one or two dimensions. This approach assumes that the conductivity tensor of the extracellular space is homogeneous throughout the tissue, and although there may be layer-specific differences in conductivity across laminated cortical structures, these do not appear to significantly affect the spatial profile of current sinks and sources (Holsheimer, 1987).

The application of CSD analysis in one dimension makes the additional assumption that there is no significant current flow orthogonal to the axis of measure, which appears to be a reasonable approximation when recording parallel to the somatodendritic axis of linearly arranged principal neurons, such as is the case in the hippocampal CA1 (Bragin et al., 1995; Csicsvari et al., 2003). Two-dimensional CSD analysis may be more appropriate for more complex anatomical structures, however, this method assumes that extracellular conductivity is equal in both dimensions. Neuronal tissue is electrically anisotropic (Nicholson and Freeman, 1975; Nicholson and Llinas, 1975), but we have found that one- and two-dimensional CSD analyses do not produce qualitatively different spatial profiles of current sinks and sources for drug-induced fast network oscillations in the hippocampus in vitro. The ability to compare both one- and two-dimensional CSD analyses using planar multi-electrode arrays may provide a greater insight into other forms of network phenomenon. However, due to the assumptions inherent in calculating current densities from the second-spatial derivative, unless detailed spatial impedance profiles are available, the amplitude of these measures should be viewed as qualitative, rather than quantitative, in nature.

The distribution of extracellular current sinks and sources during network oscillations can be calculated continuously in time from raw electrophysiological recordings (Figure 19.1B; Shimono et al., 2000), or from oscillation cycle averages (Figure 19.1F,G). Several practical issues regarding the successful calculation of CSD profiles, and the interpretation of this analysis, should be considered.

- *Electrode properties.* The construction of valid CSD profiles from planar multi-electrode arrays depends on all electrodes having essentially identical electrical properties. Spatial filtering can eliminate small differences between electrodes, and if single electrodes significantly deteriorate, it may be possible to interpolate the electrophysiological signal from surrounding electrodes. The results of such processes should be interpreted cautiously, and one-dimensional CSD analysis on a set of reliable electrode recordings is the legitimate approach.
- *Spatial resolution.* Accurate CSD analysis requires that the spatial sampling rate should be at least twice that of the highest spatial frequency component in the distribution of extracellular current sinks and sources, as given by the Nyquist theorem. For rat hippocampal slices, this is thought to demand an electrode spacing of ≤ 100 µm (Figure 19.1A; Bragin et al., 1995; Csicsvari et al., 2003), although aliasing errors can be avoided by appropriate spatial filtering when using multi-electrode arrays with interpolar distances of up to 150 µm (Shimono et al., 2000).
- *Edge effects.* The signals from the perimeter electrodes of planar multi-electrode arrays can be used for the construction of CSD profiles, but the corresponding

transformed signals are corrupted by edge effects and should be ignored for analysis and presentation (Figure 19.1A,G).

- *CSD representation*. Pseudo-color images can be used for the presentation of spatiotemporal CSD profiles (Figure 19.2; color plate; Shimono et al., 2000). The color coding for these CSD profiles is arbitrary, but traditionally warm colors have been used to represent current sources, and cool colors to represent current sinks (Bragin et al., 1995; Buzsaki, 2002; Csicsvari et al., 2003). In contrast, we have represented sinks as red and sources as blue, in order to allow intuitive comparisons with simultaneously recorded optical imaging data (see below). For optical imaging techniques, warm colors are usually used to represent excitatory events that are associated with current sinks.

In cortical slice preparations, network oscillations usually occur in a single frequency band restricted to coherently coupled anatomical subregions. However, it may be possible to generate semi-independent network oscillations in functionally connected regions of the same slice (Shimono et al., 2000), offering the possibility to study complex network interactions. The existence of multiple oscillatory current generators brings the concomitant problems of spurious signals due to volume conduction, which complicate coherence and correlation analysis. Therefore, in the future, CSD analysis with planar multi-electrode array recordings may have the additional function of isolating local currents, as is the case in vivo (e.g., Bragin et al., 1995; Csicsvari et al., 2003).

19.4   Combining Multi-Electrode Recordings with Optical Imaging Using Voltage-Sensitive Dyes

Current always flows in a complete circuit, so an active inward current in a neuronal dendrite would generate an outward current of equal magnitude at proximal and more-distal membrane sites. Although CSD analysis reveals the spatial profile of current sinks and sources, and thus provides clues to the underlying mechanisms, it does not distinguish between the active current generators and passive return currents. For example, a sink in the stratum pyramidale and source in the stratum radiatum (see Figure 19.1G), could reflect somatic excitation and/or dendritic inhibition.

One approach that we have taken to help elucidate the active events underlying network oscillations in vitro is the combination of CSD analysis with optical imaging using voltage-sensitive dyes (see Tominaga et al., 2000 for details of optical imaging technique). For the kainate-induced fast network oscillation shown in Figure 19.1 (100 nM kainate; peak frequency 21 Hz), imaging with voltage-sensitive dyes reveals that the current sinks in the stratum pyramidale are followed by depolarization in perisomatic regions of CA3 pyramidal neurons (Figure 19.2). In contrast, the corresponding sources in the stratum radiatum are not accompanied by dendritic hyperpolarization. The opposite sink/source pair precedes a hyperpolarization at perisomatic sites, without dendritic depolarization. Therefore, the active current sinks and sources driving this

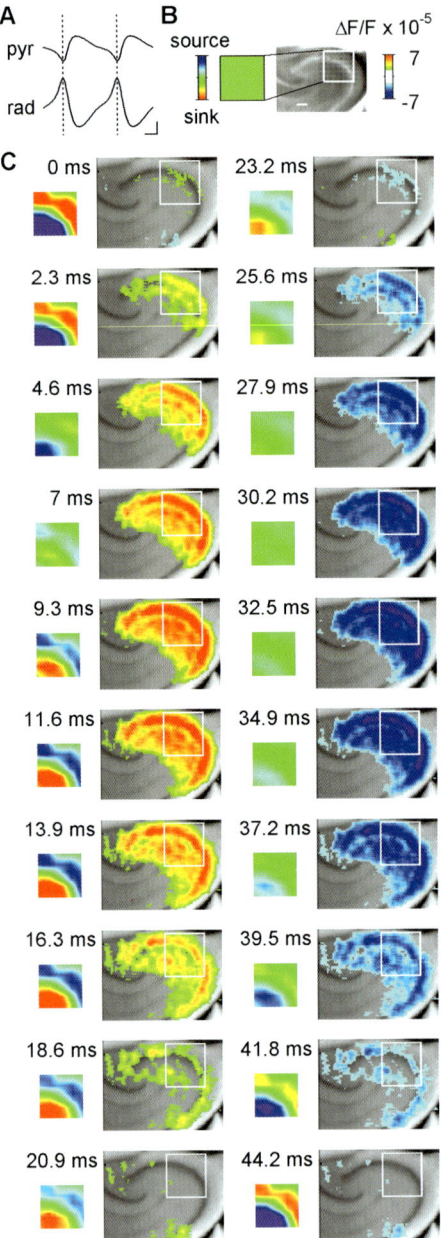

FIGURE 19.2. Combining CSD analysis with optical imaging. To reveal the cellular and synaptic mechanisms underlying network oscillations, it may be necessary to combine multi-electrode recordings with other techniques. In this example, optical imaging with voltage-sensitive dyes is used to help identify the active current sinks and sources during the kainate-induced fast network oscillation shown in Figure 19.1. For optical imaging,

kainate-induced fast network oscillation appear to be localized to the stratum pyramidale, producing predominantly passive return currents in the stratum radiatum.

As planar multi-electrode arrays record network activity from the lower surface of brain slices, simultaneous multi-site electrophysiological recordings and epifluorescence imaging can be achieved without the problems of optical distortions that would occur using conventional glass micro-electrodes or inline electrode arrays (Tominaga et al., 2001). However, there are more general issues relating to the optical imaging of network oscillations in vitro, which may be worth highlighting.

- *Dye loading.* Fluorescent optical imaging of the population activity in cortical tissue commonly involves bulk dye loading prior to experimentation. Multi-electrode probes provide a convenient and economical chamber for tissue staining. The amount of fluorescent dye required can be minimized by using silicone inserts during staining.
- *Signal synchronization.* The temporal characteristics of network oscillations are critical for their hypothesized function in cognitive processes, and therefore when combining multi-electrode recordings with other techniques, it is essential that these measurements are precisely synchronized. Coupling multi-electrode and imaging systems as "slave" and "master", or using a common synchronizing trigger input, produces jitter in the delays of acquisition onset, which may cause significant errors unless the clock rate of the slave system/s is much greater than the relevant frequency components of the physiological signal. One solution is to record an electrophysiological signal on one pixel of the imaging system, although this can lead to phase-locked bleeding into surrounding pixels. A more reliable method appears to involve recording a low-amplitude synchronizing signal, which is unrelated to the physiological signal, on one channel of the multi-electrode and imaging systems, allowing post hoc alignment. In Figures 19.1B and 19.2, channel 8 of the multi-electrode array is used to carry such a synchronization signal, as corner electrodes have no effect on two-dimensional

---

FIGURE 19.2. (*cont.*)

slices bulk-loaded with 200 μM Di-4-ANEPPS (Molecular Probes), and signals recorded using tandem-lens epifluorescence optics and CCD camera (MiCAM01, BrainVision; Sci-Media Ltd., Tokyo, Japan). (A) Peak-to-peak cycle averages of the oscillation recorded in the stratum pyramidale (pyr) and stratum radiatum (rad) (Channels 19 and 43 in Figure 19.1B). The same averaging procedure was used to improve the signal-to-noise ratio of the optical signal. Scale bars, 50 μV and 10 msec. (B) For presentation, the simultaneous two-dimensional profiles are displaced to the left, and the white box marks the area from which two-dimensional CSD profiles were constructed. The optical signal from the voltage-sensitive dye is superimposed on an image of the slice. (C) In the stratum pyramidale, current sinks (red) were followed by a depolarization (red), and current sources (blue) were followed by hyperpolarization (blue). The membrane voltage changes spread into the dendrites, but there was no apparent membrane polarization associated directly with the current sinks/sources in the distal stratum radiatum. Therefore, active current sinks and sources during this kainate-induced oscillation appear to be localized to the stratum pyramidale, producing predominantly passive return currents in the stratum radiatum.

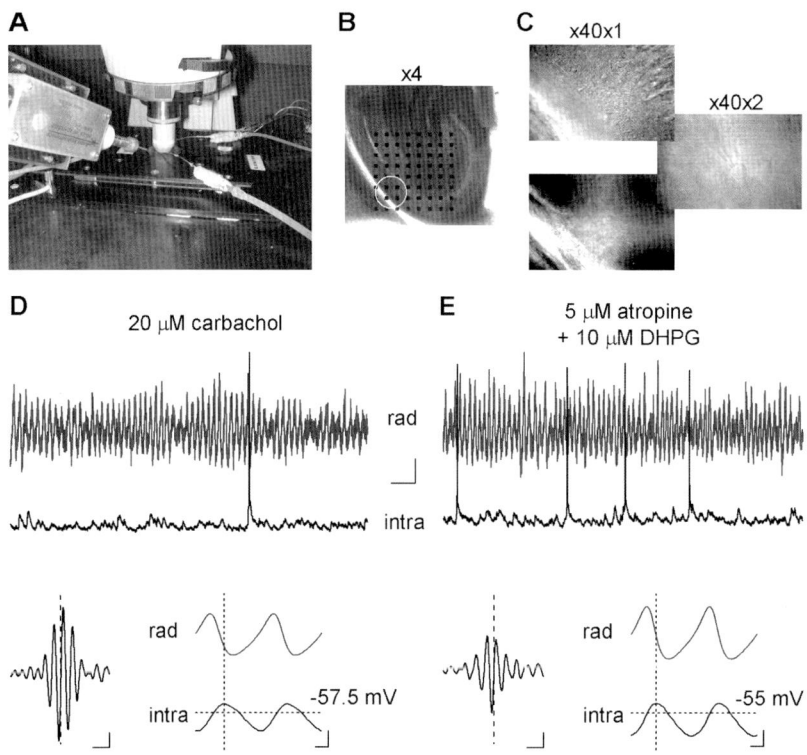

FIGURE 19.3. Whole-cell patch-clamp in conjunction with multi-electrode recordings. (A) The multi-electrode probe can be placed within an infrared-DIC microscopy patch-clamp setup. (B) This micrograph shows the positioning of a hippocampal slice mounted on a multi-electrode probe, viewed through a ×4 objective (interpolar distance of 150 μm). The area visualized in more detail in (C) has been circled. (C) When hippocampal slices are mounted on planar multi-electrode arrays, it is still possible to visualize neurons using infrared DIC microscopy (×40 objective). Focusing through the slice, one can demonstrate which electrodes from the multi-electrode array are in the same field of view, picking up the local field potential. For the example experiment shown in panels (D) to (E), fast network oscillations were recorded across the CA3 using planar multi-electrode arrays with an interpolar distance of 150 μm (Panasonic MED-P5115; Tensor Biosciences, Irvine, CA, USA). A visualized whole-cell current-clamp recording was obtained from a horizontally oriented neuron located in the CA3 stratum oriens, using the Axoclamp-2B amplifier in bridge mode. The intracellular voltage (intra) was recorded on one of the electrode channels of the multi-electrode array. This was compared with the extracellular network oscillation recorded in the stratum radiatum (rad). (D) and (E) Fast network oscillations were initially induced by activation of muscarinic receptors (20 μM carbachol; D). Subsequently, the effects of carbachol were blocked with 5 μM atropine, and fast network oscillations were induced by activation of Group I metabotropic glutamate receptors (10 μM DHPG; E). The upper figure in both panels shows an example of the simultaneously recorded extracellular and intracellular oscillations during both types of oscillation (Scale bars, 25 μV, 15 mV, and 200 msec). The lower left figure in each panel shows the cross-correlation between the intracellular and extracellular potentials (Scale bars, 0.02 and 100 msec). The lower right

CSD analysis. Of course, some acquisition programs may automatically solve these problems.

- *Signal-to-noise ratio.* The signals from optical imaging with fluorescent dyes can have a poor signal-to-noise ratio. Unless neuronal excitability is enhanced by blocking inhibition or lowering extracellular magnesium (Tsau et al., 1999; Wu et al., 1999; Bao and Wu, 2003), noise prevents the detection of network oscillations in the raw optical signal from voltage-sensitive dyes (see Tominaga et al., 2000). However, as in vitro network oscillations are often stable and persistent, it is possible to use cycle averages to improve the signal-to-noise ratio, until the underlying membrane potential oscillations are revealed (Figures 19.1F and 19.2A).

Optical imaging techniques using fluorescent dyes are becoming increasingly attractive tools with which to explore neuronal function. In combination with planar multi-electrode recordings, such techniques offer the opportunity to explore the events underlying the synchronization of large neuronal assemblies, and how these oscillations in turn modulate processing and signaling within single neurons in the network.

19.5   Combining Multi-Electrode Recordings with Patch-Clamp Recordings from Single Neurons

The cellular basis and functions of network oscillations can also be explored by combining planar multi-electrode recordings with intracellular recordings from single neurons. As CSD analysis of cortical activity primarily reveals the currents flowing in pyramidal neurons, a clear advantage of this technique would be the elucidation of the synaptic inputs driving different subclasses of interneurons within the network (Freund and Buzsaki, 1996), and the frequency and phase characteristics of the output spike patterns. It would also be possible to explore how network oscillations modulate both the dendritic processing within pyramidal cell dendrites and information transmission between pairs of neurons. Such aspects of cellular function within the network would be most easily studied with the use of visualized patch-clamp techniques, under infrared differential interference contrast (DIC) microscopy (Figure 19.3A–C; Hamill et al., 1981; Stuart et al., 1993).

FIGURE 19.3. (*cont.*)
figure in each panel shows the cycle averages for the intracellular and extracellular oscillations. The vertical dotted line shows the peak of the intracellular voltage. (Scale bars, 12 μV, 0.2 mV, and 10 ms). It is clear that the neuron shows increased depolarization during DHPG- versus carbachol-induced network oscillations. This leads to an earlier phase of the intracellular oscillation relative to extracellular oscillation, and an increased firing rate. These represent some of the methods that can be used to examine the cellular events underlying network oscillations, and how the behavior of different neuron types varies during different network states.

The major problem with visualizing slices mounted on planar multi-electrode arrays using DIC optics is reduced resolution, particularly at deeper locations, caused by the opaque electrodes and refractive properties of the circuit components embedded in relatively thick glass. However, the resulting images of individual neurons are more than adequate for somatic recordings (Figure 19.3C), although this may not be the case for dendritic recordings. The other complication that could be encountered is in grounding the patch-clamp amplifier headstage to the same bath reference electrode used for the multi-electrode array. If necessary, this can be done manually with appropriate connections at the headstage of the multi-electrode array.

An example of a whole-cell current-clamp recording performed simultaneously with planar multi-electrode field recordings is shown in Figure 19.3D,E. To facilitate the comparison between these two recordings, the intracellular voltage was recorded on one channel of the multi-electrode system. Using this technique it was possible to examine how this neuron behaved during fast network oscillations induced by both muscarinic (20 μM carbachol; Figure 19.3D) and metabotropic glutamate receptor activation (10 μM DHPG following blockade of carbachol-induced oscillations with 5 μM atropine; Figure 19.3E). Cross-correlation techniques were used to demonstrate the relation between the intracellular signal and the external oscillation (Figure 19.3D,E). Demonstrating the ability to analyze the temporal relations between intracellular and extracellular oscillations only serves to kindle interest in the aspects of network function that could be explored by combining multi-electrode recordings with the wealth of conventional single microelectrode techniques (Hajos et al., 2004; Mann et al., 2005a, b).

19.6   Conclusions and Future Perspectives

Planar multi-electrode arrays are a powerful tool to analyze network oscillations in vitro. The multi-site extracellular recording enables a detailed analysis of the spatiotemporal profile of network activity, and the planar design allows a combination with other techniques such as voltage-sensitive dye imaging and patch-clamp recording.

Acknowledgments. This research was supported by the Biotechnology and Biological Sciences Research Council (U.K.). Additional financial support from Matsushita Co., Ltd. (Japan), Alpha Med Sciences Co., Ltd. (Japan), and Pfizer Inc. (U.S.A.) is gratefully acknowledged.

References

Bao, W. and Wu, J.Y. (2003). Propagating wave and irregular dynamics: Spatiotemporal patterns of cholinergic theta oscillations in neocortex in vitro. *J. Neurophysiol.* 90: 333–341.

Boddeke, H.W., Best, R., and Boeijinga, P.H. (1997). Synchronous 20 Hz rhythmic activity in hippocampal networks induced by activation of metabotropic glutamate receptors in vitro. *Neuroscience* 76: 653–658.

Bragin, A., Jando, G., Nadasdy, Z., Hetke, J., Wise, K., and Buzsaki, G. (1995). Gamma (40–100 Hz) oscillation in the hippocampus of the behaving rat. *J. Neurosci.* 15: 47–60.

Brankack, J., Stewart, M., and Fox, S.E. (1993). Current source density analysis of the hippocampal theta rhythm: Associated sustained potentials and candidate synaptic generators. *Brain Res.* 615: 310–327.

Buzsaki, G. (2002). Theta oscillations in the hippocampus. *Neuron* 33: 325–340.

Buzsaki, G., Czopf, J., Kondakor, I., and Kellenyi, L. (1986). Laminar distribution of hippocampal rhythmic slow activity (RSA) in the behaving rat: Current-source density analysis, effects of urethane and atropine. *Brain Res.* 365: 125–137.

Charpak, S., Pare, D., and Llinas, R. (1995). The entorhinal cortex entrains fast CA1 hippocampal oscillations in the anaesthetized guinea-pig: Role of the monosynaptic component of the perforant path. *Eur. J. Neurosci.* 7: 1548–1557.

Csicsvari, J., Jamieson, B., Wise, K.D., and Buzsaki, G. (2003). Mechanisms of gamma oscillations in the hippocampus of the behaving rat. *Neuron* 37: 311–322.

Dickinson, R., Awaiz, S., Whittington, M.A., Lieb, W.R., and Franks, N.P. (2003). The effects of general anaesthetics on carbachol-evoked gamma oscillations in the rat hippocampus in vitro. *Neuropharmacology* 44: 864–872.

Ecker, C., Suckling, J.M., and Paulsen, O. (2001). Temperature dependence of cholinergically induced network oscillations in the CA1 region of the hippocampus. *Soc. Neurosci. Abstr.* 27: 47–49.

Fisahn, A., Pike, F.G., Buhl, E.H., and Paulsen, O. (1998). Cholinergic induction of network oscillations at 40 Hz in the hippocampus in vitro. *Nature* 394: 186–189.

Freund, T.F. and Buzsaki, G. (1996). Interneurons of the hippocampus. *Hippocampus* 6: 347–470.

Fries, P., Neuenschwander, S., Engel, A.K., Goebel, R., and Singer, W. (2001). Rapid feature selective neuronal synchronization through correlated latency shifting. *Nat. Neurosci.* 4: 194–200.

Gill, C.H., Soffin, E.M., Hagan, J.J., and Davies, C.H. (2002). 5-HT7 receptors modulate synchronized network activity in rat hippocampus. *Neuropharmacology* 42: 82–92.

Gillies, M.J., Traub, R.D., LeBeau, F.E., Davies, C.H., Gloveli, T., Buhl, E.H., and Whittington, M.A. (2002). A model of atropine-resistant theta oscillations in rat hippocampal area CA1. *J. Physiol.* 543: 779–793.

Hajos, N., Katona, I., Naiem, S.S., MacKie, K., Ledent, C., Mody, I., and Freund, T.F. (2000). Cannabinoids inhibit hippocampal GABAergic transmission and network oscillations. *Eur. J. Neurosci.* 12: 3239–3249.

Hajos, N., Palhalmi, J., Mann, E.O., Nemeth, B., Paulsen, O., and Freund, T.F. (2004). Spike timing of distinct types of GABAergic interneuron during hippocampal gamma oscillations in vitro. *J. Neurosci.* 24: 9127-9137.

Hamill, O.P., Marty, A., Neher, E., Sakmann, B., and Sigworth, F.J. (1981). Improved patch-clamp techniques for high-resolution current recording from cells and cell-free membrane patches. *Pflugers Arch.* 391: 85–100.

Holsheimer, J. (1987). Electrical conductivity of the hippocampal CA1 layers and application to current-source-density analysis. *Exp. Brain Res.* 67: 402–410.

Hormuzdi, S.G., Pais, I., LeBeau, F.E., Towers, S.K., Rozov, A., Buhl, E.H., Whittington, M.A., and Monyer, H. (2001). Impaired electrical signaling disrupts gamma frequency oscillations in connexin 36-deficient mice. *Neuron* 31: 487–495.

Jensen, O. and Lisman, J.E. (1996). Theta/gamma networks with slow NMDA channels learn sequences and encode episodic memory: Role of NMDA channels in recall. *Learn. Mem.* 3: 264–278.

Kocsis, B., Bragin, A., and Buzsaki, G. (1999). Interdependence of multiple theta generators in the hippocampus: A partial coherence analysis. *J. Neurosci.* 19: 6200–6212.

LeBeau, F.E., Towers, S.K., Traub, R.D., Whittington, M.A., and Buhl, E.H. (2002). Fast network oscillations induced by potassium transients in the rat hippocampus in vitro. *J. Physiol.* 542: 167–179.

Lee, D. (2002). Analysis of phase-locked oscillations in multi-channel single-unit spike activity with wavelet cross-spectrum. *J. Neurosci. Methods* 115: 67–75.

Leung, L.W. (1984). Model of gradual phase shift of theta rhythm in the rat. *J. Neurophysiol.* 52: 1051–1065.

Lisman, J.E. and Idiart, M.A. (1995). Storage of 7 +/− 2 short-term memories in oscillatory subcycles. *Science* 267: 1512–1515.

Llinas, R., Ribary, U., Contreras, D., and Pedroarena, C. (1998). The neuronal basis for consciousness. *Philos. Trans. R. Soc. Lond. B Biol. Sci.* 353: 1841–1849.

Mallat, S. (1999). *A Wavelet Tour of Signal Processing.* 2nd ed. Academic Press, San Diego.

Mann, E.O., Radcliffe, C.A., and Paulsen, O. (2005a). Hippocampal gamma-frequency oscillations: from interneurons to pyramidal cells, and back. *J. Physiol.* 562: 57–65.

Mann, E.O., Suckling, J.M., Hajos, N., Greenfield, S., and Paulsen, O. (2005b). Perisomatic feedback inhibition underlies cholinergically-induced fast network oscillations in the rat hippocampus in vitro. *Neuron* 45: 105–117.

Mitzdorf, U. (1985). Current source density method and application in cat cerebral cortex: Investigation of evoked potentials and EEG phenomena. *Physiol. Rev.* 65: 37–100.

Nicholson, C. and Freeman, J.A. (1975). Theory of current source-density analysis and determination of conductivity tensor for anuran cerebellum. *J. Neurophysiol.* 38: 356–368.

Nicholson, C. and Llinas, R. (1975). Real time current source-density analysis using multi-electrode array in cat cerebellum. *Brain Res.* 100: 418–424.

Oka, H., Shimono, K., Ogawa, R., Sugihara, H., and Taketani, M. (1999). A new planar multielectrode array for extracellular recording: application to hippocampal acute slice. *J. Neurosci. Meth.* 93:61–67.

Oppenheim, A.V. and Willsky, A.S. (1997). *Signals & Systems.* 2nd ed. Prentice-Hall, Upper Saddle River, NJ.

Phillips, C.L. and Parr, J.M. (1999). *Signals, Systems, and Transforms.* 2nd ed. Prentice-Hall, Upper Saddle River, NJ.

Proakis, J.G. and Manolakis, D.G. (1996). *Digital Signal Processing.* 3rd ed. Prentice-Hall, Upper Saddle River, NJ.

Shimono, K., Brucher, F., Granger, R., Lynch, G., and Taketani, M. (2000). Origins and distribution of cholinergically induced beta rhythms in hippocampal slices. *J. Neurosci.* 20: 8462–8473.

Singer, W. (1993). Synchronization of cortical activity and its putative role in information processing and learning. *Annu. Rev. Physiol.* 55: 349–374.

Stuart, G.J., Dodt, H.U., and Sakmann, B. (1993). Patch-clamp recordings from the soma and dendrites of neurons in brain slices using infrared video microscopy. *Pflugers Arch.* 423: 511–518.

Tominaga, T., Tominaga, Y., and Ichikawa, M. (2001). Simultaneous multi-site recordings of neural activity with an inline multi-electrode array and optical measurement in rat hippocampal slices. *Pflugers Arch.* 443: 317–322.

Tominaga, T., Tominaga, Y., Yamada, H., Matsumoto, G., and Ichikawa, M. (2000). Quantification of optical signals with electrophysiological signals in neural activities of Di-4-ANEPPS stained rat hippocampal slices. *J. Neurosci. Methods* 102: 11–23.

Torrence, C. and Compo, G. (1998). A practical guide to wavelet analysis. *Bull. Am. Meteorol. Soc.* 79: 61–78.

Towers, S.K., LeBeau, F.E., Gloveli, T., Traub, R.D., Whittington, M.A., and Buhl, E.H. (2002). Fast network oscillations in the rat dentate gyrus in vitro. *J. Neurophysiol.* 87: 1165–1168.

Traub, R.D., Whittington, M.A., Colling, S.B., Buzsaki, G., and Jefferys, J.G. (1996). Analysis of gamma rhythms in the rat hippocampus in vitro and in vivo. *J. Physiol.* 493: 471–484.

Tsau, Y., Guan, L., and Wu, J.Y. (1999). Epileptiform activity can be initiated in various neocortical layers: An optical imaging study. *J. Neurophysiol.* 82: 1965–1973.

Weiss, T., Veh, R.W., and Heinemann, U. (2003). Dopamine depresses cholinergic oscillatory network activity in rat hippocampus. *Eur. J. Neurosci.* 18: 2573–2580.

Whittington, M.A., Traub, R.D., and Jefferys, J.G. (1995). Synchronized oscillations in interneuron networks driven by metabotropic glutamate receptor activation. *Nature* 373: 612–615.

Whittington, M.A., Traub, R.D., Faulkner, H.J., Stanford, I.M., and Jefferys, J.G. (1997). Recurrent excitatory postsynaptic potentials induced by synchronized fast cortical oscillations. *Proc. Natl. Acad. Sci. U. S. A.* 94: 12198–12203.

Wu, J.Y., Guan, L., and Tsau, Y. (1999). Propagating activation during oscillations and evoked responses in neocortical slices. *J. Neurosci.* 19: 5005–5015.

Ylinen, A., Bragin, A., Nadasdy, Z., Jando, G., Szabo, I., Sik, A., and Buzsaki, G. (1995). Sharp wave-associated high-frequency oscillation (200 Hz) in the intact hippocampus: Network and intracellular mechanisms. *J. Neurosci.* 15: 30–46.

Index